Handbook of Qurʾānic Hermeneutics

De Gruyter Handbook

Handbook of Qurʾānic Hermeneutics

Vol. 3: Qurʾānic Hermeneutics from the 13th to the 19th Century

Edited by
Georges Tamer

DE GRUYTER

ISBN 978-3-11-131674-1
e-ISBN (PDF) 978-3-11-132008-3
e-ISBN (EPUB) 978-3-11-132020-5

Library of Congress Control Number: 2024940935

Bibliographic information published by the Deutsche Nationalbibliothek
The Deutsche Nationalbibliothek lists this publication in the Deutsche Nationalbibliografie; detailed bibliographic data are available on the internet at http://dnb.dnb.de.

Cover image: Calligraphy: Mr. Zuheir Elia (Erlangen, Germany). Q 3:7:
"wa-mā yaʿlamu taʾwīlahu illā Llāhu wa-r-rāsikhūna fī-l-ʿilmi"
Typesetting: Integra Software Services Pvt. Ltd.

www.degruyter.com

Contents

Introduction

Georges Tamer

The third volume of the *Handbook of Qur'ānic Hermeneutics* explores hermeneutical approaches to the Qur'ān from the thirteenth/seventh century through the beginning of the nineteeeenth/thirteenth century. In this period, Qur'ānic hermeneutics displays several characteristics. Probably the most distinctive is the preservation of approaches, methods, and interpretations from earlier major commentaries and, at the same time, adapting them to new contextual conditions. Another essential feature is that the commentaries refer to the transmission of exegetical traditions from older authorities, even if the chains of transmission have been abbreviated over time. Similar to earlier scholars of Qur'ānic hermeneutics, the majority of scholars discussed in this volume delve into traditional materials, including grammar and poetry, to elucidate certain words or expressions. Furthermore, they offer insights into the context and historical background of Qur'ānic verses. Being faithful to the established hermeneutical tradition was more appealing to the scholars featured in this volume than inventing something new. However, the preservation of traditional lines of Qur'ānic hermeneutics during the period covered in this volume coincided with an increasing dispersion and diversity of Qur'ānic scholarship after the fall of Baghdad to the Mongols and the end of the Abbasid State. Especially in regions like Khurasan and Transoxiana, Qur'ānic hermeneutics thrived, with a focus on cities such as Nishapur, Samarkand, and the border town of Balkh. Furthermore, the border towns of Marw, Rayy, and Herat in Afghanistan played a significant role in the flourishing of Qur'ānic scholarship. It is also important to note that many exegetical works have not been edited or have been lost, making it difficult to assess the Qur'ānic hermeneutics of their authors.

Qur'ānic hermeneutics as presented in this volume have multiple perspectives. In addition to conventional Sunni and Shiite hermeneutics, mystical interpretations of the Qur'ān, the interest in jurisprudential matters, and commentaries on specific themes in the Qur'ān come to the fore. Particularly striking, is the dichotomy between the traditional and mystical interpretations of the Qur'ān. The former include a holistic view of the Scripture through disciplines such as linguistics (*'ulūm al-lugha, qirā'āt* etc.), dialectical theology (*kalām*), Islamic jurisprudence (*fiqh*), and areas of knowledge that serve to contextualize certain Qur'ānic passages (*asbāb an-nuzūl, isrā'īliyyāt*, hadith, *qiṣaṣ*, etc.). Accordingly, traditional Qur'ānic hermeneutics seeks to develop interpretations utilizing specific areas of knowledge; explanations are often developed according to the expertise and preference of the respective scholar. For exegetes such

Acknowledgments: I would like to thank Mr. Ramy Abdin, Mr. Andreas Knöll, and Mr. Yaaqub Kutterer for their assistance editing the contributions of this volume. I would also like to thank Ms. Katrin Mittmann and Dr. Torsten Wollina from the publishing house Walter de Gruyter and Dr. Lydia J. White for their kind and cooperative support.

https://doi.org/10.1515/9783111320083-001

as an-Nasafī, the focus is particularly on theological issues; in al-Qurṭubī's exegesis, jurisprudential aspects are in the foreground; Ibn Kathīr views the Qur'ān through the lens of hadith literature. Although such nuances in the Qur'ānic hermeneutics become clear on closer inspection, traditional exegesis is nevertheless characterized by a holistic view which includes all or most of the above-mentioned fields of knowledge. None of the traditional Qur'ān commentators follow a simple approach based on a single field of knowledge. In addition to theological or jurisprudential aspects, linguistic and contextualizing aspects are usually investigated.

Undoubtedly, traditional exegesis also contains a certain openness towards mystical, esoteric interpretations of the Qur'ān, but esoteric Qur'ānic hermeneutics unfold especially in the works of the mystics, such as Ibn ʿArabī, as-Simnānī, Mullā Ṣadrā, and Ibn ʿAjība. Despite their nuances, explanatory variations and creative terminology, traditional Qur'ānic commentaries often contain redundancies from earlier periods. In contrast, the mystical Qur'ānic hermeneutics, by virtue of its underlying flexible interpretative potential, appears far more innovative. Representatives of the philosophy of illumination *(al-ḥikma al-ishrāqiyya)*, e.g. Mullā Ṣadrā, al-Kāshānī, and al-Lāhījī express cosmological, mystical and avicennian concepts in their exegetical writings. Alongside them, original approaches and ideas such as those of al-Biqāʿī and Ḥaydar Āmulī emerge during this period. Mystical Qur'ānic hermeneutics are therefore more varied and creative, but largely limited to Sufi or mystical topics, whereas the holistic approach in traditional Qur'ānic exegesis covers a broader spectrum of knowledge.

Equally striking is the geographical and therefore linguistic and cultural diversity of the hermeneutical approaches to the Qur'ān in this volume. The majority of exegetes in this period originate from Persian-speaking regions, where they were also active in jurisprudential scholarship. Additionally, we find Qur'ānic hermeneutics emerging in the Malay world and India. Especially with the inclusion of Persian and Ottoman Qur'ānic hermeneutics representatives, this volume pays significant attention to the differences between Sunni and Shiite interpretations.

The following overview of the scholars covered in this volume highlights their main achievements in the field of Qur'ānic hermeneutics. We will see which major commentaries and works by earlier outstanding Qur'ānic scholars they received, to what extent they were dependent on them, what they adopted from them, and what was new in their approaches. Furthermore, the interconnections between contemporaneous scholars will be examined. What is briefly outlined in the introduction will be dealt with in detail in the following chapters. The scholars examined have been arranged in chronological order based on the years of their death. The dates presented before the backslash (/) adhere to the Julian Calendar, while those after the backslash correspond to the Islamic Calendar.

Ibn ʿArabī (d. 1240/638), titled *ash-Shaykh al-Akbar*, is a prominent figure in classical Sufi literature, known for synthesizing and enhancing early Sufi doctrines. His works extend beyond Sufism, addressing various fields of knowledge, among them

Qur'ānic hermeneutics. His unique writing style blends theological reflections, philosophy, poetry, and discussions on Islamic jurisprudence. His influence on Islamic intellectual history is evident, especially in the Ottoman and Persianate worlds, where significant figures wrote commentaries on his works.

Initially living a luxurious life, Ibn ʿArabī underwent an existential crisis, leading him to devoting himself to the spiritual path. His hermeneutical method emphasizes the foundation of understanding on personal spiritual experience rather than the unquestioning adherence to concepts. Qur'ānic hermeneutics is central to Ibn ʿArabī's thought; his works are considered an immense holistic commentary on the Qur'ān. He sees the Qur'ānic revelation as a manifestation of God's self, correlating with signs in the Qur'ān, cosmos, and human beings.

Ibn ʿArabī's approach involves a twofold exploration of the Qur'ānic text, balancing its explicit and implicit aspects. He encourages approaching the Qur'ān with spiritual "illiteracy" to allow the text to provoke a comprehensive spiritual experience.[1] Perplexity, attained through a literal approach, is seen as the most perfect spiritual state, providing a unique experience of divine perfection.

Ibn ʿArabī's hermeneutical legacy remained significant, influencing later Islamic intellectual history. The so called Akbarī school emerged from his spiritual and intellectual lineage, with figures like Ṣadr ad-Dīn Qūnawī and Dawūd Qayṣarī contributing to its development. However, the diffusion of Ibn ʿArabī's ideas also sparked opposition, particularly from Ibn Taymiyya, who focused on metaphysical and theological issues rather than on his exegetical method. Nevertheless, Ibn ʿArabī's Qur'ānic hermeneutics remains a central link in the Sufi exegetical tradition, representing a milestone in Islamic intellectual history by deepening the implications of the reader's perspective and emphasizing the boundless dimension of human consciousness, as mirrored in the Qur'ān.

The Iranian Sufi scholar ʿAlā' ad-Dawla as-Simnānī (d. 1336/736), associated with the Kubrawiyya brotherhood,[2] wrote a partial *tafsīr*, focusing on the suras 1 and 52–114. Titled *Muṭṭalaʿ an-nuqaṭ wa-majmaʿ al-luqaṭ*, the work is an esoteric interpretation, emphasizing symbolic meanings over literal and historical aspects. As-Simnānī viewed the Qur'ān as an inner map which guides Sufis on a spiritual journey. His unique hermeneutics established a connection between seven major prophets in the Qur'ān and seven subtle centers within the initiate's spiritual body, representing different levels of esoteric meaning. His intellectual background, including a spiritual

1 Ibn ʿArabī's concept of illiteracy evokes the Islamic teaching of Muḥammad's illiteracy in reference to Q 7:157f. However, the term *ummī* in the Qur'ān does not mean illiterate, but "gentile," i.e., neither Jew nor Christian. In this sense it is used in plural "*ummiyyūn/ummiyyīn*" in Q 3:20, 75; 62:2. Cf. Sebastian Günther, "Ummī," in *EQ*, ed. Jane Dammen McAuliffe, accessed December 15, 2023, http://dx.doi.org/10.1163/1875-3922_q3_EQCOM_00210.

2 S. Hamid Algar, "Kubrā," in *EI*², ed. P. Bearman et al., accessed December 15, 2023, http://dx.doi.org/10.1163/1573-3912_islam_SIM_4470.

awakening during a battle, led him to abandon the Ilkhanid court and criticize the Ilkhanid rulers. As-Simnānī's *tafsīr* omits the commentary on suras 2–51, as it functioned as a continuation of an earlier, unfinished commentary by other scholars in the Kubrawiyya tradition. His hermeneutics, outlined in the introduction, focuses on the inner aspects of the Qur'ān and presents a layered theory of esoteric meanings, with seven hundred possible interpretations for each verse. The commentary's structure reflects a comprehensive system of correspondence, aligning microcosm and macrocosm through the interpretation of the Qur'ān. As-Simnānī's reliance on inspiration and the internalization of Qur'ānic meanings categorizes his work as a symbolic (*ishārī*) and inspiration-dependent (*ilhāmī*) Sufi commentary.

Al-Qurṭubī's (d. 1273/671) *al-Jāmiʿ li-aḥkām al-Qur'ān*, written under the Almohads, stands out among juridical-exegetical works for its encyclopedic nature. Focused on legal analysis, it goes beyond typical juridical-exegetical literature by incorporating grammatical analysis, reports on legal issues from earlier authorities, clarification of unclear utterances, reading variants of the Qur'ān, and references from hadiths and commentaries on theological debates. Al-Qurṭubī's unique approach demonstrates free reasoning and open-mindedness in criticizing his own Mālikiyya school of jurisprudence. Dedicating his life to exploring the Qur'ān, he presents an exegetical method that involves presenting different views, searching for relevant hadiths, and presenting legal issues in a comparative manner. He emphasizes the importance of the sunna in explaining the Qur'ān and justifies interpreting the Qur'ān through a chain of scholars, back to the Prophet, and the Qur'ān itself. Al-Qurṭubī's hermeneutics involves a *ḥaqīqa–majāz* dichotomy, prioritizing the literal meaning (*ḥaqīqa*) over figurative meanings (*majāz*). His extensive consideration of legal issues, abrogation, and the overall objectives of the sharia (*maqāṣid ash-sharīʿa*) made his commentary influential on later sources. Despite being a typical encyclopedic *tafsīr* in the Andalusian tradition, al-Qurṭubī's work stands out for its detailed analysis of Qur'ānic reading variants and his unique incorporation of Sufi terminology. A special feature of his commentary is also his inclusion of circumstances of his lifetime and examples of human experience.

Al-Baiḍāwī (d. 1291/691), an Ashʿarite theologian and Shāfiʿī jurist, was a prominent figure from Iran under Mongol Ilkhanate rule. His influential commentary, *Anwār at-tanzīl wa-asrār at-ta'wīl*, which is strongly influenced by az-Zamakhsharī's *al-Kashshāf*, earned recognition for its concise yet comprehensive approach, drawing also from various other sources. Al-Baiḍāwī, deeply rooted in the Shāfiʿī-Ashʿarī tradition, actively engaged in the incorporation of philosophical ideas into theology, reflecting the intellectual context of his time.

While concentrating on abridging and commenting on earlier works, al-Baiḍāwī's commentary covers theology, linguistics, jurisprudence (*fiqh*), principles of jurisprudence (*uṣūl al-fiqh*), history and astronomy. Noteworthy is his attention to variant readings of the Qur'ān. His most significant achievement lies, however, in masterfully condensing Sunni interpretations into a succinct text, evaluating them, and presenting his own opinions. Furthermore, al-Baiḍāwī engages with theological differences, eliminating or countering Muʿtazilī views present in az-Zamakhsharī's commentary and addressing Shiite interpretations. He refutes notions supporting the infallibility of figures like ʿAlī b. Abī Ṭālib and his family. Al-Baiḍāwī's *Anwār*, categorized as *tafsīr bi-r-raʾy*, incorporates personal reflections on Qurʾānic verses, complemented by narrations of the Prophet and his Companions. Remarkably, he retains all narrations from az-Zamakhsharī without providing chains of narrators or commenting on reliability. Furthermore, he shows caution towards Judeo-Christian traditions (*isrāʾīliyyāt*).

Abū l-Barakāt an-Nasafī (d. 1310/710) was a distinguished Ḥanafī legal scholar who authored significant legal writings, theological treatises, and a commentary on the Qurʾān in the Ḥanafī-Māturīdī tradition, titled *Madārik at-tanzīl wa-ḥaqāʾiq at-taʾwīl*. In the preface, he clarifies his motivation to provide a medium-length work that covers correct inflection, readings, tropes, symbolism, Sunni teachings, and refutation of false statements. He distances himself from the Muʿtazila and the Ashʿariyya, emphasizing the eternity of divine attributes and showcasing his commitment to the Ḥanafī-Māturīdī tradition.

An interesting point is how an-Nasafī utilized az-Zamakhsharī's Muʿtazilī Qurʾān commentary, *al-Kashshāf*, as a primary source for his *tafsīr*, despite their doctrinal differences. An-Nasafī implemented a traditionalist approach to exegesis, frequently citing hadiths without specifying chains of narrators. Moreover, he cautiously used *isrāʾīliyyāt* narratives. He also addresses legal matters, linguistic sources, variant readings, and occasions for revelation (*asbāb an-nuzūl*).

An-Nasafī's Qurʾān commentary includes *kalām* arguments to demonstrate the interconnection between his theological writings and his exegesis. The work proves useful for understanding intellectual currents in Central Asia during the Mongol occupation. By integrating various hermeneutical approaches and utilizing diverse sources, an-Nasafī's eclectic technique challenges the differentiation between conventional and contemporary Qurʾān commentaries. The incorporation of theological discussions into the commentary not only enriches the Māturīdī tradition but also contributes significantly to the expansion of Islamic intellectual history in the late Middle Ages.

Ibn Taymiyya (d. 1328/728), one of the most prominent Islamic scholars, gained fame for his role in resisting Mongol invasions and later focused on combating religious deviations. Despite facing imprisonment and other restrictions, he left a substantial theological corpus, including *Darʾ taʿāruḍ al-ʿaql wa-n-naql*, which argues for harmony between reason and revelation. Central to his vision is worship and love of God, grounded in adherence to the Sharia as outlined in foundational texts.

Ibn Taymiyya's *Muqaddima* is regarded the first systematic treatise on Qurʾānic hermeneutics. Rooted in a tradition of introductory themes found in Qurʾān commentaries, it responds to a need for general rules in understanding the Qurʾān, criticizing earlier exegesis for containing both truth and falsehood. Emphasizing the epistemological nature of his work, Ibn Taymiyya asserts the importance of knowledge based on reliable transmission from the early Muslim community (*salaf*). Thus, he elevates the interpretations of the Companions and Followers to the level of the Prophet's sunna, underlining their proximity to the Prophet and their role in transmitting Qur'ānic meanings. He assumes the existence of a consistent interpretation communicated by the Prophet and minimal conflicts among the Companions.

While Ibn Taymiyya acknowledges some irreconcilable differences in the interpretation of the Qur'ān, he attributes these mainly to inconsequential matters derived from the People of the Book (*ahl al-kitāb*), i.e. Jews and Christians. His approach involves interpreting the Qur'ān with the Qur'ān itself and turning to the sunna, Companions, and Followers if needed. He permits philological investigation when no definitive prophetic interpretation is available but strongly condemns personal opinion in Qur'ān interpretation.

An-Nīsābūrī (d. after 1329/730) made notable contributions across various disciplines, including Qur'ānic exegesis. His renowned work, *Gharā'ib al-Qur'ān wa-raghā'ib al-furqān*, reflects his broad scholarship. An-Nīsābūrī combines a rationalist approach (*at-tafsīr bi-r-ra'y*) with attention to transmitted information (*at-tafsīr bi-l-ma'thūr*) and includes esoteric exegesis (*ta'wīl*), showing a versatility that challenges modern academic specialization.

His reliance on Fakhr ad-Dīn ar-Rāzī's *at-Tafsīr al-kabīr* led him to engage in theological debates, although he did not write a separate book in philosophical theology. An-Nīsābūrī's hermeneutical approach is also noted for using science as a tool, challenging traditional specialization. His extensive sources, including az-Zamakhsharī and ar-Rāzī, shaped his comprehensive commentary.

Gharā'ib al-Qur'ān, which begins with eleven introductions on the Qur'ān and related sciences, is divided into exegetical units, each covering a verse, a part of a verse, or a group of verses, addressing alternative vocalizations (*qirā'āt*), pauses in recitation (*wuqūf*), and presenting both literal (*tafsīr*) and esoteric (*ta'wīl*) interpretations. An-Nīsābūrī integrates transmitted and rationalist knowledge, showcasing respect for diverse perspectives and utilizing hadiths to present the virtues of specific suras (*faḍā'il as-suwar*). Moreover, he employs analogies from the natural world to explain divine intent and human choices. He rejects the Mu'tazilī position on God's role in evil actions and supports a literal reading of the Qur'ān with scientific information. The collaboration between science and religious text enhances the appreciation of God's role as creator. An-Nīsābūrī employed rhetorical analysis to elucidate esoteric exegesis, guaranteeing coherence with the intended meanings of verses. His esoteric and exoteric interpretations complement each other, emphasizing God's control over natural processes.

Ibn Juzayy (d. 1340/741), a fourteenth-century scholar from Granada, authored *at-Tashīl li-'ulūm at-tanzīl*, an incomplete commentary on the Qur'ān with a comprehensive introduction on the art of *tafsīr*. He provides an advanced yet condensed work that surpasses traditional exegesis, covering various disciplines. Despite living in a politically charged era, he largely avoided political themes. Ibn Juzayy demonstrated a deep understanding of the history of Qur'ānic exegesis, acknowledging his predecessors' achievements and introducing new approaches. He criticized past exegetes for neglecting the principles of jurisprudence (*uṣūl al-fiqh*), emphasizing their importance in Qur'ānic hermeneutics. His commentary, in which he widely uses Ibn ʿAṭiyya's work, combines word-by-word interpretation with a thematic approach, occasionally departing from linear interpretation for a more systematic view. His methodological and theoretical considerations as laid out in the introduction guide the practical interpretation of the Qur'ān throughout the commentary. The work's intellectual depth, facilitated by its systematic introduction, makes it accessible for scholars, earning its reputation as an original and valuable contribution to Qur'ānic hermeneutics.

Ibn Qayyim al-Jawziyya (d. 1350/751) did not produce a complete Qur'ānic commentary but extensively referenced Qur'ānic verses in his writings. He adhered to the hermeneutical principles of his mentor, Ibn Taymiyya, adapting them to his didactic goals and promoting the Taymiyyan worldview. He composed detailed commentaries on selected short suras, such as sura 1, *al-Fātiḥa* (The Opening); sura 109, *al-Kāfirūn* (The Disbelievers); sura 113, *al-Falaq* (The Daybreak); and sura 114, *an-Nās* (Mankind). Efforts have been made to compile scattered exegetical paragraphs, creating an artificial commentary attributed to him.

Rooted in the Taymiyyan tradition, Ibn al-Qayyim prioritized the Qur'ān and hadith literature in his theology. His mentor's contextual hermeneutics greatly influenced his approach to the Qur'ānic text. The commentary on sura *al-Kāfirūn*, e.g., showcases Ibn al-Qayyim's exegetical skills and deep familiarity with Islamic sciences. While he draws on Islamic textual sources, he introduces personal reasoning with a mystical touch. Philology, linguistics, and Arabic grammar remain central to his hermeneutical approach. He asserts the uniqueness of his interpretation, claiming divine inspiration, and emphasizes the Sufi vocabulary in his independent understanding.

In works such as *aṣ-Ṣawāʿiq al-mursala ʿalā l-jahmiyya wa-l-muʿaṭṭila*, Ibn al-Qayyim systematically presents Taymiyyan theories on divine attributes, challenging the Ashʿarīs. He advocates for a literal interpretation of Qur'ānic expressions, integrating context-dependent meaning and opposing metaphorical interpretations.

Ibn Kathīr (d. 1372/774) is widely recognized for his Qur'ān commentary. He relies in his work mainly on aṭ-Ṭabarī's commentary and hadith narratives to interpret Qur'ānic verses. Ibn Kathīr's method involves naming the verse, summarizing its main idea, and explaining it using other verses, hadith narratives, sayings of companions, or jurisprudential cases associated with the verse. Furthermore, he addresses the issue of *isrāʾīliyyāt*, categorizing it into three sections based on authenticity and relevance. His work incorporates approximately 160 named sources, reflecting his extensive scholarship.

Living in Damascus under Mamluk rule, Ibn Kathīr was a salafī who differed from his teacher Ibn Taymiyya by incorporating both esoteric interpretation and delegation. His methodology follows the principles of received interpretation (*at-tafsīr bi-l-maʾthūr*), relying on the Qur'ān, Hadith, narratives from companions, classical Arabic, *isrāʾīliyyāt* and the sayings of scholars of Islamic jurisprudence. His exegesis gained approval from his contemporaries and became a primary source for later scholars. Its historical relevance is evident in its widespread influence among Muslims.

Sayyid Ḥaydar Āmulī (d. after 1385–86/787) holds a unique place for integrating Shiite esotericism, Sufi mysticism, and philosophy. He stands out as the first Shiite Imami to compose a personal commentary on the Qurʾān, emphasizing the importance of Qurʾānic hermeneutics after the occultation of the twelfth Imam.

Āmulī authored numerous works, though many are lost. His monumental, albeit unfinished, commentary on Ibn ʿArabī's *Fuṣūṣ al-ḥikam*, which was intended as a Qurʾānic commentary, consists of six introductions and a commentary on sura 1, *al-Fātiḥa*. The work draws from Ibn ʿArabī's *al-Futūḥāt al-makkiyya*, emphasizing sainthood and eschatological events. It functions not merely as a traditional *tafsir* but also as a treatise on Qurʾānic hermeneutics within the Imami tradition and mysticism.

Āmulī's concept of hermeneutics extends beyond the Qurʾān to encompass the universe and the human soul, interpreting the Qurʾānic notion of "Book" to include the universe and the soul. This expanded perspective aligns hermeneutics with the knowledge of God, the world, and humanity, uniting mystical, philosophical, and Shiite Imami perspectives. In this sense, he introduces an original interpretation of "those firmly rooted in knowledge" (Q 3:7), expanding it to include the "masters of Divine Oneness." Āmulī claims affiliation with these masters through lineage and mystical initiation, aligning himself with the Imams and Sufi traditions. Thus, his hermeneutics introduces a departure from traditional Shiite exegesis, intertwining Imami traditions with Neoplatonic philosophy and Ibn ʿArabī's mysticism. He replaces the traditional dichotomy of Imams and their enemies with a distinction between the people who possess knowledge of the outer and the people who have access to the inner meaning of the Qurʾān. Drawing from Ibn Arabi, Āmulī explores the concepts of God's self-disclosure through Divine Names and the controversial doctrine of the Unity of Existence. He distinguishes between theological and ontological oneness and places the Imams above the Prophets in spiritual and initiatory rank. Harmonizing Shiite ideas, Sufism, sharia and philosophy, Āmulī's approach lays the groundwork for a spiritual and universal interpretation of the Qurʾān, marking a significant contribution to seventeenth-century Shiite philosophy and hermeneutics.

Despite al-Biqāʿī's (d. 1480/885) reputation as a knowledgeable scholar, his lack of humility and his antagonistic attitude toward critics marginalized him in the scholarly world. To establish the validity of his Qurʾānic hermeneutics, he utilized the literary motif of "dreams and visions," narrating extraordinary experiences that supposedly aided his understanding.

Al-Biqāʿī's Qurʾānic hermeneutics was shaped by his historical context and a nostalgia for past scholars. His key methodology, "*ʿilm munāsabāt al-Qurʾān*" (the science of connections of the Qurʾān), involves understanding the aims of specific suras, identifying prerequisite elements, and evaluating their fulfillment. This method reveals the interconnectedness of Qurʾānic verses and shapes the interpretation of the text.

The differentiation between *"muḥkamāt"* (clear verses) and *"mutashābihāt"* (ambiguous verses) is central to al-Biqāʿī's hermeneutics. He argues that anthropomorphisms in the Qurʾān serve the purpose of making God known through a test for creation. Elaborating on Q 3:7, he connects the concept of *mutashābihāt* across the Qurʾān, Torah, and Gospel, viewing them as a harmonious whole. Al-Biqāʿī's hermeneutical thread integrates Qurʾānic interpretation with biblical exegesis, especially concerning the terms "fatherhood" and "sonship" in the Gospels. Incorporating the Torah and Gospel into his framework, he recognizes their legitimacy as forms of revelation. He interprets the *mutashābihāt* in the Gospels through *muḥkam* verses in the Qurʾān, emphasizing the shared revelatory core among the three Scriptures.

Jalāl ad-Dīn as-Suyūṭī, a prominent scholar and prolific author during the Mamluk era (d. 1505/911), achieved renown for his extensive contributions to Islamic sciences. As a Sunni scholar, his comprehensive body of work encompassed various branches of Islamic knowledge, making him one of the most productive polymaths of his time. Notably, his Qurʾānic encyclopedia, *al-Itqān fī ʿulūm al-Qurʾān*, stands as a comprehensive and influential work on Qurʾānic sciences.

As-Suyūṭī's impact on Qurʾānic studies is evident in works like *ad-Durr al-manthūr fī t-tafsīr bi-l-maʾthūr*, a tradition-based commentary that departs from the conventional listing of chains of transmission. Additionally, *Tafsīr al-Jalālayn*, his completion of Jalāl ad-Dīn al-Maḥallī's Qurʾānic commentary, remains widely circulated. The *al-Itqān*, his magnum opus, draws from over 352 sources, showcasing a meticulous and systematic approach.

As-Suyūṭī's methodology involved both compilation and editorial work. He skillfully integrated diverse sources, at times adopting passages verbatim, while also revising and supplementing them. Remarkably, his editorial decisions reflected personal opinions, subtly embedded between the lines. His treatment of external sources and his earlier works like *at-Taḥbīr fī ʿilm at-tafsīr* showcases this diverse processing.

While as-Suyūṭī faced accusations of presenting others' ideas as his own, such practices were common in his era. The question of his eminence versus accusations of plagiarism is complex, considering prevalent writing practices. His meticulous listing of sources in the *Itqān* indicates a conscious handling of references, aligning with the norms of his time. His influence on Qurʾānic hermeneutics and *ʿulūm al-Qurʾān* (sciences of the Qurʾān) is profound. The *Itqān* remains the standard in its genre, shaping both content and structure in the field of Qurʾānic encyclopedias. As-Suyūṭī's focus on tradition-based approaches, evident in his commentary *ad-Durr al-manthūr*, reflects the Mamluk period's preference for written sources over personal instruction. His legacy lies in his extensive and systematic contributions to Islamic sciences, particularly in the realm of Qurʾānic studies. His methodological choices, editorial approach, and the enduring influence of his works continue to shape scholarly discourse in these fields.

Abū s-Suʿūd Efendi (d. 1574/982), primarily known as a jurist, gained widespread recognition in the Ottoman Empire for his commentary titled *Irshād al-ʿaql as-salīm ilā mazāyā l-Kitāb al-karīm* in which he showcased his extensive knowledge in recitation, linguistics, rhetoric, *fiqh*, and theology. In the introduction, he outlines a vision of the Qur'ān as the key to understanding and worshiping the supreme designer of the world.

According to Abū s-Suʿūd, while the universe reflects God's power, the Qur'ān serves as a guide to unravel its secrets, removing obstacles to understanding and worship. He emphasizes the Qur'ān's role in explaining religious principles, clarifying cosmic (*takwīnī*) verses, and providing insights into the angelic world. However, he acknowledges that comprehending the profound meanings of the Qur'ān is a challenge which requires expertise.

Abū s-Suʿūd divides the history of Qur'ānic exegesis into two periods: the scholars who analyzed the Qur'ān's meanings as received from the Prophet, and later scholars who revealed its miraculous aspects. The *Irshād* combines linguistic and literary approaches, with a focus on phonology, morphology, derivation, and syntax. Although not delving deeply into issues of *fiqh* or *kalām*, he aligns with *Ahl as-Sunna*, leaning toward the Māturīdiyya in theological matters.

Drawing from diverse sources, including az-Zamakhsharī and al-Bayḍāwī, Abū s-Suʿūd interprets the Qur'ān in sequential order, analyzing each verse atomistically. His commentary emphasizes subtleties of meaning, using the theory of the poetics of sentence construction in the Qur'ān (*naẓm*) as developed by al-Jurjānī and his followers to determine linguistic nuances. While attentive to linguistic details, he doesn't engage in detailed discussions of controversial issues, maintaining a focus on the Qur'ān's linguistic and rhetorical aspects.

Mulla Ṣadrā (d. 1640/1050), a prominent figure of Islamic philosophy in Persia, is renowned for his synthesis of reason and revelation in his school of Transcendent Theosophy. Despite being recognized as a philosopher-gnostic, Ṣadrā's exceptional exegesis of the Qur'ān holds a central position in his metaphysical worldview, integrating Sufi, Shiite, theological, and philosophical traditions.

Ṣadrā criticizes commentaries focused on linguistics and dialectical theology, asserting that true knowledge of the Qur'ān surpasses these sciences. He dismisses superficial scholarship and advocates for a deeper exploration of the Qur'ān's inner reality, aligning with the Sufi methodology. Drawing from the teachings of Shiite Imams and the Sufism of Ibn ʿArabī, his hermeneutics reflects a return to the prophetic model, emphasizing direct learning from God.

Ṣadrā's sixteen works on the Qur'ān include commentaries on suras and theoretical aspects and go beyond traditional exegesis. His hermeneutics, mirroring his cosmology, recognizes three levels—rational, imaginal, and material—corresponding to intellectual, symbolic, and literal levels of interpretation. He adopts a concept of the Qur'ān that has both exoteric and esoteric dimensions, akin to Sufi hermeneutics.

Ṣadrā's method involves reading the Qur'ān in relation to the inner book of the soul, aligning outer and inner meanings. While emphasizing divine knowledge, he maintains a synthetic approach that incorporates reason, revelation, and gnosis. His commentaries follow traditional structures, beginning with philological considerations, but they reject the methods of traditional exegesis. Instead, he interprets verses through mystical insights, intertwining them with his metaphysical framework.

Unlike his more influential philosophical writings, Ṣadrā's Qur'ānic works are practice-oriented, focusing on spiritual purification, decorum, and inward conditions before engaging in interpretation. His approach is rooted in the school of Ibn ʿArabī and highlights a mystical and existential reading, emphasizing the correspondence between scripture, the book of existence, and the book of the soul. Ṣadrā's hermeneutical approach to the Qur'ān reflects a synthesis of diverse traditions with a profound emphasis on spiritual insight and the inner dimensions of the sacred text.

Al-Fayḍ al-Kāshānī (d. 1680/1090), a student of Mullā Ṣadrā, emerged as a significant scholar, especially within the Akhbārī doctrine.[3] The Akhbārīs asserted that true knowledge could only be derived from the traditions of the Imams, emphasizing the divine certitude within these traditions. Fayḍ's hermeneutical approach, influenced by Akhbārī principles, differed from typical commentaries by focusing on the Imams' traditions as the main determinants of Qur'ānic interpretation. He diverged from his Akhbārī peers on several issues, particularly in addressing conflicting traditions. Instead of resorting to abrogation or nullification, he attributed divergent traditions to the various social and political circumstances under which the Imams spoke.

In Fayḍ's Qur'ānic exegesis, the understanding of the Qur'ān is intricately connected to the Imams' traditions, surpassing grammatical nuances and linguistic aspects. His hermeneutical method maintains that only the family of the Prophet (*ahl al-bayt*) hold the correct knowledge and esoteric meanings of the Qur'ān. Fayḍ emphasized the prohibition of personal reasoning (*raʾy*) and analogy (*qiyās*) in favor of adhering to the authority of the Imams, aligning with Shiite world views in order to avoid divisions within the Shiite community.

Furthermore, Fayḍ's hermeneutical approach extends beyond mere text explication: He reads Shiite beliefs into the scripture by harmonizing the Imams' traditions with Qur'ānic contents. His hermeneutics reflects a soteriological dimension, asserting that salvation hinges on accepting the authority and traditions of the Imams. Finally, Fayḍ contends that parts of the Qur'ān explicitly mention the Imams, implying a belief in the interpolation (*taḥrīf*) or omission of certain verses.

Bahāʾ ad-Dīn al-Lāhījī (d. between 1677 and 1684/1088 and 1095), also known as Quṭb ad-Dīn Ashkiwarī, was a marginal philosopher and theologian in seventeenth-century Safavid Iran. His Qur'ānic commentary *Tafsīr Sharīf Lāhījī*, completed in 1675/1086, appears to be a traditional Shiite work but distinguishes itself by incorporating philosophical elements alongside sectarian concerns.

3 S. W. Madelung, "Akhbāriyya," in *EI*², ed. P. Bearman et al., accessed December 15, 2023, http://dx.doi.org/10.1163/1573-3912_islam_SIM_8312; Robert M. Gleave, "Akhbāriyya and Uṣūliyya," in *EI*³, ed. Kate Fleet et al., accessed December 15, 2023, http://dx.doi.org/10.1163/1573-3912_ei3_COM_0029.

Belonging to a spiritual current aiming to synthesize esoteric Imami tradition, Sufi mysticism, and peripatetic and illuminative philosophy, Lāhījī aligns with figures like Sayyid Ḥaydar Āmulī, Ibn Abī Jumhūr, and Nūr Allāh Shūshtarī. Similar to Fayḍ, his commentary follows the neo-traditional (*akhbārī*) Imami Shiite movement of the seventeenth century, reviving early Shiite hermeneutical tradition. The commentary primarily relies on traditions attributed to the Imams, notably from sources like al-Qummī and al-ʿAyyāshī, which emphasize the exclusive interpretative authority of the Imams.

The bilingual nature of his commentary, quoting Arabic verses and hadiths before translating them and commenting on them in Persian, serves a pedagogical purpose, namely to disseminate esoteric Imami tradition to Persian speakers unfamiliar with Arabic and Imami sources.

Lāhījī's hermeneutics follows two fundamental principles: the Qurʾān's dual nature (exoteric and esoteric) and the Imams as exclusive interpreters based on their immediate knowledge from the Prophet. Thus, in his commentary, he systematically refers to historical figures, especially the Prophet's family and their enemies, aligning with early Shiism's dualistic vision. Furthermore, the commentary incorporates the belief in the falsification of the Qurʾān by enemies of the Imams, supporting this notion with traditions about recovered verses. Lāhījī's commentary also introduces the concept of the imaginal world, an intermediary realm between the material and spiritual worlds, borrowed from Ishrāqī and Akbarī schools, highlighting Lāhījī's effort to harmonize scriptural tradition with philosophical and mystical thought.

Lāhījī views the Qurʾān's miraculous perfection as evidence of its resistance to falsification. His work embraces the concept of *rajʿa*, the return to life during the Mahdī's reappearance, adding an eschatological dimension to his commentary. His work contributes, albeit quietly, to the synthesis of Imami Shiite hermeneutics with Islamic philosophy in the modern period.

The seventeenth-century Malay commentator ʿAbd ar-Raʾūf b. ʿAlī l-Fanṣūrī (d. 1693/1105) holds a foundational role in Southeast Asian Qurʾānic exegesis. Undoubtedly one of his major works is the Qurʾān commentary titled *Tarjumān al-mustafīd* (The Interpreter of the Beneficial). This commentary, providing the first full Malay interpretation of the Qurʾān, aimed at facilitating understanding among ordinary people not proficient in Arabic.

ʿAbd ar-Raʾūf's commentary, influenced by the style of analytical exegesis, follows the standard order of suras and moves through the verses phrase by phrase. It does not strictly align with any chronological or thematic arrangement. While Dutch colonial writings initially associated it with al-Bayḍāwī's authoritative Arabic commentary, discrepancies with al-Bayḍāwī and usage of other sources, particularly the *Tafsīr al-Jalālayn* and al-Khāzin's commentary, indicate that *Tarjumān al-mustafīd* is an original compilation by ʿAbd ar-Raʾūf. Along with Dāʾūd ar-Rūmī, he combines efforts to transform the Qurʾānic text into a comprehensive narrative. Particularly the *Tafsīr al-*

Jalālayn contributes to in-verse commentary, while al-Khāzin's commentary provides ʿAbd ar-Raʾūf with primary material for interpolated anecdotes.

In the context of Aceh, ʿAbd ar-Raʾūf's commentary, although not explicitly Sufi, subtly refutes Ibn ʿArabī's teaching of the unity of existence (*waḥdat al-wujūd*), aligning instead with dualistic themes to reconcile Sufism and sharia. In a society recovering from polarization, the work aimed to provide clear direction in terms of doctrine and practice. It is accessible to Muslims without specialist knowledge and suitable for emerging Muslim schools, contributing to the needs of an increasingly Islamized Malay world.

In his Qurʾānic commentary, *Mirʾāt al-anwār wa-mishkāt al-asrār*, Abū l-Ḥasan al-ʿĀmilī l-Iṣfahānī (d. 1725–26/1138 or 1727–28/1140) strives to provide a complete esoteric interpretation (*taʾwīl*) of the Qurʾān, incorporating traditions attributed to the Imams for each verse. However, the project remained largely unfinished, with only three introductions and the commentary on the first and part of the second sura preserved.

The commentary, representing the Akhbārī-traditionalist theological school, is characterized more as a treatise on Qurʾānic exegesis than a commentary. This school relies solely on traditions attributed to the Imams, in contrast to the *Uṣūlī*-rationalist school, which emphasizes the authority of qualified scholars' reasoning. The first introduction outlines Shiite hermeneutics, asserting that the Qurʾān has an esoteric dimension concerning the Imams' mission and Divine Alliance. The dual vision of an outer and inner reality is presented, aligning with early Shiism. The Imams, being impeccable and infallible heirs of the Prophet, are considered the only qualified interpreters of the Qurʾān. Although the Imams are seen as authors and revealers of Qurʾānic exegesis, al-ʿĀmilī l-Iṣfahānī stresses the importance of not neglecting the exoteric content. The second introduction introduces the belief that the Qurʾān has undergone alteration following Muḥammad's death, emphasizing the esoteric aspect containing the Imams' Divine Alliance. Al-ʿĀmilī l-Iṣfahānī suggests that God's intentional concealment of esoteric truths protects them from falsification. This claim becomes a source for later Shiite criticism of the official Qurʾān. The third introduction reaffirms the esoteric nature of the Qurʾān, focusing on the Imams' Divine Alliance and leadership. The bulk of the book is a lexicon of thirteen thousand Qurʾānic terms, offering literal and esoteric explanations based on Imami traditions. The entries reflect Imamo-centrism, anti-Sunnism, and messianic themes. Finally, al-ʿĀmilī l-Iṣfahānī's work introduces the philosophical notion of God's "art of writing between the lines" in the Qurʾān, justifying the need for hermeneutics through historical and theological contexts.

Ismāʿīl Ḥaqqī l-Brūsawī (d. 1725/1137) was a versatile scholar with a strong Sufi identity. His major work, *Rūḥ al-bayān*, is a comprehensive commentary on the Qurʾān that reflects his Sufi perspective. Drawing from various sources, including traditional commentaries like *al-Kashshāf* and Sufi commentaries, al-Brūsawī crafted a commentary that combines elements of transmission, opinion, and interpretation through the Qurʾān itself.

Al-Brūsawī's approach to knowledge reflects a nuanced view. While he values reason, he emphasizes that it is valid only within its limits. Al-Brūsawī underscores the importance of certain knowledge (*yaqīn*) and the Oneness of God (*tawḥīd*), which can only be achieved by abandoning blind imitation (*taqlīd*). Regarding knowledge transmission, al-Brūsawī confines it to matters related to the sharia. He contends that true understanding comes through divine inspiration (*ilhām*) rather than mere transmission. He sees the mind as acceptable only when guided by revelation (*waḥy*), emphasizing the role of divine guidance in human understanding. He asserts that direct knowledge from a divine source to the heart is essential for understanding both the physical and metaphysical worlds.

Al-Brūsawī's Sufi identity strongly influences his Qur'ānic hermeneutics. His commentary, formed from sermons, includes Sufi material, parables, and symbolic interpretations, making it accessible to a wider public. Symbolic comments, known as *ishārī*, are carefully selected and presented in a softened manner, often validated with Qur'ānic verses and hadiths. Despite criticism for including weak and fabricated narrations, *Rūḥ al-bayān* has maintained its popularity, influencing subsequent works and serving as a valuable source for preaching and teaching activities in Arabic and Turkish.

In the realm of *tafsīr* literature and Qur'ānic hermeneutics, Shāh Walī Allāh (d. 1762/1176) holds a significant position. His impact on Muslim religious thought in South Asia is noteworthy, making him a key figure in periodizing intellectual trends. While various scholars, even those with differing views, have appropriated Walī Allāh's ideas, his work remains open to multiple interpretations.

Shāh Walī Allāh's Persian translation of the Qur'ān, *Fatḥ ar-Raḥmān*, is a crucial milestone in his larger Qur'ān project. This translation was intended to reach a broad audience, including both common people and specialists. Although it faced criticism for potentially diminishing the status of Arabic in Islamic sciences, Shāh Walī Allāh's influence, particularly through his emphasis on revisiting and preserving traditional positions on Qur'ānic exegesis, gained prominence in British India from the nineteenth century onward.

Shāh Walī Allāh's approach to Qur'ānic hermeneutics can be understood through three main aspects. First, he categorized various types of Qur'ānic narratives to help readers comprehend different forms of knowledge conveyed by the scripture. Second, he engaged with the evolution of the Muslim intellectual tradition, addressing key debates in Qur'ānic sciences (*'ulūm al-Qur'ān*). His acknowledgment of challenges in interpreting the Qur'ān led observers to label him both "traditionalist" and "modernist." Third, his hermeneutics emphasizes the divinely bestowed nature of understanding the Qur'ān. He claimed that interpreting the text is a special gift from the Almighty and kept open the possibility that selected believers could receive divine guidance in interpretive matters. His method involves understanding Arabic rhetoric and idiom, as well as relying on the sayings of early generations of Muslims.

Shāh Walī Allāh's impact on Qur'ānic hermeneutics triggered discussions on the translation of the Qur'ān into vernacular languages, challenging the traditionally Arabic exegesis in South Asia. His hermeneutics played a role in enabling Muslim scholars to grapple with emerging intellectual modernity, with his ideas being widely appropriated throughout the nineteenth and twentieth centuries.

Aḥmad b. ʿAjība (d. 1809/1223) is renowned not only for his role in the Shādhiliyya-Darqāwiyya Sufi order and his influential treatises on Sufism, but also for his comprehensive Qur'ān commentary *al-Baḥr al-madīḍ fī tafsīr al-Qur'ān al-majīd*. This work harmonizes expressive (*bi-l-ʿibāra*) and allusive (*bi-l-ishāra*) interpretations, reflecting his profound knowledge of Islamic sciences and spiritual mastery.

Ibn ʿAjība's hermeneutical method, elucidated in various works, underscores the relationship between the exoteric (divine law) and esoteric (divine reality) dimensions of the Qur'ān. He criticizes the limitation of Qur'ānic meaning to literal and historical aspects and instead advocates for both exoteric and esoteric interpretations for the reason that exoteric exegesis is relevant for scholars and Sufis, but only the latter delve into spiritual exegesis, drawing from experiential knowledge granted to saints.

The distinction between *ʿibāra* and *ishāra* is central for Ibn ʿAjība's Qur'ānic hermeneutics. *Ishāra*, allusion, holds intermediate status between clear expression (*ʿibāra*) and symbol (*ramz*). *Ishāra* serves beginners by facilitating their experience of divine presence, guiding them towards an understanding of the spiritual dimension of the Qur'ān. Advanced individuals, having merged with the divine essence, surpass the need for *ishāra*, as they are connected directly with God.

Ibn ʿAjība's commentary unfolds on three levels: linguistic, exoteric, and esoteric. He correlates these with the dimensions of the word, human being, worlds, religion, and certitude. He considers historical details secondary to ethical teachings and the sacred history of revelation with the aim of enriching spiritual insights. In his hermeneutics, the concept of *munāsaba*, the thematic interconnectedness between the suras, is central because it signifies the coherence of the Qur'ān. On the other hand, linguistic analysis clarifies difficult passages, expanding the meanings of the Qur'ānic text. Accordingly, Ibn ʿAjība employs classical commentaries, emphasizing that the exoteric and esoteric interpretations complement each other, fostering a broader audience.

According to Ibn ʿAjība, the esoteric commentary serves as a hermeneutical key, enabling readers to experience spiritual truths. *Ishāra* mediates between historical and spiritual meanings, actualizing the Qur'ānic revelation. Ibn ʿAjība's emphasis on divine reality, prophetic heritage, and theophany aligns with the Sufi tradition. His exegesis seeks to sensitize readers to the spiritual dimension, presenting the Qur'ān as a theophany and guide to divine presence. Moreover, his Qur'ānic hermeneutics provides an insightful synthesis of North African Sufism, classical Sufi exegesis, and a response to contemporary spiritual needs. His work remains a valuable source on Muslim spirituality and Sufi exegesis in the 18th and 19th centuries and beyond.

The eighteenth-century South Asian scholar Thanāʾ Allāh Pānīpatī (d. 1810/1225) was a Muslim judge with a broad knowledge of Arabic, Persian, Hindi, and Sanskrit. He engaged with various Islamic sciences and scholars, learning from important figures like Mirzā Maẓhār Jan-i Jānāṇ and Shāh Walī Allāh. His writings covered diverse subjects, including Sufism, Islamic jurisprudence, criticism on Shiism, and moral guidance. His Qurʾān commentary, *at-Tafsīr al-maẓharī* enjoyed big influence, particularly among traditionalist circles like the Deobandīs in South Asia.

The hermeneutical principles underlying Pānīpatī's Qurʾān commentary are related to three aspects. First, he appears to build on the work of Shāh Walī Allāh. Second, he critically examines traditional disciplines, contributing to an emerging genre of literature in South Asia, which aims at the reconciliation (*taṭbīq*) of competing religious insights that came from multiple disciplines of Islamic traditional sciences. Third, Pānīpatī employs a sophisticated approach to mystical commentary (*at-tafsīr al-ishārī*) within the classical tradition of Qurʾānic exegesis.

Pānīpatī shares with Shāh Walī Allāh similar hermeneutical views on the historical reports of *asbāb an-nuzūl* and the use of *isrāʾīliyyāt* and *naskh*. Both scholars strive to reconcile traditionist and deductive approaches in Qurʾānic interpretation. Pānīpatī creatively uses traditional sources, including hadiths and variant readings, to reconcile conflicting views within the Muslim intellectual tradition. His hermeneutical approach aims to root Qurʾānic interpretation in centuries-old tradition while offering a coherent understanding. His commentary is characterized as semantically polyvalent, allowing for multiple meanings. His use of classical exegesis serves the purpose of emphasizing unity within the Muslim tradition despite its diversity. Pānīpatī also integrates esoteric commentary principles, following the path outlined by earlier scholars like ash-Shāṭibī and as-Suyūṭī, aligning mystical interpretation with Qurʾānic vocabulary, human reason, and classical Arabic idiom.

The final chapter in this volume explores Qurʾānic hermeneutics during the Ottoman period, emphasizing the insufficient coverage of this era, although it can no longer be considered a "missing link". The Ottoman exegetical activities were characterized by reciting, memorizing, and understanding the Qurʾān, and depending on earlier scholars like al-Khwārizmī, ar-Rāzī, and al-Bayḍāwī.

Qurʾānic hermeneutics in the Ottoman era is categorized into four groups: full translations of the Qurʾān, translations of selected verses, supercommentaries (*sharḥ/ shurūḥ*) and glosses (*ḥāshiyas*) of classical commentaries, and translations of such commentaries. Arabic was the preferred language, but Turkish and Persian were also used. The most translated work was *Tafsīr al-Qurʾān al-karīm* by as-Samarqandī (d. 983/373).

Two main schools of Qurʾānic hermeneutics emerged in the Ottoman period: exegesis through personal reasoning (*at-Tafsīr bi-d-dirāya*) on one hand, and supercommentaries and marginal explanations. The former, rooted in classical commentaries, integrates language, theological, judicial, and Sufi ideas. The latter, involving explanations and annotations, aimed to enrich existing heritage and culture.

The Ottoman tradition of Qur'ānic hermeneutics evolved with influences from theology, Sufism, philosophy, and later, mathematics and astronomy. Ghāzī Aḥmad Mukhtār Pāshā's *Sarā'ir al-Qur'ān* is considered a pioneering work in scientific exegesis. The Ottoman exegetical impact is evident in the first Turkish commentary from modern Turkey.

This overview demonstrates the close relationship between Qur'ānic hermeneutics and the respective historical contexts in which they were developed. As has become clear, the scholars treated in this volume responded to the needs of their age and place by making Qur'ānic meanings accessible to their own societies.

Authors from different regions and academic disciplines have contributed to this volume. As each volume of the *Handbook* manifests, the history of Qur'ānic hermeneutics is marked by its diversity; its scholarly treatment must be just as diverse, as manifested in this volume and all other volumes of the *Handbook*.

Some other scholars of Qur'ānic hermeneutics who were supposed to be included in this volume are unfortunately missing. They include Muḥammad b. Ḥasan ash-Shaybānī (d. 1243/640), Abū l-Makārim Maḥmūd b. Ḥasanī Wā'iẓ (thirteenth/seventh century), Sayyid b. Ṭāwūs (d. 1266/664), Ibn an-Naqīb (d. 1298/698), Abu Ja'far b. az-Zubayr (d. 1308/708), Qutb ad-Din ash-Shīrāzī (d. 1311/710), Abū Manṣūr 'Alī b. al-Muṭṭahhir al-Ḥillī (d. 1325/726), 'Imād ad-Dīn al-Kindī (d. 1340/741), 'Alā' ad-Dīn 'Alī b. Muḥammad al-Baghdādī, known as al-Khāzin (d. 1341/741),[4] Abū Ḥayyān al-Gharnāṭī (d. 1344/745), Muḥammad b. Yūsuf Abū Ḥayyān (d. 1353/754), Aḥmad b. Yūsuf as-Samīn al-Ḥalabī (d. 1355/756), Muḥammad b. Ya'qūb ash-Shīrāzī l-Fayrūzabādī (d. 1414/816), Ibn 'Ādil al-Ḥanbalī (d. 1474–75/879), 'Abd ar-Raḥmān ath-Tha'ālibī (d. 1479/876), Hāshim b. Sulaymān al-Baḥrānī (d. 1696/1107) and Aḥmad al-Aḥsā'ī (d. 1826/1241). Although it is almost impossible to cover exhaustively all hermeneutical approaches to the Qur'ān in this or any other era of Qur'ānic scholarship, the outlined shortcomings will hopefully be corrected in a future, updated version of this volume.

4 Al-Khāzin's work is discussed in the chapter on al-Baghawī, on whom he is dependent, in Volume 2 of the *Handbook*.

Muḥyī d-Dīn b. ʿArabī

Gregory Vandamme

1 Introduction

Abū ʿAbd Allāh Muḥammad b. ʿAlī b. Muḥammad b. Aḥmad b. ʿAbd Allāh b. al-ʿArabī ṭ-Ṭāʾī l-Ḥātimī (1065–1240/560–638)[1] is certainly one of the most prominent figures of classical Sufi literature. Ibn ʿArabī's importance in the development of Sufi thought stems from his unique capacity to synthetize and enhance the doctrines and technical vocabulary of early Sufism, found in the writings of seminal figures such as Dhū n-Nūn al-Miṣrī (d. 856/245), Sahl at-Tustarī (d. 896/283) or Abū l-Qāsim al-Junayd (d. 910/298).[2] But the scope of his numerous works goes far beyond the area of Sufism; rather, as Chittick observed, Ibn ʿArabī actually addresses "every intellectual dimension of Islam" in a way that is "both intensely loyal to the tradition and exceedingly innovative."[3]

Ibn ʿArabī's sophisticated works aroused such strong enthusiasm that he was later entitled *ash-Shaykh al-Akbar* ("the greatest master") by his admirers. On the other hand, his thought-provoking works have been repeatedly placed at the center of polemical discussions about the "orthodoxy" of Sufi doctrines, often simply repeating the primary objections formulated by Taqī d-Dīn Aḥmad b. Taymiyya (d. 1328/728).[4] Despite those polemics, the influence of Ibn ʿArabī on Islamic intellectual history remained undoubtedly strong, particularly in the Ottoman and Persianate worlds, through the numerous commentaries on his works written by significant figures such as Ṣadr ad-Dīn al-Qūnawī (d. 1274/673), ʿAbd ar-Razzāq al-Kashānī (d. 1330/730), Dāwūd

1 He is usually called "Ibn ʿArabī" instead of the original "Ibn al-ʿArabī", in order to distinguish him from the Andalusian jurist, Abū Bakr b. al-ʿArabī, see Stephen Hirtenstein, "Manuscripts of Ibn ʿArabī's Works: Names and Titles of Ibn (al-)ʿArabī," *Journal of the Muhyiddin Ibn Arabi Society* 41, (2007): 109–29.

2 For a succinct but profound overview on Ibn ʿArabī's life and thought, see Claude Addas, *Ibn Arabî et le voyage sans retour* (Paris: Points, 1996), and William C. Chittick, *Ibn ʿArabi: Heir to the Prophets* (Oxford: Oneworld, 2005). These authors also wrote what are the most important scientific biographies and doctrinal synthesis on Ibn ʿArabī to this day: Claude Addas, *Ibn ʿArabī ou la quête du soufre rouge* (Paris: Gallimard, 1989), and William C. Chittick, *The Sufi Path of Knowledge: Ibn al-ʿArabī's Metaphysics of Imagination* (Albany: State University of New-York Press, 1989), and *The Self-Disclosure of God: principles of Ibn al-ʿArabi's cosmology* (Albany: State University of New-York Press, 1998). We must also mention here the important book by James W. Morris, *The Reflective Heart. Discovering Spiritual Intelligence in Ibn ʿArabī's Meccan Illuminations* (Louisville: Fons Vitae, 2005).

3 William C. Chittick, "Ibn ʿArabī" in *History of Islamic Philosophy,* eds. Seyyed Hossein Nasr and Oliver Leaman (London & New-York: Routledge, 2001), 497–98.

4 For an overview of these polemics, see Alexander D. Knysh, *Ibn ʿArabi in the Later Islamic Tradition. The Making of a Polemical Image in Medieval Islam* (Albany: State University of New-York Press, 1999).

https://doi.org/10.1515/9783111320083-002

al-Qayṣarī (d. 1350/751), ʿAbd al-Karīm al-Jīlī (d. 1412/815), and ʿAbd ar-Raḥmān al-Jāmī (d. 1492/898).[5]

2 Life and Works

Ibn ʿArabī was born in Murcia, Andalusia in 1165/560, to a prominent noble family linked to the local ruler and subsequently to the Almohad government, when they moved to Seville. After years of living the usual life of the elite youth, receiving a traditional education and enjoying various earthly pleasures, the young Ibn ʿArabī experienced an existential crisis and spiritual calling that lead him to abandon his former life in order to devote himself to the spiritual path.[6] Most of his adult life was then spent in travels, first between the Maghreb and Andalusia until 598/1201, and later through Hijaz and the Middle East, until he settled in Damascus in 620/1223, where he died in 638/1240. Ibn ʿArabī was a very prolific writer. Although more than three hundred titles were ascribed to him,[7] scholarship now considers 120 books to be a more accurate number. Among those, we find various short treatises and epistles (some gathered in various *Rasāʾil* volumes), longer works such as his famous *Fuṣūṣ al-ḥikam*, *Tarjumān al-ashwāq* and *Mawāqiʿ an-nujūm*, and a vast spiritual encyclopedia: *al-Futūḥāt al-makkiyya.*

The unique writing style of Ibn ʿArabī blends a large variety of intertwined genres: precise theological and philosophical reflections are punctuated by poetry or allusive and symbolic comments, and discussions on Islamic jurisprudence are filled with metaphysical or astrological notions. His works can be linked to a large extent with previous Sufi literature. For example, in his *Futūḥāt*, the long chapter 73 is dedicated to answering the 157 questions posed by al-Ḥakīm at-Tirmidhī (d. 930/360) in his *Khatm al-awliyāʾ*, and the first chapters of the section on the *muʿāmalāt* follow precisely the order found in the *Risāla* of ʿAbd al-Karīm al-Qushayrī (d. 1072/465).[8] This is

5 For more details on his overall influence on later Islamic thought, see the series of articles by James W. Morris "Ibn ʿArabi and His Interpreters Part I: Recent French Translations"; "Ibn ʿArabi and His Interpreters Part II: Influences and Interpretations"; "Ibn ʿArabi and His Interpreters Part II (Conclusion): Influences and Interpretations," *JAOS* 106, no. 3 (1986): 539–51; 106, no. 4 (1986): 733–56; 107, no. 1 (1987): 101–19. See also the Introduction of Michel Chodkiewicz, *Un océan sans rivage. Ibn ʿArabî, le Livre et la Loi* (Paris: Éditions du Seuil, 1992), 17–37.

6 See Ibn ʿArabī, *al-Futūḥāt al-makkiyya* 4, vols. (Beirut: Dār al-Fikr, no date), 4: 552. From now on cited: *Fut.* vol., pages.

7 See Osman Yahya, *Histoire et classification de l'œuvre d'Ibn ʿArabī. Étude critique*, 2 vols. (Damascus: Institut français de Damas, 1964).

8 See Michel Chodkiewicz, *"Miʿrâj al-kalima.* De la *Risâla Qushayriyya* aux *Futûhât Makkiyya,"* in *Reason and Inspiration in Islam. Theology, Philosophy and Mysticism in Muslim Thought. Essays in Honour of Hermann Landolt*, ed. Todd Lawson (London/New-York: I.B. Tauris/The Institute of Ismaili Studies, 2005), 248–61.

not to mention the frequent references and quotations of both renowned and obscure Sufi figures throughout his works.

In addition to classical Sufi literature, intertextuality with more peculiar texts pertaining to the Andalusian milieu in which Ibn ʿArabī grew up can also be found in his works: he avowedly wrote his *at-Tadbīrāt al-ilāhiyya* as a spiritual comment to the famous pseudo-Aristotelian *Sirr al-asrār*,[9] and he commented on passages of the *Khalʿ an-naʿlayn*, written by the proclaimed "Imam" who rebelled against the Almoravids, Aḥmad b. al-Ḥusayn b. al-Qāsī (d. 1151/546). Special attention should be given to the ambiguous case of the *Muʿtabirūn* (i.e. the proponents of a hermeneutical method called *iʿtibār*, that could be translated as "symbolic transposition"), which is sometimes considered as a distinctive movement, although it has often been simply reduced to an Andalusian expression of Sufism.[10] Its two major proponents, Abū ʿAbd Allāh Muḥammad b. Masarra (d. 931/319) and Abū l-Ḥakam b. Barrajān (d. 1141/536), are quoted by Ibn ʿArabī in several works. The latter wrote a treatise on the divine names that seems to foretell the formulation of this central doctrine in Ibn ʿArabī's works, and more notably two Qurʾānic commentaries based on the notion of *iʿtibār*, according to which both Revelation and Creation need to be interpreted in a spiritual sense. This approach will be extensively used by Ibn ʿArabī in his own Qurʾānic hermeneutics.[11] The *Muʿtabirūn* also developed an elaborate use of the science of letters (*ʿilm al-ḥurūf*) in their hermeneutics. This is already found in early Sufism, as shown in the compilation of references given by ʿAbd ar-Raḥmān as-Sulamī (d. 1021/412) in his Qurʾānic commentary.[12] An epistle attributed – probably erroneously – to Sahl at-Tustarī was used by Ibn Masarra to write his own treatise on the subject, the *Kitāb Khawāṣṣ al-ḥurūf*, in which he depicts the Arabic letters as the fundamental elements of the universe, giving a special importance to the fourteen "isolated" letters opening twenty-nine chapters of the Qurʾān.[13] A very similar cosmological use of those letters will be found in the very dense second chapter of Ibn ʿArabī's *Futūḥāt*, dedicated to the science of letters.

---

9 See Ibn ʿArabī, *at-Tadbīrāt al-ilāhiyya fī iṣlāḥ al-mamlaka al-insāniyya*, ed. Ibrāhīm al-Kayyālī (Beirut: Dār al-Kutub al-ʿIlmiyya, 2003), 20.

10 See Yousef Casewit, *The Mystics of al-Andalus: Ibn Barrajān and Islamic Thought in the Twelfth Century* (Cambridge: Cambridge University Press, 2017).

11 See Denis Gril, "La 'lecture supérieure' du Coran selon Ibn Barrağān", *AR* 47 (2000): 510–22; Denis Gril, "L'interprétation par transposition symbolique (*iʿtibār*), selon Ibn Barrağān et Ibn ʿArabī," in *Symbolisme et herméneutique dans la pensée d'Ibn ʿArabī. Actes du colloque*, ed. Bakri Aladdin (Damascus: Institut français du Proche-Orient, 2007), 147–61. See also José Bellver, "Ibn Barrağān and Ibn ʿArabī on the prediction of the capture of Jerusalem in 583/1187 by Saladin," *AR* 61, nos. 3–4 (2014): 252–86.

12 See Jean-Jacques Thibon, "L'herméneutique de Sulamī annonce-t-elle celle d'Ibn ʿArabī?," in *Symbolisme et herméneutique dans la pensée d'Ibn ʿArabī*, 21–39.

13 See Michael Ebstein, *Mysticism and Philosophy in al-Andalus. Ibn Masarra, Ibn al-ʿArabī and the Ismāʿīlī Tradition* (Leiden/Boston: Brill, 2014) 12–13 and 61–64.

The Shaykh al-Akbar regularly mentions *kalām* or *fiqh* doctrines from various schools, but usually refuses to come out in favor of one party, mostly advocating for a synthetic position without entering in a proper scholastic discussion. Although a number of textual and biographical elements seem to denote a certain attachment of Ibn ʿArabī to the Ẓāhirī school of *fiqh*, through the works of his fellow country man Abū Muḥammad ʿAlī b. Ḥazm (d. 1054/456), this attachment should be carefully qualified, as Addas recently showed.[14]

Regardless of the variety of those influences, the thought of Ibn ʿArabī should be seen overall as a synthesis and enhancement of earlier Sufi doctrines, rather than as an idiosyncratic and original appropriation of that heritage blended with miscellaneous heterogenic sources. His hermeneutical method does not sensibly differ from what we can find in the most seminal Sufi works of the previous generations, in that he stresses above all the need to base one's understanding of every reality on one's spiritual experience, rather than on the unquestioned adoption (*taqlīd*) of concepts and doctrines. The emphasis on actual spiritual realization (*taḥqīq*) is therefore at the core of Ibn ʿArabī's Qurʾānic Hermeneutics.

3 Ibn ʿArabī's Qurʾānic Hermeneutics: Approaching God through the Text

Qurʾānic hermeneutics is at the very center of Ibn ʿArabī's thought. His oeuvre, taken in its entirety, could be considered as one immense commentary on the Qurʾān, approached by every perspective at hand: metaphysics, cosmology, linguistics, poetry, and even what one might call "mythology."[15] As he himself declares: "Everything we talk about – in my gatherings and my writings – comes from the presence of the Qurʾān and its treasuries".[16]

14 Claude Addas, "The 'Radiant Way'. Ibn ʿArabī and the Interpretation of the Law. Part I: The Legacy of Ibn Hazm"; "The 'Radiant Way'. Ibn ʿArabī and the Interpretation of the Law. Part II: The Law and its Interpretation," *Journal of the Muhyiddin Ibn Arabi Society* 65 (2019): 29–57; 66 (2019): 39–76.

15 See the "mythic dialectic" analyzed in Michael A. Sells, "Ibn ʿArabī's Garden among the Flames: A Reevaluation", *History of Religions* 23, no. 4 (May, 1984): 287–315, and Michael Ebstein, "'In Truth You Are the Polytheist!': Mythic Elements in Ibn al-ʿArabī's Teachings on the Divine Names," *Intellectual History of the Islamicate World* 6 (2018): 359–87.

16 *Fut.* 3: 334.

3.1 Intertextuality between Ibn ʿArabī's Works and the Qurʾān

The intricate relation between the Qurʾānic text and Ibn ʿArabī's writings appears at least on three different levels. First, his abundant borrowing from the vocabulary and phrasing of the Revelation: his works are filled with direct quotations, and even more with allusions to Qurʾānic verses and motifs, not always obvious to a reader who is not as familiar with the Revelation as were Ibn ʿArabī and his audience. Second, his mimetic use of the perplexing structure – some might consider it an absence of structure – of the Qurʾānic text: Ibn ʿArabī's writings are far less systematically articulated than those of the Aristotelian philosophers for example, but the apparent disarray usually hides intentional cross-references and patterns, most of them directly related to the structure of the Qurʾānic text itself, as Miftāḥ and Chodkiewicz brilliantly showed.[17] Third, Ibn ʿArabī draws from the Revelation itself one of his core conceptions, that is the intrinsic correlation between the "signs" (*āyāt*) – i.e. everything that can be known – in Qurʾān, Cosmos, and Human Being itself.[18] When we take this last point into account, we understand that the nature of the commentary proposed by Ibn ʿArabī depends heavily on the correlation between the Qurʾānic text and those two analogous elements. In a certain way, he extended the approach of the *iʿtibār* by adding a third object of contemplation and symbolic transposition, which is that of the human psychological and spiritual reality in itself.

No particular work of Ibn ʿArabī could be considered as a compendium of his Qurʾānic hermeneutics, and we have to look throughout his vast literary production in order to compile his different views on the Revelation.[19] Historically authenticated *ijāzāt* indicate however that he wrote at least three Qurʾānic commentaries: *al-Jamʿ wa-t-tafṣīl fī asrār maʿānī t-tanzīl* (sixty-four exegetical volumes covering the text until the middle of sura al-Kahf (the Cave, Q 18:59)), which is unfortunately lost; the *Iʿjāz al-bayān fī t-tarjama ʿan al-Qurʾān*, which covers up to Q 2:253;[20] and the *Ishārāt al-Qurʾān fī ʿālam al-insān*, which provides elliptical allusions commenting on various verses presented in the canonical order of the chapters, and which was apparently written as an appendix to his *Kitāb at-tanazzulāt al-Mawṣiliyya*.[21] It should be noted that none of these commentaries corresponds to the so called *Tafsīr Ibn ʿArabī* widely edited in the Muslim world, which is actually ʿAbd ar-Razzāq al-Kāshānī's *Taʾwīlāt al-Qurʾān*.

---

17 See Chodkiewicz, *Un océan sans rivage*, and ʿAbd al-Bāqī Miftāḥ, *al-Mafātīḥ al-Qurʾāniyya li-Kitāb at-Tarājim wa-li-Kitāb ash-Shāhid li-Ibn ʿArabī* (Damascus: Dār Naynawā, 2019).

18 See Q 41:53: "We will show them our signs (*āyātinā*) on the horizons and in themselves (or "in their souls") (*fī anfusihim*) until it becomes clear to them that it is the Real". (Translations from the Qurʾān are mine).

19 See for instance the compilation by Maḥmūd Ghurāb, *Raḥma min ar-Raḥmān fī tafsīr wa-ishārāt al-Qurʾān min kalām ash-shaykh al-akbar Muḥyī d-Dīn b. ʿArabī*, 4 vols. (Damascus: n.e., 1989).

20 For an edition of the text, see the first volume of Ghurāb, *Raḥma min ar-Raḥmān*.

21 For an edition of the text, see ʿAbd al-Bāqī Miftāḥ, *Shurūḥ ʿalā Kitāb Ishārāt al-Qurʾān fī ʿālam al-insān* (Damascus: Dār Naynawā, 2016).

In addition to those commentaries, several other works directly or indirectly comment on the Qur'ānic text. In his *Dīwān*, a continued series of one hundred and fourteen poems are said to be inspired by "the spirits of the Qur'ānic chapters" (*arwāḥ as-suwar*).[22] The fourth section of the *Futūḥāt*, covering chapters 280 to 393, is also based on the one hundred and fourteen chapters of the Qur'ān. Each chapter of this section is dedicated to a particular spiritual "dwelling place" (*manzil*) which is linked to a Qur'ānic chapter, following a reverse order (from sura 114, *an-Nās* (Mankind) to sura one, *al-Fātiḥa* (The Opening)).[23] Unlike the poems from the *Dīwān*, the link between those chapters and the Qur'ān is not clearly stated in the section, and Ibn ʿArabī seems to expect his readers to discover by themselves the correlation. This kind of hermeneutical test seems to be part of Ibn ʿArabī's pedagogical plan to make the reader increasingly aware of the hidden intricacies of the Qur'ānic text itself: "The formalist scholar (*ʿālim ar-rusūm*) does not see (such verses) as correlated, because this belongs to the knowledge (*ʿilm*) of God. Yet we know that there is a correlation, even if it is deeply hidden".[24] Several works of Ibn ʿArabī show a secret affinity with a particular passage of the Qur'ān, as for example the hidden relation between the various theophanies described in the *Kitāb at-Tajalliyāt* and the verses of the second sura of the Qur'ān, *al-Baqara* (The Cow).[25] As shown in the works of Chodkiewicz, the correlations between different passages of Ibn ʿArabī's writings revolving around the same Qur'ānic verse reveal in the last analysis a hidden pattern in his entire oeuvre.[26]

3.2 Revelation, Manifestation and Allusion

For Ibn ʿArabī, the Qur'ānic revelation is to be considered as a particular modality of the universal Self-Manifestation of God. It is precisely named "*Qur'ān*" – from the root *q-r-'* that carries the meaning of reading or recitation, but also of collecting and gathering – because it provides the hermeneutical keys to unify all the signs (*ʿalam*, pl. *aʿlām*) composing the universe (*ʿālam*). But it is also named "*Furqān*"[27] – from the root *f-r-q* that carries the meaning of division and distinction – because it provides the distinctive criterion that allows us to distinguish between the Creator, the Creation, and the different levels of Manifestation.[28] These two aspects are considered by Ibn ʿArabī

22 Chodkiewicz, *Un océan sans rivage*, 104.

23 Ibid., 86–90

24 *Fut.* 3: 200.

25 Chodkiewicz, *Un océan sans rivage*, 109–16.

26 Ibid., 118.

27 See Q 2:185 where it is directly associated with the descent of Revelation: "The month of Ramadan is the one in which the Qur'ān was sent down as guidance for mankind (*nās*), the explanation of the guidance, and the distinguishing-criterion (*al-Furqān*)".

28 See *Fut.* 2: 88, and Chittick, *The Sufi Path of Knowledge*, 240–41.

to be linked to the classical articulation between the outward or extrinsic (*ẓāhir*) and the inward or intrinsic (*bāṭin*) aspects of every reality. The Revelation is thus *Furqān* in its extrinsic aspect, in that it is made of distinctive letters, words and formulations, and it is *Qurʾān* in its intrinsic aspect, in that it carries universal meanings pointing to a unique Reality.

The hermeneutical method of Ibn ʿArabī consists therefore in a twofold approach to the Qurʾānic text, oscillating between the scrutiny of its apparent literal form and the scrutiny of the inner experience of the reader: "God – Most-High – said: 'We will show them our signs (*āyātinā*) on the horizons and in themselves' (Q 41:53), which means 'the verses (*āyāt*) that are bestowed concerning the horizons and concerning themselves.' Every revealed verse therefore has two aspects: an aspect which they see within themselves, and an aspect which they see outside of themselves."[29] In other words, the Qurʾānic verses are providing synthetic allusions, that concern our experience of both the outside world and our own interiority.

This is why chapter 54 of the *Futūḥāt* is entirely dedicated to "knowledge of the allusions" (*maʿrifat al-ishārāt*).[30] Ibn ʿArabī explains there the difference between the exegetical approach of the "People of God" (*ahl Allāh*) and that of the "formalist scholars" (*ʿulamāʾ ar-rusūm*). According to the Shaykh al-Akbar, the first are those who keep the balance between the interior and exterior aspects of the Qurʾānic verses, whereas the second "have preferred this world to the next, and the side of Creation to the side of the Real."[31] The People of God, Ibn ʿArabī explains, name their exegetical method "allusions" (*ishārāt*)[32] rather than "exegesis" (*tafsīr*), in order to prevent the condemnations of the formalist scholars, but their respective approach to the Qurʾānic text differs even more radically, in that People of God consider that their understanding must necessarily come from the same source as Revelation itself, rather than from their own human endeavor. In short, according to Ibn ʿArabī, the understanding of the Qurʾān presented by the People of God in their commentaries (*shurūḥ*) is revealed to them, and its formulation is nothing but a rendition of their direct experience of the text. It is therefore not the fruit of an intellectual process, nor a blind imitation of the understanding of their predecessors. This is exemplified precisely in the fact that, according to Ibn ʿArabī, the technical vocabulary (*iṣṭilāḥ*) they use in their commen-

---

29 *Fut.* 1: 279.

30 Ibid., 278–81. See also Chittick, *The Sufi Path of Knowledge*, 246–50, and Kristin Zahra Sands, *Ṣūfī Commentaries on the Qurʾān in Classical Islam* (London/ New-York: Routledge, 2006) 39–41.

31 *Fut.* 1: 279.

32 This notion is already found in early Sufi sources. See for instance Abū Naṣr as-Sarrāj aṭ-Ṭūsī, *Kitāb al-Lumaʿ*, ed. ʿAbd al-Ḥalīm Maḥmūd (Cairo: Maktabat ath-Thaqāfa ad-Dīniyya, 2002) 147–48. A famous example of a Qurʾānic commentary named after this notion is Qushayrī's *Laṭāʾif al-ishārāt*. On the specificity of Sufi exegesis, see Sands, *Ṣūfī Commentaries*, and Sara Sviri, "The Countless Faces of Understanding: On *Instinbāṭ*, Mystical Listening and Sufi Exegesis," in *The Spirit and the Letter. Approaches to the Esoteric Interpretation of the Qurʾān*, eds. Annabel Keeler and Sajjad Rizvi (Oxford/ New-York: Oxford University Press/Institute of Ismaili Studies, 2016) 51–85.

taries is not a "code" that could be taught or learned. It is only when one experiences by oneself the very reality to which a precise technical term refers that one understands the allusion of the text. This peculiar conception of terminology, Ibn ʿArabī adds, is one of the characteristic features of the way of the People of God, and is nowhere to be found in other sciences.

The principle of "allusion" is actually rooted in the very nature of existence itself for Ibn ʿArabī. Those who realize the true, unique and divine nature of being are forced to acknowledge the incapacity of language to tell things as they are in themselves. Our experience of the reality of being, as well as our repeated and always frustrating attempts to formulate an adequate ontology should lead us to realize that every formulation is nothing other than an "allusion": "There is nothing in being (*wujūd*)[33] except God, whereas we are brought into existence (*mawjūdīn*). Our being is therefore by Him, and the one whose being is by another is under the rule of non-being (*ʿadam*). The allusion is therefore established, and its rule appears: the clarification (*bayān*) of what is intended by it is inevitable."[34]

3.3 Interpreting the Text as an "Illiterate"

The foundational role of allusion probably explains why Ibn ʿArabī is so reluctant about the use of the term "*taʾwīl*" and the kind of rational hermeneutical method to which it refers in his view. Contrary to the insistence of Henry Corbin on the centrality of the notion in Ibn ʿArabī's thought,[35] we find the latter considering the "active" exegetical inquiry as the lowest level of interpretation, whereas the "passive" inspiration of those who are directly taught by God is the only exegetical approach that conforms to the famous Qurʾānic verse: "No one knows its interpretation (*taʾwīl*), except God and those firmly rooted in knowledge" (Q 3:7). In Ibn ʿArabī's own words, the former "makes the Book lie down after it had been standing" while the latter "makes the Book stand up after it had been lying down".[36] Here again, we see how for Ibn ʿArabī, the spiritual experience prevails over any doctrinal formulation or belief. This ex-

33 On the notion of *wujūd* in Ibn ʿArabī and its ontological implications, see Gregory Vandamme, "L'ontologie corrélative d'Ibn ʿArabī. Métaphysique de l'être et de ses relations", in *La non-dualité: Perspectives philosophiques, scientifiques, spirituelles*, ed. Jean-Michel Counet (Louvain: Peeters, 2021) 89–102. On the legacy of the notion, see William C. Chittick, "Rūmī and *waḥdat al-wujūd*," in *Poetry and Mysticism in Islam: The Heritage of Rumi*, eds. Amin Banani, Richard Hovannisian, and Georges Sabagh (Cambridge: Cambridge University Press, 1994) 70–111.

34 *Fut.* 1: 279.

35 See the remarks in Chittick, *The Sufi Path of Knowledge*, 199–202. See also Michel Chodkiewicz, "Ibn ʿArabî dans l'œuvre d'Henry Corbin," in *Henry Corbin: philosophies et sagesses des religions du livre. Actes du colloque Henry Corbin, Sorbonne, les 6–8 novembre 2003*, eds. Mohammad Ali Amir-Moezzi, Christian Jambet, and Pierre Lory (Turnhout: Brepols, 2005), 81–91.

36 *Fut.* 2: 644. See also Chodkiewicz, *Un océan sans rivage*, 54.

plains why the only valid form of *taʾwīl* is that which comes as a direct inspiration from God, and not as the fruit of an intellectual effort.

In Ibn ʿArabī's view, everyone should address the Qurʾān as an "illiterate" (*ummī*, or literally: "as one has been conceived by one's mother"[37]), just as the Prophet who received the revelation is described in Q 7:158. But this does not mean that we should not use our rational faculties to understand the text, as long as one "does not employ one's rational examination and intellectual judgment to extract meanings and secrets which it embraces", and as long as "the heart is safe from reflective consideration."[38] In other words, to approach the Qurʾān as an "illiterate" consists in being able to suspend our rational preconceptions, in order to let the text provoke a spiritual experience that encompasses all the dimensions of the human reality.

Finally, it is also notable that, although Ibn ʿArabī considers the Qurʾān as the ultimate and all-comprehensive form of revelation, he nevertheless includes the other sacred books – at least those mentioned in the Qurʾān itself – in his hermeneutics of Revelation: "Every aspect of a verse from the Word of God – whether it be from the *Furqān*, the Torah, the Psalms (*zabūr*), the Gospel, or the Scrolls (*ṣaḥīfa*) – that can be supported by anyone who knows that language, is actually intended by God – Most-High – for that interpreter, because of his encompassing knowledge."[39] Ibn ʿArabī's use of the distinctive name "*Furqān*" in this passage seems to allude to the all-comprehensiveness of the "*Qurʾān,*" in which are included the other revelations. The *Furqān* thus designates the *Qurʾān* considered as distinct from the previous revelations, whereas the *Qurʾān* is nothing but the *Furqān* considered as comprehending them all.

4 Travelling by the Qurʾān

The inspirational understanding of the Qurʾān should be considered as being part of the Revelation itself in Ibn ʿArabī's view. In what is probably one of his earliest works, "The Book of Uncovering the Results of the Journeys" (*Kitāb al-Isfār ʿan natāʾij al-asfār*),[40] Ibn ʿArabī presents the process of revelation as an ongoing travel, in which everyone is still invited to participate.

---

37 See for instance Q 16:78.

38 *Fut.* 2: 644. See also Chittick, *The Sufi Path of Knowledge*, 235.

39 *Fut.* 2: 119.

40 Ibn ʿArabī, *Kitāb al-Isfār ʿan natāʾij al-asfār. Le dévoilement des effets du voyage*, ed. and trans. Denis Gril (Combas: L'éclat, 1994), XI–XII; and Ibn ʿArabī, *The Secrets of Voyaging. Kitāb al-Isfār ʿan natāʾij al-asfār*, trans. Angela Jaffray (Oxford: Anqa Publishing, 2016). I will refer to the Arabic text provided by A. Jaffray (from now on: *Isfār*), as it is based on Gril's edition, partially collated with the (incomplete) autograph manuscript by Ibn ʿArabī, Yusuf Aga 4859, Yusuf Aga Library, Konya (Turkey). Translations are mine.

4.1 The Qurʾānic Journey

Ibn ʿArabī's depiction of "the journey of the glorious[41] Qurʾān (*al-Qurʾān al-ʿazīz*)"[42] is founded upon the verses: "We sent it down in the Night of Determination (*laylat al-qadr*)" (Q 97:1), and "We sent it down in the blessed night" (Q 44:3), and upon a well-known tradition, according to which that descent occurred in two different times, which represent for Ibn ʿArabī two distinct modalities of the descending journey of Revelation. As considered by the traditional exegesis,[43] the Qurʾān first descended all at once (*jumlatan wāḥidatan*) upon the heaven of this world (*as-samāʾ ad-dunyā*), before descending in a splintered way (*nujūman*)[44] on the heart of Muḥammad.[45] For Ibn ʿArabī, that second modality of descent never ceased, and never will cease, as long as the Qurʾān is "recited by the tongues, covertly or openly."[46] The Qurʾān is thus still journeying since its original descent, and the second moment – or modality – of that travel did not stop with Muḥammad: "In reality, the 'Night of Determination' is still ongoing for the servant, and it is (nothing other than) his soul (*nafs*) when it has been purified and refined."[47] From Ibn ʿArabī's perspective, the process of Revelation is still occurring to everyone who participates in the journey of the Qurʾān: "The Qurʾān never ceases to be sent down. Therefore, if a man were to say: 'The Qurʾān has been sent down on me!,' he would not have lied, for the Qurʾān does not cease to travel to the hearts of those who preserve it (*qulūb al-ḥāfiẓīna lahu*)".[48]

The ongoing journey of the Qurʾān in its second modality of descent is in turn considered according to two different modalities by Ibn ʿArabī. He distinguishes between two types of descent: "the Qurʾān descending on the tongues," which remains hidden to the one who recites it "behind the veil of Nature"; and "the Qurʾān descended on the hearts," which "descends with the understanding" of its meaning. Such a conception of an inspired understanding of the Qurʾān, descending on the puri-

41 The root ʿ-*z-z* carries the polysemic sense of being "tremendous", "powerful", "beyond compare", or "glorious". I have chosen the latter because it carries a theological aspect that combines the previous meanings.

42 *Isfār*, 58–61.

43 See the various traditions in Jalāl ad-Dīn as-Suyūṭī, *al-Itqān fī ʿulūm al-Qurʾān* (Medina: Majmaʿ al-Malik Fahd li-Ṭibāʿat al-Muṣḥaf ash-Sharīf, 2005), 268–75.

44 Or, according to another transmission "*najman najman*", literally "bits by bits", or "star by star", the root *n-j-m* carrying the meaning of "star" or of "blade of grass". A. Jaffray translates it very poetically by "constellated fragments" (see *The Secrets of Voyaging*, 58), but I prefer to keep the more allusive and polysemic meaning of the Arabic term.

45 See also Q 25:32: "The disbelievers say: 'Why is the Qurʾān not revealed to him all at once?' It is so, in order for Us to strengthen thereby your heart".

46 *Isfār*, 58.

47 Ibid.

48 Ibid., 59. The word "*al-ḥāfiẓīna*" can also mean "those who memorize", and therefore actualize the journey of the Qurʾān in existence.

fied hearts of the spiritual elite, is already found in early Sufism, for example in the *Kitāb al-Lumaʿ* of Abū Naṣr as-Sarrāj (d. 988/378).[49] But, for Ibn ʿArabī, it has implications that even transcends the linguistic aspects of the text:

> The one who receives the Qur'ān on his heart knows what he recites, even though it is in another tongue than his own. He knows the meanings of what he recites, even though he does not know the meaning of the terms elsewhere than in the Qur'ān. They do not belong to his language, but he understands it when he recites them.[50]

As we can see here, this inner understanding of the meanings of the Qur'ān should definitely be linked to the method of allusion mentioned before.

4.2 The Perpetual Novelty of the Qur'ān

As a result, the Qur'ān that descends on the hearts is "always new and does not deteriorate, for it is the perpetual inspiration (*waḥy*)".[51] This perpetual novelty implies for Ibn ʿArabī that "everybody finds in (the Qur'ān) what he desires; this is why the Shaykh Abū Madyān (Shuʿayb (d. 1194/590)) said: 'The aspirant is not an aspirant unless he finds in the Qur'ān everything he desires.'"[52] Thus, the reality of the Qur'ān cannot be restricted to any of its particular understandings, for Ibn ʿArabī. Rather, it encompasses all the understandings that its ongoing descend arouses: "Every speech that does not have that encompassing generality is not the Qur'ān in reality".[53]

If Ibn ʿArabī considers the Qur'ān as "an ocean without shore,"[54] it is not because he denies the constraints of its delimited external form, nor the heritage of the exegetical and linguistic traditions on which he himself relies abundantly. Rather, it is because those delimitations are the very water of that ocean, what makes the synthetic *Qur'ān* become the distinctive *Furqān,* when it reaches the "closest heaven" of the Cosmos and the Human Being. Such an approach to the Qur'ān does not regard the text as carrying determinate meanings. This would be an immobility in what is precisely regarded as an ever-moving reality. Ibn ʿArabī's hermeneutical approach rather considers that our understanding must necessarily be renewed every time we read the text. As we have seen, the Qur'ān never ceases to descend on the hearts of those who welcome it, and this receptivity entails – among other things – that we constantly

---

49 As-Sarrāj, *Kitāb al-Lumaʿ*, 105–8.

50 *Fut.* 3: 93–94. See also Chodkiewicz, *Un océan sans rivage*, 46–47

51 *Fut.* 3: 108.

52 Ibid., 94.

53 Ibid.

54 *Fut.* 2: 582. This idiomatic expression is an allusion to Q 18:109: "Say: 'If the ocean were ink to (write down) the words of my Lord, the ocean would be exhausted before the words of my Lord, even if we brought a similar (ocean), for its aid'."

renew our approach, by getting rid of the rational preconceptions that distance us from the ongoing instant of Revelation, and by coming back to our immediate experience of the text: "Inspiration is immediacy (*al-waḥy huwa as-surʿa*)".[55] As Ibn ʿArabī declares in the *Futūḥāt*:

> The servant whose inner vision (*baṣīra*) is illuminated – i.e., the one who is guided by 'a light from his Lord' (Q 39:22) – obtains, every time he recites a verse, a new understanding, distinct from that of his previous recitation and from what he will obtain in the next recitation (. . .) The one whose understanding is the same during two successive recitations (of the Qurʾān) is a loser. The one whose understanding is renewed at every instant is a winner. As for the one who recites without understanding anything: May God have mercy on him![56]

It follows that one should never restrict the level of comprehensiveness of a verse that is indefinite in its apparent wording, as the ambivalence of the discourse is intended by the divine speech, in Ibn ʿArabī's view. Ultimately, every possible interpretation of the Revelation is intended by the Qurʾānic text itself.[57] However, those interpretations should not be limited by one's rational intellect, but should rather come as direct inspirations descending on one's heart. In other words, one should address the Qurʾānic text as a passenger on the divine journey of Revelation, rather than travelling by oneself through the text: "Whoever is put into travel will be saved, and whoever travels without being put into travel is in danger."[58] There are two types of travelers, according to Ibn ʿArabī in his *Kitāb al-Isfār*: philosophers and their like, who are travelling through the reflection of their intellects, and therefore stray from the path of the divine journey, since they have no other guide than their rational faculty; and messengers, prophets, and the most elevated among the Friends of God (*awliyāʾ*)[59], who have been "put into travel," and therefore "realized the Real".

5 Perplexity (*ḥayra*) and "Literalism"

The question remains: How are we supposed to be put into travel by the Qurʾān? The answer probably resides in what Ibn ʿArabī considers to be the most perfect spiritual state: "Perplexity" (*ḥayra*). One of the most peculiar aspects of his Qurʾānic hermeneutics is that he considers that perfection of perplexity can only be attained through a literal approach of the text.

55 *Fut.* 2: 78.

56 Ibid., 128–29. See also Chodkiewicz, *Un océan sans rivage*, 47.

57 See *Fut.* 1: 278–81, and Chodkiewicz, *Un océan sans rivage*, 65–66.

58 *Isfār*, 41.

59 *Isfār*, 47. Ibn ʿArabī names specifically here various seminal figures of early Sufism, in the following order: Sahl at-Tustarī, Abū Yazīd al-Bisṭāmī, Farqad as-Sabkhī, Junayd, and Ḥasan al-Baṣrī.

5.1 Perplexity as the Perfect Approach to the Qur'ān

According to Ibn ʿArabī, the aim of the Qurʾānic speech is to provide an experience of perplexity, which is the only way to approach the divine perfection, that can obviously not be grasped by reason. Only Revelation can inform us about divine perfection, whereas rational proofs can only bring us a half knowledge of God: his incomparability (*tanzīh*), i.e. the fact that he is ultimately other than everything. As for Revelation, it gives further information about God that is precisely "what intellect denies by its proofs", by stating his similitude (*tashbīh*) with everything that exists through his being.[60] By affirming both the transcending incomparability of God – e.g. "Glory to your Lord, the Lord who exceeds what they describe" (Q 37:180) – and his similitude with everything that exists – e.g. "Wherever you turn, there is the Face of God" (Q 2:115) – Revelation enlightens us about the perfection of Divinity. But this paradoxical perfection cannot be grasped by reason, and therefore the only real experience of its perfection we can have is precisely the perplexity of *ḥayra* brought by Revelation: "So the intellects are perplexed (*fa-ḥayyara*) and this is the divine perfection."[61]

A detailed description of the perplexing nature of the Qurʾānic speech is given by Ibn ʿArabī in the following narrative passage of his *Futūḥāt*. He depicts here how the Qurʾān deliberately blends incomparability and similitude by crossing their respective points of view:

> (God) addresses the senses and the imagination by his disengagement (*tajrīd*), to which the rational proofs point. The senses listen, so the senses and imagination are perplexed (*fa-ḥārat*), and they say: "We have nothing of him at our disposal!". And he addresses the (rational) intellects by his similitude, to which the senses and imagination point. The intellects listen, so the intellects are perplexed (*fa-ḥārat*), and they say: "We have nothing of him at our disposal!"[62]

The Qurʾān is thus the most perfect manifestation of the divine perfection that the Human Being can approach, precisely because it brings him the most perfect experience of perplexity: "(God) surpasses the perception by the intellects, the senses and imagination, and he isolates himself – glory to him! – in perfection by perplexity. So, no one knows him but himself, and no one witnesses him but himself."[63]

5.2 The Literal Approach

Such an experience of perplexity does not rise from what could be called an "esoteric" reading, based on a sophisticated interpretation (*taʾwīl*) of the Qurʾānic text that would

---

60 *Fut.* 2: 307.
61 Ibid.
62 Ibid.
63 Ibid.

pretend to unveil its internal – or implicit – (*bāṭin*) meanings.[64] Rather, Ibn ʿArabī's general approach on that matter could be surprisingly summarized as consisting in the advocacy of a scrupulous literalism that pays attention to the external and immediately apparent – or explicit – (*ẓāhir*) aspect of the text. As mentioned before, the articulation between the external and the internal – what one could call the "esoteric" and "exoteric" – dimensions of the Revelation, actually concerns the experience of the reader, and not the text itself.[65] That is to say, the implicit aspect is nothing other than the inner experience of the reader, whereas the explicit aspect is the very text itself.[66] The literal approach advocated by Ibn ʿArabī consists therefore in keeping the balance between the explicit and implicit aspects of the Qurʾānic text: "Know that God addressed human being in its totality, without giving more importance to his exteriority (*ẓāhir*) than to his interiority (*bāṭin*), nor to his interiority than to his exteriority."[67] For Ibn ʿArabī, only such a method can protect the perplexing divine perfection of the Qurʾānic text from our restrictive reading through a rational approach.[68]

It is through a close attention to the details of the literal formulation of the Qurʾān that we can let the explicit and apparent aspects of the text reveal their meanings in our immediate interiority:

> Every reality of the world is a sign orientating us towards a divine reality (. . .) Whenever God mentions the world (in the Qurʾān), pay attention to the divine name which he relates to it, and you will know which world it is. Whenever God designates himself by the singular and designates you by the plural, the verse is related to his Unicity and to your multiplicity (. . .) Whenever he designates himself by the plural, saying for example "*innā*" or "*naḥnu*" (intensive and nominal forms of "We"), this is related to God considered in regard to his Names. Whenever he speaks of you in the singular, he addresses you in regard to one of your aspects (*maʿnā*), and not to you as a whole. Therefore, know what, from yourself, is the receiver, and open your ears to his discourse.[69]

The implicit aspect of the text is thus what its most explicit aspect – consciously taken into account – provokes in the interiority of the reader, without any hermeneutical mediation.

---

64 On the "esoteric" aspect of Ibn ʿArabī's writings, see James W. Morris "Ibn ʿArabi's 'Esotericism': The Problem of Spiritual Authority," *SI* 71 (1990): 37–64.

65 This is already found in the section on the "science of the interior" (*ʿilm al-bāṭin*) of as-Sarrāj, *Kitāb al-Lumaʿ*, 43–44.

66 Such an understanding of the classical dichotomy between the *ẓāhir* and *bāṭin* aspects of the text is surprisingly absent from Annabel Keeler and Sajjad Rizvi's "Introduction" to *The Spirit and the Letter*, 1–47.

67 *Fut.* 1: 334. See also Chodkiewicz, *Un océan sans rivage*, 44.

68 In the last section of the *Kitāb al-Isfār*, Ibn ʿArabī warns against rationalization and speculative interpretation of the Qurʾān. See *Isfār*, 124–25.

69 Ibn ʿArabī, *al-ʿAbādila*, ed. ʿAbd ar-Raḥmān Aḥmad ʿAṭā (Cairo: Maktabat al-Qāhira, 1969), 42–43. See also Chodkiewicz, *Un océan sans rivage*, 56.

The "literalism" advocated by Ibn ʿArabī is not a rejection of the articulation between the internal and the external aspects of the text. Rather, it is a reminder that one should not mistake one aspect for the other, and misleadingly try to explain externally – i.e. through an interpretation of the extrinsic aspect of the text – a dimension of the reading that should always remain internal by nature – i.e. the intrinsic aspect of the reading which necessary remains in the innermost intimacy of the reader.

As Ibn ʿArabī often repeats, it is a matter of spiritual "good manners" (*adab*) to pay attention, to respect and protect the integrity of the literal aspect of the Qurʾānic text, and "the one who contradicts good manners is punished by ignorance."[70] A perfect example of those good manners resides for Ibn ʿArabī in the fact that the Qurʾān itself instructed Muḥammad to respect the intermediacy of Gabriel, who was transmitting him the Revelation in the distinctive modality of the *Furqān*, even though Muḥammad had already received it directly from God in the synthetic modality of the *Qurʾān*.[71] Following the prophetic example, one should always make sure that the distinctive aspect of the Qurʾān prevails:

> Beware of the *Qurʾān*, unless you read it as a *Furqān*. For God 'misleads many by it' – that is to say that he makes them perplexed – and 'guides many by it' – that is to say that he grants them the understanding of it as it is, with its clarification. And 'He misleads by it only the corrupt', those who exceed its limits and its tracings (*rusūm*).[72]

Ibn ʿArabī invites his readers to "travel by" the Qurʾān, but warns them about the dangers of such a journey:

> Dive into the ocean of the glorious Qurʾān if your breath is wide enough. And if not, limit yourself to examining the commentaries on its apparent (aspect), but do not dive into it, for you would perish. The ocean of the Qurʾān is deep, and if the one who dives into it does not aim at the places that are close to the shore, he will never come back to you![73]

This echoes Ibn ʿArabī's insistence on the fact that one should always be "put into travel", rather than traveling by oneself, mentioned earlier. By paying a close attention to the shores of the explicit aspect of the text, one is taken by the tides of its implicit meanings without risking sinking in the ocean of interpretation.

---

70 *Fut.* 2: 644. On the importance of *adab* in Ibn ʿArabī's hermeneutics, see Denis Gril, "Adab and Revelation ore One of the Foundations of the Hermeneutics of Ibn 'Arabi", in *Muhyiddin Ibn Arabi. A Commemorative volume*, eds. Stephen Hirtenstein and Michael Tiernan (Shaftesbury/Rockport/Brisbane: Element, 1993). See also Chittick, *The Sufi Path of Knowledge*, 231 and 237.

71 See Q 20:114: "Do not hasten with (the recitation of) the Qur'an before its revelation is completed to you, and say: 'My Lord, increase me in knowledge'!"

72 *Fut.* 4: 401. This is an allusion to Q 2:26: "God is not ashamed to present an example, (like) that of a mosquito or what is even smaller. And those who have believed know that it is the truth from their Lord. As for those who have disbelieved, they say: 'What did God intend by this example?' He misleads many thereby and guides many thereby. But He misleads not, except those who forsake".

73 *Fut.* 1: 76. See also Chodkiewicz, *Un océan sans rivage*, 43.

5.3 Perplexing Outcomes of the Literal Approach

Ibn ʿArabī's approach to the Qurʾānic text does not consider it as carrying determinate meanings. His literal approach entails that our understanding must always be renewed each time we read the text, as we have already mentioned. Ibn ʿArabī's ever-renewed literal approach to the perplexing perfection of the text brings about paradoxical and unexpected readings of the Qurʾānic vocabulary. In a passage of the *Futūḥāt* wherein he describes various kinds of sainthood (*walāya*), he surprisingly draws his spiritual typology from the etymology of several Qurʾānic terms usually understood as "disbelievers" (*kāfirūn*), "impudent" (*fujjār*), "envious" (*ḥāsidūn*) or "sorcerers" (*sāḥirūn*).[74] The verses containing those terms take thus a radically positive meaning according to this paradoxical – but always linguistically correct – "literal" reading. Ibn ʿArabī justifies this reading by explaining briefly his method: "We take every blameworthy quality in its absolute meaning, and then we determine it so that it is praiseworthy." But, he adds, this hermeneutical operation does not stem from his imaginative mind, as it is actually implied by the very wording of the Revelation: "All comes from the blessing of the Mother-Book (*umm al-kitāb*)".[75] For the Shaykh al-Akbar, the radical ambivalence of the Qurʾānic vocabulary is an aspect of its perfection. Through this paradoxical semantics, the Revelation provides to the reader – if he pays a close attention to the etymology or grammatical details of its apparent wording – the profound and unique experience of perplexity that manifests its perfection.

What has probably been the most brilliant example of Ibn ʿArabī's paradoxical hermeneutics is unfortunately lost for us, but we know that his vast Qurʾānic commentary, *al-Jamʿ wa-t-tafṣīl fī asrār maʿānī t-tanzīl*, was composed according to three points of view, in relation to the classical Sufi dichotomy of the divine reality: (1) Majesty (*jalāl*) – i.e., its intrinsic, transcendent and intimidating aspect – (2) Beauty (*jamāl*) – i.e., its extrinsic, immanent and reassuring aspect – and (3) Perfection (*kamāl*) i.e., the synthetic reunion of the two perspectives.[76] For the Shaykh al-Akbar, this is the hermeneutical approach of "those who realize the Real" (*al-muḥaqqiqūn*), those who see things "with two eyes,"[77] for they apprehend at the same time the extrinsic and intrinsic aspects of the Qurʾānic text.

74 Ibid., 134–39.

75 Ibid., 138. See also Chodkiewicz, *Un océan sans rivage*, 71–75.

76 See Chodkiewicz, *Un océan sans rivage*, 104.

77 See the conclusive chapter in Chittick, *The Sufi Path of Knowledge*, 356–80.

6 Ibn ʿArabī's Hermeneutical Legacy

The Qurʾānic hermeneutics legacy of Ibn ʿArabī remained undoubtedly significant throughout later Islamic intellectual history, thanks to the numerous influential figures deeply – although not always openly – influenced by his works, that appeared in various cultural areas of the Muslim world.

The most famous example of an exegetical work profoundly marked by the hermeneutics of the Shaykh al-Akbar is probably the *Taʾwīlāt al-Qurʾān* of ʿAbd ar-Razzāq Kāshānī, widely published today as "*Tafsīr Ibn ʿArabī.*"[78] Kāshānī's reading and use of Ibn ʿArabī's ideas stem clearly from what has usually been called the "Akbarī" school of thought, a label used for those who claimed to be part of the spiritual and intellectual lineage of the Shaykh al-Akbar, rather than a formal school with defined and coherent doctrines and methods.[79] It designates in fact a textual tradition of commentaries and original treatises revolving around the corpus and core ideas of Ibn ʿArabī. The Akbarī school should probably be considered as emerging from the circle of Ibn ʿArabī's stepson and foremost disciple, Ṣadr ad-Dīn Qūnawī, rather than from Ibn ʿArabī himself. Qūnawī wrote an extensive commentary on the *Fātiḥa*, the *Iʿjāz al-bayān fī taʾwīl umm al-Qurʾān*, that is profoundly infused with elements of Ibn ʿArabī's thought.[80]

Another pivotal figure is Dawūd Qayṣarī, who counted himself a disciple of Kāshānī.[81] His works became the classical propaedeutic literature to approach Ibn ʿArabī in the Ottoman world, especially through the influence of figures such as Muḥammad Hamza al-Fanārī (d. 1430/834), who became the first *Şeyhülislam* of the empire, and Ismaʿīl Ḥaqqī l-Brūsawī (d. 1725/1137), whose famous Qurʾānic commentary, the *Rūḥ al-bayān*, is clearly infused with Akbarī hermeneutics. It is also mainly through the introductory works of Qayṣarī that Ibn ʿArabī's ideas durably penetrated the Safavid intellectual milieu. A major figure such as Mullā Ṣadrā Shīrāzī (d. 1640/1050), who wrote several works on Qurʾānic hermeneutics and exegesis, was profoundly influenced by both Ibn ʿArabī and Qayṣarī.

In the Arab world, Ibn ʿArabī's hermeneutical ideas were also spread widely through the writings of mainstream and popular Sufis such as ʿAbd al-Wahhāb ash-Shaʿrānī (d. 1565/973) and ʿAbd al-Ghanī n-Nabūlūsī (d. 1731/1143).[82] A good example of

---

78 See Pierre Lory, *Les commentaires ésotériques du Coran selon Abd al-Razzâq al-Qâshânî* (Paris: Les deux océans, 1991), and "ʿAbd al-Razzāq al-Kāshānī", in *EI³*, ed. Kate Fleet et al., accessed on November 4, 2023. doi:http://dx.doi.org/10.1163/1573-3912_ei3_COM_22929.

79 See William C. Chittick, "The School of Ibn ʿArabī," in *History of Islamic Philosophy*, 510–23.

80 See Richard Todd, "Qūnawī's Scriptural Hermeneutics," in *The Spirit and the Letter*, 299–324. Qūnawī apparently preferred to call it simply *Tafsīr al-Fātiḥa*.

81 See Caner Dagli, *Ibn al-ʿArabī and Islamic Intellectual Culture: From Mysticism to Philosophy* (London/New-York: Routledge, 2016), 120.

82 An-Nabūlūsī considered himself as "the suckling child" of Ibn ʿArabī, who was his "milk-father". See Elizabeth Sirriyeh, *Sufi Visionary of Ottoman Damascus. Abd al-Ghani al-Nabulusi, 1641–1731* (London/New-York: Routledge, 2005) 18–38.

the discreet yet recognizable appropriation of an Akbarī hermeneutical approach by proponents of mainstream strands of Sufism can be found in the Qur'ānic commentary by the Moroccan Aḥmad b. ʿAjība (d. 1809/1224), *al-Baḥr al-madīd fī tafsīr al-Qur'ān al-majīd*, which blends the classical exegetical approaches of the genre with peculiar Sufi readings that are sometimes clearly borrowed from the Akbarī tradition.[83]

The importance of the diffusion of Ibn ʿArabī's works and ideas throughout the Muslim world is also demonstrated by the fierce reaction and opposition raised by Ibn Taymiyya and his circle, who were concerned with the growing influence exerted by the Shaykh al-Akbar after his death.[84] Interestingly enough, their attacks against his doctrines were concentrated on metaphysical and theological issues, but they rarely addressed his exegetical method per se. The polemical debates around the writings of Ibn ʿArabī that have surfaced repeatedly until today – and that are for the most part no more than repetitions of the accusations made by Ibn Taymiyya – usually evoke Ibn ʿArabī's interpretations of the Qur'ān as examples of his rhetorical justifications rather than as sources for his heretical doctrines.

Despite (or perhaps because of) the resurgence of controversies by proponents of the Wahhabi and Salafi movements, the interest in the hermeneutical method of Ibn ʿArabī did not diminish in the Modern period, and has not been restricted to the Sufi milieu. A striking example of the renewed attraction exerted by Ibn ʿArabī's hermeneutics is the study published in 1983, stemming from the PhD thesis of the famous Egyptian liberal scholar Naṣr Ḥāmid Abū Zayd (d. 2010/1431), *Falsafa at-ta'wīl: Dirāsa fī ta'wīl al-Qur'ān ʿinda Muḥyī d-Dīn b. ʿArabī*.[85] Certainly, Abū Zayd's call for a "humanistic" approach of the Qur'ānic text allowing for an "interpretational diversity" – which made him the target of Jihadist movements – was in part inspired by his study of Ibn ʿArabī's hermeneutical method.

7 Conclusion

Ibn ʿArabī's Qur'ānic hermeneutics is a central link in the chain of the Sufi exegetical tradition, and a major milestone in Islamic intellectual history. The thought of the Andalusian Shaykh synthetized brilliantly the various hermeneutical methods found in the previous literature and went into their doctrinal and practical implications in

83 On the question of Ibn ʿArabī's influence on Ibn ʿAjība, see the "Introduction" in Jean-Louis Michon, *Two Treatises on the Oneness of Existence by the Morrocan Sufi Aḥmad Ibn ʿAjība (1747–1809)* (Cambridge: Archetype, 2010), 19–34.

84 See Knysh, *Ibn ʿArabi in the Later Islamic Tradition*, 87–111. See also Michel Chodkiewicz, "Le procès posthume d'Ibn ʿArabī," in *Islamic Mysticism Contested. Thirteen Centuries of Controversies and Polemics*, eds. Frederik De Jong and Bernd Radtke (Leiden, Boston and Köln: Brill, 1999), 93–123.

85 See Michel Chodkiewicz, Review of *Falsafat al-ta'wīl: Dirāsa fī ta'wīl al-Qur'ān ʿinda Muḥyī d-Dīn b. ʿArabī*, by Naṣr Ḥāmid Abū Zayd, *SI* 60 (1984): 177–80.

greater depth. The apparent peculiarities of his Qurʾānic interpretations should be considered as the result of his particular focus on core issues of the theology of Revelation, and as developments in the fundamentals of Sufi spiritual hermeneutics at large.

His hermeneutical approach does not aim at pushing the limits of the literal implication of the text. It aims rather at deepening the implications of the reader's perspective. In other words, the Qurʾān is "an ocean without shore" for Ibn ʿArabī, in that it mirrors the boundless dimension of human consciousness.

8 Bibliography

8.1 Sources

Ghurāb, Maḥmūd. *Raḥma min ar-Raḥmān fī tafsīr wa-ishārāt al-Qurʾān min kalām ash-shaykh al-akbar Muḥyī d-Dīn b. ʿArabī*, 4 vols. Damascus: n.e., 1989.

Ibn ʿArabī, Muḥyī d-Dīn. *Al-Futūḥāt al-makkiyya*, 4 vols. Beirut: Dār al-Fikr, n.d.

Ibn ʿArabī, Muḥyī d-Dīn. *Al-ʿAbādila*, ed. ʿAbd ar-Raḥmān Aḥmad ʿAṭā. Cairo: Maktabat al-Qāhira, 1969.

Ibn ʿArabī, Muḥyī d-Dīn. *Kitāb al-Isfār ʿan natāʾij al-asfār: Le dévoilement des effets du voyage*, edited and translated by Denis Gril. Combas: L'éclat, 1994.

Ibn ʿArabī, Muḥyī d-Dīn. *At-Tadbīrāt al-ilāhiyya fī iṣlāḥ al-mamlaka al-insāniyya*, ed. Ibrāhīm al-Kayyālī. Beirut: Dār al-Kutub al-ʿIlmiyya, 2003.

Ibn ʿArabī, Muḥyī d-Dīn. *The Secrets of Voyaging: Kitāb al-Isfār ʿan natāʾij al-asfār*, trans. Angela Jaffray. Oxford: Anqa Publishing, 2016.

Miftāḥ, ʿAbd al-Bāqī. *Shurūḥ ʿalā Kitāb Ishārāt al-Qurʾān fī ʿālam al-insān*. Damascus: Dār Naynawā, 2016.

As-Sarrāj aṭ-Ṭūsī, Abū Naṣr. *Kitāb al-Lumaʿ*, ed. ʿAbd al-Ḥalīm Maḥmūd. Cairo: Maktabat ath-Thaqāfa ad-Dīniyya, 2002.

As-Suyūṭī, Jalāl ad-Dīn. *Al-Itqān fī ʿulūm al-Qurʾān*. Medina: Majmaʿ al-Malik Fahd li-Ṭibāʿat al-Muṣḥaf ash-Sharīf, 2005.

8.2 Secondary Literature

Addas, Claude. *Ibn ʿArabī ou la quête du soufre rouge*. Paris: Gallimard, 1989.

Addas, Claude. *Ibn Arabî et le voyage sans retour*. Paris: Points, 1996.

Addas, Claude. "The 'Radiant Way'. Ibn ʿArabī and the Interpretation of the Law. Part I: The Legacy of Ibn Hazm." *Journal of the Muhyiddin Ibn Arabi Society* 65 (2019): 29–57.

Addas, Claude. "The 'Radiant Way'. Ibn ʿArabī and the Interpretation of the Law. Part II: The Law and its Interpretation." *Journal of the Muhyiddin Ibn Arabi Society* 66 (2019): 39–76.

Bellver, José. "Ibn Barrağān and Ibn ʿArabī on the prediction of the capture of Jerusalem in 583/1187 by Saladin." *AR* 61, nos. 3–4 (2014): 252–86.

Casewit, Yousef. *The Mystics of al-Andalus: Ibn Barrajān and Islamic Thought in the Twelfth Century*. Cambridge: Cambridge University Press, 2017.

Chittick, William C. *The Sufi Path of Knowledge: Ibn al-ʿArabī's Metaphysics of Imagination*. Albany: State University of New-York Press, 1989.

Chittick, William C. “Rūmī and *waḥdat al-wujūd*.” In *Poetry and Mysticism in Islam: The Heritage of Rumi*, edited by Amin Banani, Richard Hovannisian, and Georges Sabagh. Cambridge: Cambridge University Press, 1994: 70–111.

Chittick, William C. *The Self-Disclosure of God: Principles of Ibn al-'Arabi's Cosmology*. Albany: State University of New-York Press, 1998.

Chittick, William C. “Ibn ʿArabī.” In *History of Islamic Philosophy*, edited by Seyyed Hossein Nasr and Oliver Leaman. London/ New-York: Routledge, 2001: 497–509.

Chittick, William C. “The School of Ibn ʿArabī.” In *History of Islamic Philosophy*, edited by Seyyed Hossein Nasr and Oliver Leaman. London/ New-York: Routledge, 2001: 510–23.

Chittick, William C. *Ibn ʿArabi: Heir to the Prophets*. Oxford: Oneworld, 2005.

Chodkiewicz, Michel. Review of *Falsafat al-taʾwīl: Dirāsa fī taʾwīl al-Qurʾān ʿinda Muḥyī d-Dīn b. ʿArabī*, by Naṣr Ḥāmid Abū Zayd. *SI* 60 (1984): 177–80.

Chodkiewicz, Michel. *Un océan sans rivage. Ibn ʿArabî, le Livre et la Loi*. Paris: Éditions du Seuil, 1992.

Chodkiewicz, Michel. “Le procès posthume d'Ibn ʿArabī.” In *Islamic Mysticism Contested. Thirteen Centuries of Controversies and Polemics*, edited by Frederik De Jon and Bernd Radtke. Leiden, Boston and Köln: Brill, 1999: 93–123.

Chodkiewicz, Michel. “Ibn ʿArabî dans l'œuvre d'Henry Corbin.” In *Henry Corbin: philosophies et sagesses des religions du livre. Actes du colloque Henry Corbin, Sorbonne, les 6–8 novembre 2003*, edited by Mohammad Ali Amir-Moezzi, Christian Jambet, and Pierre Lory. Turnhout: Brepols, 2005: 81–91.

Chodkiewicz, Michel. “*Mi'râj al-kalima*. De la *Risâla Qushayriyya* aux *Futûhât Makkiyya*.” In *Reason and Inspiration in Islam. Theology, Philosophy and Mysticism in Muslim Thought. Essays in Honour of Hermann Landolt*, edited by Todd Lawson. London/New-York: I.B. Tauris/The Institute of Ismaili Studies, 2005: 248–61.

Dagli, Caner. *Ibn al-ʿArabī and Islamic Intellectual Culture: From Mysticism to Philosophy*. London/New-York: Routledge, 2016.

Ebstein, Michael, *Mysticism and Philosophy in al-Andalus. Ibn Masarra, Ibn al-ʿArabī and the Ismāʿīlī Tradition*. Leiden/ Boston: Brill, 2014.

Ebstein, Michael. “'In Truth You Are the Polytheist!': Mythic Elements in Ibn al-ʿArabī's Teachings on the Divine Names.” *Intellectual History of the Islamicate World* 6 (2018): 359–87.

Gril, Denis. “Adab and Revelation ore One of the Foundations of the Hermeneutics of Ibn ʿArabi.” In *Muhyiddin Ibn Arabi a Commemorative volume*, edited by Stephen Hirtenstein and Michael Tiernan. Shaftesbury/Rockport/Brisbane: Element, 1993: 228–63.

Gril, Denis. “La 'lecture supérieure' du Coran selon Ibn Barrağān”, *AR* 47 (2000): 510–22.

Gril, Denis. “L'interprétation par transposition symbolique (*iʿtibār*), selon Ibn Barrağān et Ibn ʿArabī.” In *Symbolisme et herméneutique dans la pensée d'Ibn ʿArabī. Actes du colloque*, edited by Bakri Aladdin. Damascus: Institut français du Proche-Orient, 2007: 147–61.

Hirtenstein, Stephen. “Manuscripts of Ibn ʿArabī's Works: Names and Titles of Ibn (al-)ʿArabī.” *Journal of the Muhyiddin Ibn Arabi Society* 41, (2007): 109–29.

Keeler, Annabel, and Sajjad Rizvi, eds. *The Spirit and the Letter. Approaches to the Esoteric Interpretation of the Qurʾān*. Oxford/ New-York: Oxford University Press/Institute of Ismaili Studies, 2016.

Knysh, Alexander D. *Ibn ʿArabi in the Later Islamic Tradition. The Making of a Polemical Image in Medieval Islam*. Albany: State University of New-York Press, 1999.

Lory, Pierre. Les commentaires ésotériques du Coran selon Abd al-Razzâq al-Qâshânî. Paris: Les deux océans, 1991.

Lory, Pierre. “ʿAbd al-Razzāq al-Kāshānī.” In *EI*^{3}, edited by Kate Fleet et al. Brill, 2007–. Article published online 2009. Accessed on November 4, 2023. doi:http://dx.doi.org/10.1163/1573-3912_ei3_COM_22929.

Michon, Jean-Louis. *Two Treatises on the Oneness of Existence by the Morrocan Sufi Aḥmad Ibn ʿAjība (1747–1809)*. Cambridge: Archetype, 2010.

Miftāḥ, ʿAbd al-Bāqī. *Al-Mafātīḥ al-Qurʾāniyya li-Kitāb at-Tarājim wa-li-Kitāb ash-Shāhid li-Ibn ʿArabī*. Damascus: Dār Naynawā, 2019.

Morris, James W. "Ibn ʿArabi and His Interpreters Part I: Recent French Translations." *JAOS* 106, no. 3 (1986): 539–51.

Morris, James W. "Ibn ʿArabi and His Interpreters Part II: Influences and Interpretations." *JAOS* 106, no. 4 (1986): 733–56.

Morris, James W. "Ibn ʿArabi and His Interpreters Part II (Conclusion): Influences and Interpretations", *JAOS* 107, no. 1 (1987): 101–19.

Morris, James W. "Ibn 'Arabi's 'Esotericism': The Problem of Spiritual Authority." *SI* 71 (1990): 37–64.

Morris, James W. *The Reflective Heart. Discovering Spiritual Intelligence in Ibn 'Arabī's Meccan Illuminations*. Louisville: Fons Vitae, 2005.

Sands, Kristin Zahra. *Ṣūfī Commentaries on the Qurʾān in Classical Islam*. London and New-York: Routledge, 2006.

Sells, Michael A. "Ibn ʿArabī's Garden among the Flames: A Reevaluation." *History of Religions* 23, no. 4 (1984): 287–315.

Sirriyeh, Elizabeth. *Sufi Visionary of Ottoman Damascus. Abd al-Ghani al-Nabulusi, 1641–1731*. London/New-York: Routledge, 2005.

Sviri, Sara. "The Countless Faces of Understanding: On *Instinbāṭ*, Mystical Listening and Sufi Exegesis." In *The Spirit and the Letter. Approaches to the Esoteric Interpretation of the Qurʾān*, edited by Annabel Keeler and Sajjad Rizvi. Oxford/ New-York: Oxford University Press/Institute of Ismaili Studies, 2016: 51–85.

Thibon, Jean-Jacques. "L'herméneutique de Sulamī annonce-t-elle celle d'Ibn ʿArabī?." In *Symbolisme et herméneutique dans la pensée d'Ibn ʿArabī. Actes du colloque*, edited by Bakri Aladdin. Damascus: Institut français du Proche-Orient, 2007: 21–39.

Todd, Richard. "Qūnawī's Scriptural Hermeneutics." In *The Spirit and the Letter. Approaches to the Esoteric Interpretation of the Qurʾān*, edited by Annabel Keeler and Sajjad Rizvi. Oxford/New-York: Oxford University Press/Institute of Ismaili Studies, 2016: 299–324.

Vandamme, Gregory. "L'ontologie corrélative d'Ibn ʿArabī. Métaphysique de l'être et de ses relations." In *La non-dualité: Perspectives philosophiques, scientifiques, spirituelles*, edited by Jean-Michel Counet. Louvain: Peeters, 2021: 89–102.

Yahya, Osman. *Histoire et classification de l'œuvre d'Ibn 'Arabī. Étude critique*, 2 vols. Damascus: Institut français de Damas, 1964.

ʿAlāʾ ad-Dawla as-Simnānī

Giovanni Maria Martini

1 Introduction

Between 1302/701 and 1314/714 the celebrated Iranian Sufi scholar belonging to the Kubrawiyya brotherhood ʿAlāʾ ad-Dawla as-Simnānī (d. 1336/736) wrote a partial *tafsīr* consisting of an introduction and commentary on suras 1 and 52–114. Most probably the title chosen by as-Simnānī for this commentary is *Muṭṭalaʿ an-nuqaṭ wa-majmaʿ al-luqaṭ* (Place of ascent of the points and place of gathering of the gleanings), although it is also known by other names (see below). The unusual structure of this *tafsīr*, lacking the commentary of suras 2–51, is due to the fact that as-Simnānī did not aim to write a completely independent Qurʾānic commentary, but to continue a *tafsīr* that had been started by eminent Kubrawī scholars of previous generations and never completed. The work of as-Simnānī is an esoteric exegesis of the Qurʾān (*tafsīr baṭn al-Qurʾān*), in which the author disregards almost completely the literal and historical meaning, privileging a symbolic interpretation of the sacred text. According to the particular perspective adopted by Simnānī, the Qurʾān acts as a sort of inner map that speaks to the initiate on the Sufi path of invisible events taking place within himself, guiding him in the spiritual journey of purification and knowledge of God. One of the most characteristic aspects of as-Simnānī's hermeneutics is the relationship he establishes between seven major prophets present in the Qurʾān and as many subtle centers within the spiritual body of the initiate, which would correspond to at least seven levels of esoteric meaning of the sacred text. Due to its unique features, as-Simnānī's commentary is considered one of the peaks of Qurʾānic commentaries written by a Sufi scholar.

2 As-Simnānī's Life, Works and Political and Intellectual Context

ʿAlāʾ ad-Dawla as-Simnānī was a leading Sufi scholar and intellectual affiliated with the *Ṭarīqa Kubrawiyya* in Iran in the Ilkhanid era. A number of peculiar characteristics contributed to the formation of this influential and eccentric figure. One decisive element was his social background. The as-Simnānī family was one of the leading families of Persian viziers in the service of the Ilkhanid state, within which the father

https://doi.org/10.1515/9783111320083-003

and maternal and paternal uncles of ʿAlāʾ ad-Dawla occupied prominent positions.[1] ʿAlāʾ ad-Dawla grew up at court, becoming a close friend of the future Ilkhan Arghūn (r. 1284–1291/683–690). On Thursday 4 May 1284/16 Ṣafar 683, when as-Simnānī was twenty-four years old, an event occurred that would change his life forever. A spiritual crisis seized as-Simnānī while he was arrayed alongside Arghūn in a decisive battle against the latter's uncle Aḥmad Tegüder (r. 1282–1284/680–683), which would shortly thereafter contribute to the ascension to the throne of as-Simnānī's fraternal friend. The result was a sort of 'conversion,' a sudden and irreversible religious awakening that made as-Simnānī decide to abandon the mundane life of the court to devote himself completely to the strictest religious discipline. The choice of as-Simnānī, although pertinent to his personal sphere, had in fact political repercussions. It was a clear stance, also expressed in writing in his works, both against the Ilkhanid rulers, many of whom (including Arghūn) at that time had not yet converted to Islam, and against the Persian elites colluding with them, including his own family. The message of as-Simnānī, stated on paper in his works, was clear, and sounded like a declaration of war: the Islamic societies of Iran were in great danger, no compromise was possible with the idolatrous conquerors.[2] From the combination of the information handed down by the sources, it emerges that the reaction of his own family, of Arghūn, and consequently of the entire state apparatus, towards the choices of as-Simnānī mainly took the form of the restriction of his freedom of movement, which in more or less stringent forms lasted for many years, continuing intermittently even during the reigns of Arghūn's successors. It consisted of a condition of probation that forced him to live most of the time near his hometown Simnān and prevented him from leaving the territories of the Ilkhanate. Such a state of affairs had significant consequences on as-Simnānī's spiritual formation, primarily limiting his ability to set out in search of a Sufi master. This led as-Simnānī to have to compensate at least partially for the absence of a master through the study of Sufi texts, accentuating the bookish aspect of his intellectual training. The first contact with his future master Nūr ad-Dīn ʿAbd ar-Raḥmān al-Isfarāyinī (d. 1317/717) occurred through an intermediary in Simnān, where the master sent a disciple for the primary purpose of initiating as-Simnānī in 1287/685–686. As-Simnānī's first attempt to visit al-Isfarāyinī in person in Baghdad was nipped in the bud by Arghūn and ended in arrest and a period of enforced residence at court interrupted by as-Simnānī only through a daring escape. As-Simnānī had to wait some two years, until 1289/688, before he was able to travel to Baghdad to

1 In sketching as-Simnānī's life I have relied primarily on the studies of Jamal J. Elias, *The Throne Carrier of God: The Life and Thought of ʿAlāʾ ad-dawla as-Simnānī* (Albany, NY: State University of New York Press, Albany, 1995), 15–31 and Giovanni Maria Martini, *ʿAlāʾ al-Dawla al-Simnānī between Spiritual Authority and Political Power: A Persian Lord and Intellectual in the Heart of the Ilkhanate* (Leiden/Boston: Brill, 2018), 1–97, to which the reader may refer for more detailed information and references to primary sources.

2 See e.g. the passages translated in Martini, *ʿAlāʾ al-Dawla al-Simnānī*, 27–29, 83f., 443.

meet al-Isfarāyinī, and he had to do so in secret, evading police controls. To make up for the difficulty of meeting physically, al-Isfarāyinī and as-Simnānī resorted to an intense exchange of correspondence, which partially survives.[3] The practical difficulties deriving from the limitation of his freedom of movement and the impossibility of entertaining a free master-disciple relationship perhaps explains the accentuation of some particular concepts on the part of al-Isfarāyinī and as-Simnānī, such as the idea of spiritual direction by "telepathic" means that shines through in some passages of the epistolary, and the concept of "inner master" that will also play a certain role in as-Simnānī's hermeneutic theory (cf. below the concept of *ustādh ghaybī*).[4] It is reasonable to think that the high degree of originality of as-Simnānī's thought was at least partly due to the author's unusual biography. The condition of deprivation of freedom of movement did not prevent as-Simnānī from devoting himself to writing (indeed, it is likely to have been a propelling factor), nor from attracting numerous disciples around him. A probably incomplete list of as-Simnānī's writings includes more than one hundred, of which more than eighty have come down to us.[5] They range from short epistles to works of hundreds of pages. The languages used are Persian and Arabic. In this regard, one can observe a clear tendency of as-Simnānī to use Persian for the works he dedicates to specific individuals, almost always direct disciples, while Arabic is reserved mainly for wide-ranging works, addressed to the entire Islamic intellectual community and without a specific dedicatee.[6] Among the latter are some of as-Simnānī's most important and influential writings, destined to exert a lasting influence in the centuries to follow, as evidenced by the numerous manuscript copies in which they have come down to us. One of these is his partial commentary on the Qurʾān, which is also as-Simnānī's main exegetical work discussed in this article.

3 As-Simnānī's Commentary of the Qurʾān

As-Simnānī's *tafsīr* consists of an introduction and commentary on the Opening sura, which is followed by a commentary on suras 52–114, i.e., the commentary on *sūrat aṭ-Ṭūr* (The Mount) until the conclusion of the Qurʾān. The unusual structure of this *taf-*

3 Edited by Hermann Landolt, *Correspondance spirituelle échangée entre Nuroddin Esfarayeni (ob. 717/1317) et son disciple ʿAlaoddawleh Semnani (ob. 736/1336)/Murshid va murīd: Mukātabāt-i ʿAbd ar-Raḥmān Isfarāyinī bā ʿAlāʾ ad-Dawla-yi Simnānī* (Tehran/Paris: Departement d'Iranologie de l'Institut Franco-Iranien de Recherche-Librairie d'Amerique et d'Orient Adrien-Maisonneuve, 1972).

4 On this point, see Landolt, *Correspondance spirituelle*, French Introduction, 6, 13, 19–21. Cf. also Martini, *ʿAlāʾ al-Dawla al-Simnānī*, 56.

5 The most detailed list of as-Simnānī's works is in Elias, *The Throne*, 165–212, supplemented by Martini, *ʿAlāʾ al-Dawla al-Simnānī*, 461–72. The long list of works and manuscripts contained in these two indexes is in all likelihood not yet exhaustive.

6 On this point see Martini, *ʿAlāʾ al-Dawla al-Simnānī*, 122–35.

sīr, in which the discussion of suras 2–51 is absent, is neither accidental nor the result of fragmentary transmission, but due to precise reasons. It is the fact that as-Simnānī did not intend to write a completely independent Qur'ānic commentary, but to continue a *tafsīr* that had been started by eminent Kubrawī scholars of previous generations and never completed. Such an exegetical work compiled in various stages by several Sufi scholars belonging to the Kubrawiyya brotherhood, of which as-Simnānī is the author of the terminal part, is sometimes referred to for convenience as the 'Tafsīr of the Kubrawiyya' or the 'Kubrawī Tafsīr.' This appellation, although not attested in the manuscript tradition, besides having the important advantage of being simple and easily graspable, is implicitly supported by the history of the formation of the text, as well as by the manuscript tradition of the same, in which very often this *tafsīr* drafted by several authors is copied and transmitted in a unified form.

3.1 The First Part of the Kubrawī Tafsīr before as-Simnānī

The first part of the *tafsīr* of the Kubrawiyya constitutes an extremely complex philological case, on which scholars continue to be divided. It is so complex that critics have so far failed to establish with precision either the authorship (and thus the dating), the title, or the precise point at which it stops. This situation is due to the intricate manuscript tradition of the work, and to the fact that the witnesses are discordant with respect to each of these fundamental aspects. Only an accurate study of the manuscripts and the preparation of a critical edition based on serious philological criteria will perhaps be able to shed definitive light on this question.[7] Manuscripts largely agree that it was Najm ad-Dīn al-Kubrā (d. 1220/617), the eponymous founder of the *Ṭarīqa Kubrawiyya*, who has begun writing a *tafsīr* which was never completed due to his early death. According to various manuscripts, al-Kubrā's *tafsīr* stops around verse 19 of sura 51, *adh-Dhāriyāt* (The Scatterers), however there is no unanimous agreement on this point.[8] The work therefore would have been continued by another prominent Kubrawī Sufi scholar, Najm ad-Dīn ar-Rāzī Dāya (d. 1256/654). However, Dāya would also have passed away without completing the commentary. Again, it is unclear where exactly Dāya's commentary begins and ends. For example, as-Simnānī writes that Dāya would have died before he came to comment on sura 53, *an-Najm* (The Star), thus at the end of

7 The question of the authorship of the older part of the *tafsīr* of the Kubrawiyya – as well as the problem of the various titles by which it is known (on which see below) – was already posed in the 1930s by Fritz Meier, "Stambuler Handschriften dreier persischer Mystiker: ʿAin al-quḍāt al-Hamaḏānī, Naǧm addīn al-Kubrā, Naǧm ad-dīn ad-Dāja," *Der Islam* 24 (1937): 1–42, 10ff.; on the same problems see also Süleyman Ateş, "Üç müfessir bir tefsir," *Ankara Üniversitesi İlahiyat Fakültesi Dergisi* 18 (1970): 85–104, as well as the other studies cited in the following notes.

8 Cf. Meier, "Stambuler Handschriften," 10ff.; Elias, *The Throne*, 204.

sura 52, *aṭ-Ṭūr* (The Mount).[9] The doubts therefore ultimately concern the role played respectively by al-Kubrā and Dāya in the writing of the commentary, an issue on which scholars are divided between those who believe that the text is preeminently by al-Kubrā,[10] and others who believe instead that Dāya played a large and decisive role in editing the text and not at all limited to commenting on one or two suras.[11] The attribution of the first part of the commentary to Dāya as the main author has among its arguments the presence of a passage contained in the *tafsīr* of as-Simnānī in which the latter explicitly states that his work stands as a completion of the *tafsīr* of Najm ad-Dīn Dāya.[12] Taking into account as-Simnānī's privileged position within the Kubrawiyya and his direct participation in the vicissitudes of the Kubrawī *tafsīr* his testimony must

---

9 'Alā' ad-Dawla as-Simnānī, *Muṭṭala' an-nuqaṭ wa-majma' al-luqaṭ*, MS Paris, BNF, Arabe 6962, fol. 45a; cf. Najm ad-Dīn al-Kubrā and 'Alā' ad-Dawla as-Simnānī, *at-Ta'wīlāt an-Najmiyya fī t-tafsīr al-ishārī ṣ-ṣūfī, wa-yalīhi tatimmatuhu 'Ayn al-ḥayāt/al-Ta'wīlāt al-najmiyya by Najmuddīn al-Kubrā Followed by 'Ayn al-Ḥayāt by 'Alā'uddawlah al-Simnānī*, ed. Aḥmad Farīd al-Mizyadī (Beirut: Dār al-Kutub al-'Ilmiyya, 2009), 6: 54. From now on, for the sake of convenience, this work will be cited in abbreviated form as "al-Kubrā-as-Simnānī, *at-Ta'wīlāt an-Najmiyya-'Ayn al-ḥayāt*." The sixth and last volume of the printed edition, containing as-Simnānī's commentary, will be cited in abbreviated form as "as-Simnānī, *'Ayn al-ḥayāt*."

10 They include the editor of the printed edition of the text (al-Kubrā-as-Simnānī, *at-Ta'wīlāt an-Najmiyya-'Ayn al-ḥayāt*, front cover *et passim*) and, at least in part, Elias, who first mentions that the attribution of the text is debated, but later refers to it as the work of al-Kubrā (Elias, *The Throne*, 204ff.).

11 This was also the opinion of Henry Corbin and Osman Yahya, who held that the *tafsīr* of Najm ad-Dīn al-Kubrā stopped at suras 4 and 5, attributing the rest of the commentary up to sura 53 to Najm ad-Dīn Dāya (Ḥaydar Āmulī, *Jāmi' al-asrār wa-manba' al-anwār, ba inḍimām-i Risālat naqd an-nuqūd fī ma'rifat al-wujūd/La philosophie schi'ite. 1. Somme des doctrines ésotériques (Jâmi' al-asrâr), 2. Traité de la connaissance de l'être (Fî ma'rifat al-wojûd)*, ed. Henry Corbin and Osman Yahya, Tehran/Paris: Département d'iranologie de l'institut franco-iranien de recherche-Librairie d'Amérique et d'Orient Adrien-Maisonneuve, 1969, 52 of the French introduction). Other scholars who consider the first part of the *tafsīr* of the Kubrawiyya to be the work primarily or exclusively of Najm ad-Dīn Dāya are Hermann Landolt, "Review of *The Throne Carrier of God: The Life and Thought of 'Alā' ad-dawla as-Simnānī*, by Jamal J. Elias," *IJMES* 30, no. 3 (1998): 457–59, 459; Süleyman Ateş, "Üç müfessir bir tefsir;" Süleyman Ateş, *Isari tafsiri okulu* (Ankara: Ankara Üniversitesi Basımevi, 1974), 140–50; Mehmet Okuyan, *Necmuddîn Dâye ve tasavvufî tefsiri* (Istanbul: Rağbet Yayınları, 2001); cf. also W. Shpall, "A note on Najm al-Dīn al-Rāzī and the *Baḥr al-Ḥaqā'iq*," *Folia Orientalia* 22 (1981–1984): 69–80. On the complex problem of authorship of the first part of the Kubrawī *tafsīr* see Paul Ballanfat, "La prophétologie dans le 'Ayn al-hayât, tafsîr attribué à Najm al-dîn Kubrâ et. Najm-i Râzî," in *Mystique musulman: Parcours en compagnie d'un chercheur: Roger Deladrière*, ed. Geneviève Gobillot (Paris: Editions Cariscript, 2002), 171–364, 171–77.

12 *"Ya'lam ar-rajul al-muṭāli' hādhā l-kitāb al-musammā bi-"Najm al-Qur'ān," wa-huwa al-mudhayyal li-"t-Tafsīr an-Najmī" alladhī katabahu ash-shaykh al-muwaffaq Najm ad-Dīn Dāya al-Asadī ar-Rāzī – shakara Allāh sa'yahu – min awwal al-Qur'ān ilā Sūrat an-Najm (. . .) wa-alhamanī Allāh ta'ālā itmām tafsīrihi, wa-t-tafsīr (al-)maktūb bi-khaṭṭihi ash-sharīf tis' mujalladāt, wa-hādha al-mudhayyal mujallad wāḥid."* MS Paris, BNF, Arabe 6962, fol. 45a; cf. As-Simnānī, *'Ayn al-ḥayāt*, 54; cf. the observations by Corbin and Yahya in Ḥaydar Āmulī, *Jāmi' al-asrār*, 51 of the French introduction.

be held in supreme consideration. However, it is not sufficient to completely exclude Najm ad-Dīn al-Kubrā's participation in the writing of the *tafsīr*. Perhaps the most reasonable hypothesis is that the first part of the Kubrawī *tafsīr* was initiated by al-Kubrā and that this original core was later supplemented and completed by Dāya. This is the thesis held by Paul Ballanfat, who has extensively studied this complex exegetical text, of which he had begun to prepare a critical edition (currently unpublished).[13] According to this hypothesis, Dāya would have completely reworked the original core written by al-Kubrā. The latter would have been preserved in its entirety in the text that has come down to us, although following the intervention of Dāya it is practically impossible to identify with certainty the original nucleus of the text written by al-Kubrā and to establish which parts are by al-Kubrā and which by Dāya, except in some rare instances.[14] The layering and complexity of the text in question is demonstrated by numerous elements, including the presence in the commentary of long passages that are found to be identical in works by either Dāya (*Manārāt as-sā'irīn*) or al-Kubrā (*Fawātiḥ al-jamāl, al-Uṣūl al-ʿashara*), without these correspondences being acknowledged in the *tafsīr*.[15]

3.2 Authorship and Dating of as-Simnānī's *Tafsīr*

By contrast, there is no doubt about as-Simnānī's authorship of the unusual partial *tafsīr* that constitutes the final part of the *tafsīr* of the Kubrawiyya and includes the exegesis of suras 1 and 52–114. Under the name *Muṭṭalaʿ an-nuqaṭ wa-majmaʿ al-luqaṭ* (Place of ascent of the points and place of gathering of the gleanings), used within the text, the commentary is also mentioned in two other works by as-Simnānī.[16] The references made by the author of the *tafsīr* within the text to other works of as-Simnānī as

13 Ballanfat, "La prophétologie dans le *'Ayn al-hayât*," 172–77. Cf. also Paul Ballanfat, "Reality and Image in the *Tafsīr* of Kubrā and Rāzī,"*Journal of the Muhyiddin Ibn Arabi Society* 35 (2004): 75–108.

14 Ibid., 174ff.

15 On all these aspects, and in particular on the coexistence of elements and doctrines ascribed to al-Kubrā and Dāya in the first part of the Kubrawī Tafsīr, see Ballanfat's extensive and thorough study, "La prophétologie dans le 'Ayn al-hayât."

16 As-Simnānī, *Muṭṭalaʿ an-nuqaṭ wa-majmaʿ al-luqaṭ*, MS Paris, BNF, Arabe 6962, 24a; as-Simnānī, *'Ayn al-ḥayāt*, 24. In the abridged form *Muṭṭalaʿ an-nuqaṭ* the title is also cited in the introduction: MS Paris, BNF, Arabe 6962, fols. 1b–2a (cf. Paul Nwyia, "Muqaddima tafsīr al-Qur'ān li-ʿAlā' ad-Dawla as-Simnānī (736/1336)," *al-Abḥāth* 26 (1977): 141–57, 146). The problems surrounding the title of the work transpire from several places (see below). For example, we notice that the title is entirely vocalized in MS Paris, BNF, Arabe 6962, fol. 1b–2a. It is on the basis of this MS that we read the first term (m-ṭ-l-ʿ) as *muṭṭalaʿ* and not *maṭlaʿ*, as Elias does (*The Throne*, 203). Our reading of the term is supported by the use of the term *muṭṭalaʿ* that recurs in as-Simnānī's exegetical theory (on which see below). However, the doubts are far from being dispelled and more certainty can only come from a comparative examination of the various witnesses of the work. The two works in which the author mentions his commentary are: ʿAlā' ad-Dawla as-Simnānī, *Maṭāf al-ashrāf*, MS Istanbul, Süleymaniye Kütüphanesi,

being his own works is a further confirmation of this.[17] Various autobiographical references in the text, the internal coherence of the style, and the fact that this Qur'ānic commentary stands as a perfect summary presentation of as-Simnānī's thought as exposed in his other works are the definitive proof.[18]

As-Simnānī's commentary is not explicitly dated, but its writing can be indirectly placed in the first decade of the fourteenth/eighth century, and more precisely in the period from 1302/701, the date when as-Simnānī finished the treatise entitled *Mawārid ash-shawārid*, which we find quoted in the introduction to the *tafsīr*, and 1314/714, the date on which *Maṭāf al-ashrāf* was written, a short epistle in which as-Simnānī lists all the Arabic works he had written up to that time, including the one entitled *Muṭṭalaʿ an-nuqaṭ wa-majmaʿ al-luqaṭ*, i.e., his own *tafsīr*.[19]

3.3 Number of Manuscripts and Printed Edition

There are at least twenty-seven manuscripts of as-Simnānī's *tafsīr*, produced over a period of time from the first half of the fourteenth/eighth century, i.e., immediately after the author's death (d. 1336/736), to the second half of the nineteenth/thirteenth century.[20] The considerable number of manuscript witnesses and the continuity with which this commentary was copied over the centuries suggest a constant interest generated by this exegetical work. The great originality recognized in the doctrinal framework of Simnānī's *tafsīr* convinced Henry Corbin to study this text, and Simnānian thought more generally, and for the same reasons Paul Nuwyia decided to publish its introduction.[21] In spite of the importance of this exegetical work, to date there is no critical edition of as-Simnānī's *tafsīr* nor, as already mentioned, of the first part of the

Karaçelebizade 353, fols. 78b–80a, fol. 79b; ʿAlāʾ ad-Dawla as-Simnānī, *al-ʿUrwa li-ahl al-khalwa wa-l-jalwa*, ed. Najīb Māyil Hirawī (Tehran: Intishārāt-i Mawlā, 1983), 516.

17 They are the treatises entitled respectively *Mawārid ash-shawārid* (mentioned in MS Paris, BNF, Arabe 6962, fol. 3a; cf. Nwyia, "Muqaddima," 137; as-Simnānī, *ʿAyn al-ḥayāt*, 4) and *Madārij al-maʿārij* (mentioned at the end of the commentary of Q 112, MS Paris, BNF, Arabe 6962, fol. 203a; cf. as-Simnānī, *ʿAyn al-ḥayāt*, 369).

18 See e.g. the autobiographical information contained in the commentary of Q 54:9, MS Paris, BNF, Arabe 6962, fol. 43ff.; cf. as-Simnānī, *ʿAyn al-ḥayāt*, 52ff.

19 MS Paris, BNF, Arabe 6962, fol. 3a; cf. Nwyia, "Muqaddima," 137; as-Simnānī, *ʿAyn al-ḥayāt*, 4; as-Simnānī, *Maṭāf al-ashrāf*, MS Karaçelebizade 353, fol. 79b.

20 The most complete description of Simnānī's *tafsīr* manuscripts is in Elias, *The Throne*, 203–12.

21 Henry Corbin, *L'Homme de lumière dans le soufisme iranien* (Paris: Librairie de Médicis, 1971) (English translation *The Man of Light in Iranian Sufism*, trans. Nancy Pearson (Boulder: Shambala Press, 1978)); Henry Corbin, *En Islam iranien* (Paris: Gallimard, 1972), 3: 275–355; Nwyia, "Muqaddima." In fact, the importance of as-Simnānī's *tafsīr* had been noticed as early as the late 1920s by Luis Massignon, who published short extracts in *Recueil de textes inédits concernant l'histoire de la mystique en pays d'Islam* (Paris: Librarie orientaliste Paul Gauthner, 1929), 143–45.

tafsīr of the Kubrawiyya written by al-Kubrā and Dāya.[22] In 2009 the first printed edition of the whole Kubrawī *tafsīr*, including the part written by as-Simnānī, was published.[23] This edition has facilitated access to an important work of Sufi exegesis, enhancing its awareness and dissemination among a wider readership. Nonetheless, it falls short of scientific and philological standards as it relies on a single, likely late manuscript, failing to address crucial philological issues regarding the commentary, particularly authorship, and lacking critical apparatus.[24] A comparison of the printed edition with some manuscripts shows the presence of numerous and in many cases significant variants in the text, which suggests once again the usefulness that a critical edition of the work could have.

3.4 Titles of the Commentary

There is considerable uncertainty, even confusion, regarding the titles of the two parts that comprise the *tafsīr* of the Kubrawiyya, which seems not to have been known by a single name. The problem of the title(s) is strictly dependent on the complex process of formation of this commentary compiled by several authors in several stages, further exacerbated by the fact that in some manuscripts the *tafsīr* of as-Simnānī is handed down in isolated form as a separate work, while in others it is instead handed down together with the *tafsīr* by al-Kubrā-Dāya. There are titles that refer to the 'Kubrawī' *tafsīr* as a whole, titles that instead seem to refer especially to the oldest part of the *tafsīr* compiled by al-Kubrā-Dāya, and finally titles that seem to refer more precisely to the part written by as-Simnānī. Although at present doubts and uncertainties about the titling of this work, i.e. of its two parts, partly remain, it is important to note that they are hardly justifiable at least with regard to as-Simnānī's commentary, since he refers explicitly to his own *tafsīr* under the title *Muṭṭalaʿ an-nuqaṭ wa-majmaʿ al-luqaṭ* within the commentary itself, as well as in at least two other of his works, and in the shortened form *Muṭṭalaʿ an-nuqaṭ* in the introduction.[25]

22 The preparation of a critical edition of al-Kubrā-Dāya's *tafsīr* had been announced by Paul Ballanfat, "Reality and Image," 75, note 1, but was never published.

23 Al-Kubrā-as-Simnānī, *at-Taʾwīlāt an-Najmiyya-ʿAyn al-ḥayāt*.

24 Al-Kubrā-as-Simnānī, *at-Taʾwīlāt an-Najmiyya-ʿAyn al-ḥayāt*, 1:6. The editor of the edition does not provide a description of the manuscript, mentioning only that it is preserved at the Dār al-Kutub (he does not even provide the shelf mark) and that it consists of five volumes, the first four of which contain the *tafsīr* of al-Kubrā-Dāya and the fifth volume containing the *tafsīr* of as-Simnānī. From this little information it appears possible to identify the manuscript with number 15 of Elias' index (*The Throne*, 209), i.e. MS Cairo, Dār al-Kutub 6359, copied at the end of the nineteenth century (1286/1869).

25 MS Paris, BNF, Arabe 6962, fols. 1b–2a (cf. Nwyia, "Muqaddima," 146) and 24a (cf. as-Simnānī, *ʿAyn al-ḥayāt*, 24). The reference to the title of the work contained in the introduction is absent from the printed edition of the text (as-Simnānī, *ʿAyn al-ḥayāt*, 3), a demonstration of the problems arising from a non-critical edition. Under that title, the work is cited in *Maṭāf al-ashrāf*, fol. 79b and *al-ʿUrwa*, 516.

For these reasons it seems reasonable to suggest that the latter is the original name that as-Simnānī had chosen for his *tafsīr*. In one instance within the commentary as-Simnānī uses two other names related to each other, in order to designate the work, namely *Najm al-Qurʾān* and *at-Tafsīr an-najmī*.[26] The first part of the *tafsīr* of the Kubrawiyya in the manuscripts is also referred to by other titles, including *at-Taʾwīlāt an-najmiyya*, *ʿAyn al-ḥayāt*, *al-ʿAwārif* and *Baḥr al-ḥaqāʾiq*.[27] Further demonstrating the difficulty of precisely establishing the title(s) of this commentary, and the confusion that reigns in the matter, it is worth noting that in the printed edition the first part of the Kubrawī *tafsīr* is entitled *at-Taʾwīlāt an-najmiyya*, while the concluding part, the one written by as-Simnānī, bears the title *ʿAyn al-ḥayāt*.[28] It is likely that a comprehensive examination of the manuscript witnesses to the text could finally shed light on this aspect as well.

Elias suggested that by the title *Maṭlaʿ* (or *Muṭṭalaʿ*, as we proposed) *an-nuqaṭ wa-majmaʿ al-luqaṭ* as-Simnānī could refer only to the introduction and not to the entire *tafsīr* (Elias, *The Throne*, 204; Jamal J. Elias, "Ṣūfī Tafsīr Reconsidered: Exploring the Development of a Genre," *JQSt* 12 (2010): 41–55, 47f.). We do not feel comfortable with this hypothesis, which seems to be contradicted by at least two pieces of evidence: 1) the fact that the title *Muṭṭalaʿ an-nuqaṭ wa-majmaʿ al-luqaṭ* is also used by as-Simnānī within the body of the commentary (BNF, Arabe 6962, fol. 24 = as-Simnānī, *ʿAyn al-ḥayāt*, 24); 2) the fact that as-Simnānī, in mentioning the *Muṭṭalaʿ an-nuqaṭ* within *al-ʿUrwa li-ahl al-khalwa wa-l-jalwa* (p. 516) refers to a "very thick book" (*wa-huwa kitāb muṭawwal jiddan*). Examination of MS Paris, BNF, Arabe 6962, fol. 1b–2a (cf. Nwyia, "Muqaddima," 146), in which the titles are vocalized, suggests a different possibility, which seems to us worthy of consideration. That is, that the name *Muṭṭalaʿ an-nuqaṭ wa-majmaʿ al-luqaṭ* refers to the entire *tafsīr*, while that for the introduction as-Simnānī had chosen another title, directly derived from that, namely *Muttaqid muṭṭalaʿ an-nuqaṭ wa-multaqiṭ majmaʿ al-luqaṭ*. Again, there are differences in readings, as Nwyia ("Muqaddima," 146) in his edition reads *muqaddima* where we read *muttaqid*, and where other readings still are possible due to a folding of the paper that seems to conceal a letter. Elias regards the introduction and the actual commentary as two distinct exegetical works ("Simnānī (d. 1336/736) wrote two interesting works of commentary on the Qur'an"; "Simnānī's commentary on the Qur'an is more accurately seen as comprised of two distinct works written at separate times," in "Sūfī Tafsīr Reconsidered," 47 *et passim*). Although it is possible that the introduction was not written at the same time as the body of the commentary considering them two separate works seems excessive to us.

26 MS Paris, BNF, Arabe 6962, fol. 45a; cf. as-Simnānī, *ʿAyn al-ḥayāt*, 54; cf. Elias, *The Throne*, 204. It is likely that the use of these two titles in this one instance stemmed from as-Simnānī's desire to honor the shaykh Najm ad-Dīn Dāya who is mentioned immediately afterwards as the author of the first part of the Kubrawī *tafsīr* (cf. footnote 12 above).

27 See Elias, *The Throne*, 204f.

28 Al-Kubrā-as-Simnānī, *at-Taʾwīlāt an-Najmiyya-ʿAyn al-ḥayāt*, front cover *et passim*.

3.5 Relationship between as-Simnānī's *Tafsīr* and the *Tafsīr* of al-Kubrāà-Dāya

The relationship of as-Simnānī's *tafsīr* to the ancient part of the *tafsīr* of the Kubrawiyya is twofold: one of continuity and change. The fact that in the text as-Simnānī reminds the reader that Najm ad-Dīn Dāya had passed away leaving the commentary unfinished, that he was inspired by God to complete it, that he describes his own text as an "appendix" (*mudhayyal*) to the *tafsīr* that remained unfinished; and the fact that he comments from sura 52 onwards (except for the introduction and the commentary on the Opening Sura), indicate incontrovertibly his intention to link up with a pre-existing work in order to complete it.[29] This was an excellent way to underline his own belonging to a certain spiritual filiation and to sanction at the same time his own leading role in the doctrinal elaboration of the same. Other elements, instead, denote the originality of as-Simnānī's work. First of all, the presence of a theoretical introduction in which he exposes the exegetical method adopted by him. On one point it is necessary to insist in order to clarify, that is, that continuity does not mean slavish imitation, all the more unfeasible in this context for several reasons, of which the main ones seem to us the following. First, the high degree of originality of as-Simnānī's cultural and spiritual training, to which we have alluded, and which is reflected in his thought. Second, the fact that the ancient part of the *tafsīr* of the Kubrawiyya (as well as that compiled by as-Simnānī) presents a peculiar feature that makes it very difficult to even attempt to 'faithfully' resume the discourse interrupted by predecessors. This is the fact that the *tafsīr* of al-Kubrā-Dāya is virtually devoid of references to the older exegetical tradition, both Sufi and non-Sufi, thus differentiating itself from the classical tradition that even in the Sufi sphere, following the example of the *tafsīr* of as-Sulamī, usually gives great importance to the transmission of the opinions of the authorities of the previous generations.[30] The result is an extremely

29 For references and a transcription of the passage in question see footnote 12 above.

30 In al-Kubrā-Dāya's *tafsīr* among the few exceptions is a series of quotations from ancient masters added as an appendix to the commentaries on suras 12–14 (cf. al-Kubrā-as-Simnānī, *at-Ta'wīlāt an-Najmiyya-'Ayn al-ḥayāt*, 3: 349–83, 404–13 and 437–44). The fact that these quotations are brought together in bulk at the end of the commentary, and that they occur only in these three suras, suggests that they may be later interpolations. Ballanfat also points out a lengthy quotation from as-Sulāmī ("La prophétologie dans le *'Ayn al-hayât*," 173). Again, the comparison of various manuscripts or the existence of a critical edition would help to clarify this question. On the predominant role of inspiration in the writing of the *tafsīr* of the Kubrawiyya see again Ballanfat, "La prophétologie dans le *'Ayn al-hayât*," 174. The very few quotations in as-Simnānī's *tafsīr* include a passage from the *Ṭā'iyya* of Ibn al-Fāriḍ (d. 1235/632) and at least one mention of Junayd al-Baghdādī (d. 910/298) and Abū Yazīd al-Bisṭāmī (d. ca. 874/261), on which see MS Paris, BNF, Arabe 6962, fols. 52a–b, 84b, 111b; cf. Simnānī, *'Ayn al-ḥayāt*, 62, 107, 158. On as-Sulamī's two *tafsīrs*, which are almost entirely compilations of commentaries of earlier Sufis, see Süleyman Ateş, *Sülemī ve tasavvufī tefsīrī* (Istanbul: Sönmez Neşriyat, 1969), esp. 76–95; 'Abd ar-Raḥmān as-Sulamī, *Ziyādāt Ḥaqā'iq at-tafsīr/The Minor Qur'ān Commentary*

original exegesis, essentially based on the spiritual intuitions of the authors, therefore strictly personal and not replicable. In this sense, it does not seem incorrect to suggest that we are faced with a '*tafsīr ilhāmī*,' or, even, a '*tafsīr kashfī*.' This being so, it seems correct to suggest that as-Simnānī's *tafsīr* is in continuity with the unfinished *tafsīr* of al-Kubrā-Dāya just insofar as its author expounds his own exegesis applying a peculiar Sufi hermeneutics matured on the basis of his own spiritual experience that often differs from that of his predecessors, although the influences and the heritage of the Kubrawī doctrinal elaboration are evident in it. Continuity and change in the relationship between al-Kubrā-Dāya's *tafsīr* and as-Simnānī's *tafsīr* are also observed in some of the fundamental themes that innervate the work. In particular, the development of the doctrine of the subtle centers (*laṭīfa*, pl. *laṭāʾif*), corresponding to as many stages of the spiritual path, associated in turn with a phenomenon of colored photisms and the concept of 'inner prophets,' each of which is in charge of a subtle center. This latter idea allows the authors to interpret the passages narrating the stories of the prophets (*qiṣaṣ al-anbiyāʾ*) – which represent a substantial part of the Qurʾān – not only as historical events that took place or will take place at a specific time in history, but as inner events that occur in the subtle centers of each initiate during the spiritual path. Ballanfat has observed how these concepts are already present *in nuce* in the *tafsīr* (and more generally in the work) of al-Kubrā-Dāya, although not in an organized and systematic way as they appear in as-Simnānī's *tafsīr*,[31] in which these ideas and this system of correspondences come to completion in one of their highest expressions, such as to allow Corbin to state, speaking of this *tafsīr*, that "l'œuvre de Semnânî est d'une profondeur, d'une 'intériorité' incomparable avec l'œuvre de ses devanciers,"[32] and that "le *Tafsîr* de Semnânî s'atteste comme une œuvre d'une importance exceptionnelle dans la littérature mystique inspirée par l'introspection des données qorâniques; elle va jusqu'à l'extrême limite de l'intériorisation concevable."[33]

---

of Abū ʿAbd ar-Raḥmān Muḥammad b. al-Ḥusayn as-Sulamī (d. 412/1021), ed. Gerhard Böwering (Beirut: Dar el-Machreq, 1995); Gerhard Böwering, "The Major Sources of Sulamī's Minor Qurʾān Commentary," *Oriens* 35 (1996): 35–56; cf. Alan Godlas, "Ṣūfism," in *The Wiley Blackwell Companion to the Qurʾān*, ed. Andrew Rippin and Jawid Mojaddedi (Hoboken, NJ: John Wiley & Sons Ltd, 2017²), 418–29, 420.

31 Ballanfat, "La prophétologie dans le *ʿAyn al-hayât*," 171, 177 *et passim*; on the copresence of two different systems of classification of the *laṭāʾif* in the *tafsīr* of al-Kubrā-Dāya, cf. ibid., 176.

32 Henry Corbin, *En Islam iranien*, 3: 276, n90.

33 Ibid., 3: 277.

3.6 As-Simnānī's Hermeneutics

The main features, principles, and technical terminology of as-Simnānī's hermeneutics are found enunciated in the introduction to his *tafsīr*.[34] As-Simnānī in the first lines of the *muqaddima* speaks of *iṭṭilāʿ ʿalā baṭn al-Qurʾān*, "investigation of the inner (or esoteric) aspect of the Qurʾān," and of *tafsīr al-buṭūn as-sabʿa*, "exegesis of the seven inner (esoteric) aspects" of the Qurʾān, suggesting from the outset the particular perspective in which his own commentary is placed, namely a purely esoteric exegesis that does not analyze the literal and apparent meanings of the sacred text.[35] However, such an approach according to Simnānī does not imply any contradiction and no opposition between esoteric and exoteric exegesis, between occult meaning (*bāṭin*) and apparent meaning (*ẓāhir*) of the sacred text, which as-Simnānī considers not only fully valid, but even necessary to accept. The two types of exegesis for the author are entirely complementary to each other and both are necessary, as he makes explicit again in the *muqaddima*:

> Be certain that he who rejects the exegesis of the exoteric aspect of the Qurʾān (*tafsīr ẓahr al-Qurʾān*) in the world of external human events (*ʿālam al-āfāq an-nāsūtī*) is a stubborn Batinite heretic; and that he who rejects the exegesis of the exoteric aspect of the Qurʾān (*tafsīr baṭn al-Qurʾān*) in the inner celestial world of the soul (*ʿālam al-anfus al-malakūtī*) while acknowledging its exoteric exegesis is a sclerotic, a doubter,[36] and an idiot; but he who unites the exoteric and the exoteric (*jamaʿa bayna aẓ-ẓāhir wa-l-bāṭin*) he is a fulfilled Sunni Muslim.[37]

---

34 Simnānī's hermeneutics over the years has been the subject of direct or incidental consideration by various scholars. Because of their antecedence and breadth, special mention must be made of Corbin's studies. In addition to Corbin's writings cited in the preceding pages, see also "L'intériorisation du sens en herméneutique soufie iranienne," in *Eranos Jahrbuch* 26 (1957): 57–187, and "Physiologie de l'homme de lumière dans le soufisme iranien," in *Ombre et lumière* (Paris; Brussels: Desclée de Brouwer, 1961), 135–257. In more recent years the *tafsīr* has been used extensively by Elias to reconstruct Simnānī's thought in *The Throne Carrier of God* (on Simnānī's understanding of the Qurʾān see esp. pp. 107–110; on the *tafsīr* and its value has a summary of Simnānī's thought see esp. pp. 167–169 and 203–212). For an original argument on how Simnānī would have understood the very idea of "hermeneutics" see Paul Ballanfat, "Simnānī and Hermeneutics," in *The Spirit and the Letter: Approaches to the Esoteric Interpretation of the Qur'an*, ed. Annabel Keeler and Sajjad Rizvi (Oxford-New York: Oxford University Press in association with The Institute of Ismaili Studies, 2016), 345–373.

35 MS Paris, BNF, Arabe 6962, fol. 1b; Nwyia, "Muqaddima," 146; this relevant passage is missing in as-Simnānī, *ʿAyn al-ḥayāt*, 3.

36 The translation of this term is tentative. The original word is unclear and subject to variation in the manuscripts. This is m-s-b-b-h-ī in MS Paris, BNF, Arabe 6962, fol. 8b (cf. Nwyia, "Muqaddima," 151, where it is m-s-b-h-ī) and m-sh-b-h-ī in as-Simnānī, *ʿAyn al-ḥayāt*, 9. Corbin translates "obscurantiste" on the basis of another manuscript (*En Islam iranien*, 3: 302).

37 MS Paris, BNF, Arabe 6962, fol. 8b; cf. Nwyia, "Muqaddima," 151; as-Simnānī, *ʿAyn al-ḥayāt*, 9; cf. Corbin, *En Islam iranien*, 3: 302. On the complementarity and non-opposition between *tafsīr* of the exoteric and esoteric sense of the Qur'an in the Sufi context cf. Alan Godlas, "Ṣūfism," 419.

Discussing the relationship between the exegesis of the manifest and occult meanings of the Qurʾān, as-Simnānī finds the opportunity to clarify one of the most significant aspects of his own Qurʾānic commentary, as well as that of al-Kubrā-Dāya, and already pointed out above, namely the almost total absence of references to the previous exegetical tradition. This, once again, is not due to a contraposition between the two types of exegesis, but rather to their complementarity: just as it is required to adhere to the teaching of a master connected to the exegetical tradition going back to the companions of the prophet Muḥammad in the study of the apparent sense of the text, so it is necessary to adhere to the teachings of the invisible inner master (*al-ustādh al-ghaybī*) in learning the exegesis of the esoteric sense:

> He who interprets the exoteric meaning of the Qurʾān on the basis of his own individual opinion (*man fassara ẓahr al-Qurʾān bi-raʾyihi*) without having learned by direct oral teaching from an exegete (*mufassir*) whose chain of traditional transmission goes back to the Prophet's companions (*kāna isnāduhu muttaṣilan bi-ṣ-ṣaḥāba*) – that God may be pleased with them – disacknowledges (*yakfur*) because of his own ignorance most of the rules that are proper to it (*sc.* that are proper to the science of exegesis), (the science of) the causes of revelation, etc.; while he who interprets the esoteric meaning of the Qurʾān according to his own individual opinion in the absence of inspiration (*min ghayr ilhām*) (. . .) disacknowledges all the allusions proceeding from the dominical presence towards the subtleties of the subtle celestial forces and centers.[38]

And again:

> Just as the integrity of the physical external sense of hearing (*as-samʿ aẓ-ẓāhir an-nāsūtī*) is a condition for the hearer to enable him to hear the outer form of the Qurʾān (*ẓahr al-Qurʾān*) and learn the exegesis of its outer sense (*tafsīrahu aẓ-ẓāhir*) from his own visible master (*min ustādhihi ash- shahādī*) so the integrity of the hearing of the heart is a condition for the inspired (*mulham*) in hearing the inner aspect of the Qurʾān (*baṭn al-Qurʾān*) and learning the exegesis of its inner sense (*tafsīrahu al-bāṭin*) from one's invisible master (*min ustādhihi al-ghaybī*).[39]

In other words, what appears to historical scientific examination as one of the most uncommon and therefore interesting features of this exegetical work, namely its high degree of originality, was not at all perceived as such by its author. As-Simnānī actually considered his own *tafsīr* 'transmitted,' albeit in a metaphysical way, by means of the inspiration of his invisible master, and not at all the result of his own personal

---

38 MS Paris, BNF, Arabe 6962, fol. 15a; cf. Nwyia, "Muqaddima," 155–156; cf. as-Simnānī, *ʿAyn al-ḥayāt*, 16.

39 MS Paris, BNF, Arabe 6962, fol. 15a–b; cf. Nwyia, "Muqaddima," 156; cf. as-Simnānī, *ʿAyn al-ḥayāt*, 16; cf. Corbin, *En Islam iranien*, 3: 302. It is worth noting that the word I read as *ustād* (Persian for *ustādh*), "master," seems subject to variations in the manuscript tradition of the text. I read *ustād* on the basis of MS Paris, BNF, Arabe 6962, supported by the reading of Corbin (*En Islam iranien*, 3: 293, 298 *et passim*) who relied on another manuscript (MS Tehran, Kitābkhāna-yi millī-i Īrān, 1047; see *En Islam iranien*, 3: 278, n93; cf. Elias, *The Throne*, 209). However, Nwyia, in the edition of the introduction, relying on the same MS Paris, BNF, Arabe 6962, reads *isnād*. Finally, the editor of the printed edition of the text, based on the manuscript he used, reads *istinād*. This is another example of how useful, therefore desirable, a critical edition of the work could be.

lucubration (*ra'y*). In this way, the famous Sufi master confirms, albeit in an extremely original way, the attitude prevalent in the pre-modern mindset that considers only knowledge handed down by a recognized authority to be reliable, while simultaneously distrusting fully independent individual thought.

Having ascertained the importance and complementarity between *tafsīr aẓ-ẓāhir* and *tafsīr al-bāṭin*, between exoteric and esoteric exegesis of the sacred text in the thought of as-Simnānī, and the fact that he considered his own commentary, based on inspiration, a *tafsīr al-bāṭin*, it must be added that the theory on the levels of meaning of the Qur'ān of the Iranian Sufi scholar is much more complex and stratified. We learn in fact that the inner and esoteric aspect of the holy book, which until now has been considered a unit in itself as opposed to the exoteric aspect, according to as-Simnānī contains within itself seven main inner aspects which, in turn, present ten aspects each, thus reaching seventy meanings, which present ten aspects each, thus reaching seven hundred esoteric meanings of which each Qur'ānic verse is susceptible. A real inner abyss, whose multiple depths produce vertigo in the observer who faces it, and in which the initiate is called to immerse himself. It is in this context that one of the most fascinating features of as-Simnānī's hermeneutics develops. This is the idea that in the human being there are seven subtle centers (*laṭā'if*) associated with as many inner prophets, corresponding precisely to the seven main levels of esoteric exegesis of the Qur'ān.

> Be assured that the Qur'ān possesses seventy inner aspects (*li-l-Qur'ān sab'īn baṭnan*) as it has been handed down by the Prophet – peace be upon him – and I allude to what is within your capacity to consent. Know that for each of the seven subtle centers above mentioned (*al-laṭā'if as-sab'a al-madhkūra*) there are ten scion subtle centers (*al-laṭā'if al-'ashar as-salāliyya*); and that to each of these ten subtle centers there corresponds in the Qur'ān a specific disposition and a specific understanding, so that there are seventy inner senses for each of the Qur'ānic verses (*fa-yakūn sab'īn baṭnan li-kulli āya min al-āyāt*). Indeed seven hundred! Since you know for certain that each of the ten scion subtle centers possesses ten senses, (five) internal and (five) external, either in potency or in act, so that for each (secondary) subtle center there are ten understandings pertaining to the inner aspect, thus reaching seven hundred.[40]

Basically, as-Simnānī affirms that each passage of the Qur'ān on the inner plane is susceptible to an indefinite number of interpretations, corresponding to as many moments of the spiritual path. Obviously, this is not an exegesis that can be put down in writing in a paper book, but is inscribed in the invisible book of the initiate's subtle imperishable body.[41] As-Simnānī therefore in his own *tafsīr* discusses primarily the

40 MS Paris, BNF, Arabe 6962, fols. 12b–13a; Cf. Nwyia, "Muqaddima," 154; cf. as-Simnānī, *'Ayn al-ḥayāt*, 13f.

41 As-Simnānī elaborates an articulated doctrine about the subtle acquired body (*al-badan al-muktasab*) or everlasting body (*al-badan al-bāqī*), complementary to the physical body or compound body (*al-badan al-ma'jūl*) or perishable body (*al-badan al-fānī*). For more details on this aspect see Corbin, *En Islam iranien*, 3: 312–40; Elias, *The Throne*, 79ff.; for a translation of one of the many texts in

seven levels of interiorization that he considers to be the principal ones and of which the others are but further particularizations. The seven main subtle centers and the inner prophets with which they are put in correspondence by as-Simnānī in ascending order are: (1) the subtle center of the bodily receptacle (*al-laṭīfa al-qālibiyya*), which corresponds to the Adam of your being (*Ādam wujūdika*); (2) the subtle center of the vital soul (*al-laṭīfa an-nafsiyya*), which corresponds to the Noah of your being (*Nūḥ wujūdika*); (3) the subtle center of the heart (*al-laṭīfa al-qalbiyya*), which corresponds to the Abraham of your being (*Ibrāhīm wujūdika*); (4) the subtle center of the secret (*al-laṭīfa as-sirriyya*), which corresponds to the Moses of your being (*Mūsā wujūdika*); (5) the subtle center of the spirit (*al-laṭīfa ar-rūḥiyya*), which corresponds to the David of your being (*Dāwūd wujūdika*); (6) the subtle center of mystery (*al-laṭīfa al-khafiyya*), which corresponds to the Jesus of your being (*ʿĪsā wujūdika*); (7) and finally the subtle center of the True (*al-laṭīfa al-ḥaqqiyya*), which corresponds to the Muḥammad of your being (*Muḥammad wujūdika*) (s. Table 1 below).[42]

However, even 'only' seven interpretations for each Qur'ānic verse of suras 1 and 52–114 would have required as-Simnānī to write a text of enormous size. He provides an example of an esoteric sevenfold interpretation of a single verse, corresponding to the seven major subtle centers, within the Introduction, for the primary purpose of illustrating the potential development of this particular form of inspiration-based exegesis.[43] However, what happens most commonly within the Commentary is that the author provides the interpretation of the single verse or group of verses from the point of view of only one of the seven subtle centers, alternating them and referring now to this or that, and sometimes to more than one at the same time, but without going systematically into a sevenfold interpretation of the single verses. In particular, as-Simnānī suggests to the reader to interpret the numerous Koranic passages narrating stories of the prophets from the point of view of the subtle centers corresponding to the latter, as in these two examples, referring respectively to the first two subtle centers, and therefore to Adam and Noah:

> Whenever in the Book you hear the words referring to Adam, listen to them with the organ of hearing of your subtle center of the bodily receptacle (*bi-laṭīfat qābiliyyatika*) and operate your subtle center of the bodily receptacle according to what is commanded and forbidden to him (*sc.* to Adam), and meditate on the examples given to him, and be certain that the esoteric of this

---

which as-Simnānī addresses the theme of the various bodies possessed by the human being see Martini, *ʿAlāʾ al-Dawla al-Simnānī*, 395ff.

42 MS Paris, BNF, Arabe 6962, fol. 3a–b; Cf. Nwyia, "Muqaddima," 146f.; cf. as-Simnānī, *ʿAyn al-ḥayāt*, 4. It is not possible to present a detailed analysis of the doctrine of *laṭāʾif* according to as-Simnānī here. For more comprehensive discussions, please refer to Corbin, *En Islam iranien*, esp. 278–90; Corbin, *The Man of Light in Iranian Sufism*, esp. 121–30 (page numbers refer to the reprint by Omega Publications, New Lebanon (NY) 1994), and Elias, *The Throne*, esp. 85–91.

43 MS Paris, BNF, Arabe 6962, fols. 9a–12b; Cf. Nwyia, "Muqaddima," 152–54; cf. as-Simnānī, *ʿAyn al-ḥayāt*, 11–13.

> Book refers to you (*baṭn hādhā l-kitāb yataʿallaq bika*) as speech concerning the world of souls just as the exoteric corresponds to Adam in the (external) world of horizons (. . .). And whenever you hear the verses referring to Noah, listen to them with your subtle center of the vital soul (*bi-laṭīfat nafsiyyatika*) and give heed to the truth of the speech of the True One in order not to succumb to the "swarming sea" (Q 52:6, here an obvious allusion to the Flood) because of the two fires of desire and anger, and not to drown the people (*umam*, sing. *umma*, the communities of the prophets) of your inner forces (*quwā*).[44]

One of the most significant aspects of the exegesis of as-Simnānī and of his *tafsīr* is precisely the web of correspondences between subtle centers, inner prophets and levels of exegesis of the esoteric meaning of the Qurʾān that he weaves, which, by relating the initiate and the holy book, makes the Qurʾān a map and a guide on the spiritual path. This is further confirmed by additional correspondences pointed out by the author, such as the one made explicit in the Introduction and repeated several times in the commentary which associates to each of the seven subtle centers specific types of recollection (*dhikr*), ritual means par excellence in Sufism, aimed at purifying each *laṭīfa*, or each inner prophet and his relative community of believers, from its particular impurities.[45] Again, on the basis of information deduced from other works of as-Simnānī, we learn that each of the seven subtle centers is surrounded by a veil that has a specific color, whose coloring, revealing itself to the inner sight of the initiate, informs him about the stage of the spiritual path in which he comes.[46]

The doctrine of the levels of esoteric meaning of the Qurʾān described so far, which in addition to a primary distinction between an exoteric and an esoteric aspect contemplates seven levels of esoteric meaning (which can become as many as seventy, or seven hundred), despite its complexity and articulation does not at all exhaust as-Simnānī's hermeneutic theory. The Iranian Sufi scholar in fact explains that the exegesis of the esoteric sense (or rather senses) of the Qurʾān, the one with which his own *tafsīr* is primarily concerned, despite its complexity and depth, represents

44 MS Paris, BNF, Arabe 6962, fol. 4a–b; Cf. Nwyia, "Muqaddima," 147; cf. as-Simnānī, *ʿAyn al-ḥayāt*, 4f. For a more extensive and annotated translation of this passage see Corbin, *En Islam iranien*, 3: 281ff.

45 MS Paris, BNF, Arabe 6962, fols. 10a–12b and 15b–16a; Cf. Nwyia, "Muqaddima," 152–54 and 156; cf. as-Simnānī, *ʿAyn al-ḥayāt*, 11–13, 17. The doctrine of recollection is highly developed by as-Simnānī and treated by him in numerous works. The pre-eminence accorded by as-Simnānī to the silent *dhikr* seems to have contributed to the affirmation of the latter in some branches of Sufism, including the Naqshbandiyya. On recollection according to as-Simnānī see Elias, *The Throne*, Index, s.v. "dhikr" and Martini, *ʿAlāʾ al-Dawla al-Simnānī*, Index, s.v. "ḏikr."

46 As-Simnānī illustrates the doctrine of colors corresponding to the seven *laṭāʾif* in a number of works. There are some slight variations in the description of these colors depending on the texts. Corbin took great interest in this aspect. See Corbin, *The Man of Light in Iranian Sufism* and Corbin, *En Islam iranien*, 3: 289 *et passim*; A more recent discussion is in Elias, *The Throne*, esp. 135–41. For an English translation of a significant short treatise by as-Simnānī devoted specifically to this topic see Jamal J. Elias, "A Kubrawī Treatise on Mystical Visions: The *Risāla-yi nūriyya* of 'Alā' ad-dawla as-Simnānī," *MW* 83, no. 1 (1993): 68–80.

Table 1: Levels of Exoteric Exegesis and Their Correspondences.

Level of exoteric exegesis	Subtle Centre	Inner Prophet	Type of recollection	Color
First Interiority (*al-baṭn al-awwal*)	Subtle center of the bodily receptacle (*al-laṭīfa al-qālibiyya*)	Adam	Conventional formula (*adh-dhikr ar-rasmī*)	Darkness/Smoke-Grey
Second Interiority (*al-baṭn ath-thānī*)	Subtle center of the vital soul (*al-laṭīfa an-nafsiyya*)	Noah	Apprehended formula (intellectually by an authority) (*adh-dhikr at-taʿlīmī*)	Blue
Third Interiority (*al-baṭn ath-thālith*)	Subtle center of the heart (*al-laṭīfa al-qalbiyya*)	Abraham	Formula inculcated by a spiritual master (*adh-dhikr at-talqīnī*)	Red
Fourth Interiority (*al-baṭn ar-rābiʿ*)	Subtle center of the secret (*al-laṭīfa as-sirriyya*)	Moses	Formula "There is no God but God" (*lā ilāha illā Allāh*)	White
Fifth Interiority (*al-baṭn al-khāmis*)	Subtle center of the spirit (*al-laṭīfa ar-rūḥiyya*)	David	Formula of the pronoun 'He' (*adh-dhikr al-huwī*) (sound 'hā'')	Yellow
Sixth Interiority (*al-baṭn as-sādis*)	Subtle center of the mystery (*al-laṭīfa al-khafiyya*)	Jesus	Holy recollection (*adh-dhikr al-qudsī*) Transcending letter and sound (*al-munazzah ʿan al-ḥarf wa-ṣ-ṣawt*)	Luminous Black[47]
Seventh Interiority (*al-baṭn as-sābiʿ*)	Subtle center of the Truth (*al-laṭīfa al-ḥaqqiyya*)	Muḥammad	Supreme Recollection (*adh-dhikr al-aʿẓam*)	Emerald-Green/Pure Light Having no Color at All

only the second of four levels of hermeneutics of the Holy Book, above which are situated two even more universal and ineffable levels of hermeneutics to which as-Simnānī alludes in a few instances in the Commentary. That is, the sevenfold structure that governs the esoteric exegesis of the Qurʾān according to this organization represents but the development of one (the second) of four aspects of a more universal fourfold structure. These four levels are the exoteric dimension (*ẓahr*), the exoteric dimension (*baṭn*), the limit (*ḥadd*) and the point of ascent (*muṭṭala*) (s. Table 2 below).[48] Such termi-

47 Ar. *aswad nūrānī*, which Corbin also translates as "black light."
48 *Muṭṭalaʿ* is the same technical term that not surprisingly as-Simnānī seems to use in the title of his *tafsīr*: *Muṭṭalaʿ an-nuqaṭ wa-majmaʿ al-luqaṭ* (Place of ascent of the points and place of gathering of the gleanings). On the problems concerning the title of as-Simnānī's Qurʾānic commentary see above.

nology is derived from a non-canonical tradition attributed to Muḥammad that states "Not a verse of the Qur'ān is revealed but it has an exoteric dimension and an esoteric dimension, and every word has a limit, and every limit has a point of ascent." Each of these four levels of exegesis according to as-Simnānī is accomplished by a commentator

Table 2: Four Dimensions of Qur'ānic Exegesis and their Connections to Macrocosm and Microcosm according to as-Simnānī.[49]

Realm[50]	Level of Exegesis[51]	Object of Exegesis[52]	Rank of Reader[53]	Rank of Exegete[54]	Means of Exegesis[55]
Divinity (*lāhūt*)	Point of Ascent (*muṭṭala'*)	Mother of the Book (*umm al-kitāb*)	*muḥsin kāmil shahīd*	Attainer of the Secret of the Essence (*muṭṭali' 'alā sirr adh-dhāt*)	Should not comment at all
Omnipotence (*jabarūt*)	Limit (*ḥadd*)	Sublime Qur'ān (*Qur'ān 'aẓīm*)	*mu'min 'ārif rashīd*	Declarer of the Unity of God (*muwaḥḥid*)	Divine Permission (*idhn*)
Sovereignity (*malakūt*)	Esoteric Dimension(s) (*baṭn/buṭūn*) (seven inner prophets)	August Qur'ān (*Qur'ān karīm*)	*muslim sunnī sa'īd*[56]	Verifier of Truth (*muḥaqqiq*)	Inspiration (*ilhām*)
Human Realm (*nāsūt*)	Exoteric Dimension (*ẓahr*)	Glorious Qur'ān (*Qur'ān majīd*)		Exoteric exegete (*mufassir*)	Hearing from an authorized exegete (*sam'*)

49 This table, with some differences, is based on the one in Elias, *The Throne*, 109.

50 MS Paris, BNF, Arabe 6962, fols. 8b and 15a–16b; Cf. Nwyia, "Muqaddima," 151, 155–57; cf. as-Simnānī, *'Ayn al-ḥayāt*, 9, 16f.

51 MS Paris, BNF, Arabe 6962, fols. 8b and 15a–16b; Cf. Nwyia, "Muqaddima," 151, 155–57; cf. as-Simnānī, *'Ayn al-ḥayāt*, 9, 16f.

52 On this point I do not agree with Corbin, who understood the Glorious Qur'ān (*Qur'ān majīd*) as corresponding to the esoteric dimension (*En Islam iranien*, 3: 305f.), which produces a cascading slip. The passage in question is on MS Paris, BNF, Arabe 6962, fol. 15a; cf. Nwyia, "Muqaddima," 155; cf. as-Simnānī, *'Ayn al-ḥayāt*, 15f.

53 MS Paris, BNF, Arabe 6962, fol. 8b; cf. Nwyia, "Muqaddima," 151; cf. as-Simnānī, *'Ayn al-ḥayāt*, 9.

54 MS Paris, BNF, Arabe 6962, fol. 33a–b (end of *tafsīr* of Q 52); cf. as-Simnānī, *'Ayn al-ḥayāt*, 37.

55 MS Paris, BNF, Arabe 6962, fol. 33a–b (end of *tafsīr* of Q 52); cf. as-Simnānī, *'Ayn al-ḥayāt*, 37.

56 As seen above, both the viewpoint that recognizes the exegesis of the exoteric meaning alone to the exclusion of the exegesis of the esoteric meaning and, conversely, the viewpoint that recognizes the exegesis of the esoteric meaning alone to the exclusion of the exegesis of the exoteric meaning, are considered by Simnānī to be two heterodox positions.

who has obtained a specific rank applying a specific method of comment and commenting on a specific form of the sacred text, corresponding to as many epithets by which it is described in the Qurʾān. And again, each of these four levels also corresponds to the four great subdivisions of the manifestation according to a scheme rather widespread in Sufism, and with other aspects. The result is an admirable structure of correspondences that binds together and brings into sync the macrocosm and the microcosm, that is the world and man, using as a means of connection the exegesis of the Holy Book, which is thus revealed once again as a powerful tool for the elevation of the initiate on the spiritual path:

> O seeker of the inner meaning of the Qurʾān! You should first study the literal level of the Qurʾān and bring your body into harmony with its commands and prohibitions. Secondly, you should occupy yourself with purifying your inner being so that you may comprehend the hidden meaning (*baṭn*) of the Qurʾān according to the instruction of the Merciful One and the inspiration of the Holy Angel. Thirdly, you should contemplate the gnosis of its limit (*ḥadd*) in the realm of hearts. (Only then) will you be distinguished with witnessing its point of ascent (*muṭṭalaʿ*) without thought or reckoning.[57]

4 Conclusion

According to the description given by as-Simnānī himself, his exegetical work is a *tafsīr* of the esoteric meaning(s) of the Qurʾān (*tafsīr baṭn al-Qurʾān*/*tafsīr al-buṭūn as-sabʿa*). The source of authority on which as-Simnānī bases his exegesis is inspiration (*ilhām*), which is why one could also speak of a *tafsīr ilhāmī*. The terms 'Sufi *tafsīr*' and *tafsīr* '*ishārī*,' widely used in modern scholarly literature, also seem acceptable to describe as-Simnānī's exegetical work.[58] This is because the reader addressed by the Iranian scholar is eminently (perhaps it would be better to say exclusively) an initiate to Sufism;[59] and because through a complex theory of the interiorization of the meaning of the Qurʾān, combined with an articulated network of correspondences between microcosm (man), macrocosm (creation) and the Holy Book (Qurʾān), the latter is read and interpreted by as-Simnānī as a map of the inner spiritual journey. The introspective reading of the sacred text is a prominent feature shared by most of the exegetical reflections of Sufi scholars. However, in the *tafsīr* of as-Simnānī, joining an articulated plot of

57 Translation by Elias (Elias, *The Throne*, 107f). I was unable to locate this passage.
58 Regarding the use of this terminology in the modern scientific literature see e.g. Ateş, *İşârî tefsîr okulu*; Gerhard Böwering's studies on Sufi exegetical literature, esp. "Sulamī's Commentary on the Qurʾān," in *Islamic Studies presented to Charles J. Adams*, ed. W.B. Hallaq and D.P. Little (Leiden: Brill, 1991), 41–56, and Godlas, "Ṣūfism." For a critical discussion of the notion and definition of Sufi *tafsīr* see Elias, "Ṣūfī Tafsīr Reconsidered."
59 Throughout the text, the reader is systematically referred to by terms such as "wayfarer" on the spiritual path (*as-sālik*) and the like.

concepts and correspondences, such a reading assumes uncommon traits of complexity, organicity and novelty, which appear as the most significant characteristics of the work from the point of view of the contents and history of ideas, as it was already recognized by the first scholars who took an interest in this text. From the point of view of style, the Commentary is characterized by the almost complete absence of references to the previous exegetical tradition (both by Sufi scholars and by others), and in general by the scarce quotations of ancient authors (even Sufis). This gives the work a high and unquestionable degree of originality, but at the same time makes it an eccentric text and a 'special case.' These characteristics of originality and eccentricity perfectly reflect as-Simnānī's spiritual biography and thought. The lack of studies in this regard prevents us from establishing what reception and how much influence as-Simnānī's Qur'ānic commentary may have exerted on later authors. However, the relatively large number of manuscript copies of the work (twenty-seven or more) and their chronological distribution (thirteenth–nineteenth/eighth–fourteenth centuries), when compared to the number of manuscript copies in which other Sufi *tafsīr*s held in high regard by critics have come down to us, seems to suggest large and lasting interest in as-Simnānī's exegetical work.[60] A critical edition of the *tafsīr* of al-Kubrā-Dāya and as-Simnānī should be expected to be a philological challenge and a complex and lengthy undertaking, but it would certainly lead to a better knowledge of this important collective exegetical work, to a better understanding of its genesis and evolution, of the relations existing between the various phases of its composition, and of its reception. For all these reasons, it represents a significant desideratum in the field.

5 Bibliography

5.1 Sources

Al-Kubrā, Najm ad-Dīn and ʿAlāʾ ad-Dawla as-Simnānī. *At-Taʾwīlāt an-Najmiyya fī t-tafsīr al-ishārī ṣ-ṣūfī, wa-yalīhi tatimmatuhu ʿAyn al-ḥayāt/*al-Taʾwīlāt al-najmiyyah *by Najmuddīn al-Kubrā Followed by* ʿAyn al-Ḥayāt *by ʿAlāʾuddawlah al-Simnānī.* Ed. Aḥmad Farīd al-Mizyadī, 6 vols. Beirut: Dār al-Kutub al-ʿIlmiyya, 2009.

Landolt, Hermann, ed. *Correspondance spirituelle échangée entre Nuroddin Esfarayeni (ob. 717/1317) et son disciple ʿAlaoddawleh Semnani (ob. 736/1336)/Murshid va murīd: Mukātabāt-i ʿAbd ar-Raḥmān Isfarāyinī*

60 E.g. we know only one manuscript of Sulamī's minor commentary *Ziyādāt Ḥaqāʾiq at-tafsīr* (Sulamī, *Ziyādāt Ḥaqāʾiq at-tafsīr*, English Introduction); apparently only five copies of Shihāb ad-Dīn Abū Ḥafs ʿUmar b. Muḥammad as-Suhrawardī's (d. 1234/632) *Nughbat al-bayān fī tafsīr al-Qurʾān* (Yaşar Düzenli, "Şihâbuddin Sühreverdî ve Nuğbetü'l-Beyân fî Tefsîri'l-Kurân: Adlı Eserinin Tevbe Sûresine kadar Tahkîki" (PhD. diss., Marmara Üniversitesi, 1994), Arabic introduction, 15; cf. Carl Brockelman, *GALS* (Leiden: Brill, 1937), 1: 789, n4); and about eight manuscripts of Dailamī's *Futūḥ ar-raḥmān fī ishārāt al-Qurʾān* (G. Böwering, "The writings of Shams al-Dīn al-Daylamī," *Islamic Studies* 26, no. 3 (1987): 231–36, 232).

bā ʿAlāʾ ad-Dawla-yi Simnānī. Tehran/Paris: Departement d'Iranologie de l'Institut Franco-Iranien de Recherche-Librairie d'Amerique et d'Orient Adrien-Maisonneuve, 1972.
Nwyia, Paul. "Muqaddima tafsīr al-Qurʾān li-ʿAlāʾ ad-Dawla as-Simnānī (736/1336)." *Al-Abḥāth* 26 (1977): 141–57.
As-Simnānī, ʿAlāʾ ad-Dawla. *Muṭṭalaʿ an-nuqaṭ wa-majmaʿ al-luqaṭ*. MS Paris, BNF, Arabe 6962.
As-Simnānī, ʿAlāʾ ad-Dawla. *Maṭāf al-ashrāf*. MS Istanbul, Süleymaniye Kütüphanesi, Karaçelebizade 353, fols. 78b–80a.
As-Simnānī, ʿAlāʾ ad-Dawla. *Al-ʿUrwa li-ahl al-khalwa wa-l-jalwa*. Ed. Najīb Māyil Hirawī. Tehran: Intishārāt-i Mawlā, 1983.
As-Sulamī, ʿAbd ar-Raḥmān. *Ziyādāt Ḥaqāʾiq at-tafsīr/The Minor Qurʾān Commentary of Abū ʿAbd ar-Raḥmān Muḥammad b. al-Ḥusayn as-Sulamī (d.412/1021)*. Ed. Gerhard Böwering. Beirut: Dar el-Machreq, 1995.

5.2 Secondary Literature

Āmulī, Ḥaydar. *Jāmiʿ al-asrār wa-manbaʿ al-anwār, ba inḍimām-i Risālat naqd an-nuqūd fī maʿrifat al-wujūd/ La philosophie schi'ite. 1. Somme des doctrines ésotériques (Jâmi' al-asrâr), 2. Traité de la connaissance de l'être (Fî ma'rifat al-wojûd)*. Ed. Henry Corbin and Osman Yahya. Tehran-Paris: Département d'iranologie de l'institut franco-iranien de recherche-Librairie d'Amérique et d'Orient Adrien-Maisonneuve, 1969.
Ateş, Süleyman. *Sülemī ve tasavvufī tefsīrī*. Istanbul: Sönmez Neşriyat, 1969.
Ateş, Süleyman. "Üç müfessir bir tefsir." *Ankara Üniversitesi İlahiyat Fakültesi Dergisi* 18 (1970): 85–104.
Ateş, Süleyman. *Isari tafsiri okulu*. Ankara: Ankara Üniversitesi Basımevi. 1974.
Ballanfat, Paul. "La prophétologie dans le 'Ayn al-hayât, tafsîr attribué à Najm al-dîn Kubrâ et. Najm-i Râzî." In *Mystique musulmane: Parcours en compagnie d'un chercheur: Roger Deladrière*, edited by Geneviève Gobillot. Paris: Editions Cariscript, 2002: 171–364.
Ballanfat, Paul. "Reality and Image in the *Tafsīr* of Kubrā and Rāzī." *Journal of the Muhyiddin Ibn Arabi Society* 35 (2004): 75–108.
Ballanfat, Paul. "Simnānī and Hermeneutics." In *The Spirit and the Letter: Approaches to the Esoteric Interpretation of the Qur'an*, edited by Annabel Keeler and Sajjad Rizvi. Oxford-New York: Oxford University Press in association with The Institute of Ismaili Studies, 2016: 345–373.
Böwering, Gerhard. "The writings of Shams al-Dīn al-Daylamī." In *Islamic Studies* 26, no. 3 (1987): 231–36.
Böwering, Gerhard. "Sulamī's Commentary on the Qurʾān." In *Islamic Studies presented to Charles J.Adams*, edited by W.B. Hallaq and D.P. Little. Leiden: Brill, 1991:41–56.
Böwering, Gerhard. "The Major Sources of Sulamī's Minor Qurʾān Commentary." *Oriens* 35 (1996): 35–56.
Brockelman, Carl. *GALS*. 3 vols. Leiden: Brill, vol. 1, 1937.
Corbin, Henry. "L'intériorisation du sens en herméneutique soufie iranienne." *Eranos Jahrbuch* 26 (1957): 57–187.
Corbin, Henry. "Physiologie de l'homme de lumiére dans le soufisme iranien." In *Ombre et lumiére*, 135–257. Paris-Brussels: Desclée de Brouwer, 1961.
Corbin, Henry. *En Islam iranien*. 4 vols. Paris: Gallimard, 1972.
Corbin, Henry. *L'Homme de lumière dans le soufisme iranien*. Paris: Librairie de Médicis, 1971 (English translation: *The Man of Light in Iranian Sufism*, trans. Nancy Pearson. Boulder: Shambala Press. Boulder 1978. Reprint: New Lebanon, NY: Omega Publications, 1994).
Düzenli, Yaşar. "Şihâbuddin Sühreverdî ve Nuğbetü'l-Beyân fî Tefsîri'l-Kurân: Adlı Eserinin Tevbe Sûresine kadar Tahkîki." PhD. diss., Marmara Üniversitesi, 1994.
Elias, Jamal J. "A Kubrawī Treatise on Mystical Visions: The *Risāla-yi nūriyya* of ʿAlāʾ ad-dawla as-Simnānī." *MW* 83, no. 1 (1993): 68–80.

Elias, Jamal J. *The Throne Carrier of God: The Life and Thought of 'Alā' ad-dawla as-Simnānī*. Albany, NY: State University of New York Press, 1995.

Elias, Jamal J. "Ṣūfī Tafsīr Reconsidered: Exploring the Development of a Genre." *JQS* 12 (2010): 41–55.

Godlas, Alan. "Ṣūfism." In *The Wiley Blackwell Companion to the Qur'ān*, edited by Andrew Rippin and Jawid Mojaddedi. Hoboken, NJ: John Wiley & Sons Ltd, 2017^2: 418–29.

Landolt, Hermann. Review of *The Throne Carrier of God: The Life and Thought of 'Alā' ad-dawla as-Simnānī*, by Jamal J. Elias. *IJMES* 30, no. 3 (1998): 457–59.

Martini, Giovanni Maria. *'Alā' al-Dawla al-Simnānī between Spiritual Authority and Political Power: A Persian Lord and Intellectual in the Heart of the Ilkhanate*. Leiden-Boston: Brill, 2018.

Massignon, Luis. *Recueil de textes inédits concernant l'histoire de la mystique en pays d'Islam*. Paris: Librarie orientaliste Paul Gauthner, 1929.

Meier, Fritz. "Stambuler Handschriften dreier persischer Mystiker: 'Ain al-quḍāt al-Hamaḏānī, Naǧm ad-dīn al-Kubrā, Naǧm ad-dīn ad-Dāja." *Der Islam* 24 (1937): 1–42.

Okuyan, Mehmet. *Necmuddîn Dâye ve tasavvufî tefsiri*. Istanbul: Rağbet Yayınları, 2001.

Shpall, W. "A note on Najm al-Dīn al-Rāzī and the *Baḥr al-Ḥaqā'iq*." *Folia Orientalia* 22 (1981–1984): 69–80.

Abū ʿAbd Allāh al-Qurṭubī

Serdar Kurnaz

1 Introduction

One of the types of Qurʾānic exegesis is juridical exegesis, known as *Aḥkām al-Qurʾān*.[1] Shams ad-Dīn Abū ʿAbd Allāh Muḥammad b. Aḥmad b. Abī Bakr b. Farḥ[2] al-Anṣārī l-Khazrajī l-Qurṭubī's (d. 1273/671) *al-Jāmiʿ li-aḥkām al-Qurʾān* is one of these juridical-exegetical works. His work differs significantly from other *Aḥkām al-Qurʾān* works, while sharing the main concern of legal analysis of the Qurʾān: it is an encyclopedic *tafsīr* that contains grammatical analysis, reports from earlier authorities on legal issues, clarification of unclear utterances, reading variants of the Qurʾān, reports on the occasions of the revelation (*asbāb an-nuzūl*) and references from hadiths and commentaries on theological debates. Thus, al-Qurṭubī's *tafsīr* goes beyond most other juridical-exegetical works. Although being typical of its era in the writing of *tafsīr*-works in an encyclopedic fashion, it nonetheless has particular traits, which distinguished al-Qurṭubī's *tafsīr* and made it famous. In this chapter, I will highlight these features of al-Qurṭubī's *tafsīr* and locate it in the tradition of Andalusian *tafsīr*-writings. Interestingly, his *tafsīr* gives some insights into his personal life and religious attitude. Furthermore, studying Andalusian-Mālikī scholars from the eleventh to the thirteenth centuries reveals traces of the intellectual development of Northern Africa and Andalusia within the quarrels of the Almoravid (1056–1147/447–541) and Almohad (1130–1269/524–668) dynasties in their writings. Abū l-Walīd Muḥammad b. Aḥmad b. Muḥammad b. Rushd (d. 1198/595), for example, as a Mālikī scholar, was affected by the Almohadian understanding of sciences.[3] The same holds true for al-Qurṭubī: "Though he cannot be considered a supporter of the Almohads or a follower of their main ideologist Ibn Tūmart, al-Qurṭubī's writings display various features that show he was a product of the intellectual developments brought about during the reign of that dynasty."[4] Thus, after presenting al-Qurṭubī's life and works, I will try to highlight the intellectual and political atmosphere under which al-

1 Claude Gilliot, "Exegesis: Classical," In *EQ*, 2: 113f.

2 Some sources read "Farḥ" as "Faraj," see Ismāʿīl Bāshā l-Baghdādī, *Hadiyyat al-ʿārifīn asmāʾ al-muʾallifīn wa-āthār al-muṣannifīn* (Istanbul: Muʾassasat Tārīkh al-ʿArabī, 1951), 2: 129. See also Roger Arnaldez, "Al-Ḳurṭubī," in *EI²*, 5: 512.

3 Muḥammad b. Aḥmad b. Rushd, *Maßgebliche Abhandlung. Faṣl al-Maqāl*, trans. and commented by Frank Griffel (Berlin: Verlag der Weltreligionen, 2010), 75; Maribel Fierro, "The Legal Policies of the Almohad Caliphs and Ibn Rushd's *Bidāyat al-Mujtahid*." *JIS* 10, no. 3 (1999): 226–48.

4 Delfina Serrano Ruano, "al-Qurṭubī," in *EQ*, ed. Jane Dammen McAuliffe, accessed October 22, 2023: http://dx.doi.org/10.1163/1875-3922_q3_EQCOM_050504.

https://doi.org/10.1515/9783111320083-004

Qurṭubī wrote his voluminous *tafsīr*, and then present his *tafsīr* and hermeneutics. The chapter ends with a consideration of the reception of *al-Jāmiʿ*.

2 Al-Qurṭubī: His Life and Works

2.1 Life

Al-Qurṭubī's date of birth is unclear; probably lying between 1203/600 and 1213/610, during the reign of the Almohad caliph Muḥammad Nāṣir b. Abī Yūsuf Yaʿqūb b. Yūsuf b. ʿAbd al-Muʾmin al-Manṣūr (1198–1213/595–610). His place of birth was either in Córdoba or in Montoro (Ḥiṣn Munthūr),[5] 46 kilometers northwest of Córdoba. His family lived under humble circumstances: he and his father worked on the fields as hand laborers. However, he was able to study in Córdoba several disciplines, like Qurʾān, hadith, Islamic law and Arabic language, with Andalusian masters like Abū ʿĀmir Yaḥyā b. ʿAbd ar-Raḥmān b. Aḥmad b. Rabīʿ al-Ashʿarī (d. 1241/639), from whom he heard hadith directly (through *samāʿ*) in 1231/628.[6] Ibn Rabīʿ al-Ashʿarī seems to be one of his most influential masters. He also studied with Abū Jaʿfar Aḥmad b. Muḥammad b. Muḥammad al-Qaysī, known as Ibn Abī Ḥujja (d. 1246/643), who taught him the Qurʾānic reading variants in a direct lesson,[7] and Abū Sulaymān Rabīʿ b. ʿAbd ar-Raḥmān b. Aḥmad b. Rabīʿa b. Ubayy al-Ashʿarī (d. 1235/632).[8]

Al-Qurṭubī lost his father during his lifetime in Córdoba; he was killed on the field while working, caught off guard by the attack of Christian soldiers in 16 July 1230/3 Ramaḍān 627. This incident caused al-Qurṭubī to think about the legal issue of the state of people who were killed by enemies while going about their daily activities without fighting on the battlefield. Finally, after burying his father according to the ordinary ritual, he discovered later that he could have entombed him as a martyr.[9]

5 In his *tafsīr* of Q 17:45–6, al-Qurṭubī mentions that he was chased by two Christian knights and the recital of certain suras of the Qurʾān saved him from being caught, in Ḥiṣn Munthūr. Perhaps he was born and lived there in his youth; see al-Qurṭubī, *al-Jāmiʿ li-aḥkām al-Qurʾān wa-l-mubayyin li-mā taḍammanahu min as-sunna wa-āy al-furqān*, ed. ʿAbdallāh ʿAbd al-Muḥsin at-Turkī et al. (Beirut: Muʾassasat ar-Risāla, 2006), 13: 94; Mashhūr Ḥasan Maḥmūd Salmān, *al-Imām al-Qurṭubī: shaykh aʾimmat at-tafsīr* (Damascus: Dār al-Qalam, 1993), 18f.

6 Al-Qurṭubī, *al-Jāmiʿ*, 4: 220.

7 Salmān, *al-Qurṭubī*, 63–65; Al-Qurṭubī, *at-Tadhkira fī aḥwāl al-mawtā wa-l-ākhira*, ed. Ṣādiq b. Muḥammad b. Ibrāhīm (Riyadh: Dār al-Minhāj, 2004), 3: 1248; al-Qurṭubī, *al-Jāmiʿ*, 10: 434.

8 For these and further names of his masters see Serrano Ruano, "al-Qurṭubī"; Salmān, *al-Qurṭubī*, 63–70; Miftāḥ as-Sanūsī Balʿam, *al-Qurṭubī: ḥayātuhu wa-āthāruhu l-ʿilmiyya wa-manhajuhu fī t-tafsīr* (Benghazi: Manshūrāt Jāmiʿat Qāryūnus, 1998), 85f.; Tayyar Altıkulaç, "Kurtubî, Muhammed b. Ahmed," in *TDVIA*, accessed April 20, 2020, https://islamansiklopedisi.org.tr/kurtubi-muhammed-b-ahmed.

9 Al-Qurṭubī, *al-Jāmiʿ*, 5: 412f.

We know that al-Qurṭubī did not leave Córdoba before 1231/628; as mentioned above, in this year he was studying with Ibn Rabīʿ. He left the city after Ferdinand III of Castile (r. 1217–1252) conquest in 1236/633, and moved to Alexandria.

In Alexandria, al-Qurṭubī went on to study with several teachers. Among them, the most influential was Abū l-ʿAbbās Aḥmad b. ʿUmar b. Ibrāhīm b. ʿUmar al-Imām al-Anṣārī l-Qurṭubī, known as Ibn al-Muzayyin (d. 1259/657), to whom he frequently refers in his *tafsīr*.[10] He met several other Mālikī scholars like Abū Muḥammad ʿAbd al-Muʿṭī b. Muḥammad b. ʿAbd al-Muʿṭī l-Lakhmī al-Iskandarānī (d. 1240/638) and ʿAbd al-Wahhāb b. Ẓāfir b. ʿAlī b. Fattūḥ b. Abī l-Ḥasan al-Qurashī b. Rawāj (d. 1250/648)[11].

His next residence was Munyat Banī Khaṣīb, today El Minya, in Upper Egypt, where he studied with Abū l-Ḥasan ʿAlī b. Hibat Allāh al-Lakhmī, a Shāfiʿī jurist known as Ibn al-Jummayzī (d. 1251/649).[12] Biographers report an interesting anecdote involving Abū l-ʿAbbās Shihāb ad-Dīn Aḥmad b. Idrīs b. ʿAbd ar-Raḥmān al-Miṣrī l-Qarāfī (d. 1285/684), with whom he travelled to al-Fayyūm. At that time, both were regarded as masters in their fields, al-Qarāfī in law (*fiqh*). and al-Qurṭubī in *tafsīr*.[13]

It is clear from his writings that al-Qurṭubī was a Mālikī scholar[14] with obvious Ashʿarite affiliation. However, this connection to a specific school of law and theology does not refrain him from criticizing views of the eponym of the school, Abū ʿAbdallāh Mālik b. Anas b. Mālik b. Abī ʿĀmir al-Aṣbaḥī l-Yamanī (d. 795/179) and other Mālikis. Therefore, biographers and Qurṭubī-experts attest that he avoided partisanship and fanaticism.[15] For instance, he often defends scholars from other schools of law against harsh words of the Mālikī scholar Abū Bakr Muḥammad b. ʿAbdallāh b. Muḥammad b. al-ʿArabī l-Maʿāfirī (d. 1148/543),[16] so he is regarded as a critical character.[17] Among the qualities of al-Qurṭubī, besides his vast knowledge and his expertise in sciences of Qurʾān, hadith, law, eschatology and asceticism (*zuhd*), is his modest, humble and pious personality: "He appeared in public attired in a single garment and wore a

10 For example, al-Qurṭubī, *al-Jāmiʿ*, 16: 221f., where he refers to Abū l-ʿAbbās al-Qurṭubī's commentary on Abū l-Ḥusayn Muslim b. al-Ḥajjāj b. Muslim al-Qushayrī's (d. 261/875) *Saḥīḥ* with the title *al-Mufhim* (for an edition see al-Qurṭubī, *al-Mufhim li-mā ashkala min Kitāb Talkhīṣ Muslim*, ed. Muḥyī d-Dīn Dīb Mastaw et al. (Damascus/Beirut: Dār Ibn Kathīr/Dār al-Kalim aṭ-Ṭayyib, 1996). He refers to another work of his master, called *Iẓhār idbār man ajāza al-waṭy fī l-adbār* on anal intercourse, see al-Qurṭubī, *al-Jāmiʿ*, 4: 10.

11 Al-Qurṭubī, *at-Tadhkira*, 1: 341.

12 Ibid., 1: 341.

13 For these and further masters of al-Qurṭubī, see Serrano Ruano, "Al-Qurṭubī"; Salmān, *al-Qurṭubī*, 70–85.

14 He says in his *tafsīr* in commenting on the view of Mālikīs concerning the *ratio legis* (*ʿilla*) of *ribā* ('interest') in Q 2:275: "The statement of our disciples of the Mālikiyya differ in this concern," al-Qurṭubī, *al-Jāmiʿ*, 4: 389.

15 Salmān, *al-Qurṭubī*, 161; Muḥammad Ḥusayn adh-Dhahabī, *DhTM* (Cairo: Maktabat Wahba, 2000), 2: 338.

16 Salmān, *al-Qurṭubī*, 179–81.

17 Ibid., 157–85; Balʿam, *al-Qurṭubī*, 237–85.

small-cap (*taqiyya*) on his head."[18] However, he did not restrain from being critical against Sufi traditions that contradicted knowledge gained through religious texts.[19]

Interestingly, biographical sources are mostly silent on the students of al-Qurṭubī, usually referring only to his son Shihāb ad-Dīn Aḥmad and Abū Jaʿfar Aḥmad b. Ibrāhīm b. az-Zubayr ath-Thaqafī l-ʿĀṣimī l-Gharnāṭī (d. after 1300/699). We can add some more names: Ibn Ḥajar al-ʿAsqalānī mentions that Ismāʿīl b. Muḥammad b. ʿAbd al-Karīm b. ʿAbd aṣ-Ṣamad al-Kharāstānī (d. 1309/709) studied with al-Qurṭubī and, according to Ibn Rushayd al-Fihrī, the Mālikī jurist and ascetic Abū Bakr Muḥammad b. al-Maymūn al-Maymūnī l-Qasṭallānī al-Miṣrī studied Abū l-Maʿālī l-Furāwī's *al-Arbaʿūn* ('The Forty') with him. In the *Tadhkira*, al-Qurṭubī declares that he gave a copy of it to Ḍiyāʾ ad-Dīn Aḥmad b. Abī s-Suʿūd b. Abī l-Maʿālī l-Baghdādī, better known as as-Saṭrījī, on 1 October 1255/27 Shaʿbān 656, and that he gave the latter written permission to hand it on to whomever he wanted.[20]

Al-Qurṭubī died in El Minya on 28 April 1273/8 Shawwāl 671, where his tomb is still extant.[21]

2.2 Al-Qurṭubī's Works

Al-Qurṭubī's legacy is very rich. His writings "deal with *ḥadīth*, eschatology, the veneration of the prophet Muḥammad, asceticism, Qurʾānic recitation, Qurʾānic exegesis – a genre in which he excelled – and God's names *(asmāʾ)* and attributes *(ṣifāt)*."[22] The most decisive characteristic of his writings is his focus on religious texts, especially his Qurʾān-centered approach. To suggest a chronology for his writings seems impossible. However, in view of some dates in his writings, and al-Qurṭubī's own references

18 Arnaldez, "al-Ḳurṭubī," 512.

19 Serrano Ruano, "al-Qurṭubī"; Salmān, *al-Qurṭubī*, 188f., 202, for some examples of his criticism see 192–204.

20 Ibid. See also Salmān, *al-Qurṭubī*, 87–95.

21 In general, for the biographical entries see Khayr ad-Dīn az-Ziriklī, *al-Aʿlām. Qāmūs tarājim li-ashhar ar-rijāl wa-n-nisāʾ min al-ʿArab wa-l-mustaʿribīn wa-l-mustashriqīn* (Beirut: Dār al-ʿIlm li-l-Malāyīn, 2002), 5: 322; Muḥammad b. Muḥammad Makhlūf, *Shajarat an-nūr az-zakiyya fī ṭabaqāt al-Mālikiyya*, ed. ʿAbd al-Majīd al-Khayyālī (Beirut: Dār al-Kutub al-ʿIlmiyya, 2003), 1: 282; ʿUmar Riḍā Kaḥḥāla, *Muʿjam al-muʾallifīn* (Beirut: Maktabat al-Muthannā/Dār Iḥyāʾ at-Turāth al-ʿArabī, n.d.), 8: 239f.; Ṣalāḥ ad-Dīn Khalīl aṣ-Ṣafadī, *al-Wāfī bi-l-wafiyyāt*, ed. Aḥmad al-Arnaʾūṭ and Turkī Muṣṭafā (Beirut: Dār Iḥyāʾ at-Turāth al-ʿArabī, 2000), 2: 87; Jalāl ad-Dīn as-Suyūṭī, *Ṭabaqāt al-mufassirīn*, ed. ʿAlī Muḥammad ʿUmar (Kuwait: Dār an-Nawādir, 2010), 92; Muḥammad b. ʿAlī d-Dāwūdī, *Ṭabaqāt al-mufassirīn li-d-Dāwūdī* (Beirut: Dār al-Kutub al-ʿIlmiyya, 1983), 2: 69f.; Shihāb ad-Dīn Abū l-Falāḥ ʿAbd al-Ḥayy b. al-ʿImād, *Shadharāt adh-dhahab fī akhbār man dhahab*, ed. ʿAbd al-Qādir al-Arnaʾūṭ and Maḥmūd al-Arnaʾūṭ (Damascus/Beirut: Dār Ibn Kathīr, 1991), 7: 584f.; Ibrāhīm b. ʿAlī b. Farḥūn, *ad-Dībāj al-mudhahhab fī maʿrifat aʿyān ʿulamāʾ al-madhhab*, ed. Muḥammad al-Aḥmadī Abū n-Nūr (Cairo: Dār at-Turāth, n.d.), 1: 308f.

22 Serrano Ruano, "al-Qurṭubī".

to them, we have some hints for a chronology. At least for his most prominent writings, Serrano Ruano suggest the following chronology: 1) *al-Asnā*, 2) *aṣ-Ṣifāt*, 3) *at-Tadhkār*, 4) *al-Jāmiʿ* and 5) *at-Tadhkira*. She points out that al-Qurṭubī mentions the battle of ʿAyn Jālūt in 1260/658 in *at-Tadhkira*,[23] which for her is a *terminus post quem*. If we look at al-Qurṭubī's references to his own works, then this chronology seems to be problematic; Serrano Ruano does not give any other information on dating the last three works mentioned above. Salmān, on the other hand, says that al-Qurṭubī wrote *at-Tadhkār* after *al-Jāmiʿ*,[24] which would mean that this is his last work. This is, in my view, backed by his assumption that al-Qurṭubī does not quote *at-Tadhkār* in *al-Jāmiʿ*,[25] but does quote *at-Tadhkira* in *al-Jāmiʿ*[26] and *al-Jāmiʿ* in *at-Tadhkār*.[27] Salmān, however, misses that al-Qurṭubī mentions *at-Tadhkār* at least once, in his commentary on Q 38:29.[28] On the other hand, al-Qurṭubī uses *at-Tadhkira* frequently in *al-Jāmiʿ*, both early in his commentary as in Q 1:2,[29] in the middle on Q 30:55,[30] and at the end, for example, Q 108:1, even twice[31] here. His frequent use of this work in *al-Jāmiʿ* could be read as a sign for the immediate relationship between these two: once he finished *at-Tadhkira* or wrote major parts of it, he possibly began to write *al-Jāmiʿ*. This indicates that it is not probable that *at-Tadhkira* dates to after the finishing of *al-Jāmiʿ* as Serrano Ruano supposes.

It becomes even more complicated if we look into *at-Tadhkār*: Here, we find that al-Qurṭubī refers to *at-Tadhkira*[32] and, more often, to *al-Jāmiʿ*.[33] He refers to the introduction of *al-Jāmiʿ* three times,[34] twice to his entries on sura one, *al-Fātiḥa* (The Opening),[35] once to sura two, *al-Baqara* (The Cow),[36] once to sura three, *Āl ʿImrān* (The House of ʿImrān),[37] once to sura five, *al-Māʾida* (The Table),[38] and once to the beginning of sura nine, *al-Barāʾa* (The Repudiation), on the question of the absence of the

23 Al-Qurṭubī, *at-Tadhkira*, 3: 1174.
24 Salmān, *al-Qurṭubī*, 136.
25 Ibid., 135.
26 Ibid., 128f.
27 Ibid., 137, 148 (here he wants to prove the existence of another work, *al-Intihāz*; the quote also includes a reference to *al-Jāmiʿ*).
28 Al-Qurṭubī, *al-Jāmiʿ*, 18: 189
29 Ibid., 1: 211.
30 Ibid., 16: 451.
31 Ibid., 22: 520f.
32 Abū ʿAbdallāh al-Qurṭubī, *at-Tadhkār fī afḍal al-adhkār*, ed. Bashīr Muḥammad ʿUyūn. (Damascus/Ṭāʾif: Maktabat Dār al-Bayān/Maktabat al-Muʾayyad, 1986³), 71, 246.
33 For example, al-Qurṭubī, *at-Tadhkār*, 33.
34 Al-Qurṭubī, *at-Tadhkār*, 35f., 38.
35 Ibid., 228f.
36 Ibid., 50.
37 Ibid., 308.
38 Ibid., 253.

Basmala.[39] Since he refers to *al-Barā'a*, the ninth sura, at an early stage of his book *at-Tadhkār*, we can assume that he began to write it while working on *al-Jāmiʿ*, but had already written a great portion, including sura *al-Barā'a*. In *al-Jāmiʿ*, until sura *al-Barā'a*, there are several references to *at-Tadhkira*,[40] but none to *at-Tadhkār*. In *at-Tadhkira*, however, we have a single reference to *at-Tadhkār* and the introduction of *al-Jāmiʿ* together.[41] Here he possibly refers to a passage in the first third of *at-Tadhkār*,[42] or maybe the whole work. The latter seems unlikely; it appears to be a reference to a particular chapter. The allusion to *at-Tadhkār* in *at-Tadhkira*, on the other hand, refers to a later stage of *at-Tadhkār*, namely the chapter on the monotheists in hellfire, who could leave it with the help of intercession (*shafāʿa*).[43] Thus, while writing *at-Tadhkār*, more than half of *at-Tadhkira* must have existed already. Additionally, *at-Tadhkira* contains several references to *al-Jāmiʿ*, up to sura 30, *ar-Rūm* (The Byzantines); most references are in the second half of the work.[44] On the other hand, al-Qurṭubī refers in an early stage of *al-Jāmiʿ* to some later chapters of *at-Tadhkira*.[45]

Thus, we have a complicated situation concerning the chronology of his three later works here. It seems likely that at first a great portion of *at-Tadhkira* was written during his residence in Munyat Banī Khaṣīb,[46] while al-Qurṭubī began to write his *al-Jāmiʿ* afterwards. Whilst continuing *al-Jāmiʿ*, he possibly began to write *at-Tadhkār* also. Working on them at the same time, we can assume that al-Qurṭubī finished *at-Tadhkār* first, as there he referred in the last third of it to early passages located in the other two books. Since al-Qurṭubī mentions the battle of ʿAyn Jālūt in *at-Tadhkira*, which took place in 1260/658,[47] he finished it after this date. Since there are frequent references from *at-Tadhkira* until the end of *al-Jāmiʿ* and since there is a very late reference to *at-Tadhkār* in early passages of *al-Jāmiʿ*, it can be said that he completed *al-Jāmiʿ* last. If in fact it was *at-Tadhkira* that

39 Ibid., 33. Al-Qurṭubī mentions the problem of *Basmala* at the beginning of *al-Barā'a*, in the introduction (see al-Qurṭubī, *al-Jāmiʿ*, 1: 97), but he refers to the beginning of *al-Barā'a* here, where we can find the five different opinions on the reason of absence of the *Basmala*, to which he refers in *at-Tadhkār*, see al-Qurṭubī, *al-Jāmiʿ*, 10: 93–95.

40 I counted 25 references up to sura *al-Barā'a*. For this purpose, I used a different edition, which is a part of the electronic library called *al-Maktaba ash-Shāmila*; see al-Qurṭubī, *al-Jāmiʿ li-aḥkām al-Qur'ān: tafsīr al-Qurṭubī*, ed. Aḥmad al-Bardūnī and Ibrāhīm Aṭfīsh (Cairo: Dār al-Kutub al-Miṣriyya, 1964), 1: 137, 241, 378; 2: 79, 93, 96, 241; 3: 276; 4: 100f., 270, 274, 298; 5: 3; 6: 133, 136, 422; 7: 10, 42, 206, 231, 317; 8: 122, 223, 275.

41 Al-Qurṭubī, *at-Tadhkira*, 2: 962.

42 Al-Qurṭubī, *at-Tadhkār*, 78f.

43 Ibid., 71; al-Qurṭubī, *at-Tadhkira*, 2: 769.

44 Al-Qurṭubī, *at-Tadhkira*, 3: 1046; reference to Q 2:284, 2: 633; reference to *Āl ʿImrān*, 2: 780; reference to sura seven, *al-Aʿrāf* (The Heights), 2: 738; reference to *al-Barā'a*, 2: 956; reference to sura 18, *al-Kahf* (The Cave), 2: 956 and 3: 1223f.

45 The reference in Q 2:25 (see al-Qurṭubī, *al-Jāmiʿ*, 1: 362) equates to several passages in 3 of *at-Tadhkira*, see al-Qurṭubī, *at-Tadhkira*, 3: 808, 814, 929, 936, 980.

46 Al-Qurṭubī, *at-Tadhkira*, 1: 341.

47 Serrano Ruano, "al-Qurṭubī".

was produced last, then it was a small amount of the text which he wrote after the completion of *al-Jāmiʿ*. A close reading of all references in these works could shed light on the chronology of his works, including an analysis of the question whether al-Qurṭubī put some references afterwards to his works, after finishing and proofreading them.

Be this as it may, for our purpose, it is useful to categorize his writings, like Serrano Ruano suggests, into edited, non-edited, non-extant and those falsely attributed to him.[48] Since we cannot find a definitive chronology of the edited works of al-Qurṭubī, it is more appropriate to list his writings alphabetically.

2.2.1 Edited Works

1) *Al-Asnā fī sharḥ asmāʾ Allāh al-ḥusnā*,[49] writing on the beautiful names of God, which al-Qurṭubī quotes frequently in his *tafsīr*,[50] according to Salmān in *at-Tadhkira* and *at-Tadhkār*.[51] Therefore, it is considered to be one of his oldest writings. Later scholars like Abū ʿAbdallāh Shams ad-Dīn Muḥammad b. Abī Bakr b. Ayyūb az-Zuraʿī d-Dimashqī, known as Ibn Qayyim al-Jawziyya (d. 1350/751) and Abū l-Faḍl Shihāb ad-Dīn Aḥmad b. ʿAlī b. Muḥammad b. Ḥajar al-ʿAsqalānī (d. 1449/852) cite from *al-Asnā*.[52]

2) *(Al-)Jāmiʿ li-aḥkām al-Qurʾān wa-l-mubayyin li-mā taḍammanahu min as-sunna wa-āy al-furqān*[53] is al-Qurṭubī's voluminous work on Qurʾānic exegesis, which remains prominent among Muslim scholars and has attracted many recent studies. It is seen as a masterpiece of the Andalusian tradition of writing *tafsīrs*.[54] We will discuss its characteristics, features and reception below under the third heading.

48 I am following here the lists mentioned by Serrano Ruano, "al-Qurṭubī;" az-Ziriklī, *al-Aʿlām*, 322; Makhlūf, *Shajarat*, 282; Kaḥḥāla, *Muʿjam*, 239f.; aṣ-Ṣafadī, *al-Wāfī*, 82; ad-Dāwūdī, *Ṭabaqāt*, 69f.; Ibn Farḥūn, *ad-Dībāj*, 308f. Carl Brockelmann, *GAL*, 1: 529 and *GALS*, 1: 737; Salmān, *al-Qurṭubī*, 97–155; Balʿam, *al-Qurṭubī*, 133–50; Serrano Ruano, "al-Qurṭubī."

49 An example of an edition: Al-Qurṭubī, *al-Asnā fī sharḥ asmāʾ Allāh al-ḥusnā*, ed. ʿIrfān b. Salīm al-ʿAshāḥassūna ad-Dimashqī (Beirut/Sidon: Al-Maktaba al-ʿAṣriyya, 2005). Edited for the first time 1994 in Ṭanṭā: Al-Qurṭubī, *al-Asnā fī sharḥ asmāʾ Allāh al-ḥusnā*, ed. Muḥammad Ḥasan Jabal (Ṭanṭā: Dār aṣ-Ṣaḥāba, 1994).

50 I want to mention two examples, one from the beginning and one from the ending of his *tafsīr*, see al-Qurṭubī, *al-Jāmiʿ*, 1: 392 and 22: 577.

51 Salmān, *al-Qurṭubī*, 146.

52 Ibid., 147; Balʿam, *al-Qurṭubī*, 143.

53 This title is according to al-Qurṭubī's own Introduction of *al-Jāmiʿ*, 1: 8. At the end of sura *Āl ʿImrān* he quotes it as *Jāmiʿ aḥkām al-Qurʾān wa-l-mubayyin li-mā taḍammana min as-sunna wa-āy al-furqān*, see al-Qurṭubī, *al-Jāmiʿ*, 5: 492. He mentions *al-Jāmiʿ* with this latter title in *at-Tadhkār* and *at-Tadhkira* : al-Qurṭubī, *at-Tadhkār*, 253; al-Qurṭubī, *at-Tadhkira*, 2: 633. In different passages of both works he quotes *al-Jāmiʿ* abbreviated as *Jāmiʿ aḥkām al-Qurʾān*.

54 Muṣṭafā Ibrāhīm al-Mashīnī, *Madrasat at-tafsīr fī l-Andalus* (Beirut: Muʾassasat ar-Risāla, 1986), 98–101; Fahd b. ʿAbd ar-Raḥmān b. Sulaymān ar-Rūmī, *Manhaj al-madrasa al-Andalusiyya fī t-tafsīr* (Riyadh: Maktabat at-Tawba), 1997, 13f.; Serrano Ruano, "al-Qurṭubī." Mashīnī and ar-Rūmī give some

In *al-Jāmiʿ*, al-Qurṭubī quotes directly from his sources, intensively from *al-Muḥarrar al-wajīz* of ʿAbd al-Ḥaqq b. Ghālib b. ʿAbd ar-Raḥmān b. Ghālib al-Muḥāribī l-Gharnāṭī, known as Ibn ʿAṭiyya al-Andalusī (d. 1147/541) and Ibn al-ʿArabī's *Aḥkām al-Qurʾān*. Al-Qurṭubī quotes some of his sources, like Abū Jaʿfar Muḥammad b. Jarīr b. Yazīd aṭ-Ṭabarī's (d. 923/310) *Jāmiʿ al-bayān*, also via others' books[55]

The introduction to *al-Jāmiʿ*, which contains important information on Qurʾānic sciences, merits of the Qurʾān and its reading and its history, was published as an independent book.[56] Parts of it were published under the title *Faḍāʾil al-Qurʾān wa-adab at-tilāwa*[57] and *al-Wajīz fī faḍāʾil al-kitāb al-ʿazīz*.[58]

Farīd ʿAbd al-ʿAzīz al-Jundī compiled *al-Jāmiʿ* anew according to the chapters of a *furūʿ* book on Islamic law.[59] Mashhūr Ḥasan Salmān and ʿAbd al-Laṭīf ad-Dasūqī created an index of *al-Jāmiʿ*.[60] These studies on *al-Jāmiʿ* underline its ongoing prominence.

3) *Kitāb Qamʿ al-ḥirṣ bi-z-zuhd wa-l-qanāʿa*[61] is a book of forty chapters of hadith dealing with asceticism and trust in God (*tawakkul*), which al-Qurṭubī wrote before his *al-Jāmiʿ* and *at-Tadhkira* and quote in these two books. Although the above-mentioned abbreviated title frequently appears there, the complete title is unclear, since al-

---

biographical information on al-Qurṭubī and refer to his method of writing his *tafsīr* like all other sources do, but they clearly see him in the Andalusian tradition of writing *tafsīr*.

55 Some further sources of *al-Jāmiʿ*, *al-istidhkār* and *at-Tamhīd* of Abū ʿUmar Jamāl ad-Dīn Yūsuf b. ʿAbdallāh b. Muḥammad b. ʿAbd al-Barr an-Namarī (d. 1071/463), Abū l-Walīd Sulaymān b. al-Khalaf b. Saʿd at-Tūjibī al-Bājī's (d. 1081/474) *al-Muntaqā*, Abū Isḥāq Aḥmad b. Muḥammad b. Ibrāhīm ath-Thaʿlabī's (d. 1035/427) *Tafsīr* and *Iʿrāb al-Qurʾān* and *an-Nāsikh wa-l-mansūkh* of Abū Jaʿfar Aḥmad b. Muḥammad b. Ismāʿīl al-Murādī n-Naḥḥās (d. 950/338), see Serrano Ruano, "al-Qurṭubī"; al-Qurṭubī, *al-Jāmiʿ*, 1: 11–18 (the editor's introduction); Balʿam, *al-Qurṭubī*, 197–203. Al-Qurṭubī also intensively uses hadith collections, for example those of Abū ʿAbdallāh Muḥammad b. Ismāʿīl al-Bukhārī (d. 870/256), Abū Dāwūd Sulaymān b. al-Ashʿath b. Isḥāq as-Sijistānī (d. 889/275) and its commentary by Abū Sulaymān Ḥamad b. Muḥammad b. Ibrāhīm al-Khaṭṭābī (d. 998/388, an example will be given below).

56 *Muqaddimat tafsīr al-Imām al-Qurṭubī*, ed. Muḥammad Bilāl Ṭalḥa (Cairo: 1997), see Altıkulaç, "Kurtubî."

57 Edited by Aḥmad Ḥijāzī s-Saqqā, first published in Cairo 1989, see Altıkulaç, "Kurtubî." The second edition was published in Beirut, Dār al-Jīl, 1990. This book is falsely considered an independent book written by al-Qurṭubī. Serrano Ruano for example lists it at second place within the edited books of al-Qurṭubī, whereas in the Spanish article she lists it as one of al-Qurṭubī's extant but not edited books ("conservadas sin editar"), see Serrano Ruano, "al-Qurṭubī;" Serrano Ruano, "Ibn Farh al-Qurtubi," accessed April 19,2020, http://dbe.rah.es/biografias/136449/ibn-farh-al-qurtubi.

58 Claude Gilliot, "Textes arabes anciens édités en Égypte au cours des années 1992 à 1994," *MIDEO* 22 (1994), 391; al-Qurṭubī, *al-Wajīz fī faḍāʾil al-Kitāb al-ʿazīz*, ed. ʿAlāʾ ad-Dīn ʿAlī Riḍā (Cairo: Dār al-Ḥadīth, 1992).

59 Farīd ʿAbd al-ʿAzīz al-Jundī, *Jāmiʿ al-aḥkām al-fiqhiyya li-l-imām al-Qurṭubī min tafsīrihi* (Beirut: Dār al-Kutub al-ʿIlmiyya, 2005).

60 Mashhūr Ḥasan Salmān and ʿAbd al-Laṭīf ad-Dasūqī, *Kashshāf taḥlīlī li-l-masāʾil al-fiqhiyya fī tafsīr al-Qurṭubī* (Ṭāʾif: Maktabat aṣ-Ṣiddīq, 1988).

61 For an edition see al-Qurṭubī, *Qamʿ al-ḥirṣ bi-z-zuhd wa-l-qanāʿa wa-radd dhull as-suʾāl bi-l-kutub wa-sh-shafāʿa*, ed. Majdī Fatḥī s-Sayyid (Ṭanṭā: Dār aṣ-Ṣaḥāba, 1989).

Qurṭubī and later bio-bibliographical sources give different accounts of it: *Kitāb Qamʿ al-ḥirṣ bi-z-zuhd wa-l-qanāʿa wa-radd dhull as-suʾāl* 1) *bi-l-kasb wa-ṣ-ṣināʿa*, 2) *bi-l-katb*[62] *wa-sh-shafāʿa* and 3) *bi-l-kaff wa-sh-shafāʿa.*[63]

4) *Aṣ-Ṣifāt wa-mā warada fī-hā min al-āy wa-l-aḥādīth* is a book on God's attributes as they appear in Qurʾān and hadith.[64]

5) *At-Tadhkār fī afḍal al-adhkār*, which al-Qurṭubī wrote following the example of Abū Zakariyyā Yaḥyā b. Sharaf an-Nawawī's (d. 1277/676) *at-Tibyān fī ādāb ḥamalat al-Qurʾān*, contains forty chapters on Qurʾān, its merits, its recitation, memorization etc. As al-Qurṭubī mentions himself, one of his main goals was to bring forty hadiths together to memorize them, according to the recommendation of the Prophet in a hadith to learn forty hadiths.[65]

6) *At-Tadhkira fī aḥwāl al-mawtā wa-l-ākhira*, al-Qurṭubī's work on eschatology and funeral rites with 288 chapters specifically on "preparing for death, funerals, selecting land for burial, visiting graves, the permissibility of weeping, the torments of the grave, martyrdom, and the conditions of Paradise and Hell,"[66] in which he also writes about the Mahdī and the Dajjāl, the Antichrist. There are some common sources with *al-Jāmiʿ*, like Ibn al-ʿArabī's *Aḥkām al-Qurʾān* and Abū ʿUmar Ğamāl ad-Dīn Yūsuf Ibn ʿAbd al-Barr an-Namarī's (d. 1071/463) *al-Istidhkār*, but there are also specific sources on which al-Qurṭubī predominantly relied, like Abū Ḥāmid Muḥammad b. Aḥmad al-Ghazālī's (d. 1111/505) *Kashf ʿulūm al-ākhira* and Abū Muḥammad ʿAbd al-Ḥaqq b. al-Kharrāṭ al-Ishbīlī's (d. 1186/582) *al-ʿĀqiba fī dhikr al-mawt*. That *at-Tadhkira* was known to later scholars is not only attested by quotes in different *tafsīr*-works,[67] it was also the subject of commentaries and summaries.[68]

Serrano Ruano additionally mentions a certain *Khuṭab*, which is an edited work on Friday prayers and the two prayers of the two feasts (*ʿīdān*).[69]

62 Read as "*kutub*" by Serrano Ruano, "al-Qurṭubī" and by Majdī Fatḥī s-Sayyid, the editor of *Qamʿ al-ḥirṣ*.

63 Al-Qurṭubī, *al-Jāmiʿ*, 15: 287 and footnote 4; Salmān, *al-Qurṭubī*, 138–40; Serrano Ruano, "al-Qurṭubī".

64 Edited in 2001: Al-Qurṭubī, *Ṣifāt Allāh taʿālā wa-mā warada fī-hā min al-āy wa-l-aḥādīth*, ed. Sayyid b. Ibrāhīm b. Ṣādiq ʿImrān (Beirut: Dār al-Kutub al-ʿIlmiyya, 2001). The original title is as mentioned above, see al-Qurṭubī, *Ṣifāt Allāh*, 11 (the editor's introduction).

65 Al-Qurṭubī, *at-Tadhkār*, 13f.

66 Serrano Ruano, "al-Qurṭubī."

67 For example, Abū l-Fidāʾ Ismāʿīl b. ʿUmar b. Kathīr, *Tafsīr Qurʾān al-ʿaẓīm*, ed. Sāmī b. Muḥammad Salāma (Riyadh: Dār Ṭayba, 1999), 2: 307 and 5: 61; Abū Ḥafṣ Sirāj ad-Dīn ʿUmar an-Nuʿmānī, *al-Lubāb fī ʿulūm al-kitāb*, ed.ʿĀdil Aḥmad ʿAbd al-Mawjūd and ʿAlī Muḥammad Muʿawwaḍ (Beirut: Dār al-Kutub al-ʿIlmiyya, 1998), 19: 331; Abū Zayd ʿAbd ar-Raḥmān ath-Thaʿālibī, *Tafsīr ath-Thaʿālibī – al-Jawāhir al-ḥisān fī tafsīr al-Qurʾān*, ed. Muḥammad ʿAlī Muʿawwaḍ and ʿĀdil Aḥmad ʿAbd al-Mawjūd (Beirut: Dār Iḥyāʾ at-Turāth al-ʿArabī, 1997), 1: 337 and 8: 60.

68 Salmān, *al-Qurṭubī*, 133–35; Serrano Ruano, "al-Qurṭubī."

69 This book is missing in her Spanish article on al-Qurṭubī, see Serrano Ruano, "Ibn Farh al-Qurtubi." In *EQ*, she gives no information on the edition.

2.2.2 Non-Edited Extant Works

1) *Al-Miṣbāḥ fī l-jamʿ bayna l-afʿāl wa-ṣ-ṣiḥāḥ* is a book on grammatical analysis of hadith, based on Abū l-Qāʾim ʿAlī b. Jaʿfar b. al-Qaṭṭāʿ aṣ-Ṣaʿdī's (d. 1121/515; known as Ibn aṣ-Ṣiqillī) *Kitāb al-Afʿāl* and Abū Naṣr Ismāʿīl b. Ḥammād al-Jawharī's (d. before 1009/ 400) *Ṣiḥāḥ*, or *Tāj al-lugha.*
2) *At-Taqrīb li-kitāb at-Tamhīd ʿalā mā fī l-muwaṭṭaʾ min al-maʿānī wa-l-asānīd* is a commentary on Ibn ʿAbd al-Barr's *at-Tamhīd.* Ibn ʿAbd al-Barr's work is an encyclopedic commentary on Mālik's *Muwaṭṭaʾ*, in which he analyzes the chains of the hadith, provides alternative chains to ones already mentioned and categorizes and comments on them according to Islamic law.

2.2.3 Non-Extant Works

The following lists most of the non-extant works not mentioned in bio-bibliographical sources; al-Qurṭubī mentions them in his writings, on which recent scholarship sheds light. These works are:
1) *Al-Iʿlām fī maʿrifat mawlid al-muṣṭafā ʿalayhi ṣ-ṣalāt wa-s-salām*, on the celebration of the birth of the Prophet.[70]
2) *Al-Intihāz fī qirāʾat ahl al-Kūfa wa-l-Baṣra wa-sh-Shām wa-ahl al-Ḥijāz*, apparently on Qurʾānic readings, mentioned in his *Tadhkār.*[71]
3) *Kitāb al-Lumaʿ al-luʾluʾiyya fī sharḥ al-ʿishrīnāt an-nabawiyya*, which al-Qurṭubī mentions twice in *al-Jāmiʿ*,[72] is a commentary on a certain hadith work. He mentions a different title twice, and two different names for its author – the latter can be a typo: 1) *Al-Lumaʿ al-luʾluʾiyya fī sharḥ al-ʿishrīnāt an-nabawiyya* of al-Fādārī,[73] 2) *Al-Lumaʿ al-luʾluʾiyya fī sharḥ al-ʿashr bayyināt an-nabawiyya* of al-Fārābī.[74]
4) *Kitāb Sharḥ at-Taqaṣṣī*, a commentary on Ibn ʿAbd al-Barr's abridgement of Mālik's *Muwaṭṭaʾ*.

70 Al-Qurṭubī, *al-Jāmiʿ*, 18: 82. Serrano Ruano mentions this work under the title *al-Iʿlām fī-mā yajib ʿalā l-anām min maʿrifat mawlid al-muṣṭasfā ʿalayhi s-salām.*

71 Al-Qurṭubī, *at-Tadhkār*, 33; cf. Salmān, *al-Qurṭubī*, 148.

72 Salmān, *al-Qurṭubī*, 152f.

73 Al-Qurṭubī, *al-Jāmiʿ*, 13: 91.

74 Ibid., 19: 128.

5) *Manhaj al-ʿubbād wa-maḥajjat as-sālikīn wa-z-zuhhād,* a comparison and weighing about wealth and poverty.[75] Al-Qurṭubī refers to it in *al-Jāmiʿ*.[76]
6) *Al-Muqtabas fī sharḥ Muwaṭṭaʾ Mālik b. Anas,* a work on Islamic law, written comparatively, to which al-Qurṭubī refers in the *Jāmiʿ*.[77] Serrano Ruano says that "(i)t may be the same work as *al-Taqrīb,* or the *Sharḥ at-Taqaṣṣī*"[78].
7) *Urjūza jumiʿat fī-hā asmāʾ an-nabī,* a poem on the names of the Prophet, written in *rajaz* meter.

2.2.4 Falsely Attributed Works

1) *Aʿlām an-nubuwwa,* on the signs of prophecy. Salmān doubts its attribution to al-Qurṭubī.[79]
2) *Al-Iʿlām bi-mā fī dīn an-Naṣārā min al-fasād wa-l-awhām wa-iẓhār maḥāsin dīn al-Islām wa-ithbāt nubuwwat nabiyyinā Muḥammad,* is originally a work written by al-Qurṭubī's master Abū l-ʿAbbās al-Qurṭubī.[80] It is also edited as if it was written by our exegete Abū ʿAbdallāh al-Qurṭubī.
3) *Kitāb al-Aqḍiya.* Serrano Ruano notes that it is also known as *Aqḍiyat rasūl Allāh ṣallā llāh ʿalayhi wa-sallam.* "The real author was Muḥammad ibn Faraj al-Qurṭubī, who is better known as Ibn al-Ṭallāʾ (. . .); the misattribution was probably caused by the fact that the only difference between the spellings of Ibn Farḥ and Ibn al-Faraj in Arabic is one small dot."[81]
4) *Risāla fī alqāb al-ḥadīth,* a work on hadith terminology "that is, perhaps, confused with the *Qaṣīda ghazaliyya fī alqāb al-ḥadīth* ('Love poem on *ḥadīth* terminology') by Ibn Faraḥ al-Ishbīlī."[82]

Serrano Ruano points out that Brockelmann lists a *Mukhtaṣar fī faḍāʾil al-Qurʾān,* an "Abridgement of the 'Merits of the Qurʾān,'" which neither the biographers nor al-

75 Salmān, *al-Qurṭubī,* 149.
76 Al-Qurṭubī, *al-Jāmiʿ,* 18: 22
77 Salmān, *al-Qurṭubī,* 149–51. See also for example al-Qurṭubī, *al-Jāmiʿ,* 3: 374.
78 Serrano Ruano, "al-Qurṭubī."
79 Salmān, *al-Qurṭubī,* 154.
80 Samir Kaddouri, "Identificación de 'al-Qurṭubī', Autor de *al-Iʿlām bimā fī dīn al-Naṣārà min al-fasād wa-l-awhām,*" *Qanṭara* 21, no. 1 (2000): 215–19; Daniel Potthast, "Andalusische Philosophie nach Ibn Rušd: Intellektuelle Traditionen in al-Qurṭubī antichristlicher Polemik," *Der Islam* 95, no. 2 (2018): 432f.
81 Serrano Ruano, "al-Qurṭubī."
82 Serrano Ruano, "al-Qurṭubī."

Qurṭubī mentions;[83] I cannot confirm this information, since Brockelmann's *Geschichte der arabischen Litteratur* does not contain this information.[84]

2.3 Political and Intellectual Context: Andalusia in the Eleventh to Thirteenth Century

The intellectual atmosphere in which al-Qurṭubī lived and studied, both in Córdoba and different cities in Egypt, was marked by political quarrels. If we look at his lifetime in Córdoba, the city at the time was a lively and cosmopolitan city, home to prominent scholars in different disciplines, under the reign of the Almohads. The Almohads, apart from their political interests, wanted to rationalize numerous disciplines of the Islamic tradition. They supported hadith sciences, Islamic law, systematic theology, philosophy and especially Sufism (*taṣawwuf*). In contrast to their preceding political rivals, the Almoravids, a political and religious movement of the Moroccan tribe Ṣanḥāja, the Almohads were interested in rationalizing the sciences and were famous for their formula "back to the roots." Thus, one of their main goals seemed to be the reduction of different schools of law to one, or at least to rationalize the Mālikī school. Scholarship on the Almohads and the Almoravids shows that the latter were interested in conserving the traditionalist approach of the Mālikī school. This school of law was widespread in Northern Africa up to Andalusia, beginning at the western border of Egypt. Until the twelfth century, and especially under the Almoravid dynasty, the Mālikī understanding of the sciences, particularly in the realm of law, was of a traditionalist fashion. It was sufficient for them to concentrate on substantive law (*furūʿ*) without questioning their methodological and hermeneutical bases. Thus, they were not interested in Islamic legal theory. The Almohads, however, were not popular among many scholars, but their attitude towards hadith enabled a rise of hadith sciences and forced Mālikīs to reflect on their heritage. Thus, many scholars, like Abū l-Walīd Muḥammad b. Aḥmad Ibn Rushd al-Ḥafīd ("the grandson", the philosopher, d. 1198/595), tried to show that dissent among schools of law is nothing that can be avoided, a natural consequence of this science. But already under the Almoravids, notwithstanding the wish of the Almohads to rationalize the sciences, some scholars like Abū l-Walīd Muḥammad b. Aḥmad

83 Ibid. She assumes that this is an abridgement of the edited *Faḍāʾil al-Qurʾān*, "which is preceded, as in *al-Tadhkār* and *al-Jāmiʿ*, by an introduction to the merits of the Qurʾān and its experts"; In her Spanish article on al-Qurṭubī she mentions the *Faḍāʾil* as one of the non-edited books of al-Qurṭubī; see Serrano Ruano, "Ibn Farh al-Qurtubi." I could only find the edited *Faḍāʾil* of al-Qurṭubī, but this is not an independent book written by him; rather it is a part of the introduction of the *Jāmiʿ* published independently.

84 Brockelmann, *GAL*, 1: 529 and *GALS*, 1: 737.

b. Aḥmad b. Rushd al-Qurṭubī the Grandfather (d. 1126/520),[85] Abū l-Walīd Sulaymān b. Ḫalaf al-Bājī (d. 1081/474) and Ibn ʿAbd al-Barr were interested in analyzing and reflecting on different sciences rationally. This was due to their intrinsic interest and to the challenges of Ẓāhirīs like Abū Muḥammad ʿAlī b. Aḥmad b. Saʿīd b. Ḥazm al-Andalusī (d. 1064/456), and due to their studies in the eastern centers of the Muslim world. In the case of Abū l-Walīd al-Bājī, the tradition of rationalization and writing *uṣūl al-fiqh*-works, for example, was prominent. However, the reception of certain Eastern books, like *at-Talqīn* of Abū Muḥammad al-Qāḍī ʿAbd al-Wahhāb b. ʿAlī b. Naṣr al-Baghdādī (d. 1031/422), commented on by Abū ʿAbdallāh Muḥammad b. ʿAlī b. ʿUmar aṣ-Ṣiqillī l-Māzarī (d. 1151/536), also influenced the intellectual atmosphere of Andalusian scholarship towards rationalization. Especially this attitude – to question the methods used for deriving norms, to reflect on issues of Islamic law and theology comparatively, and to search for the reasons for dissenting views – seemed to shape the new scientific atmosphere under the Almohads. This would mean to accept diversity in law, which again, in contrast to the Almohadian attitude, is clearly an influence of the Almohadian challenge.[86] Al-Qurṭubī's free reasoning according to Qurʾān and sunna and his open-mindedness for criticizing his own legal school, the Mālikiyya, is therefore seen by recent scholarship on him as an Almohadian influence.[87]

The dangerous situation in Andalusia and the fall of Córdoba made al-Qurṭubī move to Egypt. For reasons we can only speculate on, he did not choose other North African centers, where Mālikīs were in the majority, but moved rather to Alexandria first. Both the western part of the Muslim world and the East were under political pressure; in 1258/656, Baghdad fell to the Mongol invasion. Al-Qurṭubī lived in Egypt during the last decades of Ayyubid reign (1175–1250/571–648) and the beginning of the reign of the Mamluk sultans (1250–1382/648–923). As the Mamluks were able to successfully defend their sovereignty against the Mongols, and the Abbasid Caliphate

85 His grandson reports that he could not resist the pressure of the traditionalist Mālikīs once he used analogy to decide on a case different than the traditional Mālikī literature: Abū l-Walīd b. Rushd, *Bidāyat al-mujtahid wa-nihāyat al-muqtaṣid*, ed. Abū Aws Yūsuf b. Aḥmad al-Bakrī (Amman: Bayt al-Afkār ad-Duwaliyya, 2007), 933; Ibn Rushd, *The Distinguished Jurist's Primer*, trans. Imran Ihsan Khan Nyazee, vol. 1 (London: Garnet, 1994), xxxiv (translator's introduction).

86 Maribel Fierro, "The Almohads (524–668/1130–1269) and the Ḥafṣids (627–932/1229–1526)," in *CHI*[2], vol. 4, *The Western Islamic World Eleventh to Eighteenth Centuries*, ed. Maribel Fierro (Cambridge: Cambridge University Press, 2011), 66–105, here 66–86; Fierro, "The Legal Policies of the Almohad Caliphs;" Maribel Fierro, "Proto-Malikis, Malikis and Reformed Malikism in al-Andalus," in *The Islamic School of Law. Evolution, Devolution, and Progress*, ed. Peri Bearman, Rudolph Peters and Frank E. Vogel (Cambridge: Harvard University Press, 2005), 56–76; Ibn Rushd, *Maßgebliche Abhandlung*, 72–74, 90–94; Serrano Ruano, "Exilio y desarraigo en la vida (y milagros) de Ibn Farḥ al-Qurṭubī, un exégeta andalusí universal." *Philologia Hispalensis* 31, no. 2 (2017): 133–48, here 134f., 142–44.

87 Canals Casassas and Serrano Ruano, "Putting Criticisms Against al-Ghazālī in Perspective: New Materials on the Interface Between Law, Rational Theology and Mysticism in Almoravid and Almohad al-Andalus (Ibn Rushd al-Jadd and al-Qurṭubī)," in *Philosophical Theology in Islam: Later Ashʿarism East and West*, ed. A. Shihadeh and J. Thiele (Leiden: Brill, 2020), 254–97, here 289f.

consolidated in Egypt, many scholars then saw a chance to live in Egypt, as the Mamluks were engaged in establishing madrasas, Islamic schools of higher education. On the other hand, Sufi traditions were also respected and became organized in various structures. Thus, Egypt became a nexus for scholarly activities and therefore seemed to be ideal for al-Qurṭubī's scientific endeavor.[88]

3 Al-Qurṭubī's *al-Jāmiʿ li-aḥkām al-Qurʾān*

3.1 Structure and Features of *al-Jāmiʿ*

Since for al-Qurṭubī, the Qurʾān is the guarantor of all the religious sciences (*jamīʿ al-ʿulūm ash-sharʿiyya*), he dedicated his whole life to exploring the Qurʾān. Therefore, he decided to write a book that contains explanations on the inflection of Qurʾānic utterances, the mention of the reading variants, the rejection of false claims on belief, the norms in the Qurʾān, reports on the revelation of the Qurʾān, reports from earlier authorities, and the elucidation of unclear passages. For this purpose, al-Qurṭubī describes his exegetical approach: he relates the different views from whom they were articulated – although he does not follow this aim on several occasions[89] – and searches for the relevant hadith in the hadith collections. The latter is important for him in order to discover the authentic reports, since, in his view, it is mostly unknown who recorded those hadith that are quoted in books on Islamic law and Qurʾānic exegesis. Further, he skips narratives and historical accounts, except those that are necessary to understand the Qurʾānic passage. His particular approach relies on juristic exegesis to derive norms from the Qurʾān, including accounts of the occasions of the revelation and clarification of unusual usage of words. If there is no ruling in a particular verse, then he seeks to provide some exegetical account without deriving norms.[90] Looking at *al-Jāmiʿ*, we can say that al-Qurṭubī more or less sticks to his plan of the composition of his *tafsīr*. The overall structure is as follows: at the beginning of each sura, al-Qurṭubī mentions whether this sura is *makkī* or *madanī*, whether there are verses

88 Arif Gezer, „Kurtubî'nin Hadis İlmindeki Yeri" (PhD diss., Ankara Üniversitesi Sosyal Bilimler Enstitüsü Temel İslam Bilimleri (Hadis) Anabilim Dalı, 2000), 22–32; Arif Gezer, "Endülüs'lü Müfessir Kurtubî, Dönemi ve Şöhret Bulma Süreci," *İstem* 14 (2009), 335–52, here 340. For further information on the Ayyubids and Mamluks during the lifetime of al-Qurṭubī: R. Stephen Humphreys, "Ayyubids, Mamluks, and the Latin East in the Thirteenth Century." *MSR* 2 (1998), 1–17; Kurt J. Werthmuller, *Coptic Identity and Ayyubid Politics in Egypt, 1218–1250* (Cairo: The American University in Cairo Press, 2010); D. De Smet et al., eds., *Egypt and Syria in the Fatimid, Ayyubid and Mamluk Eras IV: Proceedings of the 9th and 10th International Colloquium Organized at the Katholieke Universiteit Leuven in May 2000 and May 2001* (Leuven: Peeters Publishers, 2005).
89 Al-Qurṭubī, *al-Jāmiʿ*, 1: 8, footnote 4.
90 Al-Qurṭubī, *al-Jāmiʿ*, 1: 7f.

which have to be dated exceptionally,[91] and also gives accounts on the merits (*faḍāʾil*) of the particular sura. He then often mentions one or several verses together, occasionally a group in cases where they are short and closely related to each other,[92] but analyzes them verse by verse. He segments these verses; a segment can be one word, which contains several issues that must be clarified, or several legal issues (*masāʾil*).

Before he discusses the verses from a legal point of view, he accounts reports from earlier authorities, performs a grammatical analysis with the use of a great number of poetic citations,[93] and accounts about the occasions of revelation (*asbāb an-nuzūl*). While analyzing syntax, inflection and grammar, he follows the Basran school, according to Cornell,[94] and also concentrates on the variants of Qurʾānic reading. Then he lists the legal issues related to the verse. The important point here is that al-Qurṭubī mentions not only those legal issues directly related to the verse, he also tries to link every legal issue to the particular verse through even the vaguest possibility.[95] The legal discussion is mostly carried out in a comparative way; he occasionally mentions the reason for the diversity of views.[96] This feature can be read as a result of the rationalization process of Islamic law in Andalusia under the challenges of the Almohad dynasty. Thereby he tries to avoid repetition by referring to different passages of *al-Jāmiʿ*[97] or other passages of his writings for further detail.[98] However, he has a particular interest on abrogation (*an-naskh*), and tries to show the difference between seemingly similar methods, like abrogation and the specification (*at-takhṣīṣ*) of general utterances. Even if not as detailed as the legal interpretation, al-Qurṭubī also interprets many verses according to their theological implication, following primarily the Ashʿarite school.[99]

This approach of al-Qurṭubī – aside from being a complete *tafsīr* of the Qurʾān with special focus on legal interpretation – is not entirely innovative, but a rather typ-

---

91 Ibid., 6: 5.

92 Ibid., 15: 5.

93 Al-*Jāmiʿ* is therefore regarded as a rich encyclopedia of poetry, since al-Qurṭubī refers to around 3860 verses, including pre-Islamic poetries of Imruʾ al-Qays, ʿAntara b. Shaddād and Kaʿb b. Zuhayr; see Abdulmuttalip Arpa, "Şiirle İstişhâd Yöntemi Açısından Kurtubî'nin *el-Câmi' li-Ahkâmi'l-Kur'an* Tefsiri," *The Journal of Academic Social Science Studies* 6, no. 7 (2013): 98, 118.

94 Vincent J. Cornell, "*ʿIlm al-Qurʾān* in al-Andalus. The *tafsīr muḥarrar* in the Works of Three Authors," *Jusūr: The UCLA Journal of Middles Eastern Studies* 2 (1986): 63–81, here 77.

95 Serrano Ruano, "al-Qurṭubī."

96 Al-Qurṭubī, *al-Jāmiʿ*, 5: 450, 10: 324.

97 See ibid., 2: 93, 95.

98 For examples see the references to *at-Tadhkira* in *al-Jāmiʿ* mentioned above.

99 For example, on atomism, see al-Qurṭubī, *al-Jāmiʿ*, 8: 316. For *taḥsīn* and *taqbīḥ*, see al-Qurṭubī, idem, 13: 43.

ical encyclopedic[100] *tafsīr* in the Andalusian *tafsīr* tradition. A typical feature of this tradition is to focus on reading variants of the Qur'ān, which for al-Qurṭubī are so important that he has a chapter on that very question in his introduction. The rarity of considerations of *balāgha* (rhetorics) seems to be another typical feature of the Andalusian *tafsīr* tradition. Al-Qurṭubī, in true Andalusian fashion – although showing tendencies to asceticism – criticizes many Sufi practices. In contrast to other Andalusian authors, however, he makes use of a great deal of Sufi terminology, and refers to prominent esoteric exegeses like Abū Muḥammad Sahl b. ʿAbdallāh b. Yūnus b. ʿIshā b. ʿAbdallāh b. Rafīʿ at-Tustarī's (d. 896/283) *Tafsīr al-Qur'ān al-ʿaẓīm* and Abū ʿAbd ar-Raḥmān Muḥammad b. al-Ḥusayn b. Muḥammad as-Sulamī's (d. 1021/412) *Ḥaqā'iq at-Tafsīr*.[101] As Serrano Ruano notes, "(s)ingular in this respect is the inclusion of stories drawn from the apocryphal gospels and anecdotes about pious men that are not usually mentioned in other well-known *tafsīrs*."[102] Al-Qurṭubī is often criticized for using *isrā'īliyyāt*, reports that originate in the Jewish and/or Christian tradition. This is explained by the fact that it was usual to refer to these traditions, although exegetes were careful in referring to *isrā'īliyyāt* due to the ongoing threat and enmity at that time with Christian rulers in eastern and western territories of the Muslim world.

A special feature of his work is also his report on circumstances of his lifetime.[103] Commenting on Q 12:29, he complains about the lack of diligence in Egypt.[104] Moreover, he shares examples of human experience – including his own – in the art of healing warts,[105] or advice for a profitable harvest and its protection.[106]

3.2 Al-Qurṭubī's Hermeneutics

In the introduction and elsewhere in his *tafsīr*, we can discover some key elements of al-Qurṭubī's hermeneutics. For al-Qurṭubī, the Qur'ān is an eternal uncreated speech of God in clear Arabic.[107] Since the Qur'ān is transhistorical, it refers to current

---

100 He could have been influenced to write encyclopedic works by this feature of Egyptian scholarship, see Serrano Ruano, "Exilio y desarraigo", 135.

101 Abdullah Bayram, "Endülüs Tefsir Kültürünün Yorumda Yöntem ve Yaklaşımlar Açısından Analizi Kurtubî Tefsiri Örneği," *The Journal of Academic Social Science Studies* 48 (2016): 102; Mashīnī, *Madrasat at-Tafsīr*, 639–50; Serrano Ruano, "al-Qurṭubī."

102 Serrano Ruano, "al-Qurṭubī."

103 He refers to the actual practice for example in Qurṭubī, *al-Jāmiʿ*, 5: 77.

104 Qurṭubī, *al-Jāmiʿ*, 11: 325.

105 Al-Qurṭubī, *al-Jāmiʿ*, 14: 138; Balʿam, *al-Qurṭubī*, 88f. According to al-Qurṭubī, Q 20:105–7 can be used for healing warts (*tha'ālīl*, called *barāriq* in Egypt). Here al-Qurṭubī's advice is: Take three finely chopped wheat stalks, tie one end, go with it over the warts and read Q 20:105–7. Then, bury it in a moist place: The warts will be cured.

106 Al-Qurṭubī, *al-Jāmiʿ*, 20: 210.

107 Ibid., 1: 9.

events: Al-Qurṭubī says that the Qurʾānic statement "it (the war) is repulsive to you" (Q 2:216) is true in the sense that people always run away from war fearing death, even though they would die as martyrs. He connects this to Andalusia, where, he says, people fled instead of fighting, and therefore were killed and captured when Andalusia fell; a result of what they have done to themselves.[108]

Being revealed in clear Arabic, the correct inflection and grammatical analysis of the Qurʾān is essential for al-Qurṭubī. According to this attitude, the grammatical and legal interpretation of the Qurʾān does not fall under the prohibited interpretation through *raʾy* (here: 'personal reasoning'). For him, the forbidden interpretation is to suggest a view knowing that the truth is different from that or due to lack of qualification in exegesis. It is also illicit to draw the attention of the readers of the Qurʾān to a correct view, which, however, cannot be derived from the verse in question. On the other hand, it is forbidden to rely on transmission (*samāʿ*) only concerning exegetical endeavor, because this would mean that (legal) interpretation (*istinbāṭ*) would be prohibited. Since the Companions interpreted the Qurʾān and had differed in their opinions and had not heard every interpretation from the Prophet, exegesis cannot be reduced to transmission. However, we need transmissions from the earlier authorities, since many unclear or ambiguous passages can only be clarified through these transmissions. For al-Qurṭubī, to relate only to the knowledge of Arabic language does not suffice and is forbidden; transmission is the condition for correct interpretation and analysis.[109]

To legitimize and show the validity of interpreting the Qurʾān, al-Qurṭubī builds a chain from the scholars (*ʿulamāʾ*) via the Prophet to the Qurʾān via three steps: 1) The Qurʾān is the source (*aṣl*) for knowledge, on which one must contemplate in various modes (legally, theologically etc.). 2) The Sunna of the Prophet is its explanation (*bayānan lahū*), which explains unclear passages, and shows which meaning is the most probable one. 3) The scholars follow the Prophet by interpreting the Qurʾān to discover the intended meanings. Therefore, they are a special group that differs from others, and they have the duty of explaining the Qurʾān.[110] For al-Qurṭubī, their special status is attested by a hadith that says that scholars are the heirs of the Prophets.[111]

Since one has to rely on transmitted exegetical reports for correct analysis, the Sunna plays a decisive role in explaining the Qurʾān. It is obligatory for al-Qurṭubī to consider the sunna as much as the revealed Qurʾān according to the Qurʾān itself (he refers to Q 16:44, 24:63, 42:52, 59:7) and hadiths underlining that the Prophet received 'the like' of the Qurʾān. Al-Qurṭubī refers to Abū Sulaymān Ḥamd b. Muḥammad

108 Ibid., 3: 417f. Another example is his rescue from two Christian knights by reciting sura 36 and other passages from the Qurʾān, see al-Qurṭubī, *al-Jāmiʿ*, 13: 94.
109 Al-Qurṭubī, *al-Jāmiʿ*, 1: 57–59. Thus, he lists the names of authorities in *tafsīr*, see al-Qurṭubī, *al-Jāmiʿ*, 1: 60–64.
110 Ibid., 1: 6f.
111 Ibid., 5: 64.

b. Ibrāhīm b. Khaṭṭāb al-Khaṭṭābī (d. 998/388), who deduces two possible meanings from that: 1) The Prophet was given a hidden revelation, which is not recited during prayers (*al-waḥy al-bāṭin ghayr al-matluww*), besides the apparent revelation, which is recited during prayers. 2) He was given the revelation, which is recited, and was given an explanation, that means, that he was allowed to explain the Qur'ān, such as to concretize it or to add further rulings, which cannot be found in the Qur'ān. Al-Khaṭṭābī concludes from the second possibility that the sunna should be taken into account as if one was regarding the apparent wording of the Qur'ān.[112] Therefore, it is not sufficient to merely rely on the Qur'ān, or to reject a hadith, claiming that there is nothing similar to be found in the Qur'ān. This is the reason for al-Qurṭubī why early authorities said that the Qur'ān needs the sunna more than the other way around.[113]

We can assume that al-Qurṭubī holds the view that sunna is to be seen as revealed. Commenting on the phrase "except what is announced to you" (Q 5:1) concerning permitted consumption of meat, he says that not only the Qur'ān limits this permission, so does the sunna, since the sunna of the Prophet comes from the Qur'ān.[114] This could be understood to mean that al-Qurṭubī saw the sunna in line with al-Khaṭṭābī, which is not unusual in Sunni circles. Thus, he seems to allow arguing with isolated reports (*al-khabar al-wāḥid*), which are epistemologically presumptive (*ẓannī, maẓnūn*), and treats it as if it had a definitive (*qaṭʿī*) character,[115] confirmed by his argument for the validity of the abrogation of the Qur'ān by sunna, as we will see below.

Consequently, al-Qurṭubī has a special concern to prove the authenticity of hadiths to which he refers to. He does this, however, by looking for them in reliable hadith sources. A critical analysis of the hadith chains (*isnād*) is missing; he merely trusts the quality of the hadith books as attributed to them by scholars. Studies on the hadiths in *al-Jāmiʿ* show that al-Qurṭubī refers to many weak (*ḍaʿīf*), or even fictional and flat-out bogus (*mawḍūʿ*) hadiths.[116]

As he mentions in the introduction, al-Qurṭubī uses many reports from the first Muslim generations. A specific type of these are the *isrāʾīliyyāt*, which he uses occasionally, although not uncritically.[117] Another particular type of transmissions are reports on the occasions of revelation. Al-Qurṭubī refers to them not only to clarify the meaning of phrases, but also to select an opinion among competing interpretations.[118]

112 Abū Sulaymān Ḥamd b. Muḥammad al-Khaṭṭābī, *Maʿālim as-sunan wa-huwa sharḥ Sunan Abī Dāwūd*. Ed. Muḥammad Rāghib aṭ-Ṭabbāḥ (Aleppo: al-Maṭbaʿa al-ʿIlmiyya, 1932), 4: 298.

113 Al-Qurṭubī, *al-Jāmiʿ*, 1: 64–66.

114 Ibid., 7: 251.

115 Ibid., 2: 430.

116 Ibid., 1: 19–21 (the editor's introduction); Arnaldez, "al-Ḳurṭubī." For a list of hadith compilations referred to by al-Qurṭubī, see Gezer, "Kurtubî'nin Hadis İlmindeki Yeri", chapter 4.

117 Al-Qurṭubī, *al-Jāmiʿ*, 18: 215.

118 Ibid., 3: 289f.

The backbone of al-Qurṭubī's hermeneutics is the *ḥaqīqa–majāz*-dichotomy. It is common to the Muslim tradition to rely on the literal meaning of words, which should be understood according to their first coining for a specific meaning, called *ḥaqīqa*. Only an indicator or another reference in the religious texts can allow the interpreter to search for a figurative meaning (*majāz*) of a word.[119] For al-Qurṭubī, there is *majāz* in the Qur'ān,[120] but since *ḥaqīqa* is stronger than *majāz*,[121] the default meaning of an utterance is *ḥaqīqa*. Thus, without any evidence, it is forbidden to rely on *majāz*.[122] As a consequence of this dichotomy, the same is true for the default meaning of a phrase that should be understood according to the apparent meaning (*ẓāhir*) and as being general (*ʿāmm*).[123] Interestingly, even if al-Qurṭubī holds an opinion to be true, he rejects it insofar as it contradicts the apparent meaning of the quoted verse.[124] Al-Qurṭubī also tries to find the intended meaning of an utterance not only through the apparent meaning but also through referring to cohesion (*siyāq*), the inner textual context of the particular verse.[125]

Linked to that basic attitude, general utterances can only be specified and reduced through specific texts,[126] and reports on a specific occasion for a revelation do not prevent the transmitted utterance from being general.[127] Since al-Qurṭubī follows the hermeneutical standpoint of a default meaning through *ḥaqīqa* and *ẓāhir*, which should prevent arbitrary interpretation,[128] he has to define the default meaning of imperatives (*awāmir*), since he has a special concern for legal interpretation. For him, the default meaning of imperatives is an obligation (*wujūb*) and its immediate implementation (*fawr*).[129] These are two major positions among jurists. On the other hand, he negates that the grammatical form of the imperative alone can imply the obligation to repeat the obligated action.[130]

For al-Qurṭubī, the counterpart of *ẓāhir* is *mujmal*, a common dichotomy for the Shāfiʿī, Mālikī and Ḥanbalī schools. On several occasions, however, he also uses the

---

119 Robert Gleave, "Conceptions of the Literal Sense (*ẓāhir, ḥaqīqa*) in Muslim Interpretive Thought," in *Interpreting Scriptures in Judaism, Christianity and Islam. Overlapping Inquiries*, ed. Mordechai Z. Cohen and Adele Berlin (Cambridge: Cambridge University Press, 2016), 183–203.

120 Al-Qurṭubī, *al-Jāmiʿ*, 13: 336f.

121 Ibid., 5: 392.

122 Ibid., 22: 365.

123 Ibid., 5: 200, 9: 151. He also calls this *faḥwā l-khiṭāb*, in the sense of the overtly mentioned expression (*manṭūq*), see idem, 5: 200.

124 Ibid., 4: 240.

125 Ibid., 1: 378, 4: 489, 12: 219.

126 Ibid., 8: 215.

127 Ibid., 9: 192, 7: 247f.

128 Gleave attests this for various Islamic hermeneutics but shows, that it does not prevent from personal interpretation; Gleave, "Conceptions of the Literal Sense."

129 Al-Qurṭubī, *al-Jāmiʿ*, 2: 184.

130 Ibid., 5: 216.

technical terms *muḥkam* and *mutashābih* to indicate clarity and ambiguity, since the Qurʾān itself makes use of these two terms (cf. Q 3:7). For him, *muḥkam* are those utterances that can be understood through reasoning and interpretation, whereas *mutashābih* remains unclear, but is used in a very narrow sense, such as the time of the Day of Judgment (*yawm al-qiyāma*), the resurrection of Jesus, and the meaning of the so-called *al-ḥurūf al-muqaṭṭaʿa* ('broken letters').[131] Elsewhere he uses *muḥkam* in the sense of 'non-abrogated', which is one of its possible meanings: a general utterance in contrast to an abrogated (*mansūkh*) one.[132] In fact, abrogation (*naskh*) is one of the most important methods for al-Qurṭubī. According to him, it is the replacement of a Qurʾānic utterance by another, or merely abrogation without replacement, which does not imply a change in God's will and knowledge, but rather a change in the speech of God according to the interest (*maṣlaḥa*) of the people.[133] This change is possible due to his Ashʿarī view on *taḥsīn* and *taqbīḥ*, since good (*ḥusn*) and evil (*qubḥ*) do not exist, so a change in the speech would not for example permit a bad thing, and vice versa. Abrogation has different constellations: Abrogation 1) of Qurʾān by Qurʾān, or 2) of isolated report by *mutawātir* report 3) of isolated report by isolated report or 4) according to many scholars of Qurʾān by sunna, like the abrogation of Q 2:228 by the hadith 'No Testament for the Heir', in opposition to Muḥammad b. Idrīs ash-Shāfiʿī (d. 820/204). The forth constellation is valid, since for al-Qurṭubī, every ruling, either referred to by Qurʾān or sunna, comes from God; only the name of the source differs. He goes so far as to say that isolated traditions in the sunna can theoretically abrogate the Qurʾān, which some scholars like Abū l-Maʿālī Rukn ad-Dīn ʿAbd al-Malik b. ʿAbdallāh b. Yūsuf al-Juwaynī (d. 1085/478)[134] permitted.[135] Elsewhere, he gives the example of abrogation that the inhabitants of Qubāʾ accepted the turn of the *qibla* to Mecca by an isolated report, whereas their knowledge on the qibla as Jerusalem was definitive (*maqṭūʿan bihi*).[136] This is, in my view, due to his attitude towards the sunna as being part of revelation, as mentioned above. The fifth constellation of abrogation is that Qurʾān abrogates sunna, like the abrogation of the *qibla* towards Jerusalem. Abrogation is thus limited to the lifetime of the prophet, and there is no abrogation between the religious texts and consensus.[137] Since abrogation is important for al-Qurṭubī, he differentiates it from specification (*takhṣīṣ*); the specified reference (*mukhaṣṣaṣ*) is never comprised of

131 Ibid., 5: 17.

132 Ibid., 4: 487.

133 Ibid., 2: 300f., 303f., 430.

134 Abū l-Maʿālī ʿAbd al-Malik al-Juwaynī, *Kitāb at-Talkhīṣ fī uṣūl al-fiqh*, ed. ʿAbdallāh Ḥ. an-Naybānī and Sh. Aḥmad al-ʿUmarī (Beirut/Mecca: Dār al-Bashāʾir al-Islāmiyya/Maktabat Dār al-Bāz, 1996), 2: 525–29.

135 Al-Qurṭubī, *al-Jāmiʿ*, 2: 306.

136 Ibid., 2: 430f. Here he mentions al-Ghazālī as one who deemed it theoretically possible and happening during the lifetime of the Prophet; see Abū Ḥāmid al-Ghazālī, *al-Mustaṣfā min ʿilm uṣūl al-fiqh*, ed. Ḥamza b. Zuhayr Ḥāfiẓ (Medina: Sharikat al-Madīna al-Munawwara, n.d.), 2: 105f.

137 Al-Qurṭubī, *al-Jāmiʿ*, 2: 306, 430.

the specified meaning, and if it was, it would be an abrogation. Early authorities, however, used 'abrogation' figuratively for both.[138] Al-Qurṭubī's argumentation, in my opinion, is based on the assumption that the general utterance – specified by Qurʾān or sunna – does not cover the specified meaning, because sunna is revealed, too, and given to the Prophet. Thus, the general statement is already specified, but the specifier will be sent and disclosed afterwards, since for al-Qurṭubī, backed by the Mālikī position, an explanation for a revealed verse can be postponed until the need for it arises (so-called *taʾkhīr al-bayān ʿan waqt al-ḥāja*) or later than the related question.[139] Accordingly, these cases are not abrogation. Al-Qurṭubī further differentiates the removal of a ruling due to the removal of its ratio legis (*(rafʿ) li-rtifāʿ al-ʿilla (al-ḥukm)*) from removal through abrogation (*rafʿ al-ḥukm bi-n-naskh*), since abrogation is the abolishment of a ruling forever, whereas the removal of a ruling is temporary and related to the existence and absences of the *ratio legis* (*ʿilla*). As long as the *ʿilla* of a ruling returns and exists, the related ruling exists again.[140]

All these efforts are due to his attitude which holds that by existence of a religious text (*naṣṣ*) it is forbidden to pursue a personal opinion or *ijtihād*,[141] as rulings can only be determined through revelation (here *sharʿ*) and not reason, according to the impossibility of grasping good and bad through reason.[142] Consequently, al-Qurṭubī apparently follows the attitude to wait (*tawaqquf*) concerning the question of whether things are originally permitted or not, as long as there is no evidence in religious texts about their normative status,[143] and therefore, tries to resolve and harmonize competing texts, since harmonization (*al-jamʿ*) is better than a choice (*tarjīḥ*) for one of them.[144]

Linked to this attitude, al-Qurṭubī allows the consideration of implications (sg. *mafhūm*), common to Mālikī, Shāfiʿī and Ḥanbalī scholars, which he, however, calls *mafhūm al-khiṭāb* (commonly: *mafhūm al-muwāfaqa*).[145] Further, he does not reject the use of reverse conclusion (*argumentum e contrario, dalīl al-khiṭāb*).[146] Al-Qurṭubī also knows the *maqāṣid ash-sharīʿa* ('overall objectives of the Sharia'), but calls them *maṣāliḥ ash-sharīʿa*, following al-Ghazālī apparently by reducing them to the protection of religion, life, reason, progeny and property. Everything that protects these five is regarded to be an interest (*maṣlaḥa*) that should be considered for interpretation, and everything harming them is regarded to be harm (*mafsada*); its defense is also a *maṣlaḥa*.[147] Since al-Ghazālī derives at least these five goods directly from Qurʾān and

138 Ibid., 2: 305.
139 Ibid., 6: 91 and 7: 251.
140 Ibid., 14: 376.
141 Ibid., 9: 151.
142 Ibid., 13: 43.
143 Ibid., 1: 377f.
144 Ibid., 13: 142.
145 Ibid., 5: 177; 3: 48.
146 Ibid., 2: 10f. In 6: 182, he says, that the use of *dalīl al-khiṭāb* is controversial.
147 Ibid., 11: 367.

sunna,[148] al-Qurṭubī accepts their use for interpretation and flexibility, as only revelation can show what is good and bad.

3.3 Reception of al-Qurṭubī

In recent studies, a good deal of research is dedicated to al-Qurṭubī and his work a decent amount of research on al-Qurṭubī and his *al-Jāmiʿ* examines various aspects of his legal interpretations, hermeneutics, use of hadith and poetry, in studies written in Arabic, Turkish, Spanish, English, French and Bosnian.[149] There are also complete or partial translations into English, Spanish and Turkish.[150] However, al-Qurṭubī is not only prominent in recent scholarship, but was also much read and well-known from the fourteenth century onwards. He particularly influenced Mālikī jurisprudence and was quoted by several exegetes throughout the different schools of law.

---

148 Al-Ghazālī, *al-Mustaṣfā*, 2: 481–84.

149 On *al-Jāmiʿ* in general: Balʿam, *al-Qurṭubī*; Salmān, *al-Qurṭubī*; Maḥmūd Zalaṭ al-Qaṣabī, *al-Qurṭubī wa-manhajuhu fī t-tafsīr* (Cairo: Dār al-Anṣār, 1979). For example, on poems in *al-Jāmiʿ*: ʿAbd al-ʿĀl Sālim Mukarram, *ash-Shawāhid ash-shiʿriyya fī tafsīr al-Qurṭubī* (Cairo: ʿĀlam al-Kutub 1998); Arpa, "Şiirle İstişhad." Dissertations on Hadith in *al-Jāmiʿ*: Gezer, "Kurtubî'nin Hadis İlmindeki Yeri"; Farīda Zaghīshī, "Manhaj al-Qurṭubī fī t-taʿāmul maʿa l-ḥadīth wa-atharihi fī tafsīrihi li-āyāt al-Qurʾān" (PhD diss., University of Batna (Jāmiʿat al-Ḥājj Lakhdar), 2011/2012). On his views on *uṣūl al-fiqh*: Aḥmad ʿĪsā, *al-Ārāʾ al-uṣūliyya li-l-imām al-Qurṭubī min khilāl tafsīrihi* (Beirut: Dār al-Kutub al-ʿIlmiyya, 2005); Nadhīr Nabīl ʿAbd al-Ḥamīd ash-Sharāyirī, *"Manhaj al-imām al-Qurṭubī fī n-naskh dirāsatan taṭbīqiyyatan," Bingöl Üniversitesi İlahiyat Fakültesi Dergisi* 5 (2015): 165–203. An example of French scholarship is: Géraldine Jenvrin, "L'Offense à l'islam dans le *Tafsīr* d'al-Qurṭubī (m. 671/1273). Musulmans et non-musulmans en perspective," *MIDEO* 33 (2018): 95–136 (with a short description on *al-Jāmiʿ* in general). For English scholarship on al-Qurṭubī, for example: Canals Casassas and Serrano Ruano, "Putting Criticisms Against al-Ghazālī in Perspective;" Serrano Ruano, "al-Qurṭubī." An example for Spanish: Penelas Mayte, "El precepto de *al-amr bi-l-maʿrūf wa-l-nahy ʿan al-munkar* en el *Tafsīr* de *al-Qurṭubī*," in *Los almohades: Problemas y perspectivas*, ed. P. Cressier, Maribel Fierro and L. Molina (Madrid: Casa de Velázquez, 2005), 1051–73. Examples for Turkish scholarship: Bayram, "Endülüs Tefsir Kültürü;" Bayram, "Ahkâm Ayetlerini Yorumlamada Kurtubî Tefsirinin Makâsıd Açısından Analizi," *Gümüşhane Üniversitesi İlahiyat Fakültesi Dergisi* 2, no. 4 (2013): 75–95; Nefise Efe, "Kurtubî Tefsiri'nin Mukaddimesinin Tefir İlmi Açısından Değerlendirilmesi," *DEUİFD* 48 (2018): 117–67. For Bosnian: Dževad Šošić, "El-Kurtubi – slavni andaluzijski mufessir," *Zbornik Radova Fakulteta Islamskih Nauka u Sarajevu* 15 (2011): 75–89.

150 English: Aisha Bewley and Abdul Haq Bewley, *Tafsir al-Qurtubi: Classical Commentary of the Holy Qur'an* (London: Dar Al Taqwa, 2003). Partial translation of the introduction: Norman Calder, Javid Mojaddedi, and Andrew Rippin, eds. *Classical Islam. A Sourcebook of Religious Literature* (London/New York: Routledge, 2003), 98–102. For Spanish, see Zakariya Abu Mubarak Maza, *Compendio del Tafsir del Corán "Al-Qurtub,"* 10 vols. (Granada: Cdad. Musulmana Española de la Mezquita del Temor de Allāh, 2005–2013). For Turkish, see Eryarsoy M. Beşir, *Kurtubi Tefsiri – El-Câmiu li-Ahkâmi'l-Kur'ân*, 20 vols. (Istanbul: Buruç Yayınları, 2019).

If we look at biographical sources, al-Qurṭubī appears first in Abū ʿAbdallāh Shams ad-Dīn Muḥammad b. Aḥmad b. ʿUthmān adh-Dhahabī's (d. 1348/748) *Tārīkh al-Islām*, where he praises his knowledge and says that people travel to Munyat Banī Khaṣīb to study *al-Jāmiʿ*.[151] In the same century, Abū ṣ-Ṣafāʾ Ṣalāḥ ad-Dīn Khalīl b. ʿIzz ad-Dīn Aybak b. ʿAbdallāh aṣ-Ṣafadī (d. 1363/764) presents him as a great scholar. It seems that from this century onwards, almost every biographical work lists his name;[152] his contemporary Abū l-ʿAbbās Shams ad-Dīn Aḥmad b. Muḥammed b. Ibrāhīm b. Abī Bakr b. Khallikān (d. 1282/681), however, does not list al-Qurṭubī.

Al-Qurṭubī is referred to in later *tafsīr* books frequently; later authors refer to him without any specification in his name as al-Qurṭubī. There are some mentions in Abū Jaʿfar Aḥmad b. Ibrāhīm b. az-Zubayr al-Gharnāṭī's (d. 1308/708) *Milāk at-taʾwīl*, with significant changes in the wording,[153] one reference in Niẓām ad-Dīn al-Ḥasan b. Muḥammad b. Ḥusayn an-Nīsābūrī's (d. 1329/730) *Gharāʾib al-Qurʾān wa-raghāʾib al-furqān*,[154] and quotes in Abū Ḥayyān Muḥammad b. Yūsuf b. ʿAlī b. Yūsuf b. Ḥayyān al-Andalusī's (d. 1344/745) *al-Baḥr al-muḥīṭ* that are close to the wording of its source.[155] Abū l-ʿAbbās Shihāb ad-Dīn Aḥmad b. Yūsuf as-Samīn al-Ḥalabī (d. 1355/756) refers to *al-Jāmiʿ* twice.[156] Abū l-Fidāʾ ʿImād ad-Dīn Ismāʿīl b. Shihāb ad-Dīn ʿUmar b. Kathīr (d. 1373/774), however, refers more frequently to *al-Jāmiʿ*,[157] and also to *at-Tadhkira*.[158] This is not surprising, as his master Abū l-ʿAbbās Taqī d-Dīn Aḥmad b. ʿAbd al-Ḥalīm b. Majd ad-Dīn ʿAbd as-Salām b. Taymiyya (d. 1328/728) also praised al-Qurṭubī. This obviously attests the prominence of *al-Jāmiʿ*, which is occasionally compared to Abū l-Qāsim Maḥmūd b. ʿUmar b. Muḥammad az-Zamakhsharī's (d. 1144/538) *al-Kashshāf*.[159] There are numerous accounts of *al-Jāmiʿ* in *al-Lubāb* of Abū Ḥafṣ Sirāj ad-Dīn ʿUmar b. ʿAlī b. ʿĀdil

151 Adh-Dhahabī, Shams ad-Dīn Abū ʿAbdallāh, *Tārīkh al-Islām wa-wafayāt al-mashāhīr wa-l-aʿlām*, ed. ʿUmar ʿAbd as-Salām at-Tadmurī (Beirut: Dār al-Kutub al-ʿIlmiyya, 1993), 50: 74f.

152 Gezer, "Endülüs'lü Müfessir Kurtubî," 345–49.

153 Abū Jaʿfar Aḥmad b. Ibrāhīm al-Gharnāṭī, *Malāk at-taʾwīl al-qāṭiʿ bi-dhawī l-ilḥād wa-t-taʿṭīl fī tawjīh mutashābih al-lafẓ min āy at-tanzīl*, ed. ʿAbd al-Ghanī Muḥammad ʿAlī l-Fāsī (Beirut: Dār al-Kutub al-ʿIlmiyya, n.d.), 1: a) 40 b) 117, c) 138 and al-Qurṭubī, *al-Jāmiʿ*, a) 2: 138, b) 7: 250 and c) 8: 305.

154 Niẓām ad-Dīn an-Nīsābūrī, *Gharāʾib al-Qurʾān wa-raghāʾib al-furqān*, ed. Zakariyyā ʿUmayrāt (Beirut: Dār al-Kutub al-ʿIlmiyya, 1996), 6: 257. I think that this quote refers to al-Qurṭubī, *al-Jāmiʿ*, 20: 254.

155 For example, Abū Ḥayyān al-Andalusī, *al-Baḥr al-muḥīṭ fī t-tafsīr*, ed. Ṣidqī Muḥammad Jamīl (Beirut: Dār al-Fikr, 1999), a) 2: 423, b) 453, c) 545, d) 689, e) 5: 375, f) 10: 546 and al-Qurṭubī, *al-Jāmiʿ*, a) 3: 478, b) 4: 36, c) 4: 181f., d) 4: 359, e) 10: 117, f) 22: 499.

156 Abū l-ʿAbbās Shihāb ad-Dīn Aḥmad b. Yūsuf as-Samīn al-Ḥalabī, *ad-Durr al-maṣūn fī ʿulūm al-kitāb al-maknūn*, ed. Aḥmad Muḥammad al-Khirāṭ (Damascus: Dār al-Qalam, n.d.), 6: 562 and 10: 437; see al-Qurṭubī, *al-Jāmiʿ*, 11: 471 and 21: 221.

157 For example, Ibn Kathīr, *Tafsīr*, a) 1: 24, b) 1: 44, c) 1: 372; see al-Qurṭubī, *al-Jāmiʿ*, a) 1: 93f., b) 1: 72, c) 2: 274 and 281.

158 Ibn Kathīr, *Tafsīr*, 5: 61.

159 Taqī d-Dīn Abū l-ʿAbbās Aḥmad b. Taymiyya, *Majmūʿ al-fatāwā* Ed. ʿAbd ar-Raḥmān b. Muḥammad b. Qāsim (Medina: Mujammaʿ al-Malik Fahd, 2004), 13: 387f. He also refers to *at-Tadhkira*, see idem., 4: 324.

an-Nuʿmānī (d. 1373/775)[160] and some in Abū ʿAbdallāh Muḥammad b. Muḥammad b. Muḥammad b. ʿArafa at-Tūnisī's (d. 1401/803) *Tafsīr*.[161] Abū Zayd ʿAbd ar-Raḥmān b. Muḥammad b. Makhlūf ath-Thaʿālibī (d. 1471/875) in *al-Jawāhir al-ḥisān* refers to *at-Tadhkira*[162] more frequently, and even to *al-Jāmiʿ*.[163] Later scholars frequently quote from *al-Jāmiʿ*, like Abū l-Faḍl Jalāl ad-Dīn ʿAbd ar-Raḥmān b. Abī Bakr b. Muḥammad as-Suyūṭī (d. 1505/911),[164] Shams ad-Dīn Muḥammad b. Aḥmad al-Khaṭīb ash-Shirbīnī (d. 1570/977),[165] Abū s-Suʿūd Efendī (d. 1574/982),[166] Shihāb ad-Dīn Aḥmad b. Muḥammad b. ʿUmar al-Khafājī (d. 1659/1069),[167] Ismāʿīl Ḥaqqī al-Bursawī (d. 1725/1137)[168] and Abū ʿAbdallāh Muḥammad b. ʿAlī b. Muḥammad ash-Shawkānī (d. 1834/1250)[169]. We can also find authors who indirectly quote from al-Qurṭubī's *al-Jāmiʿ*, neither mentioning his name nor the title.[170] Contemporary *tafsīrs* also quote al-Qurṭubī frequently.

Evidence for the fact that appreciation of the *Jāmiʿ* has continued up to the present day can be seen in the large number of manuscripts of the work that have been preserved, the many commentaries and abridgements to which it has given rise (such as that of Sirāj ad-Dīn ʿUmar b. ʿAlī sh-Shāfiʿī, d. 1401/804), and the impressive number of modern editions.[171]

Thus, al-Qurṭubī influenced the later sources through his legal interpretation on the one hand, and on the other hand through the different opinions listed both on

---

160 See for example, an-Nuʿmānī, *al-Lubāb*, a) 1: 480, b) 1: 484, c) 20: 570; al-Qurṭubī, *al-Jāmiʿ*, a) 1: 372, b) 1: 375, c) 22: 572f.

161 Muḥammad b. Muḥammad b. ʿArafa, *Tafsīr al-imām Ibn ʿArafa*, ed. Ḥasan Munāʿī (Tunis: Markaz al-Buḥūth bi-l-Kulliyyat az-Zaytūna, 1986), 1: a) 114, b) 185, c) 330; al-Qurṭubī, *al-Jāmiʿ*, a) 1: 272, b) 1: 349, c) 2: 190f.

162 Ath-Thaʿālibī, *al-Jawāhir*, 1: 130, 337, 2: 19, 135, 3: 150, 353, 4: 79.

163 Ath-Thaʿālibī, *al-Jawāhir*, a) 2: 226, b) 3: 369; al-Qurṭubī, *al-Jāmiʿ*, a) 6: 261, b) 12, 68.

164 Jalāl ad-Dīn as-Suyūṭī, "Nawāhid al-abkār wa-shawārid al-afkār," ed. Aḥmad b. ʿAbdallāh b. ʿAlī d-Darūbī (PhD diss., Mecca: Jāmiʿat Umm al-Qurā, 2005), 3: 232; al-Qurṭubī, *al-Jāmiʿ*, 7: 252.

165 Shams ad-Dīn ash-Shirbīnī, *as-Sirāj al-munīr fī l-iʿāna ʿalā maʿrifat baʿḍ maʿānī kalām rabbinā al-ḥakīm al-khabīr* (Cairo: Būlāq, 1868), a) 1: 107, b) 1: 402, c) 4: 581; al-Qurṭubī, *al-Jāmiʿ*, a) 2: 471, b) 8: 256f., c) 22: 451.

166 Abū s-Suʿūd Efendī, *Tafsīr Abū s-Suʿūd – irshād al-ʿaql as-salīm ilā mazāyā l-kitāb al-karīm* (Beirut: Dār Iḥyāʾ at-Turāth al-ʿArabī, n.d.), a) 2: 29, b) 2: 43f.; al-Qurṭubī, *al-Jāmiʿ*, a) 5: 103, b) 5: 153f.

167 For example, Shihāb ad-Dīn Aḥmad b. Muḥammad al-Khafājī, *Ḥāshiyat ash-Shihāb al-musammā ʿInāyat al-qāḍī wa-kifāyat ar-rāḍī ʿalā tafsīr al-Bayḍāwī* (Beirut: Dār Ṣādir, n.d.), 6: 9; al-Qurṭubī, *al-Jāmiʿ*, 13: 21.

168 For example, Ismāʿīl Ḥaqqī l-Bursawī, *Rūḥ al-bayān* (Beirut: Dār al-Fikr, n.d.), 1: 201; al-Qurṭubī, *al-Jāmiʿ*, 2: 304.

169 Ash-Shawkānī's *Fatḥ al-qadīr* is full of references to *al-Jāmiʿ*. I want to mention only one here: Abū ʿAbdallāh Muḥammad ash-Shawkānī, *Fatḥ al-qadīr* (Damascus/Beirut: Dār al-Kalim aṭ-Ṭayyib, 1993), 1: 34; al-Qurṭubī, *al-Jāmiʿ*, 1: 237.

170 For example, ʿUbayd Allāh b. Aḥmad b. Abī r-Rabīʿ al-Ishbīlī, *Tafsīr al-kitāb al-ʿazīz wa-iʿrābuhu*, ed. ʿAlī b. Sulṭān al-Ḥakamī (Medina: Al-Jāmiʿa al-Islāmiyya bi-l-Madina al-Munawwara, n.d.), 370 (editor's foreword).

171 Ruano Serrano, "al-Qurṭubī."

legal and non-legal issues, which are, according to him, relevant to the related verse. It would distort the influence of *al-Jāmiʿ* to say that it influenced later exegetes to widen the scope of meanings of the verses by referring to their *ḥaqīqa* and *ẓāhir* meaning, and deriving as many rulings as possible from them; this was already a commonality among Muslim scholars in the earlier centuries. Al-Qurṭubī seems to be among those who perfectly adopted this approach in their writings and enriched its characteristics through the vast amount of *istishhād,* the analysis of Qurʾānic reading variants, legal interpretation, and the view of sunna as revelation.

4 Conclusion

We can conclude that al-Qurṭubī, with his highly critical, yet text-centered approach to Qurʾānic exegesis and special concern for legal interpretation, influenced both Qurʾānic exegesis and the legal tradition, at least the Mālikī *fiqh*. Because of his literalist approach, he widened the scope of religious scripture by regarding the sunna of the Prophet as being revealed, to derive as many norms as possible from the religious sources with special reference to the Qurʾān. With his pious attitude, he had a special concern to show that the Qurʾān should be the focal point of the life of every Muslim. Thus, for him, the Qurʾān is always up to date, as far as we can ascertain to deal with it accurately. His encyclopedic *al-Jāmiʿ* preserves many poems from the pre-Islamic era, numerous reading variants of the Qurʾān and sources of hadith that are used as the basis for deriving norms. Overall, it is a treasure for a lively interpretation of the Qurʾān, with its references to events, culture and the environment in the lifetime of al-Qurṭubī. Although being critical, al-Qurṭubī does not go beyond the borders of mainstream, literalist scholarship on Qurʾān; the work incorporates a high degree of human interpretation of God's words without being independent of the religious texts.

5 Bibliography

5.1 Sources

Abū s-Suʿūd Efendī. *Tafsīr Abū s-Suʿūd – irshād al-ʿaql as-salīm ilā mazāyā l-kitāb al-karīm*. 10 vols. Beirut: Dār Iḥyāʾ at-Turāth al-ʿArabī, n.d.

Al-Andalusī, Abū Ḥayyān. *Al-Baḥr al-muḥīṭ fī t-tafsīr*. Ed. Ṣidqī Muḥammad Jamīl. 11 vols. Beirut: Dār al-Fikr, 1999.

Al-Baghdādī, Ismāʿīl Bāshā. *Hadiyyat al-ʿārifīn asmāʾ al-muʾallifīn wa-āthār al-muṣannifīn*. 2 vols. Istanbul: Muʾassasat at-Tārīkh al-ʿArabī,1951.

Beşir, Eryarsoy M. *Kurtubi Tefsiri – El-Câmiu li-Ahkâmi'l-Kur'ân*. 20 vols. Istanbul: Buruç Yayınları, 2019.

Bewley, Aisha and Bewley, Abdul Haq. *Tafsir al-Qurtubi: Classical Commentary of the Holy Qur'an*. London: Dar Al Taqwa, 2003.

Al-Bursawī, Ismāʿīl Ḥaqqī. *Rūḥ al-Bayān*. 10 vols. Beirut: Dār al-Fikr, vol. 1, n.d.

Ad-Dāwūdī, Muḥammad b. ʿAlī. *Ṭabaqāt al-mufassirīn li-d-Dāwūdī*. 2 vols. Beirut: Dār al-Kutub al-ʿIlmiyya, vol. 2, 1983.

Adh-Dhahabī, Muḥammad Ḥusayn. *DhTM*. 3 vols. Cairo: Maktabat Wahba, vol. 2, 2000.

Adh-Dhahabī, Shams ad-Dīn Abū ʿAbdallāh. *Tārīkh al-Islām wa-wafiyyāt al-mashāhīr wa-l-aʿlām*. Ed. ʿUmar ʿAbd as-Salām at-Tadmurī, 52 vols. Beirut: Dār al-Kutub al-ʿIlmiyya, vol. 50, 1993.

Al-Gharnāṭī, Abū Jaʿfar Aḥmad b. Ibrāhīm. *Malāk at-taʾwīl al-qāṭiʿ bi-dhawī l-ilḥād wa-t-taʿṭīl fī tawjīh mutashābih al-lafẓ min āy at-tanzīl*. Ed. ʿAbd al-Ghanī Muḥammad ʿAlī al-Fāsī, 2 vols. Beirut: Dār al-Kutub al-ʿIlmiyya, vol. 1, n.d.

Al-Ghazālī, Abū Ḥāmid. *Al-Mustaṣfā min ʿilm uṣūl al-fiqh*. Ed. Ḥamza b. Zuhayr Ḥāfiẓ, 4 vols. Medina: Sharikat al-Madīna al-Munawwara, vol. 2, n.d.

Ibn ʿArafa, Muḥammad b. Muḥammad. *Tafsīr al-imām Ibn ʿArafa*. E. Ḥasan Munāʿī, 2 vols. Tunis: Markaz al-Buḥūth bi-Kulliyyat az-Zaytūna, 1986.

Ibn Farḥūn, Ibrāhīm b. ʿAlī. *Ad-Dībāj al-mudhahhab fī maʿrifat aʿyān ʿulamāʾ al-madhhab*. Ed. Muḥammad al-Aḥmadī Abū n-Nūr, 2 vols. Cairo: Dār at-Turāth, vol. 1, n.d.

Ibn al-ʿImād, Shihāb ad-Dīn Abū l-Falāḥ ʿAbd al-Ḥayy. *Shadharāt adh-dhahab fī akhbār man dhahab*. Ed. by ʿAbd al-Qādir al-Arnaʾūṭ and Maḥmūd al-Arnaʾūṭ, 10 vols. Damascus/Beirut: Dār Ibn Kathīr, vol. 7, 1991.

Ibn Kathīr, Abū l-Fidāʾ Ismāʿīl b. ʿUmar. *Tafsīr Qurʾān al-ʿaẓīm*. Ed. Sāmī b. Muḥammad Salāma, 8 vols. Riyadh: Dār Ṭayba, 1999.

Ibn Rushd, Abū l-Walīd. *The Distinguished Jurist's Primer. A Translation of Bidāyat Al-Mujtahid*, trans. Imran Ihsan Khan Nyazee, 2 vols. London: Garnet, 1994–1996.

Ibn Rushd, Abū l-Walīd. *Bidāyat al-mujtahid wa-nihāyat al-muqtaṣid*. Ed. Abū Aws Yūsuf b. Aḥmad al-Bakrī. Amman: Bayt al-Afkār ad-Duwaliyya, 2007.

Ibn Taymiyya, Taqī d-Dīn Abū l-ʿAbbās Aḥmad. *Majmūʿ al-fatāwā*. Ed. ʿAbd ar-Raḥmān b. Muḥammad b. Qāsim. 35 vols. Medina: Mujammaʿ al-Malik Fahd, 2004.

Al-Ishbīlī, ʿUbayd Allāh b. Aḥmad b. Abī r-Rabīʿ. *Tafsīr al-kitāb al-ʿazīz wa-iʿrābuhu*. Ed. ʿAlī b. Sulṭān al-Ḥakamī. Medina: Al-Jāmiʿa al-Islāmiyya bi-l-Madina al-Munawwara, n.d.

Al-Juwaynī, Abū l-Maʿālī ʿAbd al-Malik. *Kitāb at-Talkhīṣ fī uṣūl al-fiqh*. Ed. ʿAbdallāh Ḥ. an-Naybānī and Sh. Aḥmad al-ʿUmarī, 3 vols. Beirut/Mecca: Dār al-Bashāʾir al-Islāmiyya/Maktabat Dār al-Bāz, vol. 2, 1996.

Kaḥḥāla, ʿUmar Riḍā. *Muʿjam al-muʾallifīn*. 15 vols. Beirut: Maktabat al-Muthannā/Dār Iḥyāʾ at-Turāth al-ʿArabī, vol. 8, n.d.

Al-Khafājī, Shihāb ad-Dīn Aḥmad b. Muḥammad. *Ḥāshiyat ash-Shihāb al-musammā ʿInāyat al-qāḍī wa-kifāyat ar-rāḍī ʿalā tafsīr al-Bayḍāwī*. 8 vols. Beirut: Dār Ṣādir, vol. 6, n.d.

Al-Khaṭṭābī, Abū Sulaymān Ḥamd b. Muḥammad. *Maʿālim as-sunan wa-huwa sharḥ Sunan Abī Dāwūd*. Ed. Muḥammad Rāghib aṭ-Ṭabbākh, 4 vols. Aleppo: Al-Maṭbaʿa al-ʿIlmiyya, 1932.

Makhlūf, Muḥammad b. Muḥammad. *Shajarat an-nūr az-zakiyya fī ṭabaqāt al-Mālikiyya*. Ed. ʿAbd al-Majīd al-Khayyālī, 2 vols. Beirut: Dār al-Kutub al-ʿIlmiyya, vol. 1, 2003.

Maza, Zakariya Abu Mubarak. *Compendio del Tafsir del Corán "Al-Qurtubi."* 10 vols. Granada: Cdad. Musulmana Española de la Mezquita del Temor de Allāh, 2005–2013.

An-Nīsābūrī, Niẓām ad-Dīn. *Gharāʾib al-Qurʾān wa-raghāʾib al-furqān*. Ed. Zakariyyā ʿUmayrāt, 6 vols. Beirut: Dār al-Kutub al-ʿIlmiyya, vol. 6, 1996.

An-Nuʿmānī, Abū Ḥafṣ Sirāj ad-Dīn ʿUmar. *Al-Lubāb fī ʿulūm al-kitāb*. Ed. ʿĀdil Aḥmad ʿAbd al-Mawjūd and ʿAlī Muḥammad Muʿawwaḍ, 20 vols. Beirut: Dār al-Kutub al-ʿIlmiyya, 1998.

Al-Qurṭubī, Abū ʿAbdallāh. *Al-Jāmiʿ li-aḥkām al-Qurʾān: tafsīr al-Qurṭubī*. Ed. Aḥmad al-Bardūnī and Ibrāhīm Aṭfīsh, 20 vols. Cairo: Dār al-Kutub al-Miṣriyya, 1964.

Al-Qurṭubī, Abū ʿAbdallāh. *At-Tadhkār fī afḍal al-adhkār*. Ed. Bashīr Muḥammad ʿUyūn. Damascus/Ṭāʾif: Maktabat Dār al-Bayān/Maktabat al-Muʾayyad, 1986[3].

Al-Qurṭubī, Abū ʿAbdallāh. *Qamʿ al-ḥirṣ bi-z-zuhd wa-l-qanāʿa wa-radd dhull as-suʾāl bi-l-kutub wa-sh-shafāʿa*. Ed. Majdī Fatḥī s-Sayyid. Ṭanṭā: Dār aṣ-Ṣaḥāba, 1989.

Al-Qurṭubī, Abū ʿAbdallāh. *Al-Wajīz fī faḍāʾil al-kitāb al-ʿazīz*. Ed. ʿAlāʾ ad-Dīn ʿAlī Riḍā, Cairo: Dār al-Ḥadīth, 1992.
Al-Qurṭubī, Abū ʿAbdallāh. *Al-Asnā fī sharḥ asmāʾ Allāh al-ḥusnā*. Ed. Muḥammad Ḥasan Jabal, 2 vols. Ṭanṭā: Dār aṣ-Ṣaḥāba, 1994.
Al-Qurṭubī, Abū ʿAbdallāh. *Ṣifāt Allāh taʿālā wa-mā warada fī-hā min al-āy wa-l-aḥādīth*. Ed. Sayyid b. Ibrāhīm b. Ṣādiq ʿImrān. Beirut: Dār al-Kutub al-ʿIlmiyya, 2001.
Al-Qurṭubī, Abū ʿAbdallāh. *At-Tadhkira fī aḥwāl al-mawtā wa-l-ākhira*. Ed. Ṣādiq b. Muḥammad b. Ibrāhīm, 3 vols. Riyadh: Dār al-Minhāj, 2004.
Al-Qurṭubī, Abū ʿAbdallāh. *Al-Asnā fī sharḥ asmāʾ Allāh al-ḥusnā*. Ed. ʿIrfān b. Salīm al-ʿAshāḥassūna ad-Dimashqī. Beirut/Sidon: Al-Maktaba al-ʿAṣriyya, 2005.
Al-Qurṭubī, Abū ʿAbdallāh. *Al-Jāmiʿ li-aḥkām al-Qurʾān wa-l-mubayyin li-mā taḍammanahu min as-sunna wa-āy al-furqān*. Ed. ʿAbdallāh ʿAbd al-Muḥsin at-Turkī et al., 24 vols. Beirut: Muʾassasat ar-Risāla, 2006.
Al-Qurṭubī, Abū l-ʿAbbās Aḥmad b. ʿUmar. *Al-Mufhim li-mā ashkala min Kitāb Talkhīṣ Muslim*. Ed. Muḥyī d-Dīn Dīb Mastaw et al., 7 vols. Damascus/Beirut: Dār Ibn Kathīr/Dār al-Kalim aṭ-Ṭayyib, 1996.
Aṣ-Ṣafadī, Ṣalāḥ ad-Dīn Khalīl. *Al-Wāfī bi-l-wafayāt*. Ed. Aḥmad al-Arnaʾūṭ and Turkī Muṣṭafā, 29 vols. Beirut: Dār Iḥyāʾ at-Turāth al-ʿArabī, vol. 2, 2000.
As-Samīn al-Ḥalabī, Abū l-ʿAbbās Shihāb ad-Dīn Aḥmad b. Yūsuf. *Ad-Durr al-maṣūn fī ʿulūm al-kitāb al-maknūn*. Ed. Aḥmad Muḥammad al-Khirāṭ, 11 vols. Damascus: Dār al-Qalam, n.d.
Ash-Shawkānī, Abū ʿAbdallāh Muḥammad. *Fatḥ al-Qadīr*. 6 vols. Damascus/Beirut: Dār al-Kalim aṭ-Ṭayyib, 1993.
Ash-Shirbīnī, Shams ad-Dīn. *As-Sirāj al-munīr fī l-iʿāna ʿalā maʿrifat baʿḍ maʿānī kalām rabbinā l-ḥakīm al-khabīr*. 4 vols. Cairo: Būlāq, 1868/1285.
As-Suyūṭī, Jalāl ad-Dīn. "*Nawāhid al-abkār wa-shawārid al-afkār*," edited by Aḥmad b. ʿAbdallāh b. ʿAlī ad-Darūbī. 3 vols. Mecca: Jāmiʿat Umm al-Qurā, vol. 3, 2005.
As-Suyūṭī, Jalāl ad-Dīn. *Ṭabaqāt al-mufassirīn*. Ed. ʿAlī Muḥammad ʿUmar, Kuwait: Dār an-Nawādir, 2010.
Ath-Thaʿālibī, Abū Zayd ʿAbd ar-Raḥmān. *Tafsīr ath-Thaʿālibī – al-Jawāhir al-ḥisān fī tafsīr al-Qurʾān*. Ed. Muḥammad ʿAlī Muʿawwaḍ and ʿĀdil Aḥmad ʿAbd al-Mawjūd, 5 vols. Beirut: Dār Iḥyāʾ at-Turāth al-ʿArabī, 1997.
Az-Ziriklī, Khayr ad-Dīn. *Al-Aʿlām. Qāmūs tarājim li-ashhar ar-rijāl wa-n-nisāʾ min al-ʿArab wa-l-mustaʿribīn wa-l-mustashriqīn*. 8 vols. Beirut: Dār al-ʿIlm li-l-Malāyīn, vol. 5, 2002.

5.2 Secondary Literature

Arnaldez, Roger. "Al-Ḳurṭubī." In *EI*^{2}, edited by Peri J. Bearman et al., 12 vols. Leiden: Brill, vol. 5, 1986: 512.
Arpa, Abdulmuttalip. "Şiirle İstişhâd Yöntemi Açısından Kurtubî'nin *el-Câmiʾ li-Ahkâmi'l-Kur'an* Tefsiri." *The Journal of Academic Social Science Studies* 6, no. 7 (2013): 91–119.
Balʿam, Miftāḥ as-Sanūsī. *Al-Qurṭubī: ḥayātuhu wa-āthāruhu l-ʿilmiyya wa-manhajuhu fī t-tafsīr*. Benghazi: Manshūrāt Jāmiʿat Qāryūnus, 1998.
Bayram, Abdullah. Ahkâm Ayetlerini Yorumlamada Kurtubî Tefsirinin Makâsıd Açısından Analizi." *Gümüşhane Üniversitesi İlahiyat Fakültesi Dergisi* 2, no. 4 (2013): 75–95.
Bayram, Abdullah. "Endülüs Tefsir Kültürünün Yorumda Yöntem ve Yaklaşımlar Açısından Analizi Kurtubî Tefsiri Örneği." *The Journal of Academic Social Science Studies* 48 (2016): 91–113.
Brockelmann, Carl. *GALS*. 3 vols. Leiden: Brill, vol. 1, 1937.
Brockelmann, Carl. *GAL*. 2 vols. Leiden: Brill, vol. 1, 1943^2.
Calder, Norman, Jawid Mojaddedi and Andrew Rippin, eds. *Classical Islam. A Sourcebook of Religious Literature*. London/New York: Routledge, 2003.

Casassas, Canals and D. Serrano Ruano, "Putting Criticisms Against al-Ghazālī in Perspective: New Materials on the Interface Between Law, Rational Theology and Mysticism in Almoravid and Almohad al-Andalus (Ibn Rushd al-Jadd and al-Qurṭubī)." In *Philosophical Theology in Islam: Later Ashʿarism East and West*, edited by A. Shihadeh, J. Thiele. Leiden: Brill, 2020, 254–97.

Cornell, Vincent J., "*ʿIlm al-Qurʾān* in al-Andalus. The *tafsīr muḥarrar* in the Works of Three Authors." *Jusūr: The UCLA Journal of Middles Eastern Studies* 2 (1986): 63–81.

De Smet, D. et al., eds. *Egypt and Syria in the Fatimid, Ayyubid and Mamluk Eras IV: Proceedings of the 9th and 10th International Colloquium Organized at the Katholieke Universiteit Leuven in May 2000 and May 2001.* Leuven: Peeters Publishers, 2005.

Efe, Nefise. "Kurtubî Tefsiri'nin Mukaddimesinin Tefir İlmi Açısından Değerlendirilmesi." *DEUİFD* 48 (2018): 117–67.

Fierro, Maribel. "The Legal Policies of the Almohad Caliphs and Ibn Rushd's *Bidāyat al-mujtahid*." *JIS* 10, no. 3 (1999): 226–48.

Fierro, Maribel. "Proto-Malikis, Malikis and Reformed Malikism in al-Andalus." In *The Islamic School of Law. Evolution, Devolution, and Progress*, edited by Peri Bearman, Rudolph Peters and Frank E. Vogel. Cambridge: Harvard University Press, 2005, 56–76.

Fierro, Maribel. "The Almohads (524–668/1130–1269) and the Ḥafṣids (627–932/1229–1526)." In *The Western Islamic World Eleventh to Eighteenth Centuries*, edited by Maribel Fierro. vol. 4, *CHI*[2], 6 vols. Cambridge: Cambridge University Press, 2011, 66–105.

Gezer, Arif. "Kurtubî'nin Hadis İlmindeki Yeri." PhD diss., Ankara Üniversitesi Sosyal Bilimler Enstitüsü Temel İslam Bilimleri (Hadis) Anabilim Dalı, 2000.

Gezer, Arif. "Endülüs'lü Müfessir Kurtubî, Dönemi ve Şöhret Bulma Süreci." *İstem* 14 (2009): 335–52.

Gilliot, Claude. "Textes arabes anciens édités en Égypte au cours des années 1992 à 1994." *MIDEO* 22 (1994): 271–396.

Gilliot, Claude. "Exegesis: Classical." In *EQ*, edited by Jane Dammen McAuliffe, 6 vols. Leiden/Boston: Brill, vol. 2, 2002: 99–124.

Gleave, Robert. "Conceptions of the Literal Sense (*ẓāhir, ḥaqīqa*) in Muslim Interpretive Thought." In *Interpreting Scriptures in Judaism, Christianity and Islam. Overlapping Inquiries*, edited by Mordechai Z. Cohen and Adele Berlin. Cambridge: Cambridge University Press, 2016: 183–203.

Humphreys, R. Stephen. "Ayyubids, Mamluks, and the Latin East in the Thirteenth Century." *MSR* 2 (1998) 1–17.

Ibn Rushd, Muḥammad b. Aḥmad. *Maßgebliche Abhandlung. Faṣl al-Maqāl*, trans. and commented by Frank Griffel. Berlin: Verlag der Weltreligionen, 2010.

ʿĪsā, Aḥmad. *Al-Ārāʾ al-uṣūliyya li-l-imām al-Qurṭubī min khilāl tafsīrihi.* Beirut: Dār al-Kutub al-ʿIlmiyya, 2005.

Jenvrin, Géraldine. "L'Offense à l'islam dans le *Tafsīr* d'al-Qurṭubī (m. 671/1273). Musulmans et non-musulmans en perspective." *MIDEO* 33 (2018): 95–136.

Al-Jundī, Farīd ʿAbd al-ʿAzīz. *Jāmiʿ al-aḥkām al-fiqhiyya li-l-imām al-Qurṭubī min tafsīrihi.* 3 vols. Beirut: Dār al-Kutub al-ʿIlmiyya, 2005.

Kaddouri, Samir. "Identificación de 'al-Qurṭubī', Autor de *al-Iʿlām bimā fī Dīn al-Naṣārà min al-fasād wa-l-awhām*." *Qanṭara* 21, no. 1 (2000): 215–19.

Al-Mashīnī, Muṣṭafā Ibrāhīm. *Madrasat at-tafsīr fī l-Andalus.* Beirut: Muʾassasat ar-Risāla, 1986.

Mukarram, ʿAbd al-ʿĀl Sālim. *Ash-Shawāhid ash-shiʿriyya fī tafsīr al-Qurṭubī.* Cairo: ʿĀlam al-Kutub 1998.

Penelas, Mayte. "El precepto de *al-amr bi-l-maʿrūf wa-l-nahy ʿan al-munkar* en el *Tafsīr* de *al-Qurṭubī*." In *Los almohades: Problemas y perspectivas*, edited by P. Cressier, Maribel Fierro, and L. Molina. Madrid: Casa de Velázquez, 2005, 1051–73.

Potthast, Daniel. "Andalusische Philosophie nach Ibn Rušd: Intellektuelle Traditionen in al-Qurṭubī antichristlicher Polemik." *Der Islam*, 95, no. 2 (2018): 429–49.

Al-Qaṣabī, Maḥmūd Zalaṭ. *Al-Qurṭubī wa-manhajuhu fī t-tafsīr.* Cairo: Dār al-Anṣār, 1979.

Ar-Rūmī, Fahd b. ʿAbd ar-Rahmān b. Sulaymān. *Manhaj al-madrasa al-Andalusiyya fī t-tafsīr*. Riyadh: Maktabat at-Tawba, 1997.

Salmān, Mashhūr Ḥasan and ʿAbd al-Laṭīf ad-Dasūqī. *Kashshāf taḥlīlī li-l-masāʾil al-fiqhiyya fī tafsīr al-Qurṭubī*. Ṭāʾif: Maktabat aṣ-Ṣiddīq, 1988.

Salmān, Mashhūr Ḥasan Maḥmūd. *Al-Imām al-Qurṭubī: shaykh aʾimmat at-tafsīr*. Damascus: Dār al-Qalam, 1993.

Serrano Ruano, Delfina. "Exilio y desarraigo en la vida (y milagros) de Ibn Farḥ al-Qurṭubī, un exégeta andalusí universal." *Philologia Hispalensis* 31, no. 2 (2017): 133–48.

Ash-Sharāyirī, Nadhīr Nabīl ʿAbd al-Ḥamīd. "*Manhaj al-imām al-Qurṭubī fī n-naskh dirāsatan taṭbīqiyyatan*." *Bingöl Üniversitesi İlahiyat Fakültesi Dergisi* 5 (2015): 165–203.

Šošić, Dževad. "El-Kurtubi – slavni andaluzijski mufessir." *Zbornik Radova Fakulteta Islamskih Nauka u Sarajevu* 15 (2011): 75–89.

Werthmuller Kurt J. *Coptic Identity and Ayyubid Politics in Egypt, 1218–1250*. Cairo: The American University in Cairo Press, 2010.

Zaghīshī, Farīda. "Manhaj al-Qurṭubī fī t-taʿāmul maʿa l-ḥadīth wa-atharihi fī tafsīrihi li-āyāt al-Qurʾān." PhD diss., University of Batna (*Jāmiʿat al-Ḥājj Lakhdar*), 2011/2012.

5.3 Online Sources

Altıkulaç, Tayyar. "Kurtubî, Muhammed b. Ahmed." In *TDVIA*, edited by TDV İslâm Araştırmaları Merkezi. Accessed April 20, 2020. https://islamansiklopedisi.org.tr/kurtubi-muhammed-b-ahmed.

HATA file *(Historia de los autores y transmisores de al-Andalus)*, accessed October 22, 2023, https://www.eea.csic.es/red/hata/Serrano Ruano, Delfina. "al-Qurṭubī." In *EQ*, edited by Jane Dammen McAuliffe. Brill, 2001–2006. Article published online 2016. Accessed October 22, 2023. http://dx.doi.org/10.1163/1875-3922_q3_EQCOM_050504.

Serrano Ruano, Delfina. "Ibn Farh al-Qurtubi." Accessed April 19, 2020. http://dbe.rah.es/biografias/136449/ibn-farh-al-qurtubi.

Nāṣir ad-Dīn al-Bayḍāwī

Farid Suleiman

1 Introduction

Qāḍī l-Quḍāt ʿAbd Allāh b. ʿUmar b. Muḥammad b. ʿAlī Abū l-Khayr (also: Abū Saʿīd) Nāṣir ad-Dīn ash-Shīrāzī l-Bayḍāwī was an Ashʿarite theologian, Shāfiʿī jurist, historian, and exegete. He spent his life in the western part of today's Iran, especially in Shīrāz and Tabrīz, under Mongol Ilkhanate rule. Several of his writings gained widespread recognition, above all his *Tafsīr*, as attested by the (arguably unmatched) abundance of manuscript copies and glosses that exist of this work. Despite the popularity of his scholarly output, biographical data about him is scarce and at times inconsistent. His birth date has been declared unascertainable, and there is a remarkable gap between the death dates that scholars have proposed. The Arabic secondary literature prefers the date 1286/685, while Persian and Western research favors 1316/716, with the exception of Walid Saleh, who in a recently published entry on al-Bayḍāwī in the *Encyclopaedia of Islam (Third Edition)*, considers the year of 1316/719 to have the strongest support.[1] As I have argued elsewhere in detail, based on the hitherto neglected and oldest biographical note available, al-Bayḍāwī most probably died sometime after 1291/690, at the age of forty-nine.[2] There, I have also tried to show that additional, circumstantial evidence allows for the conclusion that he was born in 1243/641 and died in 1291/691.

2 His Life, Works, Political, and Intellectual Context

2.1 Life and Context

Al-Bayḍāwī was probably born in Shīrāz in the province of Fars. His *nisba* refers to a nearby town called Bayḍā, where his family lived up until the generation of his grandfather. Al-Bayḍāwī was raised during the reign of the Sunni Salghurid Atabek of Fars Abū Bakr b. Saʿd (r. 1226–1260/623–658). The Atabek rulers had up to then been tributaries to the Khwarazm Shahs, but with the beginning of Abū Bakr's reign, they became vassals of the Buddhist Mongols, namely, of Genghis Khan's third son, Ögedei (r. 1229–1241/626–639). During their military campaigns in Central Asia, the Mongols

1 Walid Saleh, "al-Bayḍāwī," in *EI*[3], ed. Kate Fleet et al. (Leiden: Brill, 2017), 5: 53–57.

2 See my article "A Note on al-Bayḍāwī's Year of Birth and Death," *JQS* (forthcoming). Recently, I have come across a master's thesis that also draws on this note: ʿAbd Allāh b. ʿUmar al-Bayḍāwī, "Mirṣād al-ifhām ilā mabādiʾ al-aḥkām," ed. Ḥasan al-Ḥusayn (Master's thesis, Muhammad Ibn Saud University, 2010), 125–33 (editor's introduction).

https://doi.org/10.1515/9783111320083-005

left behind a trail of destruction and deadly violence, which culminated in the siege of Baghdad in 1258/656 that led to the massacre of a great number of its inhabitants and the breakdown of the Abbasid caliphate. On the other hand, the province of Fars in general, and the city of Shīrāz in particular, not only remained a safe haven due to Abū Bakr's assiduous diplomatic efforts toward the Mongols, but also witnessed a time of financial and intellectual prosperity. This prompted the great Persian poet Abū Muḥammad Saʿdī (d. 1292/691), a contemporary of al-Bayḍāwī, to dedicate a eulogistic poem to Abū Bakr.

In al-Bayḍāwī's lifetime, the Mongol ruling elite appealed to several religions, especially Buddhism and Christianity; Shiite Islam was also a significant factor. Shortly after al-Bayḍāwī's death, Ghazan was the first Mongol ruler to embrace Sunni Islam. He did so likely for tactical reasons, as he might have thought that this would ease the invasion of the Sunni Mamluk Empire, which he had attempted to conquer several times with no considerable success.[3] His successors turned to Shiite Islam, which they did under the influence of one of al-Bayḍāwī's intellectual peers: the famous Shiite scholar Ibn al-Muṭahhar al-Ḥillī (d. 1325/726). As I will elaborate further, while al-Bayḍāwī's works (especially his *Tafsīr*) not only enjoyed great popularity among Sunni scholars, but also among their Shiite counterparts, there is – to my knowledge – no hint that al-Bayḍāwī took a lenient position toward Shiite theology, let alone embracing any of its elements. Rather, al-Bayḍāwī was a staunch Shāfiʿī-Ashʿarī, who, living only two generations after Fakhr ad-Dīn ar-Rāzī (d. 1210/606), not only was influenced by the process of incorporating elements of *falsafa* into *kalām,* but also actively engaged in it. Therefore, al-Bayḍāwī's scholarship is in line with the greater developments that lead to the composition of the first philosophical compendia in *kalām* in the century after him, such as those by ʿAḍud ad-Dīn al-Ījī (d. 1355/756) or Saʿd ad-Dīn at-Taftāzānī (d. 1390/792).

Another significant characteristic of al-Bayḍāwī is his interest in astronomy, which is reflected not only in his works within that field, but also in his *Tafsīr* and theological writings.[4] Again, this can be explained by the intellectual context in which al-Bayḍāwī was embedded. During the Mongol Ilkhanate, astronomy was a highly prestigious field of knowledge, and scholars like the Shiite philosopher and polymath Naṣīr ad-Dīn aṭ-Ṭūsī (d. 1274/672) – a contemporary of al-Bayḍāwī – contributed much to the advancement of the field.

Regarding the life of al-Bayḍāwī, information is scarce and at times contradictory. The Yemeni historian Bahāʾ ad-Dīn al-Janadī (d. 1332/732 or a little earlier) is the author of the earliest biographical note on him. Al-Janadī had contact with one of al-Bayḍāwī's students, the Shīrāz-based scholar Muḥammad b. Ibrāhīm az-Zanjānī (d. 1321–22/721), who wrote commentaries on several of his master's works. In 1318/718,

3 See Reuven Amitai-Preiss, "Mongol Imperial Ideology and the Ilkhanid War against the Mamluks," in *The Mongol Empire and Its Legacy*, ed. Reuven Amitai-Preiss and David O. Morgan (Leiden: Brill, 2000), 68.

4 See Robert Morrison, "Natural Theology and the Qur'an," *JQS* 15, no. 1 (2013): 1–22.

when az-Zanjānī went to Yemen and met al-Janadī, the latter asked him about al-Bayḍāwī's life and, fortunately, included the conversation in his biographical work *as-Sulūk fī ṭabaqāt al-ʿulamāʾ wa-l-mulūk*. The relevant part reads as follows:[5]

> At the time that I (al-Janadī) was informed about the excellence of his (az-Zanjānī's) master (al-Bayḍāwī), I came across passages in his works upon which I proved the accuracy of the information that reached me. I asked him (az-Zanjānī) about him (al-Bayḍāwī), and he responded: "He is ʿAbd Allāh b. ʿUmar b. Muḥammad al-Bayḍāwī. His father was the high judge of Shīrāz before him and named him Nāṣir ad-Dīn. His father was given the title 'Imām ad-Dīn.'" Then I asked him about from whom he had taken his knowledge (*ʿan man tafaqqaha*). He said: "Regarding sciences that are based on transmitted knowledge (*manqūlāt*), it was his father, and regarding those that are based on reason (*maʿqūlāt*), it was Sharīf ad-Dīn Saʿīd, the best among the scholars of Shīrāz (*awḥad ʿulamāʾ Shīrāz*)." (In terms of the name) al-Bayḍāwī, it is related to a place in the vicinity of Shīrāz, since his grandfather went from there to Shīrāz, the capital of the ruler in the province of Fars (. . .). There was no one like him (al-Bayḍāwī) in Shīrāz in terms of the number of (his) students and works. He died in the city of Tabrīz, which is part of Adharbayjān, a little after (*nayyif*) 690, when he reached the age of forty-nine.

Considering the impact of al-Bayḍāwī's scholarly input on Sunni thought, and to a considerable extent even on Shiite thought, surprisingly little is known about his life. The earliest account above presents the main elements of his biography that can be considered as well established, while later accounts only add minor and sometimes contradictory details.

2.2 Works

Al-Bayḍāwī's scholarly endeavors concentrated mostly on abridging and/or commenting on works of previous scholars. One should not consider this a shortcoming, as some researchers have done in the past.[6] Rather, al-Bayḍāwī has rightfully been held in high esteem within the Muslim tradition, as his highly eloquent and concise abridgments, among others, reflect his deep knowledge of the field to which the respective abridgment belongs. Furthermore, his efforts did not focus only on abridging texts, but also on rearranging and refining them as well as adding new information. The appreciation of al-Bayḍāwī's scholarly output started from very early on, and manifests in the vast commentary tradition that several of al-Bayḍāwī's works have spawned. In fact, within the genre of *Tafsīr*, his work *Anwār at-tanzīl*, which itself is based on the *Kashshāf* of Abū l-Qāsim Maḥmūd b. ʿUmar az-Zamakhsharī (d. 1144/538), may very well be unsurpassed in terms of the sheer number of commentaries that have been written on it.

---

5 Bahāʾ ad-Dīn al-Janadī, *as-Sulūk fī ṭabaqāt al-ʿulamāʾ wa-l-mulūk*, ed. Muḥammad al-Akwaʿ al-Ḥawālī (Sanaa: Maktabat al-Irshād, 1995), 2: 436.

6 This misjudgment is addressed by Walid Saleh, "Marginalia and Peripheries: A Tunisian Historian and the History of Qurʾanic Exegesis," *Numen* 58 (2011): 304–9.

2.2.1 Published Works

(1) *Anwār at-tanzīl wa-asrār at-taʾwīl* (*Lights of Revelation and Secrets of Interpretation*)
This commentary on the Qurʾān is al-Bayḍāwī's *magnum opus* and the work that contributed most to his reputation as a scholar. According to Ḥasan al-Ḥusayn, one of its surviving manuscripts is dated 1285/684,[7] which would then be the *terminus ad quem* for the composition of the *Tafsīr*. In section 3 below, the content and reception of this work will be examined in more detail.

(2) *Ṭawāliʿ al-anwār min maṭāliʿ al-anẓār* (*The Breaking of Dawn from the Far Horizons of Logical Reasoning*)
This work of theology, though larger than the *Miṣbāḥ* (see (8) below), is a concise treatment of Ashʿarī *kalām* at a time when it was already heavily informed by elements of *falsafa*. Among others, al-Bayḍāwī covers issues of logic, epistemology, ontology as well as the divine attributes, prophetology, and political leadership. Heidrun Eichner describes the arguments in the *Ṭawāliʿ* as "primarily based on that of the *al-Mulahkhkhaṣ fī l-ḥikma*" by Fakhr ad-Dīn ar-Rāzī.[8] The *Ṭawāliʿ* probably predates al-Bayḍāwī's *Tafsīr*, since he mentions it there.[9] It spawned several commentaries, of which the most famous is probably that by Shams ad-Dīn al-Iṣfahānī (d. 1348/749). An English translation of al-Bayḍāwī's work along with al-Iṣfahānī's commentary was published in 2002.[10]

(3) *Minhāj al-wuṣūl ilā ʿilm al-uṣūl* (*The Clear Path for Arriving at the Knowledge of the Principles*)
This short work (approximately two hundred pages in the modern edition) is an enormously influential text on *uṣūl al-fiqh* (principles of Islamic law). The work has survived in numerous manuscripts and a great number of scholars across the centuries have written commentaries on it. The text, along with the commentary of Jamāl ad-

7 Al-Bayḍāwī, "*Mirṣād al-ifhām*," 121.

8 Heidrun Eichner, "The Post-Avicennian Philosophical Tradition and Islamic Orthodoxy: Philosophical and Theological Summae in Context" (Habilitation thesis, Martin-Luther-Universität Halle-Wittenberg, 2009), 394. The work was not at my disposal, so I have taken the quoted passage from Abdelkader Al Ghouz, "Recasting al-Bayḍāwī's Eschatological Concept of Bodily Resurrection: Shams al-Dīn al-Iṣfahānī and Aḥmad al-Ījī in Comparative Perspective," *MSR* 20 (2017): 43.

9 In Faramarz Haj Manouchehri and Abbas Mosallayi-Pour, "al-Bayḍāwī," in *EI*, ed. Wilferd Madelung and Farhad Daftary (Leiden, Boston: Brill, 2013), 4: 627f., it is stated that Yaḥyā b. Sharaf an-Nawawī (d. 1277/676) refers to the *Ṭawāliʿ* in his *Tahdhīb al-asmāʾ wa-l-lughāt*, hence, al-Bayḍāwī must have composed it before an-Nawawī died. There is indeed a reference to the author of the *Ṭawāliʿ* ("*ṣāḥib aṭ-Ṭawāliʿ*") at one place; however, this is a scribal error, which should read "*ṣāḥib al-Maṭāliʿ*." The *Maṭāliʿ al-anwār* by Abū Isḥāq b. Qurqūl (d. 1173/569) is one of the main sources that an-Nawawī uses in his work.

10 ʿAbd Allāh b. ʿUmar al-Bayḍāwī and Shams ad-Dīn Iṣfahānī, *Nature, Man and God in Medieval Islam*, ed. and trans. Edwin E. Calverley and James W. Pollock, 2 vols. (Leiden: Brill, 2002).

Dīn al-Asnawī (d. 1370/772), has been part of the curriculum of prestigious Islamic universities, such as al-Azhar University. It is also frequently cited in modern Arabic textbooks on *uṣūl al-fiqh*.[11]

(4) *Al-Ghāya al-quṣwā fī dirāyat al-fatwā* (*The Ultimate Goal in Understanding Legal Opinion*)
This work (2 vols.) is an annotated abridgment of al-Ghazālī's book on Shāfiʿite jurisprudence (*fiqh*) entitled *al-Waṣīṭ fī l-madhhab* (9 vols.). While it is not possible to ascertain a precise date of composition, al-Bayḍāwī's references in this work to some of his books on principles of jurisprudence (*al-kutub al-uṣūliyya*) suggest that he wrote the *Ghāya* after the *Minhāj* and the *Mirṣād* (see (3) and (11), respectively).

(5) *Lubb al-lubāb fī ʿilm al-iʿrāb* (*The Kernel of the Hearts in the Science of Inflection*)
This is an abridgment of Ibn al-Ḥājib's (d. 1249/646) grammatical work *al-Kāfiya*. It was published twice at least, both times with the commentary of Muḥammad b. Bīr ʿAlī l-Birkilī (i.e., Birgivī; d. 1573/981) entitled *Imtiḥān al-adhkiyāʾ*.

(6) *Niẓām at-tawārīkh* (*The Order of History*)
This work on history, written in Persian, is quite insignificant compared to other works of the period, insofar as it does not contribute anything new to the subject. Hence, scholars have largely dismissed it. Charles Melville, however, has argued that the work's importance lies not in its content as such, but in the way it clearly reflects al-Bayḍāwī's intent "to steer the alien Mongol regime towards embracing the culture and tradition of their Persian subjects."[12] The first version of the work covers historical events up to its completion in 1275/674. Two continuations have been added later, the first one (at least mostly) by al-Bayḍāwī himself.

(7) *Tuḥfat al-abrār: sharh maṣābīḥ as-sunna* (*Gift of the Pious in Elucidating the Lamps of Prophetic Traditions*)
This work is a commentary of the hadith compendium *Maṣābīḥ as-sunna*, written by the Shāfiʿī scholar Abū Muḥammad al-Baghawī (d. 1122/516, 1121/515, or 1117/510). Of the 4,931 prophetic sayings that can be found therein, al-Bayḍāwī only comments on 1,599, leaving out for the most part – besides others – those that al-Baghawī himself considered to be weak (*ḍaʿīf*). Al-Bayḍāwī probably wrote his commentary before his *Tafsīr*, since he mentioned it there. The printed edition spans over three volumes and was published in 2012.

(8) *Miṣbāḥ al-arwāḥ* (*Lamp of the Souls*)
This treatise on theology spans over 150 pages in the modern edition. In structure and content, it seems to be similar to the *Ṭawāliʿ* (see (2) above). Several scholars have authored commentaries on the densely written text, among them al-Bayḍāwī himself.

11 See Monique Cardinal, "Islamic Legal Theory Curriculum: Are the Classics Taught Today?" *ILS* 12, no. 2 (2005), 241f.
12 Charles Melville, "From Adam to Abaqa: Qāḍī Bayḍāwī's Rearrangement of History," *SIr* 30 (2001), 67.

This, at least, may be inferred from the title of one of his books, *al-Īḍāḥ fī sharḥ al-miṣbāḥ*, which seems to be lost. The composition of the *Miṣbāḥ* must have been prior to that of al-Bayḍāwī's larger work on theology entitled *Ṭawāliʿ al-anwār*, since there is a reference to it there.

(9) *Muntahā l-munā fī sharḥ asmāʾ Allāh al-ḥusnā* (*The Ultimate Desire concerning the Explanation of God's Most Beautiful Names*)
Two editions have been published so far; however, Ḥasan al-Ḥusayn claims in his master's thesis that both editions are attributed falsely to al-Bayḍāwī. Al-Ḥusayn writes that he had access to a manuscript, whose authenticity is beyond any doubt, and which differs from the two published editions.[13]

(10) *Risāla fī mawḍūʿāt al-ʿulūm wa-taʿārīfihā* (*On the Subject Matter and Definitions of the Sciences*)
In this small treatise (thirteen pages), al-Bayḍāwī divides the sciences into seven main categories: linguistic/literary sciences (*ādāb*), religious sciences (*nawāmīs*), natural sciences (*al-ʿilm aṭ-ṭabīʿī*), geometry/technology (*handasa*), astronomy (*hayʾa*), music (*mūsīqāt*), and the science of character traits (*akhlāq*). He gives definitions of all of them and, in most cases, of their respective subcategories.

(11) *Mirṣād al-ifhām ilā mabādiʾ al-aḥkām* (*Pathway of Understanding to the Principles of Religious Commandments*)
This work on *uṣūl al-fiqh* is a commentary of the abridged treatise (*mukhtaṣar*) that the Mālikī ʿUthmān b. ʿUmar b. al-Ḥājib (d. 1249/646) had composed on his own work *Muntahā s-sūl wa-l-ʿamal fī ʿilmay al-uṣūl wa-l-jadal*. The *Muntahā* itself is a summary of al-Āmidī's (d. 1233/631) *al-Iḥkām fī uṣūl al-aḥkām*. There is a reference to the *Mirṣād* in al-Bayḍāwī's *Tafsīr*, which means that its composition likely preceded the *Tafsīr*. The text has been edited in 2010 as part of a master's thesis and spans over nine hundred pages.[14]

2.2.2 Unpublished Extant Works

Risālat manāzil al-qamar (*On the Lunar Mansions*)

In this very short treatise, al-Bayḍāwī lists the twenty-eight lunar mansions and offers a short explanation of each. A digital image of one of its known manuscripts is available online.[15] The treatise might have been a part of al-Bayḍāwī's lost work *al-Mukhtaṣar fī l-hayʾa*.[16]

13 See al-Bayḍāwī, "*Mirṣād al-ifhām*," 101–5 (editor's introduction).

14 See note 2 above.

15 Princeton University Library, "*Risālat manāzil al-qamar* (Selections)", https://catalog.princeton.edu/catalog/9954368663506421#view, fol. 305a-306a last accessed July 19, 2024.

16 Al-Bayḍāwī, "*Mirṣād al-ifhām*," 115 (editor's introduction).

2.2.3 Works (So Far) Only Known by Title

(1) *Al-Īḍāḥ fī uṣūl ad-dīn* (*The Elucidation of Theology*): this might be a commentary on the published work *Miṣbāḥ* (see (8) above); (2) *Tahdhīb al-akhlāq* (*Refinement of Virtues*); (3) *Sharḥ at-tanbīh*: a commentary on Abū Isḥāq ash-Shīrāzī's (d. 1083/476) work in Shāfiʿī law entitled *at-Tanbīh fī l-fiqh ash-Shāfiʿī*; (4) *Sharḥ al-Muḥaṣṣal*: a commentary on Fakhr ad-Din ar-Rāzī's (d. 1210/606) *Muḥaṣṣal afkār al-mutaqaddimīn wa-l-mutaʾakhkhirīn min al-ʿulamāʾ wa-l-ḥukamāʾ wa-l-mutakallimīn* on philosophy and theology; (5) *Sharḥ al-Maḥṣūl*: a commentary on the *Maḥṣūl* of ar-Rāzī, a work on *uṣūl al-fiqh*; (6) *Sharḥ al-Muntakhab*: a commentary on ar-Rāzī's abridgment of his own *Maḥṣūl*; (7) *Sharḥ al-Fuṣūl*: a commentary on a work written by the Shiite scholar Naṣīr ad-Dīn aṭ-Ṭūsī (d. 1274/672) that is either on theology[17] (if so, then it would probably be aṭ-Ṭūsī's *al-Fuṣūl an-Naṣīriyya*) or on astronomy;[18] (8) *Sharḥ Maṭāliʿ al-anwār*: a commentary on the logical treatise *Maṭāliʿ al-anwār* written by Sirāj ad-Dīn al-Armawī (d. 1283/682), a contemporary of al-Bayḍāwī; (9) *al-ʿAyn* (*The Eye*): a book on *tafsīr* according to Kohlberg;[19] (10) *Kitāb fī l-manṭiq* (*Book on Logic*).

3 Al-Bayḍāwī's Qurʾān Commentary, *Anwār at-tanzīl wa-asrār at-taʾwīl*, and Its Reception

3.1 The Work *Anwār*

As has been said above, there seems to be evidence that al-Bayḍāwī finished his *Anwār*[20] sometime before the year 1285/684, which is approximately seven years before his death. Considering further indications, we may assume that he did not compose the work much earlier than this date; rather, it seems that it belongs to the late period of his life. For al-Bayḍāwī states in the introduction that he had thought of composing a commentary on the Qurʾān for quite a long time, but postponed it since he had not felt competent enough to do so.[21] Secondly, al-Bayḍāwī refers to several of his major writings in his *Anwār*, which also supports a late date of composition.

Al-Bayḍāwī's *Tafsīr* is also known as *Mukhtaṣar al-Kashshāf* (*Abridgement of the* (Qurʾān commentary) *Kashshāf* (by the Muʿtazilī az-Zamakhsharī)). This, however,

---

17 See Etan Kohlberg, "al-Bayżāwī, Nāṣer-al-Dīn," in *EIr*, ed. Ehsan Yarshater, vol. 4 (London, New York: Routledge, 1989), 16.

18 See Manouchehri and Mosallayi-Pour, "al-Bayḍāwī", 629.

19 See Kohlberg, "al-Bayżāwī, Nāṣer-al-Dīn", 16.

20 Henceforth: *Anwār*.

21 ʿAbd Allāh b. ʿUmar al-Bayḍāwī, *Tafsīr al-Bayḍāwī: Anwār at-tanzīl wa-asrār at-taʾwīl*, ed. ʿAbd al-Ḥafīẓ Bayḍūn (Beirut: Dār al-Kutub al-ʿIlmiyya, 2019), 1: 22.

should not obscure the fact that al-Bayḍāwī makes use of other sources as well, most notably the Qurʾān commentaries of ar-Rāzī and Abū l-Qāsim ar-Rāghib al-Aṣfahānī (d. after 1018/409). Al-Bayḍāwī mentions neither of them explicitly; rather, he states in a general manner that he intends to present the reader with the quintessence (*ṣafwa*) that he and others before him have extracted from the teachings of the Companions of the Prophet, their students, and later scholars regarding the meaning of the Qurʾān.[22] Another issue worth noting concerns the title of al-Bayḍāwī's *Tafsīr*. Whether by coincidence or deliberately so, it bears a remarkable similarity to the title of ar-Rāzī's smaller and less-known work in Qurʾānic exegesis *Asrār at-tanzīl wa-anwār at-taʾwīl.*

Even though al-Bayḍāwī's work is relatively small (3 vols.) compared to others in the field of Qurʾānic exegesis, it still covers a wide range of issues that pertain to different disciplines, most importantly theology, linguistics, religious commandments (*fiqh*) and its principles *(uṣūl al-fiqh*), and to a lesser extent history and astronomy. Furthermore, al-Bayḍāwī's work draws attention to the implications of variant canonical readings of the Qurʾān, of which he considers those of the "eight prominent leading reciters and anomalous (*shādhdh*; or: noncanonical) readings that have been narrated from acknowledged reciters."[23]

Al-Bayḍāwī's greatest achievement in his work is that he identifies the most acknowledged interpretations within the Sunni tradition of Qurʾānic exegesis and masterly condenses them into a concise and succinct text. Moreover, he does not merely string together different interpretations, but evaluates them and concludes by stating his own opinion, which is, in matters of theology and law, most of the time in line with the teachings of the Ashʿarī and Shāfiʿī schools.

Al-Bayḍāwī either omits the places in which az-Zamakhsharī's affiliation to the theological school of the Muʿtazilī shines through in his work *Kashshāf*, or he counters them with the Ashʿarī point of view. He does so, for instance, regarding questions such as whether seeing God is possible in the hereafter, or whether angels are better creatures than prophets. To a lesser, but significant extent, al-Bayḍāwī also engages with Shiite interpretations of verses, trying to refute, for example, the view that the Qurʾān supports the infallibility of ʿAlī, Fāṭima, and their sons Ḥasan and Ḥusayn. Since al-Bayḍāwī lived in an environment in which Shiism was a significant religious and political factor, he may have felt compelled to address such issues.

The *Anwār* clearly belongs to the category of *at-tafsīr bi-r-raʾy*, that is, exegesis based on personal reflection of the Qurʾānic verses. Nevertheless, narrated sayings of the Prophet Muḥammad and his Companions also play an important role in the *Anwār*. In total, al-Bayḍāwī adduces about 1,550 narrations, of which around eighty percent go

22 Ibid.

23 Ibid. By the eight reciters al-Bayḍāwī refers to Yaʿqūb al-Ḥaḍramī and the seven reciters that have been chosen by Ibn Mujāhid (d. 936/324).

back to the Prophet.[24] Interestingly, al-Bayḍāwī has not omitted a single narration cited in az-Zamakhsharī's *Kashshāf*. He also followed him in neither providing chains of narrators, nor commenting on the reliability of the traditions, even though a significant portion of them has been classified as weak or even forged. In contrast, al-Bayḍāwī is more cautious with material from the Jewish tradition (*isrāʾīliyyāt*). He not only makes little use of it, but at times critically engages with the question of its reliability.

3.2 The Reception of the *Anwār*

It is difficult to overestimate the importance the *Anwār* had, and still has, within the Muslim commentary tradition. This is, among others, attested by the abundance of copies that have been produced throughout the centuries; indeed, as Rosemarie Quiring-Zoche says, "there is probably no library in the world that has Arabic manuscripts except that it also possesses a copy of the *Anwār*."[25] Equally impressive is the number of glosses, epitomes, and other works that somehow engage with the *Anwār*.[26] Up until today, the *Anwār* has a prominent place in academic curricula at universities in the Muslim world and is used as an important source by authors of textbooks on Qur'ānic exegesis.[27] Being of such significance, it comes as no surprise that the *Anwār* also happens to be the first Muslim Qur'ān commentary ever edited and printed in modern times.[28]

The most acknowledged glosses on the *Anwār* are those by Muḥammad b. Muṣliḥ Shaykhzāda (d. 1544/951), Shihāb ad-Dīn al-Khafājī (d. 1659/1069), Ismāʿīl al-Qūnawī (d. 1781/1195), and al-Khaṭīb al-Kāzarūnī (d. 1553/940), all of which are edited and published. Interestingly, the popularity of the *Anwār* is not confined to Sunnism. Rather, several Shiite scholars have engaged with the *Anwār*, such as the important Safavid thinker Bahāʾ ad-Dīn al-ʿĀmilī (d. 1621/1030), who has written several partial glosses on it. In the beginning of one of them, he writes that the *Anwār* "is the best and purest of

24 See Zayn ad-Dīn al-Munāwī, *al-Fatḥ as-samāwī bi-takhrīj aḥādīth tafsīr al-Bayḍāwī*, ed. Aḥmad ʿĀlim as-Salafī (Riyadh: Dār al-ʿĀṣima, 1988), 1143 (in the editor's concluding chapter).

25 Rosemarie Quiring-Zoche, "An Early Manuscript of al-Bayḍāwī's *Anwār al-tanzīl* and the Model It Has Been Copied from," in *From Codicology to Technology: Islamic Manuscripts and Their Place in Scholarship*, ed. Stefanie Brinkmann and Beate Wiesmüller (Berlin: Frank & Timme, 2009), 33.

26 Brockelmann lists eighty-six glosses and superglosses on the *Anwār*; see Carl Brockelmann, *History of the Arabic Written Tradition*, trans. Joep Lameer (Leiden, Boston: Brill, 2016), 1: 417f.

27 See note 11 above, and also ʿAbd Allāh b. ʿUmar al-Bayḍāwī, *The Lights of Revelation and the Secrets of Interpretation: Ḥizb I of the Commentary on the Qur'an by al-Bayḍāwī. Arabic edition and English translation*, ed. and trans. Gibril F. Haddad (Manchester, UK; Brunei, Darussalam: Beacon Books; UBD Press, 2016), 63 (editor's introduction).

28 The edition was prepared by Heinrich Fleischer in 1848, and reprinted in Egypt and India. Theodor Nöldeke, who did not hold the *Anwār* in high esteem, commented critically on Fleischer's choice to edit the *Anwār*; see Theodor Nöldeke, *GdQ*, (Leipzig: Dieterich, 1919^{2}), 2: 176f.

all Qur'ān commentaries."[29] Another example is Fatḥ Allāh al-Kāshānī (d. 1580/988), who based his commentary *Zubdat at-tafāsīr* on four main sources, one of them being the *Anwār*.[30]

Though the Muslim tradition held the *Anwār* in high esteem, or as Theodor Nöldeke put it, "came close to sanctifiying it,"[31] the *Anwār* was not exempt from criticism. For example, a number of scholars found fault with al-Bayḍāwī's inclusion of weak and forged narrations. More importantly, he was criticized for having unconsciously adopted some of az-Zamakhshari's views that are Muʿtazilī in nature and go against the established Ashʿarī doctrine. Several works have been written in this regard, most famously *al-Itḥāf bi-tamyīz mā tabaʿa fīhi al-Bayḍāwī ṣāḥib al-Kashshāf* by Shams ad-Dīn ad-Dāwūdī (d. 1538/945), a student of Jalāl ad-Dīn as-Suyūṭī (d. 1505/911).

4 Conclusion

Nāṣir ad-Dīn al-Bayḍāwī remains a prominent and influential Ashʿarī-Shāfiʿī scholar from thirteenth century Iran. Details on his life are scarce, even though his scholarly contribution was already acknowledged in his own lifetime. While modern scholarship has deemed his date of birth unascertainable and proposed several years of death, ranging from 1286/685 to 1319/719, hitherto largely neglected evidence suggests that he died shortly after the year 1291/690 at the age of forty-nine.

Al-Bayḍāwī mostly focused on commenting on and/or abridging works of prior scholars. He did so in a masterly fashion, which explains the hugely positive reception in later Sunni, and to a lesser extent Shiite, scholarship. In fact, his commentary on the Qur'ān, *Anwār at-tanzīl wa-asrār at-taʾwīl*, has been copied numerous times and spawned a great number of glosses, and likely remains unsurpassed by any other work in the field of Qur'ānic exegesis. It was also the first Muslim exegetical work to be edited and published in a modern edition, and holds an important place within the curricula of many Muslim universities up until today. While the work is primarily based on three earlier commentaries – those of az-Zamakhsharī, ar-Razī, and ar-Rāghib al-Aṣfahānī – it is original in the sense that it offers a very concise and succinct explanation of the Qur'ānic verses that nevertheless takes into account aspects of language, (Ashʿarī) theology, (Shāfiʿī) law, history, astronomy, and other fields of knowledge. Despite the work's popularity, it has come under criticism as well. This happened especially for two reasons. The first being

29 Cited in Maryam Rustami, "*Tafsīr-i suturg wa mufassir-i buzurg*," *Bayyināt* 12 (2006): 71. In a different gloss, preserved in manuscript form at the library *Kitabkhana-yi Majlis-i Shura-yi Milli* (#12084), al-ʿĀmilī glorifies al-Bayḍāwī, prays for his well-being in the hereafter, and praises his *Tafsīr* (f.1b).
30 See Murtażā Raḥīmī, "*Tafsīr-i Bayżāwi dar āyanih-i fiqhā-yi maẕāhib-i islāmī*," *Muṭālaʿāt-i Taqrībī-i Maẕāhib-i Islāmī* 8, no. 32 (2013): 35.
31 Nöldeke, *GdQ*, 2: 176.

al-Bayḍāwī's unsparing use of weak and forged prophetic traditions, and the second his failure to detect and remove all Muʿtazilī elements in the passages that he took from az-Zamakhsharī's *Tafsīr*.

5 Bibliography

5.1 Sources

Heinrich Fleischer edited the *Anwār* for the first time in 1848; since then, many more editions have been published. None of them are critical in the sense that they establish a *stemma codicum* of the available manuscripts. The most reader-friendly version might be a two-colored edition published in three volumes (ed. ʿAbd al-Ḥafīẓ Bayḍūn, Beirut: Dār al-Kutub al-ʿIlmiyya, 2019). Gibril Fouad Haddad's edition with English translation is based on several manuscripts and prior editions; however, so far it only covers the first *ḥizb*, which is up until Q 2:74 (*The Light of Revelation and the Secrets of Interpretation*, Manchester, UK: Beacon Books, 2016). Another noteworthy aspect of this edition is Haddad's lengthy introduction on al-Bayḍāwī, his *Tafsīr*, and its later reception.

Other partial translations cover al-Bayḍāwī's commentary on the third and the twelfth chapter of the Qur'ān:

Beeston, A.F.L., ed. *Baidawi's Commentary on Surah 12 of the Qur'an. Text, Accompanied by an Interpretative Rendering and Notes*. Oxford: Clarendon Press, 1964.

Bishop, Eric F.F., and Mohammed Kaddal, eds. *Chrestomathia Baidawiana. "The Light of Inspiration and Secret of Interpretation"; Being a Translation of the Chapter of Joseph (Surat Yusuf) with the Commentary of Nasir Id-Din al-Baidāwī*. Glasgow: Jackson, 1957.

Margoliouth, D.S., ed. *Chrestomathia Baidawiana. The Commentary of El-Baidawi on Sura III*. London: Luzac, 1894.

5.2 Secondary Literature

Abū ʿAnza, ʿAlāʾ Jamīl. "Al-Bayḍāwī wa-juhūduhu al-balāghiyya fī tafsīrihi." Master's thesis, Islamic University Gaza, 2015.

ʿAlī, Yūsuf Aḥmad. "Al-Bayḍāwī wa-manhajuhu fī t-tafsīr." PhD diss., Umm al-Qurra University, n.d.

Amitai-Preiss, Reuven. "Mongol Imperial Ideology and the Ilkhanid War against the Mamluks." In The Mongol Empire and Its Legacy, edited by Reuven Amitai-Preiss and David O. Morgan. Leiden: Brill, 2000: 21–42.

Al-Bayḍāwī, ʿAbd Allāh b. ʿUmar. "Mirṣād al-ifhām ilā mabādiʾ al-aḥkām." Ed. Ḥasan al-Ḥusayn. Master's thesis, Muhammad Ibn Saud University, 2010.

Al-Bayḍāwī, ʿAbd Allāh b. ʿUmar, and Shams ad-Dīn Iṣfahānī. Nature, Man and God in Medieval Islam. Ed. and trans. by Edwin E. Calverley and James W. Pollock, 2 vols. Leiden: Brill, 2002.

Brockelmann, Carl. History of the Arabic Written Tradition, trans. Joep Lameer. 2 vols. with 3 vols. suppl. Leiden, Boston: Brill, 2016–19.

Cardinal, Monique. "Islamic Legal Theory Curriculum: Are the Classics Taught Today?" ILS 12, no. 2 (2005): 224–72.

Eichner, Heidrun. "The Post-Avicennian Philosophical Tradition and Islamic Orthodoxy: Philosophical and Theological Summae in Context." Habilitation thesis, Martin-Luther-Universität Halle-Wittenberg, 2009.

Al Ghouz, Abdelkader. "Recasting al-Bayḍāwī's Eschatological Concept of Bodily Resurrection: Shams al-Dīn al-Iṣfahānī and Aḥmad al-Ījī in Comparative Perspective." MStR 20 (2017): 39–54.

Ḥājī, ʿAbd al-ʿAzīz. *Al-Bayḍāwī mufassiran.* Damascus: Dār al-Ḥasanayn, 2000.

Ibrahim, Lutpi. "The Theological Questions at Issue between az-Zamakhsharī and al-Bayḍāwī with Special reference to *al-Kashshāf* and *Anwār at-tanzīl*." PhD diss., University of Edinburgh, 1977.

Ibrahim, Lutpi. "Al-Bayḍāwī's Life and Works." *Islamic Studies* 18, no. 4 (1979): 311–21.

Ibrahim, Lutpi. "The Concept of *Iḥbāṭ* and *Takfīr* according to az-Zamakhsharī and al-Bayḍāwī." *WO* 11 (1980): 117–21.

Ibrahim, Lutpi. "The Question of the Superiority of Angels and Prophets between az-Zamakhsharī and al-Bayḍāwī." *AR* 28, no. 1 (1981): 65–75.

Ibrahim, Lutpi. "A Comparative Study of az-Zamakhshrī (sic) and al-Bayḍāwī about the Position of the Grave Sinners." *Islamic Studies* 21, no. 1 (1982): 55–73.

Ibrahim, Lutpi. "The Problem of the Vision of God in the Theology of az-Zamakhsharī and al-Bayḍāwī." *WO* 13 (1982): 107–13.

Idrīs, Muḥammad. "*Ihtimām mufassirī al-qarn al-hādī ʿashr bi-tafsīr al-Bayḍāwī: asbābuhu wa-maẓāhiruhu.*" *Majallat Jāmiʿat Dimashq li-l-ʿUlūm al-Iqtiṣādiyya wa-l-Qānūniyya* 29, no. 3 (2013): 489–515.

Al-Janadī, Bahāʾ ad-Dīn. As-Sulūk fī ṭabaqāt al-ʿulamāʾ wa-l-mulūk. Ed. Muḥammad al-Akwaʿ al-Ḥawālīm, 2 vols. Sanaa: Maktabat al-Irshād, 1995.

Johns, Anthony. "Exegesis as an Expression of Islamic Humanism: Approaches, Concerns and Insights of al-Bayḍāwī." *Hamdard Islamicus* 22, no. 4 (1999): 37–58.

Karimi-Nia, Morteza. "Bayżāwī, ʿAbd Allāh b. ʿUmar b. Muḥammad." In *Dānishnāma-yi jahān-i islām*, edited by Ghulām-ʿAlī Ḥaddād ʿĀdil. 27 vols. Tehran: Bunyād-i Dāʾirat al-Maʿārif-i Islāmī, vol. 5, 2000: 218–21.

Klausing, Kathrin. *Geschlechterrollenvorstellungen im Tafsīr*. Frankfurt am Main: Peter Lang, 2014.

Kohlberg, Etan. "al-Bayżāwī, Nāṣer-al-Dīn." In *EIr*, edited by Ehsan Yarshater. London, New York: Routledge, vol. 4, 1989: 15–17.

Manouchehri, Faramarz Haj, and Abbas Mosallayi-Pour. "Al-Bayḍāwī." In EI, edited by Wilferd Madelung and Daftary Farhad. Leiden, Boston: Brill, vol. 4, 2013: 623–33.

Melville, Charles. "From Adam to Abaqa: Qāḍī Bayḍāwī's Rearrangement of History." SIr 30 (2001): 67–86.

Morrison, Robert. "Natural Theology and the Qurʾan." *JQS* 15, no. 1 (2013): 1–22.

Al-Munāwī, Zayn ad-Dīn. *Al-Fatḥ as-samāwī bi-takhrīj aḥādīth tafsīr al-Bayḍāwī.* Ed. Aḥmad ʿĀlim as-Salafī. Riyadh: Dār al-ʿĀṣima, 1988.

Nöldeke, Theodor. GdQ. 3 vols. Leipzig: Dieterich, vol. 2, 1919^2.

Princeton University Library. "Risālat manāzil al-qamar (Selections)." (last accessed July 19, 2024). https://catalog.princeton.edu/catalog/9954368663506421#view, fol. 305a-306a.

Quiring-Zoche, Rosemarie. "An Early Manuscript of al-Bayḍāwī's *Anwār al-tanzil* and the Model It Has Been Copied from." In *From Codicology to Technology: Islamic Manuscripts and Their Place in Scholarship*, edited by Stefanie Brinkmann and Beate Wiesmüller. Berlin: Frank & Timme, 2009: 33–46.

Raḥīmī, Murtażā. "*Tafsīr-i Bayżāwi dar āyanih-i fiqhā-yi maẓāhib-i islāmī.*" *Muṭālaʿāt-i Taqrībī-i Maẓāhib-i Islāmī* 8, no. 32 (2013): 33–47.

Rahman, Yusuf. "Hermeneutics of al-Bayḍāwī in his *Anwār al-tanzīl wa asrār al-taʾwīl*." *IC* 71 (1991): 1–14.

Riddell, Peter. "Al-Bayḍāwī." In *The Qur'an: An Encyclopedia*, edited by Oliver Leaman. London: Routledge, 2006: 116–18.
Rustami, Maryam. "*Tafsīr-i suturg wa-mufassir-i buzurg*." *Bayyināt* 12 (2006): 66–87.
Saleh, Walid. "Al-Bayḍāwī." In *EI*[3], edited by Kate Fleet et al. 5 vols. Leiden: Brill, vol. 5, 2017: 53–57.
Saleh, Walid. "Marginalia and Peripheries: A Tunisian Historian and the History of Qur'anic Exegesis." Numen 58 (2011): 284–313.
Suleiman, Farid. "A Note on al-Bayḍāwī's Year of Birth and Death." *JQS* (forthcoming).
Van Ess, Josef. "Biobibliographische Notizen zur islamischen Theologie." *WO* 9, no. 2 (1978): 255–83.
Az-Zuḥaylī, Muḥammad. *Al-Qāḍī al-Bayḍāwī*. Damascus: Dār al-Qalam, 1988.

Abū l-Barakāt an-Nasafī

Angelika Brodersen

1 Introduction

Abū l-Barakāt an-Nasafī (died 1310/710) is primarily known as an important legal scholar and an author of legal works following the Ḥanafī *madhhab* (legal school). In addition to these works and his Qur'ān commentary *Madārik at-tanzīl wa-ḥaqā'iq at-ta'wīl* (*The Reaches of the Revelation and the Realities of Interpretation*),[1] he also authored theological treatises in the Ḥanafī-Māturīdī tradition. There is hardly any reliable information about his life, although the extant biographical sources praise him as being highly influential. Probably born in Īdāj / Khūsistān at an unknown date,[2] according to the sources he studied with Shams al-A'imma Muḥammad b. ʿAbd as-Sattār al-Kardarī (died 1244/642),[3] Ḥamīd ad-Dīn aḍ-Ḍarīr (died 1267/666/) and Badr ad-Dīn Khuwāharzāde al-Kardarī (died 1253/651). Later he taught at the Quṭbiyya as-Sulṭāniyya in Kirmān.[4] Among Abū l-Barakāt's students, Muẓaffar ad-Dīn b. as-Saʿātī (died 1294/694) and Ḥusām ad-Dīn as-Sighnāqī (died 1314/714) became particularly famous. In addition, Ṭāshköprüzāde presents a certain Muḥammad b. Muḥammad al-Jīlī, who may have been an-Nasafī's student and commented on his legal writings.[5] Abū l-Barakāt an-Nasafī traveled to Bagdad in 1310/710 and died in Rabīʿ I in that same year on his way back to Īdāj,[6] where he is buried.[7]

1 ʿAbd Allāh b. Aḥmad Abū l-Barakāt an-Nasafī, *Madārik at-tanzīl wa-ḥaqā'iq at-ta'wīl*, ed. Yūsuf ʿAlī Badīwī (Damascus/Beirut: Dār Ibn Kathīr, 2013).

2 Khayr ad-Dīn az-Ziriklī, *Al-Aʿlām* (Beirut: Dār al-ʿIlm li-l-Malāyīn, 1984), 4: 67.

3 Given the long period between al-Kardarī's death and Abū l-Barakāt's death, skepticism is more appropriate here. See, however, Sāʿid Bakdāsh in his introduction to an-Nasafī's *Kanz ad-daqā'iq* (Beirut: Dār al-Bashā'ir al-Islāmiyya, 2011), 12–13, who based on this data a lifetime of an-Nasafī from 80 or even 90 years old, and estimates his year of birth to be around 1223/620. It is more likely that Abū l-Barakāt studied with Badr ad-Dīn al-Kardarī (who died 1253/651), Shams al-A'imma Ḥumayd ad-Dīn's nephew, who was later identified with his famous uncle.

4 See Claude Gilliot, "L'exégèse du Coran en Asie Centrale et au Khorasan." *SI* 89 (1999): 129–64, at 144.

5 Aḥmad b. Muṣṭafā Ṭāshköprüzāde, *Miftāḥ as-saʿāda wa-miṣbāḥ as-siyāda fī mawḍūʿāt al-ʿulūm* (Beirut: Dār al-Kutub al-ʿIlmiyya, 1984), 2: 57.

6 Muḥammad ʿAbd al-Ḥayy al-Laknawī, *al-Fawā'id al-bahiyya fī tarājim al-Ḥanafiyya*, ed. Aḥmad az-Zaʿbī (Beirut: Dār al-Arqam Ibn Abī Arqam, 1998), 172, no. 218; and Ismāʿīl al-Baghdādī, *Hadiyyat al-ʿārifīn asmā' al-mu'allifīn wa-āthār al-muṣannifīn*, ed. Mahmut Kemal İnal and Avni Aktuç (Beirut: Dār Iḥyā' at-Turāth al-ʿArabī, 1951), 2: 1168. However, there is also other information. Aḥmad b. ʿAlī b. Ḥajar al-Asqalānī, *ad-Durar al-kāmina fī aʿyān al-mi'a ath-thāmina* (Beirut: Dār al-Jīl, 1997), 178, no. 642 calls the year 701, while Qāsim b. Quṭlūbughā, *Tāj at-tarājim fī ṭabaqāt al-Ḥanafiyya*, ed. Qāsim Muḥammad ar-Rajab (Bagdad: Maṭbaʿat al-ʿĀnī, 1962), 30, no. 86, refers to "after 710."

7 Ibn Abī l-Wafā' here refers to the biographer Tāqī d-Dīn at-Tamīmī (died 1596/1005), see al-Qurashī, ʿAbd al-Qādir b. Muḥammad b. Abī l-Wafā', *al-Jawāhir al-muḍī'a fī ṭabaqāt al-Ḥanafiyya*, ed. Muḥam-

https://doi.org/10.1515/9783111320083-006

2 The Author in his Environment

ʿAbd Allāh b. Aḥmad b. Maḥmūd Ḥāfiẓ ad-Dīn Abū l-Barakāt an-Nasafī was a descendant of the famous Makhūlī-Nasafī family, which had originated in the Transoxanian Nasaf (Persian: Nakhshab) in Sogdiana, today's Qarshī (Mongolian / Uigurian for "palace") in southern Uzbekistan. Other important representatives of this family were Abū Muṭīʿ Makḥūl b. Faḍl an-Nasafī (died 930/318), Abū l-Muʿīn an-Nasafī (died 1114/508), and Najm ad-Dīn Abū Ḥafs ʿUmar an-Nasafī (died 1142/537), author of the famous *ʿAqāʾid an-Nasafī*. Together with the Bazdawīs and ʿĪyāḍīs, this family played an important role in the dissemination of Ḥanafī theological ideas in the systematization of Abū Manṣūr al-Māturīdī (died 944/333).[8] Abū l-Barakāt an-Nasafī adhered to this direction of *kalām* (speculative theology), which forms the second pillar of Sunni confession alongside the doctrines of Abū l-Ḥasan al-Ashʿarī (died 935/324) and his followers.

The theological current we call "Māturīdiyya" consolidated itself from the eleventh century in Central Asia, particularly in the area of today's Uzbekistan, with Samarqand and Bukhārā as centers of scholarship. Supported by political circles, it spread westward, and eastward within Sunni Islam. In particular, the spread of the Turkish peoples into the geographical center of the Islamic world led not only to the strengthening of Ḥanafism in that region but also to the predominance of the Eastern traditions, especially the Transoxanian and thus the Māturīdī tradition, within Ḥanafism as a whole.[9] In Transoxania, the Sunni Qarākhānids had ruled from the end of the tenth century AD; one century later, the cities of Samarqand and Bukhārā had become thriving centers for various kinds of culture, including different theological currents that successfully coexisted.[10]

Consequently, a large body of theological writings by Māturīdī scholars was produced in the eleventh and twelfth centuries.[11] Māturīdī theology was predominant in Transoxania until the thirteenth century; later, the works were primarily passed on in Ḥanafī circles.[12]

As Madelung has pointed out, a huge migration of eastern Ḥanafīs to the western regions took place starting in the Seljuk period. This migration continued until the era

---

mad ʿAbd Allāh ash-Sharīf (Beirut: Dār al-Kutub al-ʿIlmiyya, 2005), 178, no. 642. The indication "al-Maqrīzī" in the edition of the *Jawāhir* must be based on an incorrect reading.

8 Philipp Bruckmayr, "The Spread and Persistence of Māturīdī Kalām and Underlying Dynamics." *Iran and the Caucasus* 13 (2009): 59–92, at 61.

9 For details, Wilferd Madelung, "The Spread of Māturīdism and the Turks," in *Actas do IV Congresso de Estudos Árabes e Islâmicos (Coimbra-Lisboa 1968)* (no editor indicated) (Leiden: Brill, 1971), 109–168.; Bruckmayr, "The Spread," 60–61, 64.

10 For details see Angelika Brodersen, "New Light on the Emergence of Māturīdism: Abū Shakūr al-Sālimī (5th/11th century) and his Kitāb al-Tamhīd fī bayān al-tauḥīd." *JIS* 31 (2020): 329–357.

11 An overview, for example, in Angelika Brodersen, *Between Māturīdīya and Ašʿarīya. Abū Šakūr as-Sālimī and his Tamhīd fī bayān at-tauḥīd* (Piscataway: Gorgias Press, 2018), 2, note 5.

12 Bruckmayr, "The Spread," 60–61, 64.

of Mongol rule[13] and markedly diminished the numbers of Sunni scholars in the eastern regions.[14] In other words, the political situation changed dramatically in the decades just before Abū l-Barakāt an-Nasafī lived and worked. The Mongol conquest also affected the Transoxanian region: in 1220/616, Bukhārā and Samarqand were conquered and largely destroyed. Bukhārā initially recovered relatively quickly under the successors of Genghis Khan. Thus, it was able to re-establish itself as a center for scholarship until the Mongols destroyed it again in 1273/671 and 1316/715.[15] Samarqand, in contrast, remained destroyed for about 150 years.[16]

The effects of these developments on religious life and theological disputes remain ambiguous. On the one hand, the end of the Abbasid caliphate due to the Mongol conquest of Bagdad in 1258/656, which likely occurred during the lifetime of Abū l-Barakāt an-Nasafī, meant the destruction of the center of the Islamic world and a serious challenge to the self-image of the Islamic community.[17] On the other hand, the Mongol rulers following this conquest were tolerant regarding religious matters. Genghis Khan himself had ordered tax exemption for religious leaders,[18] before the Ilkhān Ghazān converted to Islam in 1295/694 and re-established Islam as a state-supported religion, taking the name of Maḥmūd. Maḥmūd claimed adherence to the Ḥanafī school of law, and his conversion inspired numerous imitators in the Mongolian upper class.[19] Ḥanafism was thus once again the official direction in legal practice in Transoxania, ensuring its continuity as a theological tradition.

Abū l-Barakāt an-Nasafī's affiliation with this theological tradition is obvious, particularly in his *kalām* writings, in which he does not propagate original thoughts and doctrines. Rather, his theological works are very closely based on the writings of his Māturīdī family predecessors. In particular, the Creed of Najm ad-Dīn Abū Ḥafṣ ʿUmar an-Nasafī (died 1142/537),[20] on which Abū l-Barakāt commented under the title *al-Muṣaffā sharḥ al-manẓūma an-nasafiyya*,[21] and the *Tabṣirat al-adilla* of Abū l-Muʿīn

13 Madelung, "The Spread," 141.

14 Ahmed El-Shamsy, "The Social Structure of Orthodoxy," in *The Cambridge Companion to Classical Islamic Theology*, ed. Tim Winter (Cambridge: Cambridge University Press, 2008), 97–117, at 114.

15 See W. Barthold and R.N. Frye, "Bukhārā", in *EI*², ed. P. Bearman et al., accessed February 13, 2020, http://dx.doi.org/10.1163/1573-3912_islam_SIM_1508.

16 See H.H. Schaeder, C.E. Bosworth, and Yolande Crowe, "Samarḳand", in *EI*², eds. P. Bearman et al., accessed February 13, 2020, http://dx.doi.org/10.1163/1573-3912_islam_COM_0996.

17 See Monika Gronke, "Die mongolische Epoche," in *Der islamische Orient. Grundzüge seiner Geschichte*, ed. Albrecht Noth and Jürgen Paul (Würzburg: Ergon, 1998): 225–329, at 267.

18 Gronke. "Die mongolische Epoche," 266.

19 Ibid, 276–77.

20 First edited by William Cureton: Najm ad-Dīn Abū Ḥafs ʿUmar an-Nasafī, "*Al-ʿAqāʾid*," in *Pillar of the Creed of the Sunnites*, ed. William Cureton (London: Printed for the Society for the Publication of Oriental Texts, 1843), 1–5, since then countless new editions, translations and online versions.

21 See Ibn Abī l-Wafāʾ, *al-Jawāhir*, 178; al-Laknawī, *al-Fawāʾid*, 172; Ibn Quṭlūbughā, *Tāj at-tarājim*, 30.

an-Nasafī (died 1114/508)[22] obviously influenced his own *kalām* treatises to a considerable extent. In this respect, Abū l-Barakāt an-Nasafī's theological writings, namely, the *'Umdat al-'aqā'id*[23] and its elaboration in the *I'timād sharḥ 'umdat al-'aqā'id*,[24] in addition to the commentary just mentioned, can be seen as evidence of his close adherence to the Ḥanafī-Māturīdī tradition.

If Abū l-Barakāt an-Nasafī is still known and recognized today, this appreciation is based primarily on his importance as an author of and commentator on legal writings. The sources cite the following legal works:

- *al-Mustaṣfā sharḥ al-fiqh an-nāfi'*;[25]
- *al-Mustawfā sharḥ al-fiqh an-nāfi'*;[26]
- *al-Wāfī*;[27]
- *al-Kāfī sharḥ al-wāfī*;[28]
- *Kanz ad-daqā'iq*;[29]
- *Manār al-anwār*;[30]
- *Kashf al-asrār fī sharḥ al-manār*.[31]

However, contrary to what some sources claim, Abū l-Barakāt probably did not write a *Sharḥ* on al-Marghīnānī's (died 1197/593) *Hidāya*, a highly influential work on Ḥanafī *fiqh*.[32]

---

22 Edited by Claude Salamé (Damascus: Institut Français de Damas 1990–1993); and edited by Hüseyin Atay (Ankara: Diyanet Işleri Başkanlığı Yayınları 1993).

23 Alternative titles: *'Umdat al- 'aqīda, al-'Umda fī uṣūl ad-dīn*, see Ibn Abī l-Wafā', *al-Jawāhir*, 178; Ibn Quṭlūbughā, *Tāj at-tarājim*, 30; Edited by William Cureton, in: *Pillar of the Creed of the Sunnites* (London: Printed for the Society for the Publication of Oriental Texts, 1843), 1–29.

24 See al-Laknawī, *al-Fawā'id*, 172; Edited under the title *Kitāb al-I'timād fī al-i'tiqād al-musammā sharḥ al-'umda fī 'aqīda ahl as-sunna wa-l-jamā'a*, ed. A. M. Ismail (Frankfurt/Main: PhD diss., 2003). For details of this treatise see Angelika Brodersen, *Der unbekannte kalām. Theologische Positionen der frühen Māturīdīya am Beispiel der Attributenlehre* (Berlin: LIT 2014), 46–47.

25 Commentary on the legal work of Nāṣir ad-Dīn as-Samarqandī (died 1258/656), in Ibn Abī l-Wafā', *al-Jawāhir*, 178.

26 A longer commentary on the work just mentioned, in Ibn Abī l-Wafā', *al-Jawāhir*, 178; al-Laknawī, *al-Fawā'id*, 172; Ibn Quṭlūbughā, *Tāj at-tarājim*, 30, calls it *al-Manāfi' sharḥ an-nāfi'*.

27 On Ḥanafī law, in Ibn Abī l-Wafā', *al-Jawāhir*, 178; al-Laknawī, *al-Fawā'id*, 172; Ibn Quṭlūbughā, *Tāj at-tarājim*, 30.

28 Commentary on *al-Wāfī*, in Ibn Abī l-Wafā', *al-Jawāhir*, 178; al-Laknawī, *al-Fawā'id*, 172; Ibn Quṭlūbughā, *Tāj at-tarājim*, 30.

29 Abridged version of *al-Wāfī*, in Ibn Abī l-Wafā', *al-Jawāhir*, 178; al-Laknawī, *al-Fawā'id*, 172; Ibn Quṭlūbughā, *Tāj at-tarājim*, 30.

30 In al-Laknawī, *al-Fawā'id*, 172; Ibn Abī l-Wafā', *al-Jawāhir*, 178; and Ibn Quṭlūbughā, *Tāj at-tarājim*, 30 cited as *Manār fī uṣūl al-fiqh* and *Manār fī uṣūl ad-dīn*.

31 Detailed commentary on *al-Manār*, in al-Laknawī, *al-Fawā'id*, 172. Printed in 2 volumes (Beirut: Dār al-Kutub al-'Ilmiyya, 2015), no editor indicated.

32 See Sā'id Bakdāsh, Introduction to al-Nasafī's *Kanz ad-daqā'iq*, 43.

3 An-Nasafī's *Tafsīr*: The *Madārik at-tanzīl wa-ḥaqā'iq at-ta'wīl*

3.1 The Text

Why did Abū l-Barakāt an-Nasafī, famous as a jurist and a *mutakallim*, compose a commentary on the Qur'ān? He himself answers this question in the foreword to that work:

> Someone asked me for a medium-length writing about Qur'ān exegesis. This writing should include aspects of correct inflection and readings *(wujūh al-i'rāb wa-l-qirā'āt)*, as well as details on tropics and symbolism *(daqā'iq 'ilmay al-badī' wa-l-ishārāt)*. Furthermore, it should include Sunni teachings *(aqāwīl ahl as-sunna wa-l-jamā'a)* and should be free from the false statements of the innovators and the mistaken *(abāṭīl ahl al-bida' wa-ḍ-ḍalāla)*. And it should neither be tiring in length *(bi-ṭ-ṭawīl al-mumill)*, nor shamefully short *(bi-l-qaṣīr al-mukhill)*.[33]

Even if this request for a *tafsīr* is less an historically accuracy and more a standard introductory formula, its main focus is obvious: the author set out to defend Sunni teachings using linguistic and dogmatic arguments.

This preface also alerts us to the dogmatic direction of Sunnism Abū l-Barakāt had in mind. In the opening he writes:

> God is in His essence *(bi-dhātihi)* above the signs of false ideas *(ishārat al-awhām)*. His attributes are pure from being completely grasped by the intellect and the understanding *(idrāk al-'uqūl wa-l-afhām)*. His divinity exists before every created thing *(mawjūd)*, and He remains permanently after every established thing *(sarmadiyya)*. He is the King *(al-malik)*, the lights of whose glory *(subuḥāt jalālihi)* blind the eyes, and the Exalted *(al-mutakabbir)*, whose mighty sublimity drives away the thoughts *(azāhat al-afkār)*. He is the Pre-eternal *(al-qadīm)*, not similar to the things created in time *(al-ḥudthān)*, and the Sublime *(al-aẓīm)*, free from contact with a place. He is above similarity with bodies and comparability with humankind *(mushābahat al-anām)*. He is the Almighty *(al-qādir)*, who cannot be referred to by qualification *(bi-t-takyīf)*, and the Superior *(al-qāhir)*, who cannot be asked about load and burden *(at-taḥmīl wa-t-taklīf)*. He is the All-knowing *(al-'alīm)* who created man and taught him the revelation *(al-bayān)*, and the All-wise *(al-ḥakīm)* who sent the Qur'ān down as a healing for spirits and bodies *(shifā'an li-l-arwāḥ wa-l-abdān)*.[34]

In this passage, Abū l-Barakāt clearly distinguishes himself not only from the Mu'tazila, who denied the existence of God's attributes, and from anthropomorphizers, who attributed a kind of physicality to God. He also differentiates himself from the competing Sunni school known as Ash'ariyya, through his emphasis on the eternity of all divine attributes. Thus, he proves himself a staunch representative of the Ḥanafī-Māturīdī tradition within the Sunni *kalām*.

---

33 Abū l-Barakāt an-Nasafī, *Madārik at-tanzīl* 1: 24.

34 Abū l-Barakāt an-Nasafī, *Madārik at-tanzīl* 1: 23.

3.2 An-Nasafī's Sources and Hermeneutical Methods

3.2.1 Qur'ān Commentaries

A particular trait of this direction of *kalām* is the recognition of the human intellect as a means of knowing God. It is therefore surprising, although only at first glance, that Abū l-Barakāt an-Nasafī uses a Muʿtazilī Qur'ān commentary as a major source for his *tafsīr*, namely Abū l-Qāsim Maḥmūd b. ʿUmar az-Zamakhsharī's (died 1144/538) *al-Kashshāf ʿan ḥaqā'iq at-tanzīl* (*The Discoverer of Revealed Truths*). Throughout large passages of the text, Abū l-Barakāt's *Madārik at-tanzīl* appears directly as a summary of the *Kashshāf*. As a result, his *tafsīr* can also be compared with the *Anwār at-tanzīl wa-asrār at-ta'wīl* (*The Lights of Revelation and the Secrets of Interpretation*) of ʿAbd Allāh b. ʿUmar al-Bayḍāwī (died 685/1290), which in turn is a summary of the *Kashshāf* and presumably was available to Abū l-Barakāt.[35] From these two sources, he adopted linguistic and stylistic details, but omitted the specific Muʿtazilī positions from his dogma.[36]

As an example, the opening of his interpretation of Q 2:259 reads:

> *Or as the one who passed* (*aw ka-lladhī marra*), which means: 'or consider it as one who . . .'; then it is an elision and it means: 'don't you see?' Because both are words that express astonishment. Or it is a matter of agreement of meaning, not of expression. Then the completion would be: 'don't you see that it is like someone who argued with (*ḥājja*) Ibrāhīm?' Or: 'like someone who passed (*marra*).' The author of the *Kashshāf* said: the *kāf* is redundant (*zā'ida*).[37]

Here, an-Nasafī relies on az-Zamakhsharī for linguistic details. In other places, however, he clearly distances himself from the doctrines of the Muʿtazila. For example, in the exegesis of Q 6:25 he writes: "and in their ears deafness *(wa-fī ādhānihim waqran)*, i.e. heaviness that prevents hearing. And he calls it 'deafness' in the singular because it is a verbal noun *(maṣdar)*. This is our proof concerning (the doctrine of) the optimum *(al-aṣlaḥ)* against the Muʿtazila."[38]– In fact, the Muʿtazila taught that God is committed to always doing the best for people. In contrast, the Māturīdīs emphasized that God is not subject to any obligation. An-Nasafī even names prominent representatives of the Muʿtazila, for example Abū ʿAlī l-Jubbā'ī (died 303/915) in his interpretation of Q 10:67.

The editor of his *Tafsīr* names the following Qur'ān commentaries as further sources for an-Nasafī:[39]

---

35 The interpretation of sura 53, *an-Najm* (The Star) can be cited as an example of the often literal correspondence between these two *tafsīrs*.

36 See Claude Gilliot, "Exegesis of the Qur'ān: Classical and Medieval,." In *EQ*, 2: 99–124, at 112–13.

37 Abū l-Barakāt an-Nasafī, *Madārik at-tanzīl* 1: 258–259.

38 Abū l-Barakāt an-Nasafī, *Madārik at-tanzīl* 1: 497.

39 For an-Nasafī's sources as a whole, see Badīwī, introduction to *Madārik at-tanzīl*, 13–14.

- *Tafsīr Qutāda* (b. Diʿāma, died 737/118), from which he took over many topics;
- *Tafsīr Mujāhid* (b. Jabr, died 722/104), from which he took over some explanations;
- Abū Manṣūr al-Māturīdī (died 944/333): *Taʾwīlāt ahl as-sunna*, as well as its *sharḥ* by ʿAlāʾ ad-Dīn Muḥammad b. Aḥmad as-Samarqandī (died 539/1144).
- *al-Lubāb*, without further specification. This is probably a mistake on the part of the editor, which can be explained by the printing of an-Nasafī's *Tafsīr* on the margin of ʿAlī b. Muḥammad al-Khāzin's (died 1341/741) *Lubāb at-taʾwīl fī maʿānī t-tanzīl.*[40]

3.2.2 Tradition

An-Nasafī invokes the Islamic tradition by repeatedly citing hadith. He points to the *Ṣaḥīḥ al-Bukhārī, Ṣaḥīḥ Muslim,* Abū Muḥammad al-Baghawī's *Maṣābīḥ as-sunna* and some *Masānīd.* In this respect, an-Nasafī takes the approach of a traditionalist *tafsīr.* In contrast to a so-called *tafsīr bi-l-maʾthūr,* however, he does not name a chain of narrators *(isnād)* to underpin his traditional statements about the oldest generations of Islam.

In the context of the traditions that an-Nasafī obviously used, the problematic *Isrāʾīliyyāt,* i.e. narratives from Jewish or Christian tradition, should also be mentioned. An-Nasafī also cites these stories, although, as is often emphasized, he does so rather sparingly.[41] In general he does not comment on these stories, as in the exegesis of Q 27:16–35 concerning Solomon's ability to understand the language of birds. An express rejection of these texts, however, can be found, in an-Nasafī's commentary on Q 38:34, which focuses on the temptation of Solomon: "The stories about the seal (of Solomon), the demon and idolatry belong to the lies of the Jews *(fa-min abāṭīl al-yahūd).*"[42]

3.2.3 Fiqh

An-Nasafī also deals with legal issues when the text of the Qurʾān imposes them, and occasionally mentions legal writings as his sources, namely ʿAlī b. Muḥammad al-Bazdawī's (died 1089/482) *al-Mabṣūṭ fī furūʿ al-fiqh*[43] and his own *al-Kāfī sharḥ al-wāfī* and *Sharḥ al-manār.* In the context of his interpretation of the legal regulations in Q 2, for example, he first mentions the regulations in Abū Ḥanīfa b. Thābit and Abū Yūsuf

40 Carl Brockelmann, *GAL*, 2: 267.
41 Muḥammad Ḥusayn adh-Dhahabī, DhTM (Cairo: Dār al-Ḥadīth, 1961), 3: 263.
42 Gilliot, "L'exégèse," 145, and Dhahabī attribute this explicit rejection of Jewish traditions to the fact that the prophets' sinlessness is at stake here.
43 Possibly meant: al-Bazdawi's famous *Kanz al-wuṣūl ilā maʿrifat al-uṣūl.*

b. Ḥabīb to compare them with Muḥammad b. Idrīs ash-Shāfiʿī. In the course of his exegesis of Q 2:221, an-Nasafī cites Abū Isḥāq Ibrāhīm b. Muḥammad b. as-Sarī z-Zajjāj (died 922/310).[44] This kind of quotations prove that he was familiar with works from classical *fiqh*. In principle, an-Nasafī discusses various legal opinions in order to ultimately prove the correctness of the Ḥanafī dogmas.

3.2.4 Philology

An-Nasafī takes a philological approach as well. Thus, as another starting point for his Qurʾān exegesis, he relies on linguistic sources and on text examples from the philological exegesis of the Qurʾān, as first developed by al-Khalīl b. Aḥmad al-Farāhīdī (died 791/174) and Sībawayh (Abū Bashr ʿAmr b. ʿUthmān b. Qanbar al-Baṣrī, died 793/176), which primarily revolved around the Qurʾānic lexemes. For example, in the exegesis of Q 2:217, an-Nasafī names the famous grammarian from Kufa, al-Farrāʾ (Abū Zakarīyāʾ Yaḥyā b. Ziyād b. ʿAbd Allāh b. Manṣūr ad-Daylamī, died 822/207), and contrasts his theory with the opinion of the grammarians from Basra. The question here is whether a single preposition *(jārr)* can refer to several nouns *(majrūr)* or whether it must be repeated for each noun. Another example is an-Nasafī's exegesis of Q 4:81, where he mentions the phenomenon of assimilation *(idghām)*.

The editor of the *Tafsīr* was able to identify the following writings as sources for an-Nasafī:[45]

- *Kitāb Sībawayh*;
- Abū l-Baqāʾ al-ʿAkbarī (died 1219/616): *at-Tibyān fī Iʿrāb al-Qurʾān*,
- Abū Naṣr al-Jawharī (died ca 1008/303): *Tāj al-lugha wa-ṣiḥāḥ al-ʿArabiyya*,
- Abū l-Ḥasan ʿAlī b. al-Ḥusayn al-Bāqūlī ḍ-Ḍarīr (died 1148/543): *Kashf al-muʿḍilāt wa-īḍāḥ al-mushkilāt*.

Another characteristic of an-Nasafī's *Tafsīr*, which can also be classified in the field of philology, is his frequent examination of the variant readings of the Qurʾānic text. He restricts himself to the seven canonical readings, which he sometimes summarizes as the "reading from Kufa" *(kūfī)*. For example, in his explanation of Q 5:89 he writes: "*but He will impose blame upon you for what you intended (ʿaqqadtum) of oaths*, which means: for your firm commitment by an oath *(fī taʿqīdikum al-aymān)*. This is the cer-

44 For him, see Wadād al-Qāḍī, "Al-Zajjāj and Glassmaking. An expanded Range of Options in a comparative Context," in *In the Shadow of Arabic. The Centrality of Language to Arabic Culture: Studies Presented to Ramzi Baalbaki on the Occasion of His Sixtieth Birthday*, ed. Bilal Orfali (Leiden/Boston: Brill, 2011): 221–48.

45 Badīwī in Abū l-Barakāt an-Nasafī, *Madārik at-tanzīl* 1: 14.

tification *(tawthīq)*. The *kūfī* reading, with the exception of Ḥafṣ, is (to read the word ʿaqqadtum) without doubling the consonant *(bi-t-takhfīf)*."[46]

An-Nasafī also deals with the occasions for revelation *(asbāb an-nuzūl)*, for example in the context of the discussion on of the "embezzlement" *(yaghullu)* by a prophet reported in Q 3:161.

3.2.5 Theology

A dogmatic approach leads us to ask how the theological content of Abū l-Barakāt's *Tafsīr* relates to his explicit theological writings. Because like other *Mutakallimūn*, he likes to support his arguments with quotes from the Qurʾān in his *kalām* works. In fact, very often the context in which an-Nasafī uses a verse in his *Iʿtimād fī l-iʿtiqād* is reflected in his Qurʾān commentary, or vice versa. A typical example is found in an-Nasafī's interpretation of Q 7:143, where Moses' encounter with God and his request that God show himself are discussed. An-Nasafī argues that the wording of the verse in no way excludes the possibility of a vision of God, the so-called *visio beatifica*. He gives several arguments: a) Moses would never ask for the impossible. b) God does not answer: I am not visible. c) God places visibility on a condition that can be fulfilled. d) God does not blame Moses for his request, in contrast to his response to Noah in Q 11:46. These arguments are clearly taken from the *kalām*. In his *Iʿtimād*, an-Nasafī uses nearly the same arguments as the other Māturīdīs to prove that there is a possibility that believers in paradise will see God, although in a slightly different order: a); d); b); c).[47] Other passages indicate that an-Nasafī's kalām works use Qurʾānic verses that are only briefly dealt with in the *Tafsīr*. For example, he quotes Q 59:24 in the *Iʿtimād* in the context of the discussion of the concept of "creating" *(takwīn)* as a divine attribute, a concept that is central to Māturīdī theology.[48] The interpretation of the verse in his *Tafsīr*, on the other hand, is extremely short, and reflects no interest in theological issues. Seen as a whole, however, it is obvious that an-Nasafī's theological writings and his *Tafsīr* are closely related.

---

46 Abū l-Barakāt an-Nasafī, *Madārik at-tanzīl* 1: 471–472. In the Sunni tradition, the following readings are referred to as *kūfī*: Ḥamza b. Ḥabīb (died 158/775), Abū l-Ḥasan al-Kisāʾī (died 189/805), ʿĀṣim b. Abī n-Najūd (died 127/745), and Ḥafṣ b. Sulaymān (died 179/796). See Claude Gilliot, "Creation of a fixed text," in *The Cambridge Companion to the Qurʾān*, ed. Jane Dammen McAuliffe (Cambridge: Cambridge University Press, 2006): 41–57, at 50–52.

47 Here an-Nasafī refers to his *Tafsīr*. See An-Nasafī, ʿAbd Allāh b. Aḥmad Abū l-Barakāt. *Kitāb al-Iʿtimād fī l-iʿtiqād al-musammā Sharḥ al-ʿumda fī ʿaqīdat ahl as-sunna wa-l-jamāʿa*, ed. A. M. Ismail (Frankfurt/Main: PhD diss., 2003), 89.

48 Abū l-Barakāt an-Nasafī, *Kitāb al-Iʿtimād*, 77.

3.3 Text History and Reception

An-Nasafī's *Madārik at-tanzīl wa-ḥaqā'iq at-ta'wīl* has been preserved in several manuscripts. Brockelmann[49] identifies a total of six copies in London, Istanbul and Cairo. In contrast, Badīwī's edition first printed in 1995 (sixth edition 2013), which is used as the reference edition here, is based on a single copy found in Aleppo and dated 1327/727. It can also be assumed that private collections hold further copies which therefore are not accessible to the public. A first print of this text – as already mentioned, in the margin of al-Khāzin's *Lubāb* – was made in 1271/1854 in Delhi; further prints were made in 1862/1279, 1882/1300, 1886/1304, 1888/1306, 1891/1309, 1900/1318 and 1910/1328 in Cairo and Bombay.[50] In addition, Brockelmann is aware of a print in the margin of Jalāl ad-Dīn as-Suyūṭī's (died 1505/911) *al-Iklīl fī istinbāṭ at-tanzīl,* and lists some commentaries: *at-Tafsīr al-muzīl li-mughlaqāt Madārik at-tanzīl,* by ʿAbd al-Aḥad b. Isḥāq al-Qandahārī (Lahore 1904); Ilāhdād Jaunpīrī (died 1577/923, India Office 1137); and ʿAbd al-Ḥaqq al-Hindī: *al-Iklīl ʿalā Madārik at-tanzīl* (India Office 1336).[51] Today the Badīwī edition, which was first published in 1995 and has since been reprinted several times and also translated into Urdu, is the standard. The number of extant manuscripts, glosses and prints suggests that an-Nasafī's *Tafsīr* actually received considerable attention. It is said to have been used in academic teaching at the Azhar in Cairo as well as at universities in Syria, Pakistan, Iraq and Lebanon,[52] and is even quoted in the first *tafsīr* to be published in Nigeria, the *Ḍiyā' at-ta'wīl fī ma'ānī t-tanzīl* of ʿAbd Allāh b. Fūdī (died 1829).[53]

4 Conclusion

Abū l-Barakāt an-Nasafī's *Tafsīr* has proved particularly valuable as evidence for trends in Islamic intellectual history at the time of the Mongol occupation in Central Asia. Its practice of adopting linguistic passages from az-Zamakhsharī's *Kashshāf* demonstrates that text sources were used without regard for dogmatic boundaries if they were useful in any non-dogmatic way. It also appears that the distinction between traditional *(at-tafsīr bi-l-ma'thūr)* and "scientific" Qur'ān commentaries *(at-tafsīr bi-l-ʿilm)* can no longer be maintained in discussions of the time of an-Nasafī. On the one hand, his commentary contains material from the prophetic tradition, from non-Islamic sources, and from other so-called Qur'ānic sciences such as the reasons for the revela-

49 Brockelmann, *GAS,* 2: 252, no. 197.

50 Ibid, vol 2, 252; *GALS,* 2: 267; Bakdāsh in his introduction to an-Nasafī's *Kanz ad-daqā'iq,* 30.

51 Brockelmann, *GALS,* 2: 267–68.

52 Bakdāsh in his introduction to an-Nasafī's *Kanz ad-daqā'iq,* 30.

53 Andrea Brigaglie, "Tafsīr and the Intellectual History of Islam in West Africa," in *Tafsīr and Islamic Intellectual History: Exploring the Boundaries of a Genre,* ed. Andreas Görke and Johanna Pink (Oxford: Oxford University Press, 2014), 379–415, at 399.

tion and the variant readings. On the other hand, equal weight is given to considerations on the Qur'ānic language, drawing from linguistic and grammatical works. It is obvious from this that an-Nasafī was less concerned with following a certain direction of the interpretation of the Qur'ān. Rather, he combined various hermeneutic approaches to achieve his goal, which, as he stated in his introduction, was to satisfy the Sunnis, using even non-Sunni sources for this. From these findings, it can be concluded that *tafsīr* and *kalām* in Abū l-Barakāt are fundamentally closely related to one another in that theological discourses are integrated into many aspects of his Qur'ān commentary. In this respect, his *Tafsīr* is fruitful not only as a source of Qur'ān exegesis in the Māturīdī tradition, but also as a testimony to the development of Islamic intellectual history in the late Middle Ages as a whole.

5 Bibliography

5.1 Sources

Al-Asqalānī, Aḥmad b. ʿAlī b. Hajar. *Ad-Durar al-kāmina fī aʿyān al-miʾa ath-thāmina*. Ed. ʿAbd al-Wārith Muḥammad ʿAlī. Beirut: Dār al-Jīl, 1997.

Al-Baghdādī, Ismāʿīl. *Hadiyyat al-ʿārifīn asmāʾ al-muʾallifīn wa-āthār al-muṣannifīn*. Ed. Mahmut Kemal İnal and Avni Aktuç. 2 vols., Beirut: Dār Iḥyāʾ at-Turāth al-ʿArabī, 1951.

Al-Bayḍāwī, ʿAbd Allāh b. ʿUmar. *Anwār at-tanzīl wa-asrār at-taʾwīl*. Cairo: Dār al-Kutub al-Gharbiyya al-Kubrā, 1911.

Al-Bazdawī, ʿAlī b. Muḥammad. *Kanz al-wuṣūl ilā maʿrifat al-uṣul*. Ed. Sāʿid Bakdāsh. Medina: Dār as-Sirāj; Beirut: Dār al-Bashāʾir al-Islāmiyya, 2016.

Al-Khāzin, ʿAlāʾ ad-Dīn Alī b. Muḥammad. *Lubāb at-taʾwīl fī maʿānī t-tanzīl*. Ed. ʿAbd al-Ghanī ad-Daqr, Beirut: Al-Yamāma, 2003.

Al-Khāzin, Ibn Muʿādh b. Ibrāhīm al-Baghdādī. *Tafsīr al-Khāzin al-musammā Lubāb at-taʾwīl fī maʿānī t-tanzīl wa ʿalā hāmishihi Tafsīr an-Nasafī al-musammā bi-marārik at-tanzīl wa-ḥaqāʾiq at-taʾwīl li-l.imām Abī l-Barakāt ʿAbd Allāh b. Aḥmad b. Maḥmūd an-Nasafī*. 4 vols. 1899; Reprint Dār al-Kutub al-ʿArabiyya al-Kubrā, Miṣr n. d.

Al-Laknawī, Muḥammad ʿAbd al-Ḥayy. *Al-Fawāʾid al-bahiyya fī tarājim al-Ḥanafiyya*. Ed. Aḥmad az-Zaʿbī. Beirut: Dār al-Arqam Ibn Abī Arqam, 1998.

An-Nasafī, ʿAbd Allāh b. Aḥmad Abū l-Barakāt. "*ʿUmdat al-ʿaqāʾid*." In *Pillar of the Creed of the Sunnites*, edited by William Cureton. London: Printed for the Society for the Publication of Oriental Texts, 1843, 1–29.

An-Nasafī, ʿAbd Allāh b. Aḥmad Abū l-Barakāt. *Kitāb al-Iʿtimād fī l-iʿtiqād al-musammā Sharḥ al-ʿumda fī ʿaqīdat ahl as-sunna wa-l-jamāʿa*. Ed. A. M. Ismail. Frankfurt/Main: PhD diss., 2003.

An-Nasafī, ʿAbd Allāh b. Aḥmad Abū l-Barakāt. *Kanz ad-daqāʾiq*. Ed. Sāʿid Bakdāsh. Beirut: Dār al-Bashāʾir al-Islāmiyya, 2011.

An-Nasafī, ʿAbd Allāh b. Aḥmad Abū l-Barakāt. *Madārik at-tanzīl wa-ḥaqāʾiq at-taʾwīl*. Ed. Yūsuf ʿAlī Badīwī. 3 vols. Damascus/Beirut: Dār Ibn Kathīr, 2013[6].

An-Nasafī, ʿAbd Allāh b. Aḥmad Abū l-Barakāt. *Kashf al-asrār fī sharḥ al-manār*. 2 vols. Beirut: Dār al-Kutub al-ʿIlmiyya, 2015.

An-Nasafī, Abū l-Muʿīn. *Tabṣirat al-adilla*. Ed. Claude Salamé. Damascus: Institut Français de Damas, 1990–1993.

An-Nasafī, Abū l-Muʿīn. *Tabṣirat al-adilla*. Ed. Hüseyin Atay. Ankara: Diyanet Işleri Başkanlığı Yayınları, 1993.

An-Nasafī, Najm ad-Dīn Abū Ḥafs ʿUmar. *"Al-ʿAqāʾid."* In *Pillar of the Creed of the Sunnites*, edited by William Cureton. London: Printed for the Society for the Publication of Oriental Texts, 1843, 1–5.

Al-Qurashī, ʿAbd al-Qādir b. Muḥammad b. Abī l-Wafāʾ. *Al-Jawāhir al-muḍīʾa fī ṭabaqāt al-Ḥanafiyya*. Ed. Muḥammad ʿAbd Allāh ash-Sharīf. Beirut: Dār al-Kutub al-ʿIlmiyya, 2005.

Ibn Quṭlūbughā, Qāsim. *Tāj at-tarājim fī ṭabaqāt al-Ḥanafiyya*. Ed. Qāsim Muḥammad ar-Rajab. Bagdad: Maṭbaʿat al-ʿĀnī, 1962.

At-Tamīmī, Tāqī d-Dīn. *Aṭ-Ṭabaqāt as-saniyya fī tarājim al-Ḥanafiyya*. Ed. ʿAbd al-Fattaḥ Muḥammad al-Halw. Cairo: Dār ar-Rifāʿī, 1970.

Ṭāshköprüzāde, Aḥmad b. Muṣṭafā. *Miftāḥ as-saʿāda wa-miṣbāḥ as-siyāda fī mawḍūʿāt al-ʿulūm*. 2 vols., Beirut: Dār al-Kutub al-ʿIlmiyya, 1984.

Az-Zamakhsharī, Abū l-Qāsim Maḥmūd b. ʿUmar. *Al-Kashshāf ʿan ḥaqāʾiq at-tanzīl wa-ʿuyūn al-aqāwīl fī wujūh at-taʾwīl*. Beirut: Dār Iḥyāʾ at-Turāth al-ʿArabī li-ṭ-Ṭibāʿa wa-n-Nashr wa-t-Tawzīʿ, 2016.

5.2 Secondary Literature

Barthold, W. and R.N. Frye. "Bukhārā." In *EI*², edited by: P. Bearman et al., Brill, 1954–2005. Article published online 2012. Accessed February 13, 2020. http://dx.doi.org/10.1163/1573-3912_islam_SIM_1508.

Brigaglie, Andrea. "Tafsīr and the Intellectual History of Islam in West Africa." In *Tafsīr and Islamic Intellectual History. Exploring the Boundaries of a Genre*, edited by Andreas Görke and Johanna Pink. Oxford: Oxford University Press, 2014: 379–415.

Brockelmann, Carl. *Geschichte der arabischen Litteratur. Supplementbände (GALS)*. 3 vols. Leiden: Brill, vol. 2, 1937.

Brockelmann, Carl. *Geschichte der arabischen Litteratur (GAL)*. 2 vols. Leiden: Brill, vol. 2, 1949².

Brodersen, Angelika. *Der unbekannte kalām. Theologische Positionen der frühen Māturīdīya am Beispiel der Attributenlehre*. Berlin: LIT, 2014.

Brodersen, Angelika. *Between Māturīdīya and Ašʿarīya. Abū Šakūr as-Sālimī and his Tamhīd fī bayān at-tauḥīd*. Piscataway: Gorgias Press, 2018.

Brodersen, Angelika. "New Light on the Emergence of Māturīdism: Abū Shakūr al-Sālimī (5th/11th century) and his Kitāb al-Tamhīd fī bayān al-tauḥīd." *JIS* 31 (2020): 329–357.

Bruckmayr, Philipp. "The Spread and Persistence of Māturīdī Kalām and Underlying Dynamics." *Iran and the Caucasus* 13 (2009): 59–92.

Adh-Dhahabī, Muḥammad Ḥusayn. DhTM. 3 vols., Cairo: Dār al-Ḥadīth, 1961.

El-Shamsy, Ahmed. "The Social Structure of Orthodoxy." In *The Cambridge Companion to Classical Islamic Theology*, edited by Tim Winter. Cambridge: Cambridge University Press, 2008: 97–117.

Gilliot, Claude. „L'exégèse du Coran en Asie Centrale et au Khorasan." *SI* 89 (1999): 129–64.

Gilliot, Claude. "Exegesis of the Qurʾān: Classical and Medieval." In *EQ*, edited by Jane Dammen McAuliffe, 6 vols. Leiden/Boston: Brill, vol. 2, 2002: 99–124.

Gilliot, Claude. "Creation of a fixed text." In *The Cambridge Companion to the Qurʾān*, edited by Jane Dammen McAuliffe. Cambridge: Cambridge University Press, 2006: 41–57.

Gronke, Monika. "Die mongolische Epoche." In *Der islamische Orient. Grundzüge seiner Geschichte*, edited by Albrecht Noth and Jürgen Paul. Würzburg: Ergon, 1998: 225–329.

Madelung, Wilferd. "The Spread of Māturīdism and the Turks." In *Actas do IV Congresso de Estudos Árabes e Islâmicos (Coimbra-Lisboa 1968)*. Leiden: Brill, 1971: 109–68.

Al-Qāḍī, Wadād. "Al-Zajjāj and Glassmaking. An expanded Range of Options in a comparative Context." In *In the Shadow of Arabic. The Centrality of Language to Arabic Culture. Studies Presented to Ramzi Baalbaki on the Occasion of His Sixtieth Birthday*, edited by Bilal Orfali. Leiden/Boston: Brill, 2011: 221–48.

Schaeder, H.H., C.E. Bosworth, and Yolande Crowe. "Samarḳand." In *EI*², edited by P. Bearman et al., Brill, 1954–2005. Article published online 2012. Accessed February 13, 2020. http://dx.doi.org/10.1163/1573-3912_islam_COM_0996.

Le Strange, Guy. *The Lands of the Eastern Caliphate: Mesopotamia, Persia, and Central Asia*. New York: Barnes & Noble, 1905.

Az-Ziriklī, Khayr ad-Dīn. *Al-Aʿlam*. 8 vols. Beirut: Dār al-ʿIlm li-l-Malāyīn, 1984.

Ibn Taymiyya

Jon Hoover

1 Introduction

Taqī d-Dīn Aḥmad b. ʿAbd al-Ḥalīm b. Taymiyya was a prominent Ḥanbalī activist, theologian, and jurist who spent most of his life in Damascus. His influential treatise on the hermeneutics of the Qurʾān *Muqaddima fī uṣūl at-tafsīr* (Introduction to the Principles of Interpretation) will be analyzed in Section 3 below. Section 4 takes up the question of how Ibn Taymiyya's interpretive practice compares with his tradition-based hermeneutical theory, and Section 5 discusses the reception of the *Muqaddima*. The following section – Section 2 – introduces the broader context of Ibn Taymiyya's life and thought.

2 Life, Works, and Reform Vision

Ibn Taymiyya was born into a scholarly family in Ḥarrān in 1263/661. At the age of six, he and his family fled Mongol invaders and settled in Damascus, the second most important city in the Mamluk Empire after Cairo. Ibn Taymiyya rose quickly to prominence as a religious scholar, and he campaigned for jihad against three Mongol invasions of Syria from 1299/699 to 1303/702. After the Mongol threat passed, Ibn Taymiyya devoted himself to fighting currents in religious doctrines and practices that he believed weakened Muslim society. Opposition to Ibn Taymiyya's activism led to his 1306/705 imprisonment in Cairo for 18 months. Altercations with Sufis resulted in another twenty-four months in prison – sixteen in Cairo and eight in Alexandria. This was followed by three years of teaching and writing in Cairo before he returned to Damascus in 1313/712. In 1318/718 Ibn Taymiyya provoked antagonism by invalidating the divorce oaths that the Mamluk rulers used to guarantee their social commitments, and he was jailed for more than five months. Controversy over his views on grave visitation led to his incarceration in the citadel of Damascus from 1326/726 to 1328/728. The authorities stripped Ibn Taymiyya of his writing materials and books in the spring of 1328/728, and he died in prison five months later on 26 September 1328/20 Dhū l-Qaʿda 728.[1]

Ibn Taymiyya wrote several large tomes and hundreds of fatwas and treatises on the religious sciences. The scholar traveler Abū ʿAbdallāh Muḥammad b. ʿAbdallāh, known as Ibn Baṭṭūṭa (d. 1377/779), reports that Ibn Taymiyya wrote a Qurʾān com-

1 For a fuller account of Ibn Taymiyya's life and writings, see Jon Hoover, *Ibn Taymiyya* (London: Oneworld, 2019), 5–39.

https://doi.org/10.1515/9783111320083-007

mentary of around forty volumes called *al-Baḥr al-muḥīṭ* during his first imprisonment in Cairo.[2] There is however no other evidence for this,[3] and modern works presenting themselves as Ibn Taymiyya's *Tafsīr* are in fact compilations of his diverse fatwas and treatises.[4] A large portion of Ibn Taymiyya's corpus is devoted to theology, and his magnum opus *Darʾ taʿāruḍ al-ʿaql wa-n-naql* argues, against medieval Islamic philosophy and *kalām* theology, that there is no contradiction between reason properly understood and the plain sense of revelation. Ibn Taymiyya grounds this claim in his conviction that the Qurʾān contains not only indicative reports but also rational arguments, and he allows engaging the terminology and argumentation of his opponents by analogy to the permission to translate the Qurʾān and the hadith into other languages.[5]

At the core of Ibn Taymiyya's vision for reform is worship and love of God, understood ethically as obedience to God's law (*sharīʿa*). Worshipping and loving the one true God is first and foremost a matter of adhering to the theological doctrines, ritual practices, moral acts, and dispositions of the heart that God has prescribed. The rulings of God's law are found most definitively in the foundational texts of the religion – the Qurʾān, the sunna of the Prophet, and the dicta of the early Muslims (*salaf*) – not in the consensus of the Sunni legal schools, and certainly not in the theological and ritual innovation (*bidʿa*) of later generations.[6] The textualist impulse in Ibn Taymiyya's reformist thought is readily apparent in his *Muqaddima*.

---

2 Abū ʿAbdallāh Muḥammad b. Baṭṭūṭa, *Riḥlat Ibn Baṭṭūṭa*, ed. Muḥammad ʿAbd al-Munʿim al-ʿIryān (Beirut: Dār Iḥyāʾ al-ʿUlūm, 1987), 1: 111. Ibn Baṭṭūṭa's report on Ibn Taymiyya is translated in Donald P. Little, "Did Ibn Taymiyya Have a Screw Loose?" *SI* 41 (1975): 93–111 (95–96).

3 For discussion of whether Ibn Taymiyya wrote a full Qurʾān commentary, see Walid A. Saleh, "Ibn Taymiyya and the Rise of Radical Hermeneutics: An Analysis of *An Introduction to the Foundations of Qurʾānic Exegesis*," in *Ibn Taymiyya and His Times*, ed. Yossef Rapoport and Shahab Ahmed (Karachi: Oxford University Press, 2010), 123–162 (155, no. 5).

4 See for example Ibn Taymiyya, *Tafsīr shaykh al-Islām Ibn Taymiyya*, ed. Iyād b. ʿAbd al-Laṭīf b. Ibrāhīm al-Qaysī, 7 vols. (Ad-Dammām: Dār Ibn al-Jawzī, 2011–2012).

5 Ibn Taymiyya, *Darʾ taʿāruḍ al-ʿaql wa-n-naql*, ed. Muḥammad Rashād Sālim, 11 vols. (Riyadh, Jāmiʿat al-Imām Muḥammad b. Suʿūd al-Islāmiyya, 1991). On *Darʾ* see Carl Sharif El-Tobgui, *Ibn Taymiyya on Reason and Revelation: A Study of* Darʾ taʿāruḍ al-ʿaql wa-l-naql (Leiden: Brill, 2020); and Jon Hoover with Marwan Abu Ghazaleh Mahajneh, "Theology as Translation: Ibn Taymiyya's Fatwa Permitting Theology and Its Reception into His Averting the Conflict between Reason and Revealed Tradition (*Darʾ Taʿāruḍ Al-ʿAql Wa l-Naql*)," *MW* 108, no. 1 (2018): 40–86. M. Sait Özervarli very aptly describes Ibn Taymiyya's theology as "Qurʾānic rational theology" in: "The Qurʾānic Rational Theology of Ibn Taymiyya and His Criticism of the *Mutakallimūn*," in *Ibn Taymiyya and His Times*, ed. Yossef Rapoport and Shahab Ahmed (Karachi: Oxford University Press, 2010), 78–100.

6 For elaboration see Hoover, *Ibn Taymiyya*, 41–88.

3 Ibn Taymiyya's *Muqaddima fī uṣūl at-tafsīr*

Ibn Taymiyya appears to be the first in the Islamic tradition to write a systematic standalone treatise on the hermeneutics of the Qurʾān.[7] However, the treatise is not without precedent. It stands in a tradition of common themes found in introductions to Qurʾān commentaries. These include emphasis on the importance of knowing the Qurʾān's meaning, advocacy for interpretation through prophetic hadith, argument against using personal opinion (*raʾy*) in interpretation, criticism of other interpreters, and explanation of interpretive methodology.[8] Ibn Taymiyya's text is of modest length, at eighty-three pages in the widely-cited edition of ʿAdnān Zarzūr, and comprises an introduction and six sections (*fuṣūl*). It lacked a formal title in the manuscript tradition, and Zarzūr credits the Ḥanbalī mufti of Damascus Muḥammad Jamīl ash-Shaṭṭī (d. 1959/1379) for giving the work its modern title *Muqaddima fī uṣūl at-tafsīr* when printing the treatise for the first time in 1936.[9]

The date of the *Muqaddima* is uncertain. Ibn Taymiyya reports that he wrote the treatise from memory (*min imlāʾ al-fuʾād*),[10] which leads Walid Saleh to suggest that

7 Saleh, "Ibn Taymiyya," 125. Saleh provides a full description and analysis of Ibn Taymiyya's text, as does Farid Suleiman, "Ibn Taymīyas Theorie der Koranexegese," in *Koran Exegese als "Mix and Match": Zur Diversität aktueller Diskurse in der tafsīr-Wissenschaft*, ed. Abbas Poya (Bielefeld: Transcript, 2017), 15–43. Jane Dammen McAuliffe, "Quranic Hermeneutics: The Views of al-Ṭabarī and Ibn Kathīr," in *Approaches to the History of the Interpretation of the Qurʾān*, ed. Andrew Rippin (Oxford: Clarendon, 1988; reprint Piscataway, NJ: Gorgias Press, 2013), 46–62 (56–58), analyses the last two sections of Ibn Taymiyya's treatise as they appear in Ibn Kathīr's *Tafsīr*.

8 Karen Bauer, "Justifying the Genre: A Study of Introductions to Classical Works of *Tafsīr*," in *Aims, Methods and Contexts of Qur'anic Exegesis (2nd/8th—9th/15th c.)*, ed. Karen Bauer (Oxford: Oxford University Press, 2013), 39–65.

9 Ibn Taymiyya, *Muqaddima fī usūl at-tafsīr*, ed. ʿAdnān Zarzūr, 2nd printing (Kuwait: Dār al-Qurʾān al-Karīm, 1972, 1st printing 1971), 24; Ibn Taymiyya, *Muqaddima fī uṣūl at-tafsīr*, ed. Jamīl ash-Shaṭṭī (Damascus: Maṭbaʿat at-Taraqqī, 1936). The treatise is also printed without title in Ibn Taymiyya, *Majmūʿ fatāwā shaykh al-Islām Aḥmad b. Taymiyya* (hereafter *MF*), ed. ʿAbd ar-Raḥmān b. Muḥammad b. Qāsim and Muḥammad b. ʿAbd ar-Raḥmān b. Muḥammad (Riyadh: Maṭābiʿ ar-Riyāḍ, 1961–1967), 13: 329–75. Page references to *Muqaddima* will be first to Zarzūr's edition and then to *MF*. On ash-Shaṭṭī, see also Ahmed El Shamsy, *Rediscovering the Islamic Classics: How Editors and Print Culture Transformed an Intellectual Tradition* (Princeton, NJ: Princeton University Press, 2020), 188. The *Muqaddima* has been translated into English, German, and French: Ibn Taymiyyah, *An Introduction to the Principles of Tafseer*, trans. Muhammad Abdul Haq Ansari (Birmingham: Al-Hidaayah, 1993); Jane McAuliffe, "Ibn Taymiya: Treatise on the Principles of Tafsir," in *Windows on the House of Islam: Muslim Sources on Spirituality and Religious Life*, ed. John Renard (Berkeley: University of California Press, 1998), 35–43 (Sections Five and Six only); Ibn Taymiyya, *Einführung in die Methodologie der Qurʾānexegese*, trans. Elsayed Elshahed (Riyadh: Imam Muḥammad ibn Saud Universität, Abt. für Kultur und Publikation, 2000); Ibn Taymiyya, *Déontologie de l'exégèse coranique dans l'enseignement d'Ibn Taymiyya*, trans. Muhammad Diakho (Paris: Dar Albouraq, 2014).

10 Ibn Taymiyya, *Muqaddima*, 35; *MF*, 13: 331.

he composed the work "during his last stint in jail, without access to his notes and books."[11] Farid Suleiman provides evidence to support this conjecture. Ibn Taymiyya's foremost student Shams ad-Dīn Muḥammad b. Abī Bakr az-Zarʿī, known as Ibn Qayyim al-Jawziyya (d. 1350/751), writes, "At the end of his life, (Ibn Taymiyya) sent me a guide concerning (Qurʾān) interpretation (*qāʿida fī t-tafsīr*) in his own handwriting, and on the back of it were some poetic verses."[12] Ibn al-Qayyim then quotes the verses. Ibn Taymiyya's main biographer Shams ad-Dīn b. ʿAbd al-Hādī (d. 1344/744) also quotes the same verses, introducing them as follows: "Verses (of poetry) have been found in the handwriting of the sheikh, which he composed in the citadel." Suleiman concludes from this evidence that the *Muqaddima* dates to Ibn Taymiyya's final imprisonment in the citadel of Damascus.[13] Younus Mirza is not aware of the statements of Ibn al-Qayyim and Ibn ʿAbd al-Hādī, and he argues against Saleh for an earlier date. Mirza observes that the first four parts of the *Muqaddima* differ somewhat in style from the latter two parts. Additionally, the latter two parts cite full transmission chains for traditions, which, Mirza maintains, Ibn Taymiyya could not have recalled in prison without books. Mirza therefore concludes that Ibn Taymiyya probably wrote at least the last two parts of the *Muqaddima* before his last imprisonment. However, Ibn Taymiyya did in fact have access to books during his final imprisonment. As noted above, he was only dispossessed of his books, as well as his writing instruments, five months before he died. That aside, Mirza also draws attention to the fact that ash-Shaṭṭī, the editor of the 1936 edition of *Muqaddima*, reports that he found the work in a collection copied in 1312–1313/712.[14] Mirza thus dates the *Muqaddima* to earlier than that. If Ibn Taymiyya did write the treatise in prison, Mirza reasons, it would have been during his incarcerations in Egypt, that is, sometime between 1306/705 and 1310/709.[15] In that case perhaps the "rule concerning interpretation" mentioned by Ibn al-Qayyim is not the same treatise as the *Muqaddima*. Yet, no firm conclusions on an early date can be drawn without a fuller investigation of the manuscript tradition.

Ibn Taymiyya explains in the introduction to the *Muqaddima* that he wrote the work to respond to a request for "general rules" (*qawāʿid kulliyya*) to understand the Qurʾān. Guidance was needed because "the books written on *tafsīr* are full of the lean and the fat, and clear falsehood and plain truth."[16] Ibn Taymiyya also signals the fundamentally epistemological character of the treatise by explaining that certain knowledge (*ʿilm*) is based on "reliable transmission (*naql*) from someone protected from

11 Saleh, "Ibn Taymiyya," 124.

12 Shams ad-Dīn ʿAbdallāh Muḥammad b. Qayyim al-Jawziyya, *Madārij as-sālikīn*, ed. Nāṣir b. Sulaymān as-Saʿwī, et al. (Riyadh: Dār as-Sumayʿī, 2011), 2: 1330.

13 Suleiman, "Ibn Taymīyas Theorie," 19.

14 Ibn Taymiyya, *Muqaddima*, ed. ash-Shaṭṭī, 2.

15 Younus Y. Mirza, "Ibn Taymiyya as Exegete: Moses' Father-in-Law and the Messengers in *Sūrat Yā Sīn*," *JQS* 19, no. 1 (2017): 39–71 (42–43).

16 Ibn Taymiyya, *Muqaddima*, 33; *MF*, 13: 329.

error or a statement for which there is a known proof (*dalīl*)."[17] Ibn Taymiyya focuses in the *Muqaddima* on securing transmission from the early Muslim community (*salaf*) as the foundational source of knowledge for what the Qur'ān means. The treatise does not explore the methods that the *salaf* themselves used to interpret the Qur'ān, nor does it detail an approach to reading the Qur'ān independently of what the *salaf* said it meant. Instead, it advances what Walid Saleh calls a "radical hermeneutics" that pits the definitive authority of the *salaf* over against the philological mode of the mainstream Sunni *tafsīr* tradition.[18]

To appreciate the uniqueness of the *Muqaddima*, it is useful first to examine the ostensibly similar but essentially different approach of Muḥammad b. Jarīr aṭ-Ṭabarī (d. 923/310). In the introduction to his vast Qur'ān commentary, aṭ-Ṭabarī outlines three categories of knowledge concerning the Qur'ān's interpretation.[19] The first category is known only by transmission from the Prophet. For aṭ-Ṭabarī the Prophet's interpretative authority encompasses God's command and prohibition, the fixed punishments (*ḥudūd*), inheritance shares, and other legal rulings (*aḥkām*) found in the Qur'ān. While aṭ-Ṭabarī does not say so explicitly, this is the domain of the jurists (*fuqahā'*) who turn to the sunna of the Prophet recorded in hadith reports to ascertain the religious law (*fiqh*). Aṭ-Ṭabarī's second category includes interpretations known only to God, such as the times of future events and especially the Day of Resurrection. The third category is philological. It is the interpretation by those who know the language of the Qur'ān. For example, aṭ-Ṭabarī explains, it takes only knowledge of Arabic to understand that the verse, "Do not work corruption in the land" (Q 2:11) means avoiding whatever causes harm. The Prophet may also have issued specific rulings linked to this verse, but the general meaning is accessible to those who know the language. This third category is the domain of the philologist who produces exegesis of the Qur'ān by drawing on linguistics and Arabic poetry. Aṭ-Ṭabarī cautions that philology-based interpretations should fall within the bounds of interpretations found among the Companions (*ṣaḥāba*) of the Prophet, their Followers (*tābiʿūn*), and other early scholars of the Muslim community.[20] However, the Companions and Followers do not carry prophetic authority in aṭ-Ṭabarī's exegetical practice. Their diverse interpretations simply form the set of opinions within which the exegete should work.[21]

---

17 Ibn Taymiyya, *Muqaddima*, 33; *MF*, 13: 329.

18 Saleh, "Ibn Taymiyya," 126–27, 148, and passim.

19 Muḥammad b. Jarīr aṭ-Ṭabarī, *Tafsīr aṭ-Ṭabarī: Jāmiʿ al-bayān ʿan ta'wīl āy al-Qur'ān*, ed. ʿAbd Allāh b. ʿAbd al-Muḥsin at-Turkī (Cairo: Dār al-Hijr, 2001), 1: 68–89; English abridged translation: aṭ-Ṭabarī, *The Commentary on the Qur'ān*, trans. J. Cooper, vol. 1 (Oxford: Oxford University Press, 1987), 32–40. For a more detailed analysis of this passage, see McAuliffe, "Quranic Hermeneutics," 49–50.

20 Aṭ-Ṭabarī, *At-Tafsīr*, 1: 68–70, 87–89.

21 For the *salaf* in the mainstream Sunni *tafsīr* tradition and the approaches of Ibn Taymiyya and aṭ-Ṭabarī, see Mustafa Shah, "Introduction," in *Tafsīr: Interpreting the Qur'ān*, ed. Mustafa Shah (London: Routledge, 2013), 1: 1–157 (3–7, 17–20, 27–31).

In the *Muqaddima* Ibn Taymiyya commends aṭ-Ṭabarī for transmitting the exegetical traditions of the *salaf* and calls his commentary "one of the greatest and most extensive of the tradition-based commentaries (*at-tafāsīr al-ma'thūra*)."[22] Yet, in contrast to aṭ-Ṭabarī, Ibn Taymiyya revives a marginal and strongly traditionalist current in the *tafsīr* tradition that elevates the Qur'ān interpretations of the Companions and the Followers to the level of the Prophet's sunna.[23] Ibn Taymiyya begins the first section of the *Muqaddima*, "It is obligatory to know that the Prophet made the meanings (*ma'ānī*) of the Qur'ān evident to his Companions just as he made its words (*alfāẓ*) evident to them"[24] Likewise, the Companions transmitted the Prophet's interpretations to the Followers. Ibn Taymiyya claims that there was minimal conflict over the interpretation of the Qur'ān among the Companions because of their proximity to the Prophet. There was more disagreement among the Followers, but nothing like the divergence that occurred thereafter. Ibn Taymiyya furthermore draws an explicit comparison between transmission of the Qur'ān's interpretation and transmission of the Prophet's sunna: "The Followers received the *tafsīr* of the Companions just as they received knowledge of the sunna from them."[25] The Companions and Followers' interpretations of the Qur'ān are prophetic knowledge on the same order as their knowledge of the Prophet's normative statements and practice transmitted in *ḥadīth* reports. Whereas aṭ-Ṭabarī limits the Prophet's authoritative interpretation of the Qur'ān to matters of religious law, Ibn Taymiyya extends it to the Qur'ān as a whole.

Ibn Taymiyya assumes that the Prophet communicated one consistent and coherent interpretation of the Qur'ān. He thus devotes Section Two of the *Muqaddima* to arguing that disagreements among the Companions and Followers are few.[26] He writes, "Difference over interpretation (of the Qur'ān) among the *salaf* is small (. . .). Most of their actual difference is a matter of (indicating) the same thing in different ways (*ikhtilāf tanawwu'*), not outright opposition (*ikhtilāf taḍādd*)."[27] Ibn Taymiyya cites many examples to make his point. Two will suffice here. In the first he observes that some interpret the expression "Straight Path" (*aṣ-ṣirāṭ al-mustaqīm*, Q 1:6) to be the Qur'ān, while others interpret it to mean the religion of Islam. Ibn Taymiyya explains that the two interpretations agree because Islam is the same as following the Qur'ān. Yet others interpret the "Straight Path" as the way of servitude, obedience to God and his messenger, or the sunna of the Prophet and the Consensus of the Commu-

22 Ibn Taymiyya, *Muqaddima*, 90; *MF*, 13: 361 (without *ma'thūra*). Ibn Taymiyya also commends aṭ-Ṭabarī in *Muqaddima*, 79–80; *MF*, 13: 355.
23 Suleiman, "Ibn Taymīya's Theorie," 24–32; Walid A. Saleh, "Preliminary Remarks on the Historiography of Tafsīr in Arabic: A History of the Book Approach," *JQS* 12 (2010): 6–40 (especially 21–31).
24 Ibn Taymiyya, *Muqaddima*, 35–38; *MF*, 13: 331–33 (extent of first section; quotation on 35; 331).
25 Ibid., 337–38; *MF*, 13: 332–33.
26 Ibid., 38–55; *MF*, 13: 333–44.
27 Ibid., 38; *MF*, 13: 333.

nity. According to Ibn Taymiyya, these expressions all refer to the same thing.[28] In a second example, Ibn Taymiyya notes that Companions sometimes differ over the occasion of revelation for a Qurʾānic verse. One Companion says it was revealed to deal with one situation, but a second Companion says it was revealed to deal with another. Ibn Taymiyya resolves the contradiction by suggesting that the same verse was revealed twice.[29]

While Ibn Taymiyya maintains that the core message of the Qurʾān is clear and consistent, he does allow that irreconcilable differences in Qurʾān interpretation still exist, and he seeks to explain their origins in the following two sections of the *Muqaddima*. Ibn Taymiyya writes in the introduction to Section Three, "Certain knowledge is either authenticated transmission or verified inferential reasoning (*al-ʿilm imma naql muṣaddaq wa-imma istidlāl muḥaqqaq*)."[30] Implied in this statement is Ibn Taymiyya's claim mentioned above that revealed tradition and correct reasoning do not contradict,[31] and Kristin Sands links the *Muqaddima*'s mention of inferential reasoning to the rational argumentation often found in Ibn Taymiyya's interpretive practice.[32] In the *Muqaddima* however he shows no interest in the character or content of correct rational inference, nor in demonstrating that the Qurʾān interpretations of the *salaf* agree with reason. He only seeks to establish that differences of interpretation arise from unverifiable transmission (Section Three) or erroneous inferential reasoning (Section Four).[33]

In Section Three, Ibn Taymiyya implies that real differences in Qurʾān interpretation among the *salaf* concern only inconsequential matters transmitted from the People of the Book (*ahl al-kitāb*, i.e. Jews and Christians). He divides transmission into 1) that which is known to be either authentic (*ṣaḥīḥ*) or weak (*ḍaʿīf*) by virtue of proofs established by God and 2) that whose authenticity or weakness cannot be known for lack of proof. The first category involves matters of religious importance to Muslims, whereas the second is of no religious value and includes exegetical traditions concerning things like the size of Noah's Ark and the color of the dog belonging to the Sleepers of the Cave (Q 18:18, 22). According to Ibn Taymiyya, traditions of this kind were taken from the People of the Book, and they cannot be judged true or false because the Prophet said, "If the People of the Book speak to you, neither accept what they say as true nor call them liars lest they speak to you in truth and you call them

28 Ibid., 41–43; *MF*, 13: 336–37.

29 Ibid., 48–49; *MF*, 13: 340.

30 Ibid., 55; *MF*, 13: 344.

31 Suleiman, "Ibn Taymīyas Theorie," 20–24.

32 Kristin Zahra Sands, *Ṣūfī Commentaries on the Qurʾān in Classical Islām* (London: Routledge, 2006), 144.

33 Ibn Taymiyya, *Muqaddima*, 55–78; *MF*, 13: 344–54 (Section Three), and 79–93; *MF*, 13: 355–63 (Section Four).

liars, or they speak to you falsely and you accept what they say as true."[34] Later in the *Muqaddima*, in Section Five, Ibn Taymiyya clarifies that the Prophet permits transmission of such "Israelite traditions" (*al-aḥādīth al-isrāʾīliyya*) – commonly known as *isrāʾīliyyāt* – and that they may be cited as corroborating evidence. However, they may not be used to establish points of doctrine, either because Islamic revelation has already shown them to be true or false, or because they speak to matters of no religious import.[35]

Ibn Taymiyya also dedicates part of Section Three to establishing that Qurʾān interpretations ascribed solely to the Followers constitute reliable prophetic knowledge. These exegetical traditions lack a chain of transmission going back to the Prophet through a Companion, and they usually make no claim even to report the interpretations of the Prophet himself. Nonetheless, Ibn Taymiyya equates such traditions to *mursal* reports from the Followers about the Prophet's battles (*maghāzī*). A *mursal* report assumes that an unnamed Companion served as an intermediary transmitter from the Prophet to a Follower. Ibn Taymiyya then maintains that a *mursal* tradition reported from multiple Followers without collusion is authentic and true, whether it deals with the Prophet's battles or the interpretation of the Qurʾān. Moreover, Ibn Taymiyya places exegetical traditions on a par with *mursal ḥadīth* reports of the Prophet's sunna, which both he and most mainstream Sunni jurists accept as probative in matters of religious law.[36]

At the end of Section Three, Ibn Taymiyya takes aim at alleged purveyors of erroneous exegetical traditions. He accuses Abū Isḥāq ath-Thaʿlabī (d. 1035/427), ʿAlī b. Aḥmad al-Wāḥidī (d. 1076/468), and Abū l-Qāsim Maḥmūd b. ʿUmar az-Zamakhsharī (d. 1144/538) of transmitting many forged hadith reports on the virtues of the Qurʾānic suras, and he criticizes especially ath-Thaʿlabī for transmitting hadith without any discrimination whatsoever: "He was a man gathering wood at night, transmitting whatever he found in the books of commentary (*tafsīr*), whether authentic, weak, or forged."[37] Ibn Taymiyya adds that ath-Thaʿlabī's student al-Wāḥidī may have been better than his teacher at Arabic but was even further from soundness in following the *salaf*. Ath-Thaʿlabī had synthesized a vast array of Islamic culture and tradition into an encyclopedic commentary that gained wide influence in the following centuries.[38] Ibn Taymiyya sought to proscribe the polyvalence of this commentary tradition and

34 Ibid., 55–58; *MF*, 13: 344–46 (quotation, 57; *MF*, 13: 345).

35 Ibid., *Muqaddima*, 98–101; *MF*, 13: 366–67.

36 Ibn Taymiyya, *Muqaddima*, 58–65; *MF*, 13: 346–50. On juristic views of *mursal ḥadīth*, see Mohammad Hashim Kamali, *Principles of Islamic Jurisprudence* (Cambridge, UK: The Islamic Texts Society, 2003[3]), 108–10.

37 Ibn Taymiyya, *Muqaddima*, 75–78 (quotation on 76); *MF*, 13: 354.

38 Walid A. Saleh, *The Formation of the Classical Tafsīr Tradition: The Qurʾān Commentary of al-Thaʿlabī (d. 427/1035)* (Leiden: Brill, 2004).

narrow it down to one single coherent and purified vision of what the *salaf* said the Qur'ān meant.

The fourth section of the *Muqaddima* examines differences in Qur'ān interpretation arising from inferential reasoning (*istidlāl*). Among the causes of such differences are exegetes who impose erroneous doctrines onto the words of the Qur'ān. Ibn Taymiyya cites Muʿtazilīs like ʿAbd al-Jabbār b. Aḥmad (d. 1025/415) and az-Zamakhsharī as the main culprits here. He also criticizes Shiites who interpret the Qur'ān esoterically, as when they read, "What are they asking about? About the great news" (Q 78:1–2), to refer to the first Shiite Imam ʿAlī b. Abī Ṭālib. Ibn Taymiyya claims that differences caused by inferential reasoning emerged only after the first three generations following the Prophet, and he commends fifteen Qur'ān commentators for transmitting the interpretations of the early Muslims and safely avoiding those later difficulties. Among them are ʿAbd ar-Razzāq aṣ-Ṣanʿānī (d. 827/211), aṭ-Ṭabarī, and Ibn Abī Ḥātim ar-Rāzī (d. 938/327).[39] Ibn Taymiyya warns that deviating from the path of the *salaf* only ends in error and innovation: "Whoever turns away from the doctrine of the Companions, the Followers, and their (Qur'ān) interpretation (*tafsīr*) to whatever differs from that is one who errs in that. Indeed, he is an innovator (*mubtadiʿ*), even if he is an independent thinker (*mujtahid*) forgiven for his error."[40] With this Ibn Taymiyya leaves no room for interpretation that breaks with the *salaf*.

The final two sections of the *Muqaddima* discuss methods for interpreting the Qur'ān, or, more specifically, acquiring knowledge of what the Qur'ān means. Ibn Taymiyya begins, "The best method in this is to interpret the Qur'ān by the Qur'ān because what is ambiguous in one place has been interpreted in another, and what is condensed in one place has been elaborated in another. If that does not work for you, then turn to the sunna because it explicates the Qur'ān and makes it clear."[41] Commentators prior to Ibn Taymiyya rarely mentioned interpretation of the Qur'ān by the Qur'ān,[42] and Ibn Taymiyya says no more about it here than this. He does not analyze the hermeneutical process of judging which parts of the Qur'ān interpret others. He also does not reflect on how exegesis of the Qur'ān by the Qur'ān can lead to diverse interpretations, including interpretations falling outside the bounds set by the sunna of the Prophet and the dicta of the *salaf*.[43] Saleh speculates that Ibn Taymiyya is not at this point opening the door to philology or modern forms of textual analysis. Rather,

---

39 Ibn Taymiyya, *Muqaddima*, 79–93; *MF*, 13: 355–63.

40 Ibid., *Muqaddima*, 91; *MF*, 13: 361.

41 Ibid., *Muqaddima*, 93; *MF*, 13: 363.

42 Sohaib Saeed Bhutta, "Intraquranic Hermeneutics: Theories and Methods in Tafsir of the Qur'an through the Qur'an" (PhD diss., School of Oriental and African Studies, University of London, 2017), 22–51, surveys hermeneutical discussions of interpretation of the Qur'ān by the Qur'ān in the *tafsīr* tradition and states that Ibn Taymiyya's *Muqaddima* was "the earliest source I found to address the topic of intraquranic exegesis directly" (24, see also 47–50).

43 See Musāʿid aṭ-Ṭayyār, *Sharḥ Muqaddima fī uṣūl at-tafsīr li-Ibn Taymiyya* (Ad-Dammām: Dār Ibn al-Jawzī, 2007²), 271–78, for a discussion of issues raised by interpretation of the Qur'ān by the Qur'ān.

he places the Qurʾān first, followed by the sunna, in order to replicate the first two sources in the well-established hierarchy that Sunni jurists set out for ascertaining the law: the Qurʾān, the sunna, the Consensus (*ijmāʿ*) of the Muslim community, and analogy (*qiyās*).[44]

Ibn Taymiyya continues in the *Muqaddima*, "If we do not find the interpretation in the Qurʾān or the sunna, we resort for that to the sayings of the Companions because they are the most knowledgeable about that."[45] Then, after noting the virtues of the Companions ʿAbdallāh b. Masʿūd (d. ca. 652–654/32–33) and ʿAbdallāh b. ʿAbbās (d. ca. 687–688/68) as Qurʾān commentators, he writes, "If you do not find the interpretation in the Qurʾān or the sunna, and you do not find it among the Companions, many of the leading scholars have turned for that to the sayings of the Followers."[46] Given Ibn Taymiyya's view that the Prophet transmitted the interpretation of the Qurʾān to the Companions and the Companions to the Followers, the sayings of the Companions and the Followers are tantamount to an extension of the Prophet's sunna, the second source in the juristic hierarchy. After this, Ibn Taymiyya briefly considers how to proceed if the views of the Followers are inconclusive: one may turn "to the language of the Qurʾān or the sunna, or the whole of the language of the Arabs or the sayings of the Companions."[47] With this Ibn Taymiyya permits philological investigation when no definitive prophetic interpretation is available.

Ibn Taymiyya ends the *Muqaddima* by censuring "personal opinion" (*raʾy*): "As for interpretation of the Qurʾān by mere personal opinion, it is forbidden (*ḥarām*)."[48] The only legitimate interpretation of the Qurʾān is that of the *salaf*, and the *salaf* themselves, Ibn Taymiyya explains, did not engage in interpretation of matters about which they had no knowledge.[49] All Qurʾān interpretation touching on core religious matters goes back to the Prophet.

4 Ibn Taymiyya's Hermeneutics in Practice

Ibn Taymiyya's *Muqaddima* outlines a sequence of steps for interpreting the Qurʾān. First, interpret the Qurʾān by the Qurʾān itself. If that does not work, turn to the sunna for an interpretation, then the Companions, and then the Followers. If all else fails, resort to philology. Musāʿid aṭ-Ṭayyār comments that Ibn Taymiyya's *Muqaddima*, despite all appearances, cannot be taken literally to prescribe a particular sequence of

44 Saleh, "Ibn Taymiyya," 145.
45 Ibn Taymiyya, *Muqaddima*, 95; *MF*, 13: 364.
46 Ibid., 95; *MF*, 13: 364.
47 Ibid., 105; *MF*, 13: 370.
48 Ibid., 105; *MF*, 13: 370.
49 Ibid., 105–115; *MF*, 13: 370–75.

interpretive steps: "No interpretation (*tafsīr*) is sequenced in this way, not even the interpretation that Shaykh al-Islām (Ibn Taymiyya) wrote."[50] Not much work has been done to compare Ibn Taymiyya's hermeneutical theory with his practice,[51] and space does not permit extensive discussion here.[52] It will suffice to examine one short fatwa that Younus Mirza describes as putting Ibn Taymiyya's "qur'anic hermeneutic in crisis."[53]

Ibn Taymiyya's fatwa addresses the question whether Isaac or Ishmael was Abraham's intended sacrifice.[54] This appears to be the only place in his corpus where he discusses this issue.[55] The Qur'ānic account of the event (Q 37:101–111) does not name the son who was to be sacrificed, and the Qur'ān does not identify the intended sacrifice elsewhere either. The *salaf* and the classical exegetes were divided over the matter, with aṭ-Ṭabarī arguing that the intended sacrifice was Isaac.[56] Ibn Taymiyya is fully aware of this situation. Nevertheless, he asserts of the intended sacrifice, "It is definitive that it was Ishmael. That is what the Book, the sunna, and well-known proofs support, and it is what the Torah possessed by the People of the Book proves."[57]

Ibn Taymiyya cites his evidence from the Torah first. The Torah commands Abraham to sacrifice his only or firstborn son (Genesis 22:2). This only son was Abraham's firstborn Ishmael, according to Ibn Taymiyya, but the People of the Book corrupted the Torah account by adding in the name Isaac.[58] Mirza observes that Ibn Taymiyya facilely ignores his prescribed first step of interpreting the Qur'ān by the Qur'ān by starting with the Bible.[59] It could also be that Ibn Taymiyya mentions his evidence from the Torah first merely to corroborate the Qur'ān and the sunna before going into the substance of his interpretation.

The Torah aside, Ibn Taymiyya turns to philology. As Mirza points out, Ibn Taymiyya copies without acknowledgement several lines from the commentary of az-Zamakhsharī, whom he censures in the *Muqaddima*, on the Qur'ānic verse, "So, we gave him the glad tidings of a forbearing boy" (Q 37:101).[60] Az-Zamakhsharī's commen-

50 Aṭ-Ṭayyār, *Sharḥ*, 269.

51 Elliott A. Bazzano, "Ibn Taymiyya, Radical Polymath, Part 2: Intellectual Contributions," *Religion Compass* 9, no. 4 (2015): 117–39 (120–21), reflects briefly on the state of the research on this question.

52 For thorough discussion of Ibn Taymiyya's hermeneutic principles in theology, see Farid Suleiman, *Ibn Taymiyya und die Attribute Gottes* (Berlin: de Gruyter, 2019), 178–227; and El-Tobgui, *Ibn Taymiyya on Reason and Revelation*, 179–226.

53 Younus Y. Mirza, "Ishmael as Abraham's Sacrifice: Ibn Taymiyya and Ibn Kathīr on the Intended Victim," *Islam and Christian-Muslim Relations* 24, no. 3 (2013): 277–98 (284).

54 The untitled fatwa begins *'an ad-dabīḥ min walad Khalīl Allāh Ibrāhīm . . .*, *MF*, 4: 331–36.

55 Mirza, "Ishmael," 284.

56 See Mirza, "Ishmael," 283–84, for an account of aṭ-Ṭabarī's argumentation.

57 Ibn Taymiyya, *MF*, 4: 331.

58 Ibid., *MF*, 4: 331–32.

59 Mirza, "Ishmael," 285.

60 Ibn Taymiyya, *MF*, 4: 332, lines 6–10; Mirza, "Ishmael," 285–86.

tary explains that the boy was male, that he would reach maturity, and that he was forbearing. This sets the stage for Ibn Taymiyya to claim that Ishmael was the intended sacrifice on philological grounds. I will note only his first four arguments. First, Ibn Taymiyya explains that the announcement of the son to be sacrificed and the ensuing ordeal (Q 37:101–11) come just before the announcement of Isaac's birth (Q 37:112). The implication is that Ishmael must have been the intended sacrifice since Isaac was only born after the event. Second, the sacrifice story does not occur elsewhere in the Qur'ān, but Isaac's birth is announced in other verses, none of which mention an intended sacrifice. Third, the Qur'ān also calls the intended sacrifice a forbearing boy, which is a fitting description of Ishmael in the circumstances, whereas the Qur'ān calls Isaac a "knowledgeable boy" (Q 15:53). Fourth, both Abraham and Sarah were informed of the birth of Isaac (Q 11:71–72), whereas Abraham alone received the announcement of Ishmael, the intended sacrifice. Sarah was not involved. Ibn Taymiyya adds that this fits with authentic hadith reports of Sarah growing jealous when Hagar gave birth to Ishmael and of Abraham, Hagar, and Ishmael then travelling to Mecca where Abraham was commanded to sacrifice.[61]

Following Mirza, I have identified Ibn Taymiyya's interpretative method here as philology – Ibn Taymiyya's method of last resort – but it could just as well be called interpretation of the Qur'ān by the Qur'ān, as it builds its argument from within the narrative and semantic world of the Qur'ān. We might also even wonder whether Ibn Taymiyya deems the question of the intended sacrifice to be religiously inconsequential; the *salaf* would presumably have spoken with a unified voice had it been a core issue. Ibn Taymiyya unfortunately does not reflect on his methodology in the fatwa, but comparing this fatwa to the *Muqaddima* does highlight flexibility in his interpretive practice and ambiguity in his hermeneutical theory.

5 Reception of Ibn Taymiyya's *Muqaddima fī uṣūl at-tafsīr*

The significance of the *Muqaddima* lies to a large extent in its reception. Several works of pre-modern traditionalist Qur'ān scholarship quote parts of it. Ismāʿīl b. ʿUmar b. Kathīr (d. 1373/774) copied the last two sections of the *Muqaddima* into the introduction of his large Qur'ān commentary, the *Tafsīr*, without attribution.[62] Ibn Kathīr's exegetical practice in the *Tafsīr* follows Ibn Taymiyya's *Muqaddima* in ground-

61 Ibid., *MF*, 4: 332–35.

62 Ismāʿīl b. ʿUmar b. Kathīr, *Tafsīr al-Qur'ān al-ʿaẓīm*, ed. Sāmī b. Muḥammad as-Salāma (Riyadh: Dār aṭ-Ṭayba, 1997; reprint 1999), 1: 7–14 (main body of text; differs slightly from published versions of the *Muqaddima*); English translation: Ibn Kathīr, *The Exegesis of the Grand Holy Qur'an*, trans. Muhammad Mahdi ash-Sharif, 4 vols. (Beirut: Dar al-Kotob al-Ilmiyah, 2006), 10–17.

ing the meaning of the Qur'ān firmly in prophetic hadith and prophetic knowledge transmitted from the Companions and the Followers. Ibn Kathīr nonetheless retains elements of the mainstream philological method, and he shuns the rational argumentation that Ibn Taymiyya sometimes uses to defend his views in other works.[63]

Additionally, Badr ad-Dīn az-Zarkashī (d. 1392/794) and Jalāl ad-Dīn as-Suyūṭī (d. 1505/911) draw on the *Muqaddima* in their encyclopedias of Qur'ānic sciences. Az-Zarkashī quotes the key statements on interpreting the Qur'ān by the Qur'ān and then the sunna and the sayings of the Companions from the *Muqaddima*, Section Five. After that, he advises resort to rational methods and does not mention the Followers.[64] As-Suyūṭī quotes the same key statements on interpreting the Qur'ān by the Qur'ān, the sunna, and the Companions, and then adds an abridgement of the first four sections of the *Muqaddima*.[65] Muḥammad b. Muḥammad Murtaḍā z-Zabīdī (d. 1791/1205) quotes the *Muqaddima* material found in as-Suyūṭī with minor omissions into his *Itḥāf as-sāda*, a commentary on the *Iḥyā' 'ulūm ad-dīn* of Abū Ḥāmid al-Ghazālī (d. 1111/505).[66] It is likely that az-Zabīdī copied directly from as-Suyūṭī's work.

In the modern period the *Muqaddima*, published as a standalone text, has played a key role in inspiring tradition-based revivalism and reformism. For centuries, the philological Qur'ān commentaries of az-Zamakhsharī, Fakhr ad-Dīn ar-Rāzī (d. 1210/606), and Nāṣir ad-Dīn al-Bayḍāwī (d. 1319/719)[67] had been mainstays of traditionalist seminary training in much of the Islamic world. Then, the publication of aṭ-Ṭabarī and Ibn Kathīr in the early twentieth century and Ibn Taymiyya's *Muqaddima* in 1936 contributed to a shift in the dominant interpretative paradigm from philology to prophetic tradition. Aṭ-Ṭabarī, despite his philological commitments, was subsumed under the category "tradition-based interpretation" (*at-tafsīr bi-l-ma'thūr*), which was now the new gold standard in Qur'ānic hermeneutics.[68] Ibn Kathīr supplanted al-Bayḍāwī as the most popular Qur'ān commentary, and it retains this position in the early twenty-first century, with

---

63 Younus Y. Mirza, "Was Ibn Kathīr the 'Spokesperson' for Ibn Taymiyya? Jonah as a Prophet of Obedience," *JQS* 16, no. 1 (2014): 1–19; Saleh, "Ibn Taymiyya," 153.

64 Badr ad-Dīn az-Zarkashī, *Al-Burhān fī 'ulūm al-Qur'ān*, ed. Muḥammad Abū l-Faḍl Ibrāhīm (Cairo: Maktabat Dār at-Turāth, 1957; 1984^3), 2: 175–76. Az-Zarkashī quotes without attribution to Ibn Taymiyya.

65 Jalāl ad-Dīn as-Suyūṭī, *al-Itqān fī 'ulūm al-Qur'ān*, ed. Markaz ad-Dirāsāt al-Qur'āniyya (Medina: Mujamma' al-Malik Fahd, 2005), 6: 2274–75 (beginning of Section 5), 6: 2277–84 (Sections 1–4) (Chapter 78).

66 Muḥammad b. Muḥammad Murtaḍā z-Zabīdī, *Itḥāf as-sāda al-muttaqīn bi-sharḥ Iḥyā' 'ulūm ad-dīn* (Cairo: al-Maṭba'a al-Maymaniyya, 1893–1894; reprint, Beirut: Mu'assasat Tārīkh al-'Arabī, 1994), 4: 537–39. I owe this reference to El Shamsy, *Rediscovering the Islamic Classics*, 58; El Shamsy suggests that az-Zabīdī quotes almost the entire *Muqaddima*, but he in fact leaves out large blocks of text.

67 Al-Bayḍāwī's death date has been a matter of debate. For explanation of the date given here, see Walid Saleh, "al-Bayḍāwī," in *EI*^{3}, ed. Kate Fleet et al., accessed May 12, 2020, http://dx.doi.org/10.1163/1573-3912_ei3_COM_25267.

68 Saleh, "Preliminary Remarks," 10–16, 24–35.

translations into several major languages.[69] The primacy of Ibn Kathīr signals not only a transition in Qur'ānic hermeneutics but also the modern mobilization of the Qur'ān as a book of religious guidance and belief. In the premodern period such matters were largely the domain of jurisprudence (*fiqh*), theology (*kalām*), and Sufism, while the Qur'ān served mainly liturgical and aesthetic functions. Against this backdrop, modern reformers and revivalists have foregrounded the Qur'ān as the primary authoritative voice of Islamic guidance to unify the Muslim community and displace premodern scholasticisms, with Ibn Taymiyya's hermeneutics providing the theoretical foundation.[70]

Ibn Taymiyya's "interpretation of the Qur'ān by the Qur'ān" found at the beginning of Section Five in the *Muqaddima* has been subject to diverse understandings. It has inspired not only tradition-based interpretation but also a spectrum of modernist and feminist approaches, some of which eschew prophetic hadith and exegetical traditions entirely and derive broad principles such as gender egalitarianism from intratextual readings of the Qur'ān.[71] Modernists and reformists also cast the *isrāʾīliyyāt* traditions into question. Whereas Ibn Taymiyya had merely stripped the *isrāʾīliyyāt* of religious consequence, his views were now deployed in the twentieth century to reject the *isrāʾīliyyāt* corpus entirely in the interest of a more exclusive focus on Qur'ānic narrative.[72]

6 Conclusion

Ibn Taymiyya turned to the Qur'ān as a book of belief and guidance to circumvent the religious authority structures of his day and to ground his own vision of what Islam properly taught. Interpretation of the Qur'ān permeates his writings, and, even if he did not write a line-by-line commentary, his entire corpus is tantamount to a reflection on what the Qur'ān means. Ibn Taymiyya's influential treatise on hermeneutical theory, the *Muqaddima*, reflects his wider reformist concern to ground Islamic doctrine and practice in the Qur'ān, the sunna, and the *salaf* over against later religious developments that he rejects. Modern revivalists have deployed the treatise to champion tradition-based interpretation (*at-tafsīr bi-l-ma'thūr*) of the Qur'ān and expungement of Israelite traditions (*isrāʾīliyyāt*) from Qur'ān commentary. However, closer

69 Johanna Pink, *Muslim Qur'ānic Interpretation Today: Media, Genealogies and Interpretive Communities* (Sheffield: Equinox, 2019), 35, 49–61.

70 Pink, *Muslim Qur'anic Interpretation*, 14–34.

71 Saeed, "Intraquranic hermeneutics," 159–177; Aysha A. Hidayatullah, *Feminist Edges of the Qur'an* (Oxford, UK: Oxford University Press, 2014), 87–88, quotes Ibn Taymiyya's *Muqaddima* to help elucidate the hermeneutics of modernist Fazlur Rahman (d. 1988/1408) and several contemporary feminist interpreters.

72 Roberto Tottoli, *Biblical Prophets in the Qur'ān and Muslim Literature* (Richmond, UK: Curzon, 2002), 170–88.

attention to Ibn Taymiyya's interpretive practice shows that he makes more space for philology and the use of Biblical material than the modern revivalist appropriation of his *Muqaddima* might suggest.

7 Addendum

The above article was written before publication of the important new edition of the *Muqaddima* produced by Sāmī b. Muḥammad b. Jād Allāh (Riyadh: Dār al-Muḥdath, 2022). Ibn Jād Allāh argues in his introduction (pp. 5–63) that the final two sections of the *Muqaddima*—Sections 5 and 6—should not be ascribed to Ibn Taymiyya but to his student Ibn Kathīr. Ibn Jād Allāh notes differences in style and content between Sections 1–4 and Sections 5–6, including the use of full transmission chains for traditions in the latter two sections, and he argues that Sections 5–6 better reflect Ibn Kathīr's style than Ibn Taymiyya's. Furthermore, no manuscripts have yet been found that present Sections 1–6 together as one treatise and date to earlier than ash-Shaṭṭī's 1936 printed edition. As Ibn Jād Allāh explains, ash-Shaṭṭī himself notes that he based his edition on an ancient manuscript of the *Muqaddima* that lacked part of Section 4, and so he filled in the gap from another source. Ibn Jād Allāh was unable to find either of ash-Shaṭṭī's sources. Nonetheless, he suggests through extensive textual analysis that ash-Shaṭṭī's second source was probably the abridgement (*Mukhtāṣar*) of the massive anthology *al-Kawākib ad-darārī* by the Ḥanbalī scholar Abū l-Ḥasan ʿAlī b. al-Ḥuṣayn b. ʿUrwa (d. 1433–34/837). Ibn ʿUrwa copies Sections 1–4 of Ibn Taymiyya's *Muqaddima* into *al-Kawākib ad-darārī* and then Sections 5–6 found in the introduction to Ibn Kathīr's *Tafsīr*, the latter section of which Ibn ʿUrwa attributes to Ibn Kathīr himself. The attribution to Ibn Kathīr, however, is not found in the *Mukhtāṣar*, from which ash-Shaṭṭī was probably copying. The upshot of Ibn Jād Allāh's analysis is that Sections 5–6 of Ibn Taymiyya's *Muqaddima* as it is known today were most likely composed first by Ibn Kathīr in the introduction to his *Tafsīr* and much later appended to the *Muqaddima* by ash-Shaṭṭī from the *Mukhtaṣar* of Ibn ʿUrwa's *al-Kawākib ad-darārī*. If further investigation shows Ibn Jād Allāh to be correct, much modern scholarship on Ibn Taymiyya's Qurʾān hermeneutics will need to be revised, especially the notion that he articulated a hermeneutic of interpreting the Qurʾān by the Qurʾān.

8 Bibliography

8.1 Sources

Ibn Baṭṭūṭa, Abū ʿAbdallāh Muḥammad. *Riḥlat Ibn Baṭṭūṭa*. Ed. Muḥammad ʿAbd al-Munʿim al-ʿIryān, 2 vols. Beirut: Dār Iḥyāʾ al-ʿUlūm, 1987.

Ibn Kathīr, Ismāʿīl b. ʿUmar. *Tafsīr al-Qurʾān al-ʿaẓīm*. Ed. Sāmī b. Muḥammad as-Salāma, 8 vols. Riyadh: Dār aṭ-Ṭayba, 1997; reprint 1999.

Ibn Kathīr, Ismāʿīl b. ʿUmar. *The Exegesis of the Grand Holy Qur'an*, trans. Muhammad Mahdi ash-Sharif. 4 vols. Beirut: Dar al-Kotob al-Ilmiyah, 2006.

Ibn Qayyim al-Jawziyya, Shams ad-Dīn ʿAbdallāh Muḥammad. *Madārij as-sālikīn*. Ed. Nāṣir b. Sulaymān as-Saʿwī et al., 6 vols. Riyadh: Dār as-Sumayʿī, 2011.

Ibn Taymiyya, Taqī d-Dīn Aḥmad. *Muqaddima fī uṣūl at-tafsīr*. Ed. Jamīl ash-Shaṭṭī. Damascus: Maṭbaʿat at-Taraqqī, 1936.

Ibn Taymiyya, Taqī d-Dīn Aḥmad. *Majmūʿ fatāwā shaykh al-Islām Aḥmad b. Taymiyya*. Ed. ʿAbd ar-Raḥmān b. Muḥammad b. Qāsim and Muḥammad b. ʿAbd ar-Raḥmān b. Muḥammad, 37 vols. Riyadh: Maṭābiʿ ar-Riyāḍ, 1961–1967. (abbreviated MF)

Ibn Taymiyya, Taqī d-Dīn Aḥmad. *Muqaddimat at-tafsīr*. In *Majmūʿ fatāwā shaykh al-Islām Aḥmad b. Taymiyya*. Ed. ʿAbd ar-Raḥmān b. Muḥammad b. Qāsim and Muḥammad b. ʿAbd ar-Raḥmān b. Muḥammad, 37 vols. Riyadh: Maṭābiʿ ar-Riyāḍ, 1961–1967, vol. 13: 329–75.

Ibn Taymiyya, Taqī d-Dīn Aḥmad. *Muqaddima fī uṣūl at-tafsīr*. Ed. ʿAdnān Zarzūr. 2nd printing. Kuwait: Dār al-Qurʾān al-Karīm, 1972. 1st printing 1971.

Ibn Taymiyya, Taqī d-Dīn Aḥmad. *Darʾ taʿāruḍ al-ʿaql wa-n-naql*. Ed. Muḥammad Rashād Sālim, 11 vols. Riyadh: Jāmiʿat al-Imām Muḥammad b. Suʿūd al-Islāmiyya, 1991.

Ibn Taymiyya, Taqī d-Dīn Aḥmad. *Tafsīr shaykh al-Islām Ibn Taymiyya*. Ed. Iyād b. ʿAbd al-Laṭīf b. Ibrāhīm al-Qaysī, 7 vols. Ad-Dammām: Dār Ibn al-Jauzī, 2011–2012.

Ibn Taymiyya, Taqī d-Dīn Aḥmad. *Muqaddima fī uṣūl at-tafsīr*. Ed. Sāmī b. Muḥammad b. Jād Allāh. Riyadh: Dār al-Muḥdath, 2022.

Ibn Taymiyya, Taqī d-Dīn Aḥmad. "Treatise on the Principles of Tafsir," trans. Jane Dammen McAuliffe. In *Windows on the House of Islam: Muslim Sources on Spirituality and Religious Life*, edited by John Renard. Berkeley: University of California Press, 1998: 35–43. (Translation of Sections Five and Six of the *Muqaddima*).

Ibn Taymiyya, Taqī d-Dīn Aḥmad. *Déontologie de l'exégèse coranique dans l'enseignement d'Ibn Taymiyya*, trans. Muhammad Diakho. Paris: Dar Albouraq, 2014.

Ibn Taymiyya, Taqī d-Dīn Aḥmad. *An Introduction to the Principles of Tafseer*, trans. Muhammad Abdul Haq Ansari. Birmingham: Al-Hidaayah, 1993.

Ibn Taymiyya, Taqī d-Dīn Aḥmad. *Einführung in die Methodologie der Qurʾānexegese*, trans. Elsayed Elshahed. Riyadh: Imam Muḥammad ibn Saud Universität, Abt. für Kultur und Publikation, 2000.

As-Suyūṭī, Jalāl ad-Dīn. *Al-Itqān fī ʿulūm al-Qurʾān*. Ed. Markaz ad-Dirāsāt al-Qurʾāniyya, 7 vols. Medina: Mujammaʿ al-Malik Fahd, 2005.

Aṭ-Ṭabarī, Muḥammad b. Jarīr. *The Commentary on the Qurʾān*, trans. J. Cooper. Vol. 1. Oxford: Oxford University Press, 1987.

Aṭ-Ṭabarī, Muḥammad b. Jarīr. *Tafsīr aṭ-Ṭabarī: Jāmiʿ al-bayān ʿan taʾwīl āy al-Qurʾān*. Ed. ʿAbd Allāh b. ʿAbd al-Muḥsin at-Turkī, 26 vols. Cairo: Dār al-Hijr, 2001.

Az-Zabīdī, Muḥammad b. Muḥammad Murtaḍā. *Itḥāf as-sāda al-muttaqīn bi-sharḥ Iḥyāʾ ʿulūm ad-dīn*. 10 vols. Cairo: al-Maṭbaʿa al-Maymaniyya, 1893–1894; reprint, Beirut: Muʾassasat Tārīkh al-ʿArabī, 1994.

Az-Zarkashī, Badr ad-Dīn. *Al-Burhān fī ʿulūm al-Qurʾān*. Ed. Muḥammad Abū l-Faḍl Ibrāhīm, 4 vols. Cairo: Maktabat Dār at-Turāth, 1957; 1984[3].

8.2 Secondary Literature

Baraka, Ibrāhīm. *Ibn Taymiyya wa-juhūduhu fī t-tafsīr*. Beirut: al-Maktaba al-Islāmiyya, 1984.

Bauer, Karen. "Justifying the Genre: A Study of Introductions to Classical Works of *Tafsīr*." In *Aims, Methods and Contexts of Qur'anic Exegesis (2nd/8th – 9th/15th c.)*, edited by Karen Bauer. Oxford: Oxford University Press, 2013: 39–65.

Bazzano, Elliott A. "Ibn Taymiyya, Radical Polymath, Part 2: Intellectual Contributions." *Religion Compass* 9, no. 4 (2015): 117–39.

El Shamsy, Ahmed. *Rediscovering the Islamic Classics: How Editors and Print Culture Transformed an Intellectual Tradition*. Princeton, NJ: Princeton University Press, 2020.

El-Tobgui, Carl Sharif. *Ibn Taymiyya on Reason and Revelation: A Study of* Darʾ taʿāruḍ al-ʿaql wa-l-naql. Leiden: Brill, 2020.

Ḥarīz, Sāmī Muḥammad Hishām. *Ibn Taymiyya mufassiran wa-Muqaddimatuhu fī uṣūl at-tafsīr: ʿarḍ wa-dirāsa*. Amman: Dār al-Bidāya, 2007.

Hidayatullah, Aysha A. *Feminist Edges of the Qur'an*. Oxford, UK: Oxford University Press, 2014.

Hoover, Jon, and Marwan Abu Ghazaleh Mahajneh. "Theology as Translation: Ibn Taymiyya's Fatwa Permitting Theology and Its Reception into His Averting the Conflict between Reason and Revealed Tradition (*Darʾ Taʿāruḍ Al-ʿAql Wa l-Naql*)." *MW* 108, no. 1 (2018): 40–86.

Hoover, Jon. *Ibn Taymiyya*. London: Oneworld, 2019.

Kamali, Mohammad Hashim. *Principles of Islamic Jurisprudence*. Cambridge, UK: The Islamic Texts Society, 2003^3.

Little, Donald P. "Did Ibn Taymiyya Have a Screw Loose?" *SI* 41 (1975): 93–111 (95–96).

McAuliffe, Jane Dammen. "Quranic Hermeneutics: The Views of al-Ṭabarī and Ibn Kathīr." In *Approaches to the History of the Interpretation of the Qur'ān*, edited by Andrew Rippin. Oxford, UK: Clarendon Press, 1988. Reprint, Piscataway, NJ: Gorgias Press, 2013: 46–62.

Mirza, Younus Y. "Ishmael as Abraham's Sacrifice: Ibn Taymiyya and Ibn Kathīr on the Intended Victim." *Islam and Christian-Muslim Relations* 24, no. 3 (2013): 277–98.

Mirza, Younus Y. "Was Ibn Kathīr the 'Spokesperson' for Ibn Taymiyya? Jonah as a Prophet of Obedience." *JQS* 16, no. 1 (2014): 1–19.

Mirza, Younus Y. "Ibn Taymiyya as Exegete: Moses' Father-in-Law and the Messengers in *Sūrat Yā Sīn*." *JQS* 19, no. 1 (February 2017): 39–71.

Özervarli, M. Sait. "The Qur'ānic Rational Theology of Ibn Taymiyya and His Criticism of the *Mutakallimūn*." In *Ibn Taymiyya and His Times*, edited by Yossef Rapoport and Shahab Ahmed. Karachi: Oxford University Press, 2010: 78–100.

Pink, Johanna. *Muslim Qur'ānic Interpretation Today: Media, Genealogies and Interpretive Communities*. Sheffield: Equinox, 2019.

Saeed Bhutta, Sohaib. "Intraquranic Hermeneutics: Theories and Methods in Tafsir of the Qur'an through the Qur'an." PhD diss., School of Oriental and African Studies, University of London, 2017.

Saleh, Walid A. *The Formation of the Classical Tafsīr Tradition: The Qur'ān Commentary of Al-Thaʿlabī (d. 427/1035)*. Leiden: Brill, 2004.

Saleh, Walid A. "Ibn Taymiyya and the Rise of Radical Hermeneutics: *An Analysis of An Introduction to the Foundations of Qur'ānic Exegesis*." In *Ibn Taymiyya and His Times*, edited by Yossef Rapoport and Shahab Ahmed. Karachi: Oxford University Press, 2010: 123–162.

Saleh, Walid A. "Preliminary Remarks on the Historiography of Tafsīr in Arabic: A History of the Book Approach." *JQS* 12 (2010): 6–40.

Saleh, Walid. "al-Bayḍāwī." In EI^3, edited by Kate Fleet et al. Brill, 2007–. Article published online 2017. Accessed on May 12, 2020. http://dx.doi.org/10.1163/1573-3912_ei3_COM_25267

Sands, Kristin Zahra. *Ṣūfī Commentaries on the Qur'ān in Classical Islam*. London: Routledge, 2006.

Shah, Mustafa. "Introduction." In *Tafsīr: Interpreting the Qur'ān*, edited by Mustafa Shah, 4 vols. London: Routledge, vol. 1, 2013: 1–157.
Suleiman, Farid. "Ibn Taymīyas Theorie Der Koranexegese." In *Koranexegese Als »Mix and Match« Zur Diversität Aktueller Diskurse in Der Tafsīr-Wissenschaft*, edited by Abbas Poya. Bielefeld, Germany: Transcript Verlag, 2017: 15–43.
Suleiman, Farid. *Ibn Taymiyya und die Attribute Gottes*. Berlin: de Gruyter, 2019.
Aṭ-Ṭayyār, Musāʿid. *Sharḥ Muqaddima fī uṣūl at-tafsīr li-Ibn Taymiyya*. Ad-Dammām: Dār Ibn al-Jawzī, 2007[2].
Tottoli, Roberto. *Biblical Prophets in the Qur'ān and Muslim Literature*. Richmond, UK: Curzon, 2002.

Niẓām ad-Dīn an-Nīsābūrī l-Qummī

Robert Morrison

1 Introduction

Al-Ḥasan b. Muḥammad b. al-Ḥusayn Niẓām ad-Dīn al-Aʿraj an-Nīsābūrī l-Qummī (d. after 1330/730) made significant contributions to a number of disciplines, including astronomy, astrology, mathematics, Arabic grammar, Arabic rhetoric, and, of course, *tafsīr*. An-Nīsābūrī's *tafsīr*, entitled *Gharāʾib al-Qurʾān wa-raghāʾib al-Furqān* (The Wonders of the Qurʾān and the Desiderata of the Proof; GQ hereafter) remains his most famous work and his most widely available printed text. GQ has been published in stand-alone editions and in the margins of the 1905 Būlāq edition of Abū Jāʿfar aṭ-Ṭabarī's (d. 923/310) *Jāmiʿ al-Bayān fī tafsīr al-Qurʾān* (The Comprehensive Exposition of the Exegesis of the Qurʾān).[1] In modern scholarship, Guy Monnot's "Le verset du trône" likewise paired translations of aṭ-Ṭabarī and an-Nīsābūrī's exegeses of Q 2:255.[2] An-Nīsābūrī's combination of a rationalist approach to exegesis (*at-tafsīr bi-r-raʾy*) with attention to transmitted information about the *Qurʾān* (*at-tafsīr bi-l-maʾthūr*), along with his inclusion of esoteric exegesis (*taʾwīl*),[3] explains why he has left different impressions with different modern scholars.[4] The only other text by an-Nīsābūrī that has been printed is his commentary on the *Shāfiya* (*Sharḥ Shāfiyat Ibn al-Ḥājib*) of ʿUthmān b. al-Ḥājib (d. 1249/646) on morphology.[5]

GQ reflected an-Nīsābūrī's breadth of learning. Not only did an-Nīsābūrī's comments draw on the fields in which he wrote discrete works, but GQ has also attracted scholarly attention for how an-Nīsābūrī's esoteric interpretations were informed by Sufism, a field in which an-Nīsābūrī did not write a discrete work.[6] An-Nīsābūrī's stated dependence on Fakhr ad-Dīn ar-Rāzī's (d. 1210/606) *at-Tafsīr al-kabīr* (The Grand Tafsīr), also known as *Mafātīḥ al-ghayb* (The Keys of the Unseen), led him, in GQ, to enter debates within *ʿilm al-kalām* (philosophical theology) even though he did

1 Modern stand-alone printed editions include ed. Zakariyyā ʿUmayrāt (Beirut: Dār al-Kutub al-ʿIlmiyya, 1996) and ed. Ibrāhīm ʿAṭwa ʿAwaḍ (Cairo: Muṣṭafā l-Bābī l-Ḥalabī, 1962). There is also an 1864 lithograph printed in Tehran.

2 Guy Monnot, "Le verset du trône," *MIDEO* 15 (1982): 119–44.

3 In other contexts, *taʾwīl* can be a synonym for *tafsīr*.

4 Cf. Mahmoud M. Ayoub, *The Qur'an and its Interpreters* (Albany: State University of New York Press, 1984), 1: 6; Robert Morrison, *Islam and Science: The Intellectual Career of Niẓām al-Dīn al-Nīsābūrī* (Abingdon, Oxon: Routledge, 2007); and ʿĀdil Rāḍī r-Rifāʿī, *al-Athar al-balāghī fī tafsīr an-Nīsābūrī* (Amman: Dār Majdalāwī li-n-Nashr wa-t-Tawzīʿ, 2012).

5 Niẓām ad-Dīn an-Nīsābūrī, *Sharḥ Shāfiyat Ibn al-Ḥājib al-musammā bi-Sharḥ an-Niẓām*, ed. ʿAlī sh-Shamlāwī (Beirut: Shams al-Mashriq, 1992).

6 Ayoub, *The Qur'an and its Interpreters*, 1: 6.

https://doi.org/10.1515/9783111320083-008

not write a separate text in that discipline.[7] An-Nīsābūrī has received a great deal of modern scholarly attention for his use of science as a hermeneutic tool in GQ.[8] An-Nīsābūrī's decision to pursue multiple exegetical approaches in GQ challenges readers formed in the modern era of academic specialization.

2 Biography

Conversely, an-Nīsābūrī's own intellectual formation was, broad. He was born in Nishapur into a Shiite family with roots in Qom, perhaps around 1270/668.[9] On the basis of an-Nīsābūrī's compositions as well as other records of his activities in the first decade of the fourteenth century, we can conclude that he received a *madrasa* education that comprised not only Qur'ān, *fiqh*, *ḥadīth*, and Arabic grammar, but also the fields of *'ilm al-kalām*, mathematics and *'ilm al-hay'a* (astronomy).[10] Transformations in the field of *'ilm al-kalām* in the thirteenth century account for this breadth of education. Following the career of Fakhr ad-Dīn ar-Rāzī, theologians (*mutakallimūn*) started to include disciplines such as astronomy and Avicennan philosophy in their *summae* and appropriated those disciplines' own technical terms.[11] A debate, in which an-Nīsābūrī participated, among *madrasa*-educated scholars over the nuances of the presentation of astronomy in *'ilm al-kalām* ensued. That debate spilled over into GQ.[12] Marginalia on manuscripts of an-Nīsābūrī's work, as well as reports from later scholars, indicate that his texts were studied in *madrasas* as part of a tradition of religious scholarship.[13] Thus, internal and external evidence shows that GQ, with its manifold named and unnamed sources, is the product of the post-classical (post-1200/600) period of Islamic thought in which

7 Niẓām ad-Dīn an-Nīsābūrī, *Gharā'ib al-Qur'ān wa-raghā'ib al-furqān* (printed in margins of aṭ-Ṭabarī, *Jāmi' al-bayān*) (Beirut: Dār al-Ma'rifa, 1992; reprint of the 1905–11 Būlāq edition), 1: 5–6.

8 Ahmad S. Dallal, "Science and the Qur'ān," in *EQ*, ed. Jane Dammen McAuliffe (Leiden: Brill, 2004), 4: 540–58. For a modern scholar of GQ who did not pay much attention to an-Nīsābūrī's use of science, see Ḥusayn Muḥammadkhānī, *Nīshābūrī va tafsīr-i Gharā'ib al-Qur'ān* (Tehran: Khānah-i Kitāb, 2014), 19. There, Muḥammadkhānī described an-Nīsābūrī's approach to *tafsīr* as mixed (*mukhtaliṭ*), though Muḥammadkhānī devoted only three pages of his book to the role of science and natural philosophy in GQ.

9 Morrison, *Islam and Science*, 7.

10 Ibid., 9–17.

11 The classic article about these developments is A. I. Sabra, "Science and Philosophy in Medieval Islamic Theology: The Evidence of the Fourteenth Century," in *ZGAIW* 9 (1994): 1–42. For information about ar-Rāzī's pivotal position in these developments, see Heidrun Eichner, "The Post-Avicennian Philosophical Tradition and Islamic Orthodoxy. Philosophical Theological Summae in Context" (Habil. thesis., Martin-Luther-Universität Halle-Wittenberg, 2009).

12 Robert Morrison, "What *Was* the Purpose of Astronomy in Ījī's *Kitāb al-Mawāqif fī 'ilm al-kalām*?" in *Tabriz after the Mongol Conquest*, ed. Judith Pfeiffer (Leiden: Brill, 2013), 201–29.

13 Morrison, *Islam and Science*, 17–19.

many fields were understood to be relevant to understanding God, revelation, and creation.

An-Nīsābūrī's three major astronomy and astrology texts come from the first decade or so of the fourteenth century.[14] He completed all three of these texts at the Ilkhanid Court in Tabriz, at which he had arrived by 1304/703. He had begun his commentary *Taʿbīr at-Taḥrīr* (The Interpretation of the Recension) on Naṣīr ad-Dīn aṭ-Ṭūsī's (d. 1274/672) recension of the Arabic translations of Ptolemy's (fl. second century CE) *Almagest* before coming to Tabriz. Occasional comments in *at-Taʿbīr*, which an-Nīsābūrī completed in 1305/704, show that he was already thinking, early in his career, about how astronomy could inform religious scholarship. Next, an-Nīsābūrī completed *Kashf-i ḥaqāʾiq-i Zīj-i Īlkhānī* (The Investigation of the Truths of the Ilkhanid Astronomical Handbook with Tables) a commentary on aṭ-Ṭūsī's *Zīj-i Īlkhānī* (The Ilkhanid Astronomical Handbook with Tables) in 1308–9/708. An-Nīsābūrī devoted most of *Kashf-i ḥaqāʾiq* to explain the details of the mathematical computation of the planets' positions in longitude and latitude. He also elucidated theoretical astronomy in Persian for those who could not read Arabic, the dominant language of theoretical astronomy at the time in Islamic societies. Like many authors of texts in the *zīj* (astronomical handbook with tables) genre, an-Nīsābūrī included in *Kashf-i ḥaqāʾiq* a lengthy section on astrological forecasting because he believed that God controlled events on earth through astral forces. In 1311/711, an-Nīsābūrī finished his final major astronomy text, *Tawḍīḥ at-Tadhkira* (Elucidation of the *Tadhkira*), a commentary on aṭ-Ṭūsī's *at-Tadhkira fī ʿilm al-hayʾa* (Memoir on Astronomy). *At-Tawḍīḥ* analyzed the innovative physical models for planetary motions that scholars affiliated with the observatory in Maraghah, near Tabriz, proposed in order to resolve the many significant physical inconsistencies of Ptolemy's *Almagest*. *At-Tawḍīḥ* incorporated excursuses about how creation and regular natural processes are evidence for God's power and wisdom. An-Nīsābūrī's astronomy teacher Quṭb ad-Dīn ash-Shīrāzī (d. 1311/710) had also remarked on the theological value of studying astronomy in his own astronomy texts.

An-Nīsābūrī cultivated a variety of patrons. He dedicated *Taʿbīr at-Taḥrīr*, as well as a 1310/709 copy of *Kashf-i ḥaqāʾiq*, to Saʿd ad-Dīn as-Sāwajī, a Ilkhanid co-chief minister with whom ash-Shīrāzī had been acquainted as well.[15] As-Sāwajī was executed by Rashīd ad-Dīn Ṭabīb (c. 1318/d. c. 718), the other co-chief minister, in 1312/712. An-Nīsābūrī dedicated *at-Tawḍīḥ* to ʿAlī b. Maḥmūd al-Yazdī, a friend of Rashīd ad-Dīn. By the second decade of the fourteenth century, powerful figures recognized an-Nīsābūrī's knowledge in fields besides astronomy. Rashīd ad-Dīn's *Asʾila wa ajwiba* (Questions and Answers), which al-Yazdī edited, assembled questions from a number of scholars at Tabriz on a variety of matters and provided Rashīd ad-Dīn's responses. An-Nīsābūrī's three recorded questions, submitted by 1311/710, treated theological

14 Ibid., 148–51.
15 Ibid., 65 and 149f.

matters.[16] He was present at a 1311/711 assembly, organized by the Ilkhanid Sultan Öljaytü (d. 1316/716), of all the scholars who submitted questions. *Sharḥ Shāfiyat Ibn al-Ḥājib* (The Commentary on Ibn al-Ḥājib's *Shāfiya*), a text on morphology, dates from the same decade or perhaps the next.[17] Some manuscripts of *Sharḥ Shāfiyat Ibn al-Ḥājib* contain a dedication to Majd ad-Dīn Ismāʿīl b. Yaḥyā b. Ismāʿīl al-Fālī (d. 1355/756) who was a judge in Shiraz. The second decade of the fourteenth century was the most likely time for the composition of an-Nīsābūrī's commentary on as-Sakkākī's *Miftāḥ al-ʿUlūm* (Key of the Sciences), a text on rhetoric, and *ar-Risāla ash-Shamsiyya fī l-ḥisāb* (The Solar Epistle on Arithmetic), which was dedicated to Rashīd ad-Dīn's son, Shams ad-Dīn.[18] Scholarship on Ilkhanid Tabriz has found that patrons encouraged the inter-disciplinary interests of an-Nīsābūrī and numerous other scholars.[19] GQ was the final work an-Nīsābūrī authored. The Būlāq edition of GQ ends with a transcription of the colophon of some manuscripts which inform us that an-Nīsābūrī inspected (or annotated) the text in late November-early December 1329 (the beginning of Ṣafar 729) in Dawlatabad, India.[20]

3 Hermeneutical Method

The many renowned texts that an-Nīsābūrī cited at the end of GQ are the foundations of his wide-ranging approach to *tafsīr*.[21] As his most significant sources, he named two *tafsīr*s, Abū l-Qāsim az-Zamakhsharī's (d. 1144/538) *al-Kashshāf* and ar-Rāzī's *at-Tafsīr al-kabīr,* that are hard to categorize.[22] Muḥammad b. Abū Yazīd Tayfūr as-Sajāwandī's (d. 1165/560) *Kitāb al-Wuqūf* (The Book of Pauses) was an-Nīsābūrī's source for *wuqūf,* pauses in *qurʾān*ic recitation. His dependence on as-Sājawandī was not total; for example, an-Nīsābūrī disagreed with him over why a pause was permitted between the two instances of the verb *hamma* in Q 12:24.[23] An-Nīsābūrī's sources for *asbāb an-nuzūl* (reasons for revelation) were Majd ad-Dīn b. al-Athīr's (d. 1210/606) *Jāmiʿ al-uṣūl li-aḥādīth ar-Rasūl* (The Compendium of Principles for the Ḥadīth of the

16 For the text of the questions, see ibid., 41–43.

17 Ibid., 151.

18 Ibid., 152–54.

19 Josef van Ess, *Der Wesir und seine Gelehrten* (Wiesbaden: Franz Steiner, 1981).

20 Morrison, *Islam and Science*, 152; and an-Nīsābūrī, *GQ*, 30: 229.

21 An-Nīsābūrī, *GQ*, 30: 222f. For a summary of an-Nīsābūrī's sources, see Mājid Zakī al-Jallād, *an-Nīsābūrī wa-manhajuh fī t-tafsīr* (Amman: Dār al-Fikr li-ṭ-Ṭibāʿa wa-n-Nashr wa-t-Tawzīʿ, 2000), 47f.

22 See Andrew Lane, *A traditional Muʿtazilite Qurʾan commentary: the Kashshāf of Jār Allāh al-Zamakhsharī (d. 538/1144)* (Boston/Leiden: Brill, 2006) for a book that explores the renowned rationalist commentator az-Zamakhsharī's use of *ḥadīth*. Though ar-Rāzī's *at-Tafsīr al-kabīr* is known for philosophically informed exegesis, an-Nīsābūrī mentioned it and *al-Kashshāf* as sources for *ḥadīth*.

23 Al-Jallād, *an-Nīsābūrī*, 127f.

Apostle), which doubled as a source for *ḥadīth*, and ʿAlī b. Aḥmad al-Wāḥidī's (d. 1076/468) *al-Basīṭ fī t-tafsīr* (The Plain Way in Exegesis). On matters of *fiqh*, an-Nīsābūrī drew on ʿAbd al-Karīm ar-Rāfiʿī's (d. 1226/623) *Sharḥ al-Wajīz* (The Commentary on the Compendium), a commentary on *al-Wajīz* by the Shāfiʿī luminary Abū Ḥāmid al-Ghazālī (d. 1111/505). An-Nīsābūrī acknowledged the work of Najm ad-Dīn ar-Rāzī Dāya (d. 1256/654) as a source for esoteric interpretations in GQ, but in that field and others, scholarship on an-Nīsābūrī has found that he was informed throughout GQ by sources that he did not name at the end of GQ.[24] Because an-Nīsābūrī formatted his comments in a consistent manner, in the following paragraphs I describe the different dimensions of an-Nīsābūrī's hermeneutic approach in the order in which they appear in GQ.

The unit of commentary in GQ was generally a group of verses.[25] After presenting the passage to be interpreted, an-Nīsābūrī reviewed the alternative vocalizations (*qirāʾāt*) and the necessary and permissible pauses (*wuqūf*) in recitation. He acknowledged ten readings of the *Qurʾān*, favoring the reading of Abū Ḥātim as-Sijistānī (d. 869/255). Then, he entered into the exoteric (*tafsīr*) and then esoteric (*taʾwīl*) interpretations of that passage. All this transmitted knowledge was often marshaled by an-Nīsābūrī to answer questions posed in fields such as rhetoric, law, and *ʿilm al-kalām*. For instance, alternative vocalizations impinged on other matters, such as nuanced understandings of God's role in the natural world.[26] Q 7:54 explained that God created "the sun, and the moon, and the stars subservient by His command."[27] The Arabic word translated by 'subservient', *musakhkharāt*, could be read in either the nominative (*musakhkharātun*) or accusative (*musakhkharātin*) cases. If *musakhkharāt* were read in the accusative case, then God would have created the sun, moon, and stars in a state in which they were subservient or subjugated, and creation would have occurred through God's command. If *musakhkharāt* were read in the nominative case, then *musakhkharāt* would be the predicate of the sun, moon, and the stars and those celestial bodies would be subservient because they followed God's command. The format of each unit of commentary in GQ indicates an-Nīsābūrī's respect for and concern with both transmitted and rationalist knowledge.

Ḥadīth was an important source of information for the virtues of various *sūra*s (*faḍāʾil as-suwar*). An-Nīsābūrī explained that he chose to set aside the *ḥadīth* reports

24 For example, al-Jallād found that an-Nīsābūrī relied on several other *tafsīr*s besides the ones he listed in the conclusion, see al-Jallād, *an-Nīsābūrī*, 48–60. See those pages for other sources, in areas such as *ḥadīth*, rhetoric, history, and grammar, which an-Nīsābūrī also did not list in the conclusion to his *tafsīr*.

25 Khaled Troudi, "Qurʾānic Hermeneutics with Reference to Narratives: A Study in Classical Exegetical Traditions" (Ph.D. Dissertation, University of Exeter, 2012), 175. An-Nīsābūrī's tendency to comment on groups of verses distinguished GQ from ar-Rāzī's *at-Tafsīr al-kabīr*.

26 Al-Jallād, *an-Nīsābūrī*, 109f.

27 All translations of the *Qurʾān* come from *The Koran Interpreted*, trans. A. J. Arberry (New York: Macmillan, 1955).

that Zamakhsharī had provided in the *Kashshāf* on the virtues of *sūras* because other critics had shown those reports to be false.[28] Az-Zamakhsharī's famous and influential linguistically informed comments did not save him from an-Nīsābūrī's unfavorable judgment of his choice of *ḥadīth*! Reports from earlier figures were also a weighty source for an-Nīsābūrī's explanations of legal rules (*al-aḥkām al-fiqhiyya*). Q 4:43 contained the command to perform ablutions with sand (*tayammum*) when water was not available. The verse reads:

> O believers, draw not near to prayer when you are drunken until you know what you are saying, or defiled– unless you are traversing a way– until you have washed yourselves; but if you are sick, or on a journey, or if any of you come from the privy, or you have touched women, and you can find no water, then have recourse to wholesome dust and wipe your faces and your hands; God is All-pardoning, All-forgiving.

An-Nīsābūrī considered the narrow question at hand to be two different possible interpretations of the word *lāmastum* (you have touched).[29] The first possibility, which the companion Ibn Masʿūd (d. c. 653/d. c. 32), ʿAbd Allāh b. ʿUmar b. al-Khaṭṭāb (d. 693/73), Muḥammad b. Idrīs ash-Shāfiʿī (d. 820/204) and others favored and which an-Nīsābūrī chose (*kamā huwa muqtaḍā l-lugha*), was to interpret *lāmastum* as the meeting of the two people's skin, perhaps through intercourse but also in other ways (*aw bi-ghayrih*). The second possibility, preferred by ʿAbd Allāh b. ʿAbbās (d. 686–8/68), Qatāda b. Diʿāma (d. c. 735/d. c. 117), Abū Ḥanīfa (d. 767/150) and Shiites, was to interpret *lāmastum* as an allusion (*kināya*) to only intercourse because the *Qurʾān* elsewhere used *massa* (to touch) as a metaphor for intercourse and not for other contact.

An-Nīsābūrī's adherence to the Shāfiʿī *madhhab*, reflected by his dependence throughout *GQ* on the legal positions found in ar-Rāfiʿī's *Sharḥ al-Wajīz*, has raised questions about his Shiite loyalties.[30] In fact, al-Jallād has argued that an-Nīsābūrī was Sunni.[31] Yet on a verse that broached the controversial issue of temporary (*mutʿa*) marriage, a Shiite practice that typically elicited the scorn of Sunnis, an-Nīsābūrī's comments were balanced. The relevant portion of Q 4:24, reads, "Such wives as you enjoy thereby, give them their wages apportionate." An-Nīsābūrī first presented the arguments against *mutʿa* marriage and then the arguments for it. He concluded that ʿUmar b. al-Khaṭṭāb's (d. 644/23) proscription of the practice meant that ʿUmar believed that *mutʿa* marriage had been abrogated during Muḥammad's lifetime even though the majority of reports (*akthar ar-riwāyāt*) related that Muḥammad permitted the

---

28 Al-Jallād, *an-Nīsābūrī*, 95. For al-Jallād's observation that an-Nīsābūrī was usually more interested in the contents of the report than in the chains of transmission, see ibid., 97.

29 An-Nīsābūrī, *GQ*, 5: 58.

30 Morrison, *Islam and Science*, 47f. and 112.

31 See al-Jallād, *an-Nīsābūrī*, 25–31 and 23f. for the argument that an-Nīsābūrī's upbringing in Nishapur exerted more influence upon his religious outlook than his birth in Qom.

practice at the farewell pilgrimage.[32] But an-Nīsābūrī omitted ar-Rāzī's analytic rebuttal of the arguments in favor of *mut'a* marriage in which ar-Rāzī concluded that the once licit practice had been abrogated.[33] Since an-Nīsābūrī's presentation of the arguments against and for *mut'a* marriage tracked ar-Rāzī's, the omission of the rebuttal is striking.

There were certainly places in *GQ* where an-Nīsābūrī did not adopt typically Shiite positions, such as in his comments on Q 5:54 ("God will assuredly bring a people He loves"), a verse which previous Shiite commentators had understood as a reference to ʿAlī b. Abī Ṭālib (d. 661/40).[34] An-Nīsābūrī remarked that the Companions would have been apostates for overlooking a clear reference to ʿAlī, had the passage indeed referred to ʿAlī.[35] Ar-Rāzī had taken the same position and had added that the passage was actually a reference to the first caliph Abū Bakr.[36] In contrast, an-Nīsābūrī concluded his remarks simply with *Allāh aʿlam* (God is most knowing). A clear indication of respect for ʿAlī is found in the closing words of *GQ*, where an-Nīsābūrī singled out ʿAlī, named as God's *walī* (friend), for praise.[37] Thus, an-Nīsābūrī, given his connection to Qom, might be described as a non-sectarian Shiite whose legal positions resulted from his choice of teachers, above all the Shāfiʿī Quṭb ad-Dīn ash-Shīrāzī.[38] An-Nīsābūrī was one of many Shiites to identify as a Shāfiʿī.[39]

As an-Nīsābūrī was educated within a *madrasa* curriculum that had adopted the study of the new *ʿilm al-kalām*, which incorporated fields such as astronomy and natural philosophy into speculation about God, it is not surprising that he used such fields to make sense of the theological questions posed by Q 2:6–7. The verse reads, "As for the unbelievers, alike it is to them whether thou hast warned them or hast not warned them, they do not believe. God has set a seal on their hearts and on their hearing, and on their eyes is a covering, and there awaits them a mighty chastisement." An-Nīsābūrī's comments are best understood in the context of ar-Rāzī's exegesis. In *at-Tafsīr al-kabīr*, he used psychology, a rationalist analysis of the soul, to argue that God could lead humans to sin by creating not only the capability but also the motivations to misbehave.[40] The observed existence of variations in these motivations simply reflected God's role in creating humans' different characters. Ar-Rāzī concluded that a recognition of God's total control over our behavior was valuable be-

32 An-Nīsābūrī, *GQ*, 5: 18f.

33 Fakhr ad-Dīn ar-Rāzī, *at-Tafsīr al-kabīr* (Beirut: Dār Iḥyāʾ at-Turāth al-ʿArabī, n.d.), 10: 53f.

34 Al-Jallād, *an-Nīsābūrī*, 26f.

35 An-Nīsābūrī, *GQ*, 6: 166.

36 Ar-Rāzī, *al-Tafsīr al-kabīr*, 12: 22.

37 An-Nīsābūrī, *GQ*, 30: 229.

38 A list of arguments for why an-Nīsābūrī was Shiite can be found in Muḥammadkhānī, *Nīshābūrī*, 6–11.

39 Devin Stewart, *Islamic Legal Orthodoxy: Twelver Shiite Responses to the Sunni Legal System* (Salt Lake City: The University of Utah Press, 1998), 102–9.

40 Robert Morrison, "Science and Theodicy," in *Nature and Scripture in the Abrahamic Religions*, ed. Jitse M. van der Meer and Scott Mandelbrote (Leiden: E.J. Brill, 2008), 249–72, at 262–64.

cause such a recognition restrained us from being overconfident in our own knowledge. An-Nīsābūrī, though, drew on Ibn Sīnā's theory of vision to argue confidently that the *Qur'ān* was speaking directly when it said that God might interfere with unbelievers' hearts and eyes so as to lead to the unbelievers' eventual punishment.[41] Since vision did not improve when many people stood together viewing the same object, Ibn Sīnā had concluded that vision was not due to rays emitted from the eye. Rather, individuals' eyes perceived forms emitted from visible objects. Since God created both the Sun, whose light facilitated vision, and blind people, differences in humans' visual acuity should be attributed to God.[42] Then, an-Nīsābūrī argued that the more noble heart perceived through a light more refined than visible light.[43] Since God could obstruct and even eliminate vision, God could seal the heart.

Subsequently, an-Nīsābūrī used analogies from the natural world to rationalize why God would not intend everyone to benefit from His guidance, which is a consequence of God sealing the hearts of some. Just as blind people might not know what the sun looks like, evildoers might not know any other way of action. Conversely, humans could not choose good if the bad choice were unavailable. Natural philosophy's account of causation gave God a role in humans' sins. Anything God created, such as metal, could be used for good (e.g. building) or bad purposes (e.g. murder). God's place in the chain of causes that led to evil actions did not mean that God was arbitrary or unmerciful. Rather, the use of a sword for evil indicated the swordsman's distance from God. In this way, an-Nīsābūrī rejected the *Muʿtazilī* position that attributing to God a role in humans' evil actions would compromise theodicy. Rather, an-Nīsābūrī added information which reinforced Rāzī's conclusion, in *at-Tafsīr al-kabīr*, that God could, in fact, seal one's heart. In this case, philosophy and psychology has supported a literal reading of the *Qur'ān*.

An-Nīsābūrī and ar-Rāzī agreed that scientific and philosophic information about creation would enhance the reader's appreciation of God's role as creator. There were cases, however, in which an-Nīsābūrī's presentation of the scientific material accepted scientists' explanations on their own terms whereas ar-Rāzī directly questioned the certainty of the scientists' conclusions. For example, with regard to Q 2:29 ("It is He who created for you all that is in the earth, then he lifted Himself to heaven and levelled them seven heavens; and He has knowledge of everything"), an-Nīsābūrī and ar-Rāzī advanced competing explanations for the location of the fixed stars, the stars that made up the constellations. Because there would have to be a single translucent orb (a hollow sphere), divided into partial orbs, to move each of the Sun, Moon, and the five known planets, the Qur'ān's mention of "seven heavens" forced interpreters to consider whether there could be more than seven orbs. An additional orb, i.e. an

41 An-Nīsābūrī, *GQ*, 1: 143 and Morrison, "Science and Theodicy," 265f.

42 Morrison, "Science and Theodicy," 268.

43 An-Nīsābūrī, *GQ*, 1: 143.

eighth heaven, might be necessary to move the fixed stars. Ar-Rāzī suggested that the fixed stars might be located below the moon's orb, but above the atmosphere, or even distributed among the orbs that account for the planets' motions.[44] Both of those proposals placed some of the constellations closer to the earth than some of the planets, with the effect that some planets, instead of appearing to travel against the background of the zodiacal constellations, would be obscured at times by the zodiacal constellations. Yet observations showed that the planets always moved in front of the zodiacal signs! Ar-Rāzī's comments implied that the astronomers' observations were less certain than astronomers believed them to be.

In contrast, an-Nīsābūrī suggested locating the fixed stars on the outermost part of the orb that moves Saturn, the outermost planet.[45] That exegesis of Q 2:29 cohered with the direct sense of 'seven heavens' *and* with cutting-edge astronomy, as ash-Shīrāzī had propounded the same theory in his astronomy text entitled *al-Tuḥfa ash-shāhiyya* (The Royal Gift). Ar-Rāzī undercut the power of science to explain God's creation by considering theories that astronomers could not take seriously. He wrote, "Why is it not conceivable that it be said that the planets move on their own, without them being fixed in another body? This possibility (*iḥtimāl*) cannot be undermined except by undermining the omnivolent (*al-mukhtār*)."[46] An-Nīsābūrī and ar-Rāzī agreed that the effect, in this case the planets' and fixed stars' motions, was evidence for the cause, but they disagreed over the reliability of astronomers' explanation for how God produced the effect. Since the only way astronomy could explain the planets' motions was through orbs, when ar-Rāzī proposed that the planets moved on their own, he rejected the astronomy of his time. Without orbs, a naturalistic explanation of the planets' motions was not possible. An-Nīsābūrī explained in his comments on Q 57:1 that the goal of demonstrations was to learn more about God.[47] When ar-Rāzī undermined the reliability of fields such as astronomy he undercut the theological conclusions that one might draw from such fields.

Ar-Rāzī and an-Nīsābūrī disagreed about how to weigh the epistemological claims of astronomy also in their comments on Q 17:12. The verse reads: "We have appointed the night and the day as two signs; then We have blotted out the sign of the night."[48] They agreed that blotting "out the sign of the night" could refer to the blemish (*kalaf*) visible in the face of the moon and considered the possibility that the blotting out was due to the Angel Gabriel placing his wing over the face of the moon at God's command. Both commentators recognized that the blotting out could also be interpreted as a reference to the phases of the moon. They both considered the philosophers' (*al-falāsifa*) explanation that there were bodies in the face of the moon that were either

44 Morrison, *Islam and Science*, 115f.
45 An-Nīsābūrī, *GQ*, 1: 211.
46 Ar-Rāzī, *at-Tafsīr al-kabīr*, 2: 156.
47 Morrison, *Islam and Science*, 116.
48 Ibid., 119f.

less luminescent or less reflective than the rest of the moon, which would lead to a visible blemish.[49] These less luminescent bodies would be embedded in certain parts of the moon just as the visible planets were fixed in their translucent orb. While ar-Rāzī and an-Nīsābūrī agreed on the range of possible explanations, they disagreed about the theological significance of the interpretations. Ar-Rāzī concluded that the placement of these small bodies in certain parts of the moon rather than others, like the placement of the planets in the orbs, was unintelligible without the unfathomable actions of an omnipotent creator (*al-fāʿil al-mukhtār*). An-Nīsābūrī chose to emphasize the underlying intelligibility of natural processes as evidence for God's existence, power, and wisdom.

Though GQ was the most broadly informed of pre-modern commentaries about philosophy and astronomy, an-Nīsābūrī had other hermeneutic concerns. From the outset of GQ, an-Nīsābūrī paid a great deal of attention to the Qurʾān's rhetorical strategies. For example, an-Nīsābūrī noticed that in Q 1:5 (*iyyāka naʿbud wa-iyyāka nastaʿīn*; Thee only we serve; to Thee alone we pray for succor) the object preceded the verb twice.[50] The object's preceding the verb drew attention to God and indicated the worshipper's intention.[51] In his comments on the same *sūra*, Nīsābūrī explained why *ar-Raḥmān* (the all-merciful) preceded *ar-Raḥīm* (the all-compassionate) in Q 1:3 (*ar-Raḥmān ar-Raḥīm*). According to the pre-Islamic Arabs, *raḥmān* connoted multitudes of grace which justified its preceding *ar-Raḥīm*.[52] ʿAbd al-Qāhir al-Jurjānī (d. 1078/471) influenced an-Nīsābūrī's interest in the literary device of precedence as well as an-Nīsābūrī's theory of overall literary coherence.[53] Given the close links between *ʿilm al-kalām* and the science of rhetoric in az-Zamakhsharī's *al-Kashshāf*, it is not surprising that an-Nīsābūrī also explored the ontological relationship between those two divine names (the all-merciful, the all-compassionate) and God.[54]

An-Nīsābūrī used rhetorical analysis to clarify the different ways in which esoteric exegesis (*taʾwīl*) could be relevant.[55] Esoteric exegesis was necessary for a complete understanding of the Qurʾān, but an-Nīsābūrī was extremely concerned that his *taʾwīlāt* might depart from the desired meaning (*murād*) of the verse.[56] Despite that reservation, an-Nīsābūrī commented that the phrase "the keys of the unseen (*mafātīḥ al-ghayb*)" in Q 6:59 referred to the rational sciences (*al-ʿulūm al-ʿaqliyya*) that the ex-

49 Ibn Sīnā, *ash-Shifāʾ, aṭ-Ṭabīʿiyyāt: al-Samāʾ wa-l-ʿālam*, ed. Ibrāhīm Madkūr and Maḥmūd Qāsim (Cairo: General Egyptian Book Organization, 1969), 37ff.

50 Ar-Rifāʿī, *al-Athar*, 72–75.

51 An-Nīsābūrī, *GQ*, 1: 92.

52 Ar-Rifāʿī, *al-Athar*, 82 and an-Nīsābūrī, *GQ*, 1: 67f.

53 Ar-Rifāʿī, *al-Athar*, 286. See Troudi, *Qurʾānic Hermeneutics*, 175 for how an-Nīsābūrī's theory of literary coherence was influential on _al-Biqāʿī (d. 1480/885).

54 An-Nīsābūrī, *GQ*, 1: 59.

55 Morrison, *Islam and Science*, 127f.

56 Al-Jallād, *an-Nīsābūrī*, 90.

ceptional could use to unlock esoteric interpretations.[57] An-Nīsābūrī's identification of clear (*muḥkam*) and ambiguous (*mutashābih*) passages through the science of rhetoric informed his decisions about the relationship between a verse's direct sense, a verse's exoteric interpretation, and a verse's esoteric interpretation.[58] Passages were ambiguous, in special need of *ta'wīl*, when rational (*'aqlī*) evidence militated against the plain sense of the passage.[59] If the evidence that favored an esoteric interpretation were semantic, then the passage would no longer be ambiguous![60]

Q 18:86 ("He (Dhū l-Qarnayn) found it (the sun) setting in a muddy spring") was such a passage that rational analysis found to be ambiguous. An-Nīsābūrī adduced much more evidence from astronomy than ar-Rāzī had in order to argue that a celestial body many times the size of the earth could not fit in a spring.[61] An-Nīsābūrī pointed out that the Qur'ān's phrase 'found it setting' rather than 'set' meant that the text was conveying Dhū l-Qarnayn's (Alexander The Great) impression of what happened, and not necessarily a direct report of what did happen. Moreover, the sun setting in the spring was not a figure of speech. In the *ta'wīl*, an-Nīsābūrī explained that the lower or nether world (*'ālam as-suflā*) was where the sun of the human soul sets.[62] The spring was the material world. Here *ta'wīl* superseded an exoteric interpretation (*tafsīr*) of the direct sense of the Qur'ān because the direct and metaphorical senses of the passage made no sense.

There were cases in which an-Nīsābūrī's analysis of the Qur'ān's rhetorical techniques aided his exoteric exegesis. An-Nīsābūrī examined the use of a simile (*tashbīh*) in Q 36:39 which reads: "And the moon– we have determined it by stations (*manāzil*, sing. *manzil*), till it returns like an aged palm-bough." An-Nīsābūrī's *tafsīr* of the verse explained why the word 'aged' (*qadīm*) could not apply directly to the moon. Rather, he explained that an aged palm-bough would become thin, twisted, and yellow. All those words described the new moon.[63] Since the Qur'ān could not be speaking directly in this passage, an-Nīsābūrī's *ta'wīl*, which held that attaining truth was the final station (*manzil*) of the human soul, should also be taken seriously precisely because even the exoteric interpretation diverged from the passage's direct sense.[64] The presence of indirect speech in a passage did not mean setting aside exoteric exegesis.

There were other cases where an-Nīsābūrī's esoteric interpretations functioned in concert with the direct sense of a passage. For instance, an-Nīsābūrī explored in his *tafsīr* of Q 6:76–78

---

57 An-Nīsābūrī, *GQ*, 7: 160.
58 Morrison, *Islam and Science*, 50.
59 Al-Jallād, *an-Nīsābūrī*, 138.
60 Morrison, *Islam and Science*, 128.
61 Ibid., 128f.
62 An-Nīsābūrī, *GQ*, 16: 26.
63 Ibid., 23: 17. See also ar-Rafā'ī, *al-Athar*, 145.
64 An-Nīsābūrī, *GQ*, 23: 20.

> When night outspread over him he saw a star and said, 'This is my Lord.' But when it set he said, 'I love not the setters.' When he saw the moon rising, he said, 'This is my Lord,' but when it set, he said, 'If my Lord does not guide me I shall surely be of the people gone astray.' When he saw the sun rising, he said, 'This is my Lord; this is greater!' But when it set he said, 'O my people, surely I am quit of that you associate'

how Abraham's experience with astrology and his reflection on celestial phenomena led him to a profound belief in a single God.[65] In the *taʾwīl* of the same verse, an-Nīsābūrī expounded upon the hidden meaning of the rising and setting of heavenly bodies. When the moon set, Abraham's desire for mystical, divine guidance increased. Without that guidance, Abraham would have gone the way of his father Āzar. An-Nīsābūrī commented that when the sun rose and set, "there remained only God (*al-wājib al-ḥaqq*). Some identify the star with sense perception and the moon with imagination and the sun with conjecture and reason. This means these three perceptive powers are weak, possessing finite power as the guardian (*al-mudabbir*) of the universe dominates them and exercises power over them."[66] In this case, an-Nīsābūrī's esoteric and exoteric interpretations of the same passage reinforced each other as they both pointed to God's full control over natural processes and rejected the existence of other deities.

4 The Reception of an-Nīsābūrī's Work

Three types of evidence indicate that GQ has always been an-Nīsābūrī's best-known work. First, the titles of the three modern monographs on an-Nīsābūrī in Arabic and Persian that I have cited in this chapter indicate their sole focus on GQ. Second, bio-bibliographic sources such as *adh-Dharīʿa ilā taṣānīf ash-Shīʿa* and *Rawḍāt al-Jannāt* pay the most attention to GQ of all of an-Nīsābūrī's works. Third, even modern scholarship on an-Nīsābūrī that has studied his work on science has done so by bringing his scientific texts into conversation with GQ. Modern reprints of GQ reflect a continuing interest in an-Nīsābūrī's exegesis of the Qur'ān. In the centuries that followed an-Nīsābūrī's death, his astronomy texts were widely studied. There is a report of *at-Tawḍīḥ* being studied in the Bahmani sultanate, located in the Deccan, where an-Nīsābūrī finished GQ, by the end of the fourteenth century.[67] In the mid-fifteenth century, at Samarqand, where astronomy was studied in a *madrasa* that Ulugh Bey founded, Fatḥ Allāh ash-Shirwānī (d.c. 1486/891) praised both *at-Tawḍīḥ* and *Taʿbīr at-Taḥrīr*, which he had stud-

65 Ibid., 7: 175–77.

66 An-Nīsābūrī, *GQ*, 7: 180.

67 Morrison, *Islam and Science*, 152 and an-Nīsābūrī, *GQ*, 30: 229.

ied with Qāḍī Zādah Rūmī (d.c. 1432/835) who had written a gloss on *Taʿbīr at-Taḥrīr*.[68] Ash-Shirwānī also referred to *Kashf-i ḥaqāʾiq-i Zīj-i Īlkhānī*.[69] The modern Shiite reception of GQ is related to an-Nīsābūrī's reputation for scientific proficiency and the ongoing role of science in Islamic scholarship. In the 2007 *Durūs-i hayʾat va dīgar rishtahhā-i riyāḍī* (Lessons on astronomy and other mathematical subjects), the author, Ayatollah Āmulī, cited GQ (apud Q 1:4) as a source for the definition of a day and as an authoritative source (apud Q 18:86) for the sphericity of the earth.[70] No other *tafsīr* has combined fluency in cutting-edge science with a serious engagement with transmitted knowledge.

5 Bibliography

5.1 Sources

The Koran Interpreted. Translated by A. J. Arberry. New York: Macmillan, 1955.

An-Nīsābūrī, Niẓām ad-Dīn. *Gharāʾib al-Qurʾān wa-raghāʾib al-Furqān*. Ed. Ibrāhīm ʿAṭwa ʿAwaḍ. Cairo: Muṣṭafā l-Bābī l-Ḥalabī, 1962.

An-Nīsābūrī, Niẓām ad-Dīn. *Sharḥ Shāfiyat Ibn al-Ḥājib al-musammā bi-Sharḥ an-Niẓām*. Ed. ʿAlī sh-Shamlāwī. Beirut: Shams al-Mashriq, 1992.

An-Nīsābūrī, Niẓām ad-Dīn. *Gharāʾib al-Qurʾān wa-raghāʾib al-furqān* (printed in margins of aṭ-Ṭabarī, *Jāmiʿ al-bayān*), 30 vols. in 12. Beirut: Dār al-Maʿrifa, 1992; reprint of the 1905–11 Būlāq edition.

An-Nīsābūrī, Niẓām ad-Dīn. *Gharāʾib al-Qurʾān wa-raghāʾib al-Furqān*. Ed. Zakariyyā ʿUmayrāt. Beirut: Dār al-Kutub al-ʿIlmiyya, 1996.

Ar-Rāzī, Fakhr ad-Dīn. *At-Tafsīr al-kabīr*. 32 vols in 16. Beirut: Dār Iḥyāʾ at-Turāth al-ʿArabī, n.d.

5.2 Secondary Literature

Āmulī, Ḥasan Zādah. *Durūs-i hayʾat va dīgar rishtahhā-i riyāḍī*. Qom: Markaz-i Intishārāt-i Daftar-i Tablīghāt-i Islāmī, 2007.

Ayoub, Mahmoud M. *The Qur'an and its Interpreters*, 2 vols. Albany: State University of New York Press, 1984.

Dallal, Ahmad S. "Science and the Qurʾān." In *EQ*, edited by Jane Dammen McAuliffe. 6 vols. Leiden: Brill, vol. 4., 2004: 540–58.

Eichner, Heidrun. "The Post-Avicennian Philosophical Tradition and Islamic Orthodoxy. Philosophical Theological Summae in Context." Habil. thesis. Martin-Luther-Universität Halle-Wittenberg, 2009.

Ibn Sīnā. *Ash-Shifāʾ, aṭ-Ṭabīʿiyyāt: al-Samāʾ wa-l-ʿālam*. Ed. Ibrāhīm Madkūr and Maḥmūd Qāsim. Cairo: General Egyptian Book Organization, 1969.

68 Morrison, *Islam and Science*, 17f.

69 Scott G. Trigg, "From Samarqand to Istanbul: Astronomy and Scientific Education in the Commentaries of Fatḥallāh al-Shirwānī" (Ph.D. Dissertation, University of Wisconsin-Madison, 2016), 132.

70 Ḥasan Zādah Āmulī, *Durūs-i hayʾat va dīgar rishtahhā-i riyāḍī* (Qom: Markaz-i Intishārāt-i Daftar-i Tablīghāt-i Islāmī, 2007), 1: 74f. (definition of a day) and 279 (sphericity of the earth). See also an-Nīsābūrī, *GQ*, 1: 73.

Al-Jallād, Mājid Zakī. *An-Nīsābūrī wa-manhajuh fī t-tafsīr*. Amman: Dār al-Fikr li-ṭ-Ṭibāʿa wa-n-Nashr wa-t-Tawzīʿ, 2000.

Lane, Andrew. *A traditional Muʿtazilite Qurʾan commentary: The Kashshāf of Jār Allāh al-Zamakhsharī (d. 538/1144)*. Boston/Leiden: Brill, 2006.

Monnot, Guy. "Le verset du trône." *MIDEO* 15 (1982): 119–44.

Morrison, Robert. *Islam and Science: The Intellectual Career of Niẓām al-Dīn al-Nīsābūrī*. Abingdon, Oxon: Routledge, 2007.

Morrison, Robert. "Science and Theodicy." In *Nature and Scripture in the Abrahamic Religions*, edited by Jitse M. van der Meer and Scott Mandelbrote. Leiden: E.J. Brill, 2008: 249–72.

Morrison, Robert. "What *Was* the Purpose of Astronomy in Ījī's *Kitāb al-Mawāqif fī ʿilm al-kalām*?" In *Tabriz after the Mongol Conquest*, edited by Judith Pfeiffer. Leiden: Brill, 2013: 201–29.

Muḥammadkhānī, Ḥusayn. *Nīshābūrī va tafsīr-i Gharāʾib al-Qurʾān*. Tehran: Khānah-i Kitāb, 2014.

Ar-Rifāʿī, ʿĀdil Rāḍī. *Al-Athar al-balāghī fī tafsīr an-Nīsābūrī*. Amman: Dār Majdalāwī li-n-Nashr wa-t-Tawzīʿ, 2012.

Sabra, A. I. "Science and Philosophy in Medieval Islamic Theology: The Evidence of the Fourteenth Century." In *ZGAIW* 9 (1994): 1–42.

Stewart, Devin. *Islamic Legal Orthodoxy: Twelver Shiite Responses to the Sunni Legal System*. Salt Lake City: The University of Utah Press, 1998.

Trigg, Scott G. "From Samarqand to Istanbul: Astronomy and Scientific Education in the Commentaries of Fatḥallāh al-Shirwānī." Ph.D. Dissertation, University of Wisconsin-Madison, 2016.

Troudi, Khaled. "Qurʾānic Hermeneutics with Reference to Narratives: A Study in Classical Exegetical Traditions." Ph.D. Dissertation, University of Exeter, 2012.

Van Ess, Josef. *Der Wesir und seine Gelehrten*. Wiesbaden: Franz Steiner, 1981.

Muḥammad b. Aḥmad b. Muḥammad b. Juzayy al-Kalbī

Ralf Elger

1 Introduction

Ibn Juzayy was a scholar and exegete from Granada. He wrote *at-Tashīl li-ʿulūm at-tanzīl* (The Simplified Approach to the Knowledge of Revelation), a commentary on the Qurʾān with an extensive introduction that presents the art of *tafsīr* (exegesis).[1] Ibn Juzayy often refers to earlier exegetes, but also adopts original approaches. While the *Tashīl* was not widely received until the twentieth century, it is now well studied and commented upon in Islamic academic literature, though not in Western scholarship.

2 Life, Works, Political and Intellectual Context

The only independent biography of Ibn Juzayy is contained in the biographical collection *al-Iḥāṭa fī akhbār Gharnāṭa*,[2] written by his student, Lisān ad-Dīn b. al-Khaṭīb (d. 1374/776).[3] It says: Ibn Juzayy was born into a learned family. He received a good scholarly education in Granada. From a young age, he served as a *khaṭīb* (preacher) in the great mosque of the city, and he was an exceptionally good teacher. Ibn Juzayy seems to have spent his life only in Granada and did not undertake any major travels. No major events concerning him are reported until he died in battle near the city of Ṭarīf (now Tarifa) on October 30, 1340 (Jumādā I, 741).

Besides some poems,[4] all works of Ibn Juzayy are about the disciplines of the learned curriculum, theology, law, hadith, grammar and also Qurʾān. A comprehensive overview of his works is not in place[5], but some remarks shall be made: Ibn Juzayy tended to write quite briefly. For example, in his two books on theology, *an-Nūr*

1 The text has ca. 1000 pages in the edition of Abū Bakr Saʿdāwī (Sharja, UAE: Al-Muntadā l-Islāmī 2012). This edition is used in this article.

2 Lisān ad-Dīn b. al-Khaṭīb, *al-Iḥāṭa fī akhbār Gharnāṭa*, ed. Yūsuf ʿAlī Ṭawīl (Beirut: Manshūrāt Muḥammad ʿAlī Bayḍūn, 2003), 3: 10–13.

3 Cynthia Robinson, "Lisān al-Dīn IBN al-KHAṬĪB," in *Essays in Arabic literary biography, 1350–1850*, ed. Joseph E. Lowry and Devin J. Stewart (Wiesbaden: Harrassowitz, 2009), 159–73; Ralf Elger, "Adab and historical memory. The Andalusian poet/politician Ibn al-Khaṭīb presented in Aḥmad al-Maqqarī's (1577–1632/986–1041/ 'Nafḥ aṭ-ṭīb'," *WI* 42 (2002): 289–306.

4 These are described in: Fernando Velázquez Basanta and Celia del Moral Molina, "Los Banu Ŷuzayy. Una familia de juristas e intelectuales granadinos del siglo XIV. I: Abū l-Qāsim Muḥammad Ibn Ŷuzayy," *Miscelanea de estudios arabes y hebraicos* 45 (1996): 189–97.

5 See the bibliography of this article.

https://doi.org/10.1515/9783111320083-009

al-mubīn fī qawāʿid ʿaqāʾid ad-dīn, and juridical method, *Taqrīb al-wuṣūl ilā ʿilm al-uṣūl*, he refrains from lengthy expositions of the scholarly tradition and provides little evidence for his statements. The texts give the impression that they were intended for school use. And indeed, the *Taqrīb*, Ibn Juzayy says,[6] he wrote as primer for his son Abū ʿAbd Allāh Muḥammad b. Juzayy (d. 1356/757).[7] Ibn Juzayy's *Kitāb al-Anwār as-saniyya fī l-alfāẓ as-sunniyya*, was intended[8] for another son, Abū Bakr Muḥammad b. Juzayy (d. 1383/785).[9] It represents a basic set of hadiths for a young, budding scholar. Now, one might think that Ibn Juzayy's *Tashīl* was also meant for beginners, since its title means "to make something easy." But this is deceptive, for the *Tashīl* presupposes great knowledge in different disciplines. And it goes beyond the *tafsīr* tradition in many respects. Ibn Juzayy wanted to write something advanced, albeit in a short, condensed form.

Most remarkably, the *Tashīl* is not mentioned in the *Iḥāṭa* among Ibn Juzayy's works.[10] Several reasons are conceivable for this. The *Tashīl* was Ibn Juzayy's last work before he died in the disastrous battle of Tarifa. Perhaps his manuscript was not completed at the time and therefore was not circulated. Thus, Ibn al-Khaṭīb might not have known it or did not consider the exegetical material Ibn Juzayy had produced as a completed book.[11]

How did Ibn Juzayy relate to his social environment in Granada?[12] Not much is known. Only a few things can be said about his political position. Ibn Juzayy had no court office though this would not have been uncommon for a scholar in Granada. Ibn Juzayy's pupil Ibn al-Khaṭīb had close contacts to the court. Also, Ibn Juzayy's three sons had official positions. Abū ʿAbd Allāh Muḥammad b. Juzayy worked in the court of Granada and later in Fes where he penned the autobiography of Abū ʿAbd Allāh Muḥammad b. Baṭṭūṭa. The second son, Abū Bakr Muhammad b. Juzayy after service to

---

6 Muḥammad b. Aḥmad b. Muḥammad b. Juzayy al-Kalbī (693–741H), *Taqrīb al-wuṣūl ilā ʿilm al-uṣūl*, ed. ʿAbd Allāh Muḥammad Jubūrī (Ammān: Dār an-Nafāʾis, 2002), 88.

7 See Ralf Elger "Ibn Juzayy, Abū ʿAbdallāh," in *EI*^{3}, ed. Kate Fleet et al. Brill, 2007–, article published online 2020, accessed on June 23, 2022, https://referenceworks.brillonline.com/entries/encyclopaedia-of-islam-3/ibn-juzayy-abu-abdallah-COM_30849?s.num=153&s.start=140.

8 Muḥammad b. Aḥmad b. Muḥammad b. Juzayy al-Kalbī (693–741H), *Kitāb al-Anwār as-saniyya fī l-alfāẓ as-sunniyya*, ed. Muḥammad Shāyib Sharīf al-Jazāʾirī (Beirut: Dār al-Kutub al-ʿIlmiyya, 1971), 17.

9 Fernando Velázquez Basanta, "Ibn Ŷuzayy al-Kalbī, Abū Bakr," in *Biblioteca de al-Andalus*, ed. Jorge Lirola Delgado, Puerta Vílchez, and José Miguel (Almería: Fundación Ibn Tufayl de Estudios Árabes, 2009), 6: 196–203 (no. 1414).

10 Some other works are mentioned by Ibn al-Khaṭīb, al-Iḥāṭa, 3: 12. But they are not extant today.

11 ʿAlī Muḥammad az-Zubayrī, *Ibn Juzayy wa-manhajuhu fī t-tafsīr. Dirāsa mushaba ʿan al-Imām al-mufassir al-Andalusī sh-shahīd Ibn Juzayy wa-tawḍīḥ mufaṣṣal li-manhajihi fī tafsīrihi at-Tashīl li-ʿulūm at-tanzīl* (Damascus: Dār al-Qalam, 1987), 222.

12 See for Granada in general: Adela Fábregas, ed., *The Nasrid Kingdom of Granada between East and West (Thirteenth to Fifteenth Centuries)* (Leiden: Brill 2020).

the Granada court chose the life of a scholar. The third son, Abū Muhammad ʿAbd Allāh b. Juzayy (d. 1408/810), wrote a book about horses for the Granada court.[13]

Ibn Juzayy kept clear of politics in general. Just one political subject relevant to his writings comes to mind, that is *jihād*, the struggle against the infidels. This was a serious problem of the state of Granada in the first half of the fourteenth century, when Ibn Juzayy lived. In fact, his life spanned between two important battles. In 1295/695, two years after his birth, the Sultan fought in the Battle of Ḥiṣn al-Lawz (Castle of the Almond Trees, now the town of Iznalloz) about thirty-five km north of Granada and won a great victory against Castile. The second battle, at Tarifa in 1340/741, ended in defeat and cost Ibn Juzayy's life and that of many other notables.

The constant struggle between the Christians and Muslims made several Andalusian scholars intensely deal with jihad. Aḥmad b. Ibrāhīm b. az-Zubayr (d. 1308/708),[14] a teacher of the Ibn Juzayy, wrote a (now lost) work entitled: *as-Sabīl ar-rashād fī faḍl al-jihād* (The Way of Good Behavior, about the Favor of jihad).[15] Ibn Juzayy did not prepare a monograph on this subject, but deals with it in chapter seven of his law book *al-Qawānīn al-fiqhiyya*. In the *Tashīl*, he comments on jihad in a passage on *nāsikh* and *mansūkh* (abrogating and abrogated verses in the Qurʾān). In this, he presents a number of passages that advocate "peaceful coexistence with the disbelievers and enduring their vices" (*musālamat al-kuffār wa-l-ʿafū ʿanhum*). The first is in Q 2:83: *Wa-qūlū li-n-nās ḥusnan* ("speak to people in a good way"). Then in Q 2:191 it says: *Lā taqtulūnahum* ("do not fight them"). He quotes many other similar commandments and then adds that all of them are overridden by these two passages: *Fa-qtulū l-mushrikīna* ("fight the pagans," Q 9:5) and *kutiba ʿalaykum al-qitāl* ("the fight is prescribed for you," Q 2:216). Ibn Juzayy gives no further commentary on this subject.

Another field of activity that Ibn Juzayy could have chosen was Sufism. But as far can be derived from the biographical sources, Ibn Juzayy was not active in any *ṭarīqa* (Sufi order). It is unknown how well he knew Andalusian Sufis[16] such as Ibn Barrajān (d. 1141/536) from Seville who wrote a famous Qurʾān commentary.[17] In the *Tashīl*, Ibn Juzayy refers to the Sufi exegesis of the Persian mystic Muḥammad b. al-Ḥusayn as-Sulamī (d. 1021/412), called *Ḥaqāʾiq at-tafsīr* (The truths of exegesis). In his commentary on the phrase *al-ḥamdu li-llāh* in sura one, *al-Fātiḥa* (The Opening), Ibn Juzayy says that the term *ḥamd* (praise, thanks) is similar to the term *shukr* and continues:

13 See Teresa Isabel Sobrado Galanes, "Traducctión y estudio del Matla de Ibn Yuzavy: sobre rasgos y características del caballo" (PhD Diss., Madrid, 2015).

14 J. Lirola Delgado and E. Navarro i Ortiz, "Ibn al-Zubayr, Abū Ǧaʿfar," in *Biblioteca de al-Andalus*, 6: 327–339; Ibn al-Khaṭīb, *al-Iḥāṭa*, 1: 72–75.

15 Ibn al-Khaṭīb, *al-Iḥāṭa*, 1: 73.

16 See for Andalusian Sufism: Yousef Casewit, *The Mystics of al-Andalus: Ibn Barrajān and Islamic Thought in the Twelfth Century* (Cambridge: Cambridge University Press, 2016).

17 See Gerhard Böwering and Yousef Casewit, eds., *A Qurʾān Commentary by Ibn Barrajān of Seville (d. 536/1141). Īḍāḥ al-ḥikma bi-aḥkām al-ʿibra (Wisdom Deciphered, the Unseen Discovered)* (Leiden: Brill, 2015).

> *Shukr* has three degrees. The degree of the *ʿawāmm* (common people) is *ash-shukr ʿalā n-niʿam* (giving thanks for benefits from God). The degree of the *khawāṣṣ* (notables, here probably Sufis) is *ash-shukr ʿalā n-niʿam wa-n-niqam* (giving thanks for the good and the bad things that God makes happen) and for every situation one finds oneself in. The degree of *khawāṣṣ al-khawāṣṣ* (elite of the elite) is to lose sight of the benefits by seeing only the benefactor (God).[18]

This is an adaptation of as-Sulamī, who also lists three levels of people who praise and thank God.[19]

Ibn Juzayy does not often refer to as-Sulamī. Thus, when he speaks about another Sufi term, *taqwā* (piety), in the commentary on Q 2:2, no similarity to *Ḥaqāʾiq at-tafsīr* is apparent. The same is true of the term *dhikr* (Q 2:152). Ibn Juzayy had information about Sufi terminology and thought from sources other than as-Sulamī, like the manual of al-Qushayrī, which also deals with the terms *shukr*, *taqwā* etc. But direct connections are not apparent. The Sufi passages make up only a small part of the *Tashīl*, which is by no means an allegorical commentary, a *taʾwīl*. Ibn Juzayy makes this clear: *Taʾwīl*, he says, is an interpretation that does not follow the obvious meaning of the text, as the *tafsīr* does. And *tafsīr* is what he wanted to offer.[20]

3 The Intellectual Background of Ibn Juzayy's Hermeneutical Approach

Now to Ibn Juzayy's scholarly environment. From among his teachers,[21] only one shall be dealt with who had a significant position in Qurʾānic exegesis, the above mentioned Aḥmad b. Ibrāhīm b. az-Zubayr. Ibn Juzayy refers to him in the *Tashīl* as the "seal of the scholars of the Qurʾān in Andalus."[22] This is probably not an exaggeration, for the importance of Ibn az-Zubayr is also underlined by Ibn Juzayy's contemporary, the Andalusian-born scholar Abū Ḥayyān al-Gharnāṭī (d. 1344/745). Abū Hayyān studied under Ibn az-Zubayr before eventually immigrating to Egypt and writing the great exegesis *al-Baḥr al-muḥīṭ* (The Comprehensive Ocean).[23]

Two works on Qurʾānic exegesis were written by Ibn az-Zubayr. The first is *Milāk at-taʾwīl* (The Foundation of Exegesis), an appendix to *Durrat at-tanzīl* (The Pearl of Revelation) by al-Khaṭīb al-Iskāfī (d. 1029/420). Al-Iskāfī presents similar formulations

18 Ibn Juzayy, *Tashīl*, 48.

19 Muḥammad b. al-Ḥusayn as-Sulamī, *Ḥaqāʾiq at-tafsīr*, ed. Sayyid ʿUmrān (Beirut: Dār al-Kutub al-ʿIlmiyya, 2001), 33.

20 Ibn Juzayy, *Tashīl*, 7.

21 See the list in Ibn al-Khaṭīb, *al-Iḥāṭa*, 3: 11.

22 Ibn Juzayy, *Tashīl*, 12.

23 Abū Ḥayyān al-Gharnāṭī, *al-Baḥr al-muḥīṭ*, ed. ʿĀdil Aḥmad ʿAbd al-Mauǧūd, Beirut: Dār al-Kutub al-ʿIlmiyya, 2001, 1: 21.

and repetitions in the Qur'ān and comments on their functions. Ibn az-Zubayr adds more verses and his own comments. Ibn Juzayy mentions Ibn az-Zubayr's book several times in the *Tashīl* when discussing homonymous phrases in the Qur'ān.
One example concerns Q 2:125–126:

> Remember We made the House a place of assembly for men and a place of safety; and take ye the station of Abraham as a place of prayer; and We covenanted with Abraham and Ismail, that they should sanctify My House for those who compass it round, or use it as a retreat, or bow, or prostrate themselves (therein in prayer) (125) And remember Abraham said: "My Lord, make this a City of Peace . . ." (126)

Q 14:34–35 says:

> And He giveth you of all that ye ask for. But if ye count the favors of Allah, never will ye be able to number them. Verily, man is given up to injustice and ingratitude. (34) Remember Abraham said: "O my Lord! make this city one of peace and security: and preserve me and my sons from worshipping idols." (35)

Under discussion is verse Q 2:126: *Ijʿal hādhā baladan āminan* ("make this a city of safety") and Q 14:35: *Ijʿal hādhā l-balada āminan* ("make this city one of peace"). Ibn az-Zubayr asks why *balad* has a definite article in sura 14, but not in 2. The reason is that in sura 2 it is preceded by *idh jaʿalnā l-bait mathābatan li-n-nās*, which means "we made the house a place of refuge." And then God is asked to make this said place a place of peace. Sura 14, *Ibrāhīm* (Abraham), refers again to that same place, which was yet already established. Therefore, it is logical that the speaker, Abraham, says: "the place."[24] Whether this explanation is plausible or not, Ibn Juzayy accepts it.

In his second exegetical work, *al-Burhān* (The Proof), Ibn az-Zubayr argues that there is a certain order and interrelationship (*tartīb* or *tanāsub*) within and throughout the Qur'ānic suras. He claims that his book was the first monograph on this subject. This is confirmed by Muḥammad b. ʿAbd Allāh az-Zarkashī (d. 1392/794), who devotes a chapter to the subject in his great work *al-Burhān fī ʿulūm al-Qur'ān* (The Proof in the Knowledge of the Qur'ān).[25]

Ibn az-Zubayr's *Burhān* was also known to Abū Ḥayyān al-Gharnāṭī. In commenting on the passage *dhālika l-kitāb* ("this Book") from Q 2:2, he quotes from it the statement that these words respond to *ihdinā* ("Guide us!") from sura one, *al-Fātiḥa* (The Opening).[26] However, Abū Ḥayyān does not often refer to *tanāsub*. In the *Tashīl*, the subject is not mentioned at all. This is somewhat surprising. Perhaps Ibn az-Zubayr's

---

24 Aḥmad b. Ibrāhīm b. az-Zubayr, *Milāk at-taʾwīl*, ed. Saʿīd al-Fallāḥ (Beirut: Dār al-Gharb al-Islāmī, 1983), 50; Ibn Juzayy, *Tashīl*, 82.
25 Muḥammad b. ʿAbd Allāh az-Zarkashī, *al-Burhān fī ʿulūm al-Qur'ān*, ed. Muḥammad Abū l-Faḍl Ibrāhīm (Cairo: Dār at-Turāth, (199-?)), 1: 35–62. In fact, it is one of the first chapters, which may be due to the importance az-Zarkashī attached to the subject.
26 Abū Ḥayyān al-Gharnāṭī, *al-Baḥr*, 1: 61.

ideas on the *tanāsub* seemed too far-reaching. After all, they amount to considering the Qur'ān as an integrated text and are thus in some way opposed to the verse-by-verse exegesis. One can only speculate about this.

Ibn az-Zubayr is not the most important source for the *Tashīl*. This position holds firstly *al-Muḥarrar al-wajīz fī tafsīr al-kitāb al-ʿazīz* of the Andalusian scholar Abū Muḥammad ʿAbd al-Ḥaqq b. ʿAṭiyya (date of death not clear, ca. 1147/541), secondly the *Tafsīr al-Kashshāf* of Maḥmūd az-Zamakhsharī (d. 1144/538).[27] The first of the two was an obvious choice, since Ibn ʿAṭiyya was an accepted authority in al-Andalus. The appeal to az-Zamakhsharī, on the other hand, was delicate. Ibn Juzayy was aware that his exegeses contained Muʿtazilite, i.e. "heterodox" thoughts, and he states explicitly, that one must take just the useful from az-Zamakhsharī and omit the other.[28]

Ibn Juzayy also used other authors, although less frequently. He presents them in a passage of the *Tashīl* that describes the history of *tafsīr* since the very first Muslim generation. Included are among many others Abū Jaʿfar aṭ-Ṭabarī's *tafsīr* and Abū Muḥammad b. Qutayba's (d. 889/276) *Gharīb al-Qur'ān*. Several Maghrebinians and Andalusians appear, like Abū l-Ḥakam al-Ballūṭī (d. 966/355), Makkī b. Abī Ṭālib (d. 1045/437), and Abū Bakr b. al-ʿArabī (d. 1148/543). Then come az-Zamakhsharī and Ibn ʿAṭiyya and finally the youngest author of the list, Abū Faḍl b. al-Khaṭīb, as Ibn Juzayy calls him. He is better known as ar-Rāzī (d. 1209/606). Ibn Juzayy praises ar-Rāzī's *at-Tafsīr al-kabīr*, but says that some abridgement would do it good.[29]

The list shows that Ibn Juzayy was well versed in the history of *tafsīr*. He knew the value of what his predecessors had accomplished. However, his work is not only a compilation from previous exegeses, but also brings quite new approaches.

3.1 Exegetical-Hermeneutical Principals

That Ibn Juzayy was not content to follow old paths is especially evident in the two introductions to the *Tashīl*.[30] The first one (pp. 1–20 in the Sharja edition) introduces to various forms of Qur'ān-related areas of knowledge. The second (pp. 21–44) is a glossary of words with their meanings from the Qur'ān. The rest is the commentary. (p. 45–1024).

27 The two are also the principal sources of Abū Ḥayyān al-Gharnāṭī. See ibid., 1: 21.

28 Adday Hernández, "Qur'anic studies in al-Andalus. An overview on the state of research on qirā'āt and tafsīr," *JQS* 19 (2017): 74–102, points to the fact that many manuscripts of the two *Tafsīrs* exist in Spain.

29 Ibn Juzayy, *Tashīl*, 12.

30 The originality of the two introductions is pointed out by Ṭāriq b. Aḥmad Fāris, *ʿUlūm al-Qur'ān ʿinda al-imām Ibn Juzayy al-Kalbī wa-atharuhā fī tafsīrihi* at-Tashīl li-ʿulūm at-tanzīl (Riyadh: Kursī l-Qur'ān al-Karīm wa-ʿUlūmihi, Jāmiʿat al-Malik Saʿūd, 2016–2017), 14.

Ibn Juzayy's first introduction consists of ten chapters (*bāb*, pl. *abwāb*):[31] *Bāb* 1) The revelation (*nuzūl al-Qurʾān*) and the composition of the Qurʾānic text. *Bāb* 2) The revelations in Mecca and Medina. *Bāb* 3) The general aim (*maqṣūd*) of the Qurʾān is "the call to the creation to serve God (*ʿibāda*) and to come into the religion of God." – The *ʿibāda* consists of belief in the correct principles of faith (*uṣūl al-ʿaqāʾid*) and obeying the rules of action (*aḥkām al-ʿamal*). To make people serve God, the Qurʾān uses the two means of *tarhīb* and *targhīb*. That is, on the one hand, it threatens those who are not willing to serve God, and on the other hand, it arouses the desire to serve Him through promises.

The Qurʾān contains seven classes of statements (*maʿānī* or *ʿulūm*, both terms mean the same here) detailing what its general purpose is: 1) *Rubūbiyya* (statements explaining the concept of godhead). 2) *Nubuwwa* (explanation of what prophecy is). 3) *Maʿād* (the hereafter). 4) *Aḥkām* (the standards that people should obey). 5) *Waʿd* (the promise of a favorable judgment at the Last Judgment). 6) *Waʿīd* (the threat of a negative judgment). 7) *Qiṣaṣ* (knowledge of the prophetic stories in the Qurʾān). Ibn Juzayy discusses the phenomenon of repetition (*tikrār*) of the *qiṣaṣ*. It serves, he argues, to emphasize different details in different places, depending on the purpose of the account. Another reason is that the *qiṣaṣ* are rendered in a detailed form (*iṭnāb*) in some places and in a reduced, elliptical manner (*ījāz*) in others. This shows the *fasāḥa* (stylistic mastery) of the Qurʾān.[32]

Bāb 4) The twelve *funūn al-ʿulūm* (disciplines of interpretation) are: 1) *Tafsīr*, the main exegetical discipline, to which all the following contribute. 2) *Qirāʾāt* (readings), 3) *aḥkām* (norms), 4) *naskh* (abrogation), 5) *ḥadīth* (traditions from the Prophet and his companions), 6) *qiṣaṣ* (stories about prophets 7) *taṣawwuf* (Sufism). Twelve Sufi terms from the Qurʾān are dealt with in the course of interpretation. They are *shukr* (thanksgiving), *taqwā* (piety), *ṣabr* (perseverance), *tauḥīd* (confession of the oneness of God), *maḥabba* (love), *tawakkul* (trust in God), *murāqaba* (spiritual contemplation), *khawf* (fear), *rajāʾ* (hope), *tawba* (repentance), and *ikhlāṣ* (devotion). 8) *Uṣūl ad-dīn* (the basic rules of theology), 9) *uṣūl al-fiqh* (the basic rules of jurisprudence) 10) *lugha* (semantics, word meaning). The topic is taken up again below in the second introduction. 11) *Naḥw* (syntactic rules), 12) *bayān* (rhetoric).

Bāb 5) Twelve areas of disagreement among exegetes: 1) Different readings of the Qurʾān. 2) Differences in the vocalizations of words (*iʿrāb*) that cause differences in syntax. 3) Differences over word meanings (*maʿnā l-kalima*). 4) Differences over identical word forms (*lafẓ*) with different meanings. 5) The question of whether a statement about a category of people refers to all of them or only to some (*ʿumūm* or *khuṣūṣ*). 6) Does a statement apply absolutely or restrictedly? (*iṭlāq* or *taqyīd*). 7) Should the exegete ascribe a real meaning or a metaphorical meaning to certain words (*ḥaqīqa* or *majāz*). 8) Is there an ellipsis in certain places (*iḍmār*) or not (*istiqlāl*). 9) There is the possibility that a word

31 See for a commentary on this first introduction Musāʿid b. Sulaymān Ṭayyār, *Sharḥ muqaddimat at-Tashīl li-ʿulūm at-tanzīl li-Ibn Juzayy* (Dammam: Dār Ibn al-Jawzī, 2009–2010).

32 Ibn Juzayy, *Tashīl*, 6.

is integrated into a sentence that is not necessary for the meaning of the sentence (*iḥtimāl kawn al-kalima zāʾida*). Whether this option was used, however, is often debatable. 10) Do parts of the sentence come before or after their normal position, so that the exegete must understand the sentence in a different form than that given? (*iḥtimāl ḥaml al-kalām ʿalā t-tartīb wa-ʿalā t-taqdīm wa-t-taʾkhīr*). 11) Is one norm overridden by another norm or not (*mansūkh* or *muḥkam*)? 12) Differences over which *tafsīr* statements of the Prophet and the Companions are relevant to the interpretation of certain passages.

Twelve ways to resolve the disputes and indicate the best variant of interpretation (*wujūh at-tarjīḥ*): 1) Problematic passages in the Qurʾān should be interpreted on the basis of other but analogous passages (*tafsīr baʿḍ al-Qurʾān bi-baʿḍ*). 2) If there is a suitable hadith from the Prophet pointing to a particular interpretation, it settles the dispute. 3) The opinion of the majority of exegetes is to be followed in certain cases. 4) If there is a suitable interpretation by the Companions of the Prophet, it is to be preferred over a deviating opinion by the scholars. 5) The exegete should understand the sentence in accordance with the normal usage of the Arabic language (*kalām al-ʿarab*). 6) The exegete primarily assumes that the sentence to be interpreted has a normal sentence structure (*siyāq al-kalām*), with no deviations from the normal flow of words. 7) A plausible meaning (*mutabādir ilā dh-dhihn*) is to be assumed. 8) It is better to assume a non-metaphorical use of a word than a metaphorical one (*taqdīm al-ḥaqīqa ʿalā l-majāz*). 9) A statement is to be interpreted in a general rather than a specific sense (*taqdīm al-ʿumūm ʿalā l-khuṣūṣ*). 10) A statement should be interpreted in such a way that it is given an absolute binding force and not only a limited binding force (*al-iṭlāq ʿalā t-taqyīd*). 11) Normally, the exegete should not assume the existence of an ellipsis (*al-istiqlāl ʿalā l-iḍmār*). 12) The sentence is to be interpreted as if the order of its elements is the usual one (*ḥaml al-kalām ʿalā tartībihi*), except in cases where a preposition or postposition of a word is more plausible.

Bāb 6) A chronologically ordered list of exegetes (*mufassirūn*), beginning with the first Muslim generation and ending with Fakhr ad-Din ar-Rāzī (d. 1209/606).

Bāb 7) Abrogated and abrogating verses. *Bāb* 8) On the readings of the Qurʾān. *Bāb* 9) *Mawāqif* (The use of pause forms of words in reciting the Qurʾān). *Bāb* 10) *Al-Faṣāḥa wa-l-balāgha wa-adawāt al-bayān* (Stylistic features of the Qurʾān, with a list of rhetorical devices used). *Bāb* 11) *Iʿjāz* (the inimitability of the Qurʾān). *Bāb* 12) *Faḍāʾil* (the excellence of the Qurʾān).

This is a long list of topics that Ibn Juzayy addresses. He did this in order to fill the gap which the existing *tafsīrs* left that covered only a few of the relevant hermeneutical disciplines. Ibn ʿAṭiyya is a good case for comparison. His introduction covers the "merits of the Qurʾān" (*faḍl al-Qurʾān*) (pp. 36ff.), *tafsīr al-Qurʾān* (40–42), the seven readings (43–48), the collection of the Qurʾān (49f.), foreign words in the Qurʾān (51), inimitability (*iʿjāz*) (52), the concise style (*ījāz*), and (54) the names of the Qurʾān, of the suras and verses (56–58). Ibn Juzayy brings many more subjects.[33] Indeed, this first introduction to the *Tashīl* is like a brief predecessor to az-Zarkashī's very detailed work on the *ʿulūm al-Qurʾān*, the disciplines of exegesis.

Ibn Juzayy's quest for methodical penetration of the exegetical practice is evident in his definition of the term *tafsīr*:

> The meaning of *tafsīr* is *sharḥ al-Qurʾān* (the explanation of the Qurʾān) and *bayān maʿnāhu* (clarification of its meaning) and *al-ifṣāḥ bi-mā yaqtaḍīhi* (making clear what follows from the words of the Qurʾān) – whether in the *naṣṣ* (absolutely clear passages), in the *ishāra* (allusions to a hidden meaning), or in the *faḥwā* (implied meaning).[34]

That means: For understanding of the Qurʾān, it is a prerequisite to know the discipline of *uṣūl al-fiqh*, from which the terms in this definition are taken. Ibn Juzayy criticizes that many exegetes in the past have not applied the *uṣūl al-fiqh*, although mastery of their methods contributes to the understanding of the Qurʾānic text and provides criteria for deciding between different interpretations. He wants to give the *uṣūl* their deserved place in Qurʾānic hermeneutics.

In addition to the terms in the above definition of *tafsīr*, other useful concepts can be drawn from the *uṣūl*, Ibn Juzayy explains.[35] These are the categories *ẓāhir* (obvious sense, but not entirely clear), *mujmal* (ambiguous sense) and *mubayyan* (clarified sense). Also the classification of statements: *ʿāmm* (a general statement), *khāṣṣ* (a specified statement), *muṭlaq* (an absolutely valid statement) and *muqayyad* (a qualified statement). All these terms come from *uṣūl al-fiqh*, says Ibn Juzayy. This is also true for various categories of speech, *faḥwā l-khiṭāb* (the deeper meaning of speech), *laḥn al-khiṭāb* (actually grammatically incorrect speech, but here it means "speech that uses ellipses"), *dalīl al-khiṭāb* (what the speech points to). Also from the *uṣūl*, one can learn what the *shurūṭ an-naskh* are (how abrogated and abrogating statements come about). Further, one can learn the *wujūh at-taʿāruḍ* (types of discrepancies in interpretation) and *asbāb al-khilāf* (reasons for disagreement among exegetes).

These topics also appear in Ibn Juzayy's monograph entitled *Taqrīb al-wuṣūl ilā ʿilm al-uṣūl*, which was certainly written before *Tashīl*. In this work, he shows in detail how *uṣūl al-fiqh* is helpful for understanding the Qurʾān. He also describes the rules that should be applied to deal with the differences between exegetes.[36] One rule is labeled *ash-sharʿī ʿalā l-ʿaqlī* (the sharia is preferred over the ratio). Corresponding to this in the *Tashīl* is the advice to "follow the sayings of the Prophet." The rule *al-ʿurfī ʿalā l-lughawī* (the argument from tradition beats the linguistic argument) from *Taqrīb*

33 See Fāris, *ʿUlūm al-Qurʾān ʿinda al-imām Ibn Juzayy*, 14f.

34 Ibn Juzayy, *Tashīl*, 7.

35 Ibid., 9.

36 Muḥammad b. Aḥmad b. Muḥammad b. Juzayy al-Kalbī (693–741H), *Taqrīb al-wuṣūl ilā ʿilm al-uṣūl*, ed. ʿAbd Allāh Muḥammad Jubūrī (ʿAmmān: Dār an-Nafāʾis, 2002), 175ff.

al-wuṣūl has as its counterpart in the *Tashīl* the advice to follow the "majority of the exegetes" and the "sayings of the companions."

One can see that Ibn Juzayy does much to establish *uṣūl al-fiqh* as the methodological basis for *tafsīr*. Of course, other exegetes were no strangers to the terminology listed above. Ibn ʿAṭiyya uses it in his *tafsīr* sometimes, but he does not offer a systematic methodology based on *uṣūl al-fiqh*. That is the contribution of Ibn Juzayy.

The second introduction is an alphabetically ordered[37] list of words with explications, which Ibn Juzayy conceived as an appendix to the *fann al-lughāt* (semantics) from the first introduction where he explains how important it is for the exegete to know the correct meanings of Qurʾānic words. This second introduction is the practical implementation in the form of a small dictionary. The first lemma is *āya*. Ibn Juzayy says that this word has two meanings, a "sign" or "proof" (*burhān*) as well as a "verse" in the Qurʾān. It is followed by *atā* which means "he came." A word with the same root is contained in the Qurʾānic passage *waʿduhu maʾtiyyan* ("Verily, His promise is eminent," Q 19:61) The next word is *ʿātā*, which means "he gave". And from it comes *wa-l-muʿṭūna z-zakāta* ("and the givers of zakāt," Q 4:162). The next word is *abā*, "He avoided." Here Ibn Juzayy gives no further explanation. Some other words which Ibn Juzayy explains: The word *umm* has two meanings: Mother and origin, as in *umm al-qurā* (origin of the towns). The word *rabb* has four meanings: The lord, the master, the owner of something, and the one who sets a thing in order.

Ibn Juzayy argues that the list saves space, because it includes words that appear several times in the Qurʾān. The explanation of these words is generally given only once in the list. Then the interpretation at the various places in the Qurʾān where they occur can refer to the list. This is not a bad idea, one might say. However, he himself points out that despite the entry in the list, a word in the course of the *tafsīr* occasionally needs to be explained because it depends on the context, which meaning the word has straight. So the space saving effect does not work quite well.[38]

In conceiving the list, was Ibn Juzayy following a tradition? There is no model in the *tafsīr*-genre.[39] A better candidate are the works dealing with *gharīb al-Qurʾān* (the difficult words of the Qurʾān). Ibn Juzayy refers to authors such as Ibn Qutayba and Makkī b. Abī Ṭālib.[40] Yet these two did not arrange their lists alphabetically, but placed the words according to the order of the verses and suras. Now it could be that Ibn Juzayy has extracted his alphabetical list from these works or he used other *gharīb* works that have an alphabetical order like those of Abū Bakr as-Sijistānī (d. ca.

37 See for more details of the alphabetical order az-Zubayrī, *Ibn Juzayy wa-manhajuhu*, 629ff.

38 Az-Zubayrī, *Ibn Juzayy wa-manhajuhu fī t-tafsīr. Dirāsa mushaba ʿan al-Imām al-mufassir al-Andalusī sh-shahīd Ibn Juzayy wa-tawḍīḥ mufaṣṣal li-manhajihi fī tafsīrihi* at-Tashīl li-ʿulūm at-tanzīl (Damascus: Dār al-Qalam, 1987), 634ff.

39 Fāris, *ʿUlūm al-Qurʾān ʿinda al-imām Ibn Juzayy*, 15.

40 Makkī b. Abī Ṭālib al-Qaysī, *Tafsīr al-mushkil min gharīb al-Qurʾān al-ʿaẓīm ʿalā l-ījāz wa-l-ikhtiṣār*, ed. Hudā aṭ-Ṭawīl al-Marʿashlī (Beirut: Dār an-Nūr al-Islāmī, 1988).

942/330)[41] and ar-Rāghib al-Iṣfahānī (d. 1108/502).[42] The problem with all these works, however, is that they want to explain difficult words whereas Ibn Juzayy presents words that occur several times in the Qur'ān. A difficult word that occurs only once is excluded. In view of this, the question remains open as to exactly which model Ibn Juzayy was referring to.

Turning now to the commentary section of the *Tashīl*: It consists largely of a word-by-word interpretation. But sometimes Ibn Juzayy abandons this and applies a thematic approach. This is evident in the passage on the *isti'ādha* (the formula *a'ūdhu bi-llāh* = "I seek forgiveness in god") (p. 45f.) and the *basmala* ("In the name of god, the merciful"). (p. 46f.) Each of the two sections consists of ten *fawā'id* (topics or subjects of discussion). The following commentary on sura one, *al-Fātiḥa* (The Opening), proceeds similarly, with twenty *fawā'id* (p. 48–51). This section will now be considered as an example.

The first *fā'ida* deals with rules for reciting the *Fātiha* in ritual prayer (*ṣalāt*). The jurists Mālik b. Anas and Muḥammad b. Idrīs ash-Shāfi'ī say it is *wājib* (obligatory), explains Ibn Juzayy, but not Abū Ḥanīfa. Ibn Juzayy brings the respective arguments for both positions. The second *fā'ida* is about whether the implied meaning at the beginning of the *Fātiḥa* is *qūlū: al-ḥamdu li-llāh* ("say: Praise be to Allah, the Sustainer of the worlds"), that is, that the command *qūlū* ("say") is omitted from the scripture. Ibn Juzayy makes arguments for the different positions without expressing his own opinion. He adds that there is definitely an ellipsis (*iḍmār al-qawl*) in verse 4: *Iyāka na'budu* ("Thee we serve"). This is because it is not likely that God is saying, "Thee do we serve," but that the command "say" is implied. Thus, the implicit meaning is, "Say, Thee do we serve."

The third *fā'ida* deals with the word *al-ḥamd*. The fourth *fā'ida* treats *shukr* as the first of the twelve Sufi terms that Ibn Juzayy recognizes in the Qur'ān. In the fifth *fā'ida*, he arrives at the phrase *al-ḥamdu li-llāhi rabbi l-'ālamīn* as a whole. In the sixth he discusses *rabb*, in the seventh *mālik*, and in the following *fawā'id* he follows the course of the text. In the *fawā'id* 13 and 14, a digression appears again. In 15 there is a commentary on *aṣ-ṣirāṭ*. Ibn Juzayy then follows the course of the text until *fā'ida* 19. In 20, he brings a general interpretation of the *Fātiḥa*, saying that this sura contains all the meanings of the Qur'ān as if it were an abridged copy of it. A list of these meanings has already been given in the introduction, *bāb* 3, he adds. The concluding section (*khātima*) describes the rule of saying *amīn* (amen) after reading the *Fātiḥa*.

Ibn Juzayy does not explain why he uses the *fā'ida* structure. Surely, he had in mind the model of Fakhr ad-Dīn ar-Rāzī. Ar-Rāzī built his entire *tafsīr* on thematic chapters, many of which digress quite far from the Qur'ānic text, such as Ibn Juzayy's *fā'ida* on *shukr*. But Ibn Juzayy did not use the systematic structure to the extent that his predecessor did.

---

41 Abū Bakr Muḥammad as-Sijistānī, *Kitāb Gharīb al-Qur'ān al-musammā bi-Nuzhat al-qulūb*, no ed. given (Cairo: Maṭba'at as-Sa'āda, 1907).

42 Abū l-Qāsim ar-Rāghib al-Iṣfahānī, *al-Mufradāt fī gharīb al-Qur'ān*, ed. Ṣafwān 'Adnān Dāwūdī (Mekka/Riyadh: Maktabat Nizār Muṣṭafā l-Bāz, 1997).

The commentary on sura two, *al-Baqara* (The Cow), begins as a word-by-word *tafsīr* until Ibn Juzayy comes to the term *muttaqīn* ("those who fear Allah") in verse 2. This gives him the opportunity to introduce the second Sufi term (after *shukr* in the *Fātiḥa*), *taqwā* (fear of God), a word with the same root as *muttaqīn*. To do this, he uses a digression divided into three *fuṣūl* (chapters, digressions), which together are about two pages long. In the next digression, on the word *dhikr* (remembrance) in Q 2:152, Ibn Juzayy enumerates three types of *dhikr*: *dhikr* with the heart (*qalb*), with the tongue (*lisān*) and with both. He adds several *dhikr*-formulas. In later suras, digressions on other Sufi terms – or rather, terms understood by Ibn Juzayy from a Sufi perspective – follow: *ṣabr* (endurance), *tawḥīd* (declaration of the oneness of God), etc.

Another digression, in Q 2:40, concerns the Banū Isrāʾīl (Israelites):[43] Sometimes God addresses them with gentleness and tells them what benefits He has brought them. Sometimes God frightens them, sometimes He brings arguments against their disbelief. Ibn Juzayy enumerates from the entire Qurʾān ten verses in which the benefits are mentioned. This is followed by ten verses about the bad deeds of the Israelites and so on. This passage of half a page is reminiscent of an analogous passage in ar-Rāzī's commentary, but is less extensive.[44] What has been said makes clear how Ibn Juzayy sometimes abandons linear interpretation in favor of a more systematic approach. In general, however, he follows the word-by-word exegesis.

In the course of interpretation, Ibn Juzayy often makes a choice between different possibilities of interpretation. This topic, called *tarjīḥ*, will occupy the last part of this article. One question is: Does Ibn Juzayy apply the rules he presents in the introduction in the commentary section? Another question: To what extent does he differ from other exegetes? This investigation can, of course, only be exemplary. Of all the rules for *tarjīḥ* that Ibn Juzayy presented in his first introduction, two will be considered. First, the rule stating that *ʿumūm* is preferable to *khuṣūṣ*. That is, a statement in the Qurʾān is preferably to be understood as general rather than specific. Second, *istiqlāl* is preferable to *iḍmār*, i.e., the exegete should preferably assume that a sentence has no omitted and implied elements.

First case: In Q 2:3, the word *yunfiqūn* ("they spend") raises the question of what exactly is spent. There are three alternative answers, *zakāt* (obligatory dues), *taṭawwuʿ* (voluntary dues), or generally all kinds of dues. That would be the *ʿumūm*-solution. Ibn Juzayy may have relied on Ibn ʿAṭiyya for this listing. The latter also lists several alternatives, but does not come to a decision. Ibn Juzayy, on the other hand, makes it clear that *ʿumūm* is preferable because there is nothing in the text demanding a *takhṣīṣ*, i.e., to determine the nature of the expenditure.

---

43 Ibn Juzayy, *Tashīl*, 64.

44 Fakhr ad-Dīn Muḥammad ar-Rāzī, *at-Tafsīr al-kabīr wa-mafātīḥ al-ghayb*, ed. Muḥammad Riḍwān ad-Dāya (Beirut: Dār al-Fikr al-Muʿāṣir, 1990), 3: 30ff.

In the commentary on sura three, *Āl-ʿImrān* (The House of ʿImrān), verse 32, Ibn ʿAṭiyya says that *qul aṭīʿū llāha wa-r-rasūl* (Say: "Obey Allah and His Messenger") is addressed to the Christians of the city of Najran. Ibn Juzayy accepts this, but he adds, *wa-maʿnāhā ʿalā l-ʿumūm fī jamīʿ an-nās* ("and the sentence has a general tenor and includes all people"). Ibn Juzayy also argues for a *ʿumūm* reading in the commentary on the beginning of sura four, *an-Nisāʾ* (Women): *Yā ayyuhā n-nāsu ttaqū rabbakum* ("O you people, worship your patron") is "a statement with a general meaning" (*khiṭāb ʿalā l-ʿumūm*). No such remark is found in the work of Ibn ʿAṭiyya.

Regarding sura five, *al-Māʾida* (The Table Spread), verse 38, Ibn ʿAṭiyya presents specifications for the rule of cutting off a thief's hand. This rule is only applied under certain conditions, such as the stolen thing being above a certain value. Ibn Juzayy accepts this and says, "The general meaning of the verse applies to every thief" (*ʿumūm al-āya yaqtaḍī kull sāriq*). But, he adds, the jurists introduced conditions for cutting off the hand with which they specified the general meaning. Ibn Juzayy himself lists eleven conditions in his work on applied jurisprudence *al-Qawānīn*. Again, Ibn Juzayy refers to the rule *ʿumūm ʿalā l-khuṣūṣ* (the reading that the statement is general is preferred over the specified reading) in the context of the *tafsīr*. But he accepts the juridical specification.

The last example of the *ʿumūm* vs. *khuṣūṣ* question. Sura 9, *at-Tawba* (Repentance), verse 12, states: "But if they break their oaths according to their covenant, and mock you for the sake of your faith, then fight the chiefs of unbelief; for their oaths are nothing to them, that they may thus be subdued." Here the question arises as to who are "the masters of disbelief" (*aʾimmat al-kufr*). Nothing is said about this by Ibn ʿAṭiyya and az-Zamakhsharī. Ibn Juzayy states that some exegetes pointed to contemporaries of Muḥammad, such as Abū Jahl, Umayya b. Khalaf, and others. But he argues that most of these people were dead before the revelation of the verse. Therefore, it is better to assume that all "advocates of unbelief" are meant. Again, Ibn Juzayy adopts his rule of preferring *ʿumūm* to *khuṣūṣ*.

The concept of ellipsis appears frequently in the commentary section of the *Tashīl*. Ibn Juzayy uses for it the term *iḍmār*, familiar from his rendering of *uṣūl al-fiqh*, and often *ḥadhf*, a term that does not appear in the introduction. The two seem to be synonyms in Ibn Juzayy's usage. One example has already been discussed, the beginning of the *Fātiḥa*: *al-ḥamdu li-llāh* ("praise is due to God"). Exegetes said that *qul* ("say") is absent but implied. This is discussed by ar-Rāzī, who does not accept *iḍmār* for a number of reasons. Ibn Juzayy, on the other hand, leaves this question open and thus differs from his predecessor in this case.

Another example is sura 12, *Yūsuf* (Joseph), verse 82: "Ask the city in which we have been and the caravan with which we have returned, and (you will find) that we are indeed telling the truth." Ibn ʿAṭiyya discusses whether the tenor of this sentence is "ask the people of the city" and the "people of the caravan," where "people" is elliptical (*ḥadhf*). On the other hand, some exegetes say that the city and the caravan are metaphors (*majāz*) for the people who make up the city and the caravan. While Ibn

ʿAṭiyya does not decide between the different positions, Ibn Juzayy chooses the first solution because the metaphorical reading of the Qurʾān should be avoided. Discussions like this crop up again and again in *Tashīl*. Ibn Juzayy often follows tradition, but also has his own ideas.

3.2 Historical Importance

The reception of the *Tashīl* was rather weak for a long time. One reason was given above: It can be assumed that the *Tashīl* was not quite finished at Ibn Juzayy's death. Probably it remained so for a time. Then, it would be interesting to know when a finished copy was made and when it became the basis of the subsequent copying tradition. One way to find out about this is to review the existing manuscripts. Today around ten copies are known, a not very high number compared to other classical *tafsīrs*. The interesting question is which one is the oldest of these copies. For the moment this position has the manuscript from al-Khizāna al-Ḥasaniyya (Bibliothèque Royale), Rabat, Morocco, nr. 2390, written 1633/1024.[45] It is hard to belief that this is the first one that was written. There may be an older one, but this is hard to say given the fact that some manuscripts are not dated. Perhaps older copies are lost.

Whatever the case, there was some reception of the *Tashīl* before the age of printing. One case is an undated manuscript from the library of the Nāṣiriyya Brotherhood in Tamghrut, Morocco, entitled *Taṣfiyat al-qulūb fī l-wuṣūl ilā ḥaḍrat ʿallām al-ghuyūb*. In the edition of the text, it was attributed to Ibn Juzayy and the English translators Hamza Yusuf and Musa Furber also consider it to be a work of his. But this is doubtful, for several reasons. The first is that no such title is mentioned in Ibn al-Khaṭīb's *Iḥāṭa* or in other biographies of Ibn Juzayy. Now one might counter that Ibn al-Khaṭīb could have overlooked it. The second argument is more important. The manuscript itself does not clearly name Ibn Juzayy as the author. Rather, it appears to be an excerpt from the *Tashīl*, written by a Sufi scholar who wanted to prove that Sufism was derived from the Qurʾān. He collected all the passages that Ibn Juzayy wrote about the twelve Sufi terms in the *Tashīl*.

Reception became more intense in the twentieth century, beginning with the oldest edition in Cairo in 1936 or 1937. Published research on the *Tashīl* began in 1978.

45 This is mentioned in the edition prepared by Muḥammad b. Saydī Muḥammad Maulāy (Kuwait: Dār aḍ-Ḍiyāʾ, 2013), 1: 12f.

4 Conclusion

The *Tashīl* was in its time an original effort to provide a comprehensive account of the exegetical disciplines. Further it demonstrated how the methodological and theoretical considerations of Ibn Juzayy could be used in the practical interpretation of the Qurʾān. Going back and forth between the systematical introduction and the *tafsīr* itself is made easy. In this respect the work is aptly called *Tashīl*. Its intellectual depth though is not at all easy to explore.

The text of Ibn Juzayy did not find much interest by later Muslim scholars. The reasons are not known. One may be that the precarious political situation in Andalusia obstructed the further flourishing of an exegetical tradition. It took until the twentieth century that a broad reception of the *Tashīl* began. Today, it is well studied by academic bodies in Islamic countries, perhaps because it is rather compact and thus can be easily interpreted in its entirety, in contrast to the many multi-volume works the exegetical tradition produced. The *Tashīl* would need further study also in Western research.

5 Bibliography

5.1 Sources

5.1.1 *Tashīl* – Editions (selection)

Ibn Juzayy, Abū ʿAbd Allāh Muḥammad. *At-Tashīl li-ʿulūm at-tanzīl*. No editor named. Cairo: al-Maktaba at-Tijāriyya al-Kubrā, 1936–1937.

Ibn Juzayy, Abū ʿAbd Allāh Muḥammad. *At-Tashīl li-ʿulūm at-tanzīl*. Ed. Muḥammad ʿAbd al-Munʿim al-Yūnusī and Ibrāhīm ʿAṭwa ʿAwaḍ. Cairo: Dār al-Kutub al-Ḥadītha, 1973.

Ibn Juzayy, Abū ʿAbd Allāh Muḥammad. *At-Tashīl li-ʿulūm at-tanzīl*. Ed. Lajnat Taḥqīq at-Turāth. Beirut: Dār al-Kitāb al-ʿArabī, 1973.

Ibn Juzayy, Abū ʿAbd Allāh Muḥammad. *At-Tashīl li-ʿulūm at-tanzīl*. Ed. ʿAbd Allāh al-Khālidī. Beirut: Dār al-Arqam 1995.

Ibn Juzayy, Abū ʿAbd Allāh Muḥammad. *At-Tashīl li-ʿulūm at-tanzīl*. Ed. Muḥammad Sālim Hāshim. Beirut: Dār al-Kutub al-ʿIlmiyya, 1995 and 2015.

Ibn Juzayy, Abū ʿAbd Allāh Muḥammad. *At-Tashīl li-ʿulūm at-tanzīl*. Ed. Riḍā Faraj al-Hamāmī. Sidon/Beirut: al-Maktaba al-ʿAṣriyya, 2003.

Ibn Juzayy, Abū ʿAbd Allāh Muḥammad. *At-Tashīl li-ʿulūm at-tanzīl*. Ed. Abū Bakr b. ʿAbd Allāh Saʿdāwī. Sharja, UAE: Al-Muntadā al-Islāmī 2012.

Ibn Juzayy, Abū ʿAbd Allāh Muḥammad. *At-Tashīl li-ʿulūm at-tanzīl*. Ed. Muḥammad b. Saydī Muḥammad Maulāy. 4 vols. Kuwait: Dār aḍ-Ḍīyāʾ, 2013.

5.1.2 Manuscripts

The following manuscripts are mentioned in the edition of Maulāy, 1: 12f

Al-Khizāna al-Ḥasaniyya (Bibliothèque Royale), Rabat, Morocco, nr. 2390, 214 fol., written 1024 h.
Al-Khizāna al-Ḥasaniyya (Bibliothèque Royale), Rabat, Morocco, no. 11111, written 1198 h., incomplete.
Al-Khizāna al-Malikiyya (= al-Khizāna al-Ḥasaniyya ?), no. 1197.
Al-Khizāna al-ʿĀmma, Rabat, Morocco, nr. 627/dal, 281 fol.
Al-Maʿhad al-ʿĀlī li-d-Dirāsāt wa-l-Buḥūth al-Islāmiyya (Mauretania), no. 1809.

The following are mentioned in the edition of Saʿdāwī, 1030

Dār al-Kutub al-Waṭaniyya, Tunis, no 11115, written 1206 h.
Markaz al-Malik Fayṣal li-l-Buḥūth wa-d-dirāsāt al-Islāmiyya, Riyadh, no. 10771, written 1082 h.
Markaz al-Malik Fayṣal li-l-Buḥūth wa-d-Dirāsāt al-Islāmiyya, Riyadh, no. 12802, written 1184 h.

Other manuscripts

Jāmiʿat al-Malik Saʿūd (King Saud University), no. 71.
Staatbibliothek Berlin, Landberg 85, catalogue Ahlwardt, no. 877. Ahlwardt says that the copy is from c. 1000 / 1591. But he has to proof for this.

5.1.3 Other Works of Ibn Juzayy in Editions

Only some of Ibn Juzayy's extant works are dated. A chronological order is thus excluded

Ibn Juzayy, Abū ʿAbd Allāh Muḥammad. *Kitāb al-Anwār as-saniyya fī l-alfāẓ as-sunniyya*. Ed. Muḥammad Shāyib Sharīf al-Jazāʾirī. Beirut: Dār al-Kutub al-ʿIlmiyya, 1971.

Ibn Juzayy, Abū ʿAbd Allāh Muḥammad. *Taqrīb al-wuṣūl ilā ʿilm al-uṣūl*. Ed. Muḥammad A. Farkūs. Cairo: Dār al-Aqṣā, 1990.

Ibn Juzayy, Abū ʿAbd Allāh Muḥammad. *Taqrīb al-wuṣūl ilā ʿilm al-uṣūl*. Ed. ʿAbd Allāh Muḥammad Jubūrī. Ammān: Dār an-Nafāʾis, 2002.

Ibn Juzayy, Abū ʿAbd Allāh Muḥammad. Al-Mukhtaṣar al-bāriʿ fī qirāʾat Nāfiʿ. Ed. Fatḥī l-ʿUbaydī. Aleppo: Dār ar-Rifāʿī, 2004.

Ibn Juzayy, Abū ʿAbd Allāh Muḥammad. *Al-Anwār fī nasab āl an-nabī l-mukhtār*. Ed. Mahdī r-Rajāʾī. Qum: Marʿashi Library 2010.

Ibn Juzayy, Abū ʿAbd Allāh Muḥammad. *Al-Qawānīn al-fiqhiyya fī talkhīṣ madhhab al-Mālikiyya wa-t-tanbīh ʿalā madhhab ash-Shāfiʿiyya wa-l-Ḥanafiyya wa-l-Ḥanbaliyya*. Ed. Mājid Ḥamawī. Beirut: Dār Ibn Ḥazm, 2013.

Ibn Juzayy, Abū ʿAbd Allāh Muḥammad. *An-Nūr al-mubīn fī qawāʿid ʿaqāʾid ad-dīn*. Ed. Nizār Ḥammādī. Hawally, Kuwait: Dār aḍ-Ḍiyāʾ li-n-Nashr wa-t-Tawzīʿ; Tūnis: Dār al-Imām Ibn ʿArafa, 2015.

Ibn Juzayy, Abū ʿAbd Allāh Muḥammad. *The Clear Light Fundamentals of Religious Beliefs by Ibn Juzayy Al-Kalbī*, trans. Fahim Hoosen. N.p. (South Africa): Dār al-Ḥikma publications, 2017.

Ibn Juzayy, Abū ʿAbd Allāh Muḥammad. *ʿAqīdah and ʿIbādāt*. Vol. 1, Al-Qawānīn al-fiqhiyya. *The* Judgments *of* Fiqh, trans. Asadullah Yate, 2 vols. Bradford: Diwan Press, 2019.

5.1.4 Other Sources

Anon. *Taṣfiyat al-qulūb fī l-wuṣūl ilā ḥaḍrat ʿallām al-ghuyūb*. Ed. Munīr Q. Būdashīsh. Casablanca: n.p., 1998.

Böwering, Gerhard, and Yousef Casewit, eds. *A Qur'ān Commentary by Ibn Barrajān of Seville (d. 536/1141)*. Īḍāḥ al-ḥikma bi-aḥkām al-ʿibra (*Wisdom Deciphered, the Unseen Discovered*). Leiden: Brill, 2015

Al-Gharnāṭī, Abū Ḥayyān. *Al-Baḥr al-muḥīṭ fī tafsīr al-Qur'ān al-ʿaẓīm*. Ed. ʿĀdil Aḥmad ʿAbd al-Mawǧūd. 9 vols. Beirut: Dār al-Kutub al- ʿIlmiyya, 2001.

Ibn ʿAṭīya, ʿAbd al-Ḥaqq. *Al-Muḥarrar al-wajīz fī tafsīr al-kitāb al-ʿazīz*. Ed. ʿAbd as-Salām ʿAbd ash-Shāfī Muḥammad. Beirut: Dār al-Kutub al-ʿIlmiyya, 2001.

Ibn Juzayy, ʿAbd Allāh. *Maṭla' al-yumn wa-l-iqbāl fī intiqā' kitāb al-ihtifāl*. Ed. Muḥammad al-ʿArabī l-Khaṭṭābī. Beirut: Dār al-Gharb al-Islāmī, 1986.

Ibn al-Khaṭīb, Lisān ad-Dīn. *Al-Iḥāṭa fī akhbār Gharnāṭa*: Ed. Yūsuf ʿAlī Ṭawīl. 4 vols. Beirut: Manshūrāt Muḥammad ʿAlī Bayḍūn, 2003.

Ibn az-Zubayr, Abū Jaʿfar Aḥmad. *Milāk at-ta'wīl al-qāṭiʿ bi-dhawī l-ilḥād wa-t-taʿṭīl fī tawjīh al-mustashābih al-lafẓ min āy at-tanzīl*. Ed. Saʿīd al-Fallāḥ. Beirut: Dār al-Gharb al-Islāmī, 1983.

Ibn az-Zubayr, Abū Jaʿfar Aḥmad. *Al-Burhān fī tartīb suwar al-Qur'ān*. Ed. Saʿīd b. Jumʿa al-Fallāḥ. Dammam: Dār Ibn al-Jauzī, 2007.

Al-Iṣfahānī, Abū l-Qāsim ar-Rāghib. *Al-Mufradāt fī gharīb al-Qur'ān*. Ed. Ṣafwān ʿAdnān Dāwūdī. Mekka/Riyadh: Maktabat Nizār Muṣṭafā l-Bāz, 1997.

Al-Iskāfī, Muḥammad. *Kitāb Durrat at-tanzīl wa-ghurrat at-ta'wīl fī bayān al-āyāt al-mutashābihāt fī kitāb Allāh al-ʿazīz*. Cairo: Maṭbaʿat Muḥammad Maṭar al-Warrāq, 1909.

Al-Qaysī, Makkī b. Abī Ṭālib. *Tafsīr al-mushkil min gharīb al-Qur'ān al-ʿaẓīm ʿalā l-ījāz wa-l-ikhtiṣār*. Ed. Hudā ṭ-Ṭawīl al-Marʿashlī. Beirut: Dār an-Nūr al-Islāmī, 1988.

Ar-Rāzī, Fakhr ad-Dīn Muḥammad. *At-Tafsīr al-kabīr wa-mafātīḥ al-ghayb*. Ed. Muḥammad Riḍwān ad-Dāya. 33 vols. Beirut: Dār al-Fikr al-Muʿāṣir, 1990.

Ar-Rāzī, Fakhr ad-Dīn Muḥammad. *The Fātiḥa*. Vol. 1, *The Great Exegesis. Al-Tafsīr al-kabīr*, trans. Sohaib Saeed. Cambridge: The Royal Aal al-Bayt Institute für Islamic Thought and The Islamic Texts Society, 2018

As-Sijistānī, Abū Bakr Muḥammad. *Kitāb Gharīb al-Qur'ān al-musammā bi-Nuzhat al-qulūb*. No ed. given. Cairo: Maṭbaʿat as-Saʿāda, 1907.

As-Sulamī, Muḥammad b. al-Ḥusayn. *Ḥaqā'iq at-tafsīr. Tafsīr al-Qur'ān al-ʿazīz*. Ed. Sayyid ʿUmrān. Beirut: Dār al-Kutub al-ʿIlmiyya, 2001.

Az-Zamakhsharī, Maḥmūd b. ʿUmar. *Tafsīr al-Kashshāf ʿan ḥaqā'iq at-tanzīl wa-ʿuyūn al-aqāwil fī wujūh at-ta'wīl*. Ed. Khalīl Ma'mūn Shīḥā. Beirut: Dār al-Maʿrifa, 2009.

Az-Zarkashī, Muḥammad b. ʿAbd Allāh. *Al-Burhān fī ʿulūm al-Qur'ān*. Ed. Muḥammad Abū l-Faḍl Ibrāhīm. 4 vols. Cairo: Dār at-Turāth, (199-?).

5.2 Secondary Literature

5.2.1 Secondary Literature for the *Tashīl*

Barrāk, ʿAbd ar-Raḥmān b. Nāṣir. *At-Taʿlīqāt ʿalā l-masā'il al-ʿaqadiyya fī Kitāb at-Tashīl li-ʿulūm at-tanzīl li-Ibn Juzayy, m. 741h*. Riyadh: Mu'assasat Waqf ash-Shaykh ʿAbd ar-Raḥmān b. Nāṣir al-Barrāk etc., 2018.

Baraka, ʿAbd ar-Raḥmān Musʿad ʿAlī. *Ibn Juzayy wa-manhajuhu fī t-tafsīr min khilāl kitābihi at-Tashīl li-ʿulūm at-tanzīl*. Sabha, Libya: Kullīyat al-Ādāb wa-t-Tarbiya, Jāmiʿat as-Sabhā, 1994.

Clarke, Abdassamad. *The story of stories on Sayyiduna Yusuf. Based on the Tafsir of Ibn Juzayy al-Kalbi.* London: Ta-Ha, 1999.

Elmalı, Abdurrahman. *İbn Cüzey el-Kelbī el-Ğırnatī'nin Kitābu't-teshīl li ulūmi't-tenzīl adlı tefsirinin tahlili. Tefsir metodu üzerine bir deneme*. Urfa: Harran Üniversitesi İlahiyat Fakültesi Vakfı Yayınları, 1998.

Fāris, Ṭāriq b. Aḥmad. ʿUlūm al-Qurʾān ʿinda al-imām b *Juzayy al-Kalbī wa-atharuhā fī tafsīrihi at-Tashīl li-ʿulūm at-tanzīl*. Riyadh: Kursī l-Qurʾān al-Karīm wa-ʿUlūmihi, Jāmiʿat al-Malik Saʿūd, 2016–2017.

Furber, Musa. *Ibn al-Juzay's Sufic exegesis. Selections from Kitāb al-Tashīl li-ʿulūm al-Tanzīl*. N.p.: Islamosaic, 2016.

Al-Ghamādī, Muḥammad. "Tarjīḥāt b. Juzayy al-Kalbī fī tafsīrihi. ʿArḍ wa-munāqasha, min awwal sūrat al-Fātiḥa ilā nihāyat sūrat al-Baqara." Unpublished PhD thesis, University Umm al-Qurā, 2006 (Later works in the same series treat the following suras).

Jumaylī, Ṣāliḥ Ḥaydar. *Ibn Juzayy al-Kalbī naḥwiyyan fī Kitāb at-Tashīl li-ʿulūm at-tanzīl*. Bagdad: Dīwān al-Waqf as-Sunnī, Markaz al-Buḥūth wa-d-Dirāsāt al-Islāmiyya, 2012.

Muḥammad, Muḥammad ʿAbd ar-Raḥīm. *Manhaj Ibn Juzayy al-Kalbī l-Andalusī fī tafsīr al-Qurʾān al-karīm*. Al-Azhar Cairo: Dār aṭ-Ṭibāʿa al-Muḥammadiyya, 1989.

Ar-Rifāʿī l-Juhanī, Sāmī b. Musāʿid. *Takhrīj al-aḥādīth wa-l-āthār fī kitāb at-tashīl li-ʿulūm at-tanzīl li-bn Juzayy al-Kalbī (t 741h)*. Mecca: Jāmiʿat Umm al-Qurā, 1997.

As-Sabt, Khālid b. ʿUthmān. *Sharḥ Kitāb at-Tashīl li-Ibn Juzayy*, 2018. Online with commentary and sound.

Ṭayyār, Musāʿid b. Sulaymān. *Sharḥ muqaddimat at-Tashīl li-ʿulūm at-tanzīl li-Ibn Juzayy*. Dammam: Dār Ibn al-Jawzī, 2009–2010.

Yusuf, Hamza. *Refinement of the heart*. N.p.: Sandala Productions 2016. Lectures on CD.

Az-Zubayrī, ʿAlī Muḥammad. *Ibn Juzayy wa-manhajuhu fī t-tafsīr. Dirāsa mushaba ʿan al-Imām al-mufassir al-Andalusī sh-shahīd Ibn Juzayy wa-tawḍīḥ mufaṣṣal li-manhajihi fī tafsīrihi at-Tashīl li-ʿulūm at-tanzīl*. Damascus: Dār al-Qalam, 1987.

5.2.2 General Secondary Literature

Arcas Campoy, Maria. "Un resumen de la historia de al-Andalus del alfaqui granadino Abu-l-Qasim B. Yuzayy, siglo XIV." *Miscelanea de Estudios Arabes y Hebraicos* 36 (1987): 157–63.

Basanta, Fernando Velázquez. "Ibn Ŷuzayy al-Kalbī, Abū ʿAbd Allāh." In *Biblioteca de al-Andalus*, edited by Jorge Lirola Delgado, Puerta Vílchez, and José Miguel. 7 vols. Almería: Fundación Ibn Tufayl de Estudios Árabes, vol. 6, 2009: 180–95 (no. 1412).

Basanta, Fernando Velázquez. "Ibn Ŷuzayy al-Kalbī, Abū Bakr." In *Biblioteca de al-Andalus*, edited by Jorge Lirola Delgado, Puerta Vílchez, and José Miguel. 7 vols. Almería: Fundación Ibn Tufayl de Estudios Árabes, vol. 6, 2009: 196–203 (no. 1414).

Basanta, Fernando Velázquez. "Ibn Ŷuzayy al-Kalbī, Abū Muḥammad." In *Biblioteca de al-Andalus*, edited by Jorge Lirola Delgado, Puerta Vílchez, and José Miguel. 7 vols. Almería: Fundación Ibn Tufayl de Estudios Árabes, vol. 6, 2009: 203–14 (no. 1415).

Basanta, Fernando Velázquez. "Ibn Ŷuzayy al-Kalbī, Abū l-Qāsim." In *Biblioteca de al-Andalus*, edited by Jorge Lirola Delgado, Puerta Vílchez, and José Miguel. 7 vols. Almería: Fundación Ibn Tufayl de Estudios Árabes, vol. 6, 2009: 214–30 (no. 1416).

Basanta, Fernando Velázquez, and Celia del Moral Molina. "Los Banū Ŷuzayy. Una familia de juristas e intelectuales granadinos del siglo XIV. I: Abū l-Qāsim Muḥammad Ibn Ŷuzayy." *Miscelanea de Estudios Arabes y Hebraicos* 45 (1996): 161–201.

Casewit, Yousef. *The Mystics of al-Andalus: Ibn Barrajān and Islamic Thought in the Twelfth Century.* Cambridge: Cambridge University Press, 2016.

Elger, Ralf. “Adab and historical memory. The Andalusian poet/politician Ibn al-Khaṭīb presented in Aḥmad al-Maqqarī’s (986/1577-1041/1632) ‘Nafḥaṭ-ṭīb’.” *WI* 42 (2002): 289–306.

Elger, Ralf. “Ibn Juzayy, Abū ʿAbdallāh.” In *EI*[3], edited by Kate Fleet et al. Brill, 2007–. Article published online 2020. Accessed on June 23, 2022. https://referenceworks.brillonline.com/entries/encyclopaedia-of-islam-3/ibn-juzayy-abu-abdallah-COM_30849?s.num=153&s.start=140.

Fábregas, Adela, ed. *The Nasrid Kingdom of Granada between East and West (Thirteenth to Fifteenth Centuries)*. Leiden: Brill 2020.

Hernández, Adday. “Qurʾanic studies in al-Andalus: An overview on the state of research on qirāʾāt and tafsīr.” *JQS* 19 (2017): 74–102.

Lirola Delgado, J., and E. Navarro i Ortiz. “Ibn al-Zubayr, Abū Ŷaʿfar.” In *Biblioteca de al-Andalus*, edited by Jorge Lirola Delgado, Puerta Vílchez, and José Miguel. 7 vols. Almería: Fundación Ibn Tufayl de Estudios Árabes, vol. 6, 2009: 327–39.

Robinson, Cynthia. “Lisān al-Dīn IBN al-KHAṬĪB.” In *Essays in Arabic literary biography, 1350–1850*, edited by Joseph E. Lowry and Devin J. Stewart. Wiesbaden: Harrassowitz, 2009: 159–73.

Sobrado Galanes, Teresa Isabel. “Traducción y estudio del Matla de Ibn Yuzavy: sobre rasgos y características del caballo.” PhD Diss., Madrid, 2015.

Ibn Qayyim al-Jawziyya

Miriam Ovadia

1 Introduction

Although the numerous works of the prolific Ḥanbalī scholar of Mamluk Damascus Ibn Qayyim al-Jawziyya (d. 1350/751; henceforth: Ibn al-Qayyim) cover nearly all Islamic disciplines, he did not write a single work exclusively dedicated to Qurʾānic exegesis in the form of a complete commentary (*tafsīr musalsal*). However, Ibn al-Qayyim's vast literary output is abundant in exegetical references to Qurʾānic verses. In his writings, Ibn al-Qayyim validated his argumentations by means of citing Qurʾānic verses and clarifying their intended meaning within the context of a particular discussion.[1] Still, Ibn al-Qayyim's exegetical contribution has not yet received much focus in Western research.[2] Even if not originally placed within the literary arrangement of the *tafsīr* genre, examining his deliberations on the Qurʾān is bound to shed light on his overall hermeneutical approach to the Qurʾānic text. As previous scholarship has already proven, in his theological and jurisprudential output Ibn al-Qayyim embraced the key hermeneutical principles set by his renowned mentor, Taqī d-Dīn Aḥmad b. Taymiyya (d. 1328/728).[3] However, rather than merely reiterating his mentor's ideas, Ibn al-Qayyim skillfully reshaped them to fit his own objectives. These objectives were to transmit knowledge in a didactical manner and to popularize the Taymiyyan worldview.[4] Along these lines, this chapter will demonstrate that Ibn al-Qayyim's approach to Qurʾānic verses is a more coherent version of Ibn Taymiyya's interpretation of the Qurʾān.

The chapter begins with a short biography of Ibn al-Qayyim, which serves to describe the intellectual background of his scholarly activities (section II). The following

1 In a few sporadic cases, Ibn al-Qayyim concentrated on interpreting particular passages of the Qurʾān, thus producing clearly defined *tafsīr* units; Birgit Krawietz, "Ibn Qayyim al-Jawzīyah: His Life and Works," *MSR* 10, no. 2 (2006): 36.

2 In contrast, the topic has been addressed in the Arab world in studies such as: ʿAbd al-Fattāḥ Lāshīn, *Ibn Qayyim al-Jawziyya wa-ḥissuhu al-balāghī fī tafsīr al-Qurʾān* (Beirut: Dār ar-Rāʾid al-ʿArabī, 1982); Bakr ʿAbdallāh Abū Zayd, *at-Taqrīb li-ʿulūm Ibn al-Qayyim* (Riyadh: Dār al-ʿAṣīma, 1996²), 36–40; Ṣabrī l-Mutawallī, *Manhaj ahl as-sunna fī tafsīr al-Qurʾān al-karīm: Dirāsa mawḍūʿiyya li-juhūd Ibn al-Qayyim at-tafsīriyya* (Cairo: Maktabat Zahrāʾ ash-Sharq, 2002); Muḥammad ad-Dusrī and Muḥammad b. ʿAbdallāh al-Qaḥṭānī, *Ikhtiyārāt Ibn al-Qayyim wa-tarjīḥātuhu fī t-tafsīr*, 2 vols. (Riyadh: Jāmiʿat al-Malik Suʿūd, 2018).

3 Livnat Holtzman, "Ibn Qayyim al-Jawziyya," *EI*³, forthcoming.

4 Irmeli Perho, "Man Chooses his Destiny: Ibn Qayyim al-Jawziyya's Views on Predestination," *Islam and Christian-Muslim Relations* 12, no. 1 (January 2001): 62f.; Caterina Bori and Livnat Holtzman, "A Scholar in the Shadow," in *A Scholar in the Shadow: Essays in the Legal and Theological Thought of Ibn Qayyim al-Ǧawziyyah*, eds. Caterina Bori and Livnat Holtzman (Rome: Instituto per l'Oriente C.A. Nallino, 2010): 25.

https://doi.org/10.1515/9783111320083-010

section (III) outlines Ibn al-Qayyim's preoccupation with Qur'ān exegesis in his available works (or modern compilations attributed to him), while taking a closer look at selected parts of his commentary of sura 109, *al-Kāfirūn* (The Disbelievers). To present Ibn al-Qayyim's understanding and use of practical commentary terminology, this section also offers a preliminary overview of certain features of the *tafsīr* tradition. This preliminary overview enables us to adequately evaluate the affinity between Ibn al-Qayyim's hermeneutical approach and the Taymiyyan model of Qur'ānic hermeneutics. Thereafter, in section IV, we feature Ibn al-Qayyim's singular reshaping of his mentor's fundamentals which resulted in a rationalized version of the traditionalistic exegetical methods. The chapter concludes with an overall assessment of the multifaceted and nuanced approach Ibn al-Qayyim fashioned with respect to the Qur'ānic text.

2 Biography

Ibn al-Qayyim was born in 1292/691 in the village *az-Zura'*, located south of Damascus, where his family moved to when he was still an infant. In Damascus, he was given a very thorough education and trained from an early age to become a professional Ḥadīth scholar (*muḥaddith*). At the age of twenty-one Ibn al-Qayyim met the forceful thinker Ibn Taymiyya; captivated by Taymiyya's controversial teachings, he soon became his most fervent disciple. It should be noted here that the city of Damascus grew to be the primary center of Ḥanbalī learning after Baghdad fell into Mongols hands in 1258/656. Under the rule of the Mamluk Sultanate, religious scholars (*'ulamā'*) from all four schools of Islamic law (*madhāhib*, sing. *madhhab*) were equally permitted to hold official positions; therefore, political interests played an inevitable role in the scholarly activities of the *'ulamā'*.[5] Affiliated with the Ḥanbalī school, Ibn Taymiyya promoted a distinct type of traditionalistic doctrines which stirred the intellectual setting of Mamluk Damascus.[6] Indeed, Ibn Taymiyya stood in clear opposition to Shāfi'ī scholars who upheld later-Ash'arī theology of rational-traditionalistic tendency, mostly following the philosophical theology of Fakhr ad-Dīn ar-Rāzī (d. 1209/606). Ibn al-Qayyim soon became the most vigorous advocate of the Taymiyyan standpoints which emphasized the doctrines of the first generations of Islam, namely the *salaf* (lit. pious ancestors), and his

5 This was the state of public affairs in the Mamluk Sultanate after the reform of the judicial system introduced by the Sultan Baybars in 1264–65/633. Accordingly, each *madhhab* had its formal representation in the form of an appointed chief judge (*qāḍī l-quḍāt*), as well as other administrative institutions and religious colleges (*madrasa*, pl. *madāris*); Yaacov Lev, "Symbiotic Relations: Ulama and the Mamluk Sultans," *MSR* 13, no. 1 (2009): 1–26.

6 Although Ibn Taymiyya considered himself to be Ḥanbalī, it is questionable whether his theology and legal opinions complied with the school's traditional teachings and methods of argumentation. See: Abdul Hakim I. Al-Matroudi, *The Hanbali School of Law and Ibn Taymiyyah: Conflict or Conciliation* (London: Routledge, 2006).

profound commitment to his master and the Taymiyyan ideology greatly shaped the course of his career. The most formative event in Ibn al-Qayyim's life was his imprisonment in the Citadel of Damascus together with his mentor in 1326/726; among Ibn Taymiyya's close circle of disciples (*jamāʿa*), he was the only one who was imprisoned. The direct reason for this imprisonment was because Ibn al-Qayyim preached against visiting Abraham and the Prophet's graves, which he considered a forbidden and un-Islamic ritual. After Ibn Taymiyya's death in prison in 1328/728, Ibn al-Qayyim was released and immediately accepted as a full member of the Damascene scholarly milieu. He was appointed to three academic positions, the most esteemed of which was a teaching position in al-Madrasa aṣ-Ṣadriyya, a prominent school for the Ḥanbalī community in the city. An active exponent of the Taymiyyan theological and legal views, Ibn al-Qayyim was further persecuted by the Shāfiʿī Chief Judge of Damascus Taqī d-Dīn as-Subkī (d. 1355/756) who rebuked him directly in two treatises. However, Ibn al-Qayyim's political activity continued throughout his prosperous career of public debates (*munāẓarāt*) and extensive writing, in which he systemized and proliferated unconventional Taymiyyan opinions as well as spread anti-Ashʿarī propaganda.[7]

3 Works Containing *tafsīr*

As noted above, direct engagements with Qurʾānic exegesis can be found in Ibn al-Qayyim's allusions to Qurʾān verses scattered irregularly throughout his various works. However, Ibn al-Qayyim did compose a consecutive commentary of a few short *sūras* in full. These are *sūras* 1, *al-Fātiḥa* (The Opening), 109, *al-Kāfirūn* (The Disbelievers), 113, *al-Falaq* (Daybreak) and 114, *an-Nās* (People). These commentaries can be considered singular *tafsīr*-units within respective works that Ibn al-Qayyim authored;[8] this is probably why modern publishers have also printed these units separately from the works in which they initially appear, as independent commentary titles of those *sūras*.[9] In addi-

7 Holtzman, "Ibn Qayyim al-Jawziyya"; Bori and Holtzman, "A Scholar in the Shadow," 13–44.

8 Ibn al-Qayyim's commentary of sura 1, *al-Fātiḥa* (The Opening) appears in his Ṣūfī compendium *Madārij as-sālkīn bayna manāzil iyyāka naʿbudu wa-iyyāka nastaʿīn* (Stages of the Travelers between "It is You we worship, and it is You we ask for help," paraphrasing Q 1:5). This unit was published as *Tafsīr sūrat al-Fātiḥa*, ed. Muḥammad Ḥāmid al-Fiqī (Cairo: Dār al-Jihād, ca. 1979). Another work was published later entitled *Tafsīr al-muʿawwidhatayn* (Commentary of the Two (Chapters of) Refuge). Here, Ibn al-Qayyim added commentary to the two final chapters of the Qurʾān in his work *Badāʾiʿ al-fawāʾid* (The Amazing Benefits); *Tafsīr al-muʿawwidhatayn*, ed. Sayyid Ibrāhīm (Cairo: Dār al-Ḥadīth, 1989). A third publication gathers Ibn al-Qayyim's commentary on *al-muʿawwidhatayn* together with his remarks on sura *al-Kāfirūn*, all found in *Badāʾiʿ al-fawāʾid*; *Tafsīr suwar al-Kāfirūn wa-l-muʿawwidhatayn*, ed. Muḥammad Ḥāmid al-Fiqī (Cairo: Maṭbaʿat as-Sunna al-Muḥammadiyya, 1949).

9 Krawietz, "Ibn Qayyim al-Jawzīyah: His Life and Works," 35–37, 47f.; Bakr ʿAbdallāh Abū Zayd, *Ibn Qayyim al-Jawziyya: ḥayātuhu, athāruhu, mawāriduhu* (Riyadh: Dār al-ʿĀṣima li-n-Nashr wa-t-Tawzīʿ, 2002²), 232f.

tion to his *tafsīr*-units, Ibn al-Qayyim's work of linguistic-thematic exegesis concentrating on the subject of oaths (*aqsām*; sing. *qasam*) in the Qur'ān, entitled *at-Tibyān fī l-īmān bi-l-Qur'ān* (Explaining the Belief (in Oaths) in the Qur'ān), is also notable.[10] Interestingly enough, in the past few decades, and particularly since the 1990s, several attempts have been made to collect the entirety of dispersed exegetical paragraphs throughout Ibn al-Qayyim's writings (together with the fully interpreted *sūras*) and piece them together to create an almost complete – albeit artificial – commentary of the Qur'ān ascribed to Ibn al-Qayyim. These synthetic *tafsīr* works which are based on Ibn al-Qayyim's hermeneutical efforts, differ in volume and quality. The first of these pseudo-commentaries was published in 1948 under the title *at-Tafsīr al-qayyim* (The Valuable Commentary), a play-on-words of its alleged author's name.[11] Additional publications attempted to reconstruct a more comprehensive and refined *tafsīr* out of Ibn al-Qayyim's writings, such as the books *Badā'i' at-tafsīr* (The Amazements of *tafsīr*)[12] and *aḍ-Ḍaw' al-munīr 'alā t-tafsīr* (The Brilliant Light on *tafsīr*).[13] The most recent reconstructed commentaries attributed to Ibn al-Qayyim vary in style and size. From these commentaries we note *al-Intifā' bi-l-Qur'ān li-Ibn Qayyim al-Jawziyya* (Benefiting from the Qur'ān by Ibn Qayyim al-Jawziyya) which is a mere 280 pages,[14] and the set of 22 electronic volumes of *Maṣābiḥ at-tafāsīr al-Qur'āniyya: al-jāmi' li-tafsīr Ibn Qayyim al-Jawziyya* (Spotlights of Qur'ānic Commentaries: A Collection of Ibn Qayyim al-Jawziyya's Commentary).[15] Considering the high demand for Ibn al-Qayyim's writings, we may expect additional attempts to compile the ultimate "commentary" out of his oeuvre.

---

10 The book was first published in 1903/1321 under the title of *Kitāb al-Aqsām, al-musammā bi-t-Tibyān* (Mecca: al-Maṭba'a al-Amīriyya). It is available today in multiple editions assembled by various publishers across the Arab world, mostly under the inaccurate title of *at-Tibyān fī aqsām al-Qur'ān*. As clarified in the critical introduction to the edition made by Bakr Abū Zayd's project, Ibn al-Qayyim titled his work *at-Tibyān fī l-īmān bi-l-Qur'ān*. This is the edition used in this paper: *at-Tibyān fī l-īmān bi-l-Qur'ān*, ed. 'Abdallāh b. Salām al-Baṭāṭī and Bakr b. 'Abdallāh Abū Zayd (Mecca: A'lām al-Fawā'id, 2008), 29–31.

11 Ibn Qayyim al-Jawziyya, *at-Tafsīr al-qayyim li-l-imām Ibn al-Qayyim*, complied by Muḥammad Uways an-Nadawī, ed. Muḥammad Ḥāmid al-Fiqī (Beirut: Dār al-Kutub al-'Ilmiyya, 1948).

12 Ibn Qayyim al-Jawziyya, *Badā'i' at-tafsīr: al-jāmi' li-tafsīr al-imām Ibn Qayyim al-Jawziyya*, ed. Yusayrī s-Sayyid Muḥammad, 5 vols. (Riyadh: Dār Ibn al-Jawzī, 1993). This title displays wordplay, alluding to the authentic work Ibn al-Qayyim composed entitled *Badā'i' al-fawā'id* (The Amazing Benefits). A shortened edition of *Badā'i' at-tafsīr*, revised by Ṣāliḥ Aḥmad ash-Shāmī, was published in 2005/1427 in three volumes. The book can also be found in a single compendium under the title *al-Badā'i' fī 'ulūm al-Qur'ān*, ed. Yusayrī s-Sayyid Muḥammad (Beirut: Dār al-Ma'rifa, 2003).

13 Ibn Qayyim al-Jawziyya, *aḍ-Ḍaw' al-munīr 'alā t-tafsīr – min kutub al-imām al-muḥaddith al-mufassir al-faqīh . . . al-ma'rūf bi-Ibn Qayyim al-Jawziyya*, ed. 'Alī l-Ḥamad al-Muḥammad aṣ-Ṣāliḥī, 6 vols. (Dakhna/Riyadh: Mu'assasat an-Nūr in collaboration with Maktabat Dār as-Salām, 1994).

14 Ibn Qayyim al-Jawziyya, *al-Intifā' bi-l-Qur'ān li-Ibn Qayyim al-Jawziyya*, ed. Jamīl Ibrāhīm Ḥabīb (Beirut: Dār al-Arqam, 2008).

15 Ibn Qayyim al-Jawziyya, *Maṣābiḥ at-tafāsīr al-Qur'āniyya: al-jāmi' li-tafsīr Ibn Qayyim al-Jawziyya*, ed. 'Abd ar-Raḥmān al-Qammash, 22 vols. (2016). The source has not been published in print and is only available as an e-book (.bok file) through the online program al-Maktaba al-Shamela. *Maṣābiḥ at-*

For the current purpose of exploring Ibn al-Qayyim's hermeneutical approach to the Qur'ān, I will concentrate on selected passages of Ibn al-Qayyim's engagement with Qur'ānic verses as they appear in his original works rather than the modern reassembled compilations of these texts, which deserve a separate investigation. In my opinion, an unmediated study of Ibn al-Qayyim's approach in light of the literary context in which it was conceived will allow for a more accurate grasp of his hermeneutics. This point becomes crucial when bearing in mind the underlying theological-political tensions at the time of Ibn al-Qayyim's writing.[16] From this perspective, it is evident that the interpretations Ibn al-Qayyim provides to Qur'ānic verses often polemicize with various trends of Islamic rationalism, from the ultra-rationalistic Muʿtazila to the rational-traditionalistic Ashʿariyya. Emanating from the Taymiyyan example of Islamic traditionalism, Ibn al-Qayyim's foremost commitment in his theology was to the canonical scriptures of the Qur'ān and Ḥadīth literature. In contrast, Islamic rationalism was more tolerant to using methods and argumentations outside of Islamic textual sources, such as dialecticism and philosophy.[17] The different epistemologies of traditionalism and rationalism also manifested with relation to the interpretation of the Qur'ānic text. This aspect is clearly demonstrated in Ibn al-Qayyim's original works, where his interpretation entails a combination of tradition-based exegesis (*tafsīr bi-l-ma'thūr*) and reason-based exegesis (*tafsīr bi-r-ra'y*).

A suitable stepping stone for evaluating Ibn al-Qayyim's hermeneutical approach is his *tafsīr*-unit of sura *al-Kāfirūn* (Q 109) located in his work *Badā'iʿ al-fawā'id* (The Amazing Benefits).[18] *Badā'iʿ al-fawā'id* is a major work of grammar, in which Ibn al-Qayyim delves into linguistic questions in order to 'benefit' from them and gain a clearer understanding of the language of the Qur'ān and Ḥadīth for doctrinal matters (*ʿaqīda*) and jurisprudence (*fiqh*). In this grammatical framework, Ibn al-Qayyim explicates the function of the Arabic word *mā* when used as a relative pronoun (*mā l-mawṣūla*; in the

---

tafāsīr can also be accessed via the following link: https://ketabonline.com/en/books/104346/read?page=1.

16 This observation is hardly surprising. Throughout the evolution of Qur'ānic exegesis, various historical circumstances brought about a politicization of the exegetical traditions which were meant to serve the worldview and practical aims of the respective commentator. Qur'ānic exegesis thus became one of the scholarly platforms for expressing disagreements over theologically controversial issues. In that manner, early exegetists of sectarian schools of thought (like the Shiʿa or the Muʿtazila), who were active alongside the mainstream traditionalists, were more inclined towards an allegorical interpretation of the Qur'ān; an esoteric meaning supposedly conveyed in the text could be shaped to better suit the exegetist's non-mainstream opinions; Hussein Abdul-Raof, *Schools of Qur'anic Exegesis: Genesis and Development* (New York: Routledge, 2010), 55–63, 63–82.

17 See: Binyamin Abrahamov, *Islamic Theology: Traditionalism and Rationalism* (Edinburgh: Edinburgh University Press, 1998).

18 Ibn Qayyim al-Jawziyya, *Badā'iʿ al-fawā'id*, ed. ʿAlī b. Muḥammad al-ʿImrān and Bakr b. ʿAbdallāh Abū Zayd (Mecca: Dār ʿĀlam al-Fawā'id li-n-Nashr wa-t-Tawzīʿ, 2003), 1: 230–49. This section appears also in: Ibn Qayyim al-Jawziyya, *Tafsīr suwar al-Kāfirūn wa-l-muʿawwidhatayn*, 4–15.

meaning of *al-lādhī*):[19] on its own the particle *mā* conveys extreme obscurity (*ism mubham fī ghāyat al-ibhām*), therefore it must be followed by a subordinate clause which clarifies its meaning. Ibn al-Qayyim first sustains a direct linkage between the vocalized sound of the word *mā* and its signification of a broad meaning, suggesting semantic symbolism:[20] *mā* ends with the letter *alif* as a long vowel which extends the sojourn of the word in the speaker's mouth, as it is compatible with the extended generic meaning it conveys (*al-madd wa-l-ittisāʿ fī hawāʾ al-fam mushākila li-ttisāʿ maʿnāhā fī l-ajnās*). If one wishes to use *mā* in order to express a specific meaning, the word itself transforms and becomes shorter by exchanging the elongating *alif* for the consonant *nūn* that carries no vowel (*nūn sākina*), hence creating the word *man* (lit.: 'who'). In this manner, the pronunciation of the word shortens to counterbalance its short, specific meaning (*fa-ṣāra qaṣran li-l-lafẓ muwāzinan li-qaṣr al-maʿnā*).[21] Ibn al-Qayyim then cites several Qurʾānic verses in which, he stresses, the word *mā* fundamentally appears in its obscure and generic meaning;[22] an insightful person who possesses a spiritual taste under-

---

19 According to Wright, "the conjunctive noun *al-lādhi* may be used either substantively or adjectively. In the former case, it includes the idea of a person or a thing, that is to say, it is equivalent to the substantive *man* and *mā*, when they are definite (*mawṣūla*), *he who, that which*. In the latter case, it agrees, like any other adjective, with its antecedent, which is always as definite substantive in gender, number and *case*, and thusly markedly differs from the relative pronouns of the Indo-German languages"; William Wright, *A Grammar of the Arabic Language*, trans. Carl Paul Caspari and ed. with numerous additions and corrections by W. Wright (Beirut: Librairie du Liban, 1996[3]), 2: 320.

20 Sound symbolism was never a mainstream branch of knowledge and thus rather marginal for most Arabic grammarians. However, the topic was discussed with some connection to the question of the origin of language, for instance, by Abū l-Fatḥ Ibn Jinnī (d. 1002/392) in his work *al-Khaṣāʾiṣ*. Ibn al-Qayyim did not always accept Ibn Jinnī's approach (for example, concerning the origin of language and *majāz*; see footnote 58 henceforth); however, it seems that Ibn al-Qayyim adopted Ibn Jinnī's ideas regarding sound symbolism as a natural, instinctive occurrence which relates to human sensorial capacity. Elsewhere in *Badāʾiʿ al-fawāʾid* Ibn al-Qayyim mentions that when he pointed the phenomenon out to Ibn Taymiyya, the latter shared his amazement of the subject. See: Ibn Qayyim al-Jawziyya, *Badāʾiʿ al-fawāʾid*, 166, 189; Kees Versteegh, "Sound Symbolism," in *Encyclopedia of Arabic Language and Linguistics Online*, eds. Lutz Edzard and Rudolf de Jong, accessed on 06 September 2020, http://dx.doi.org/10.1163/1570-6699_eall_EALL_COM_0315.

21 Ibn Qayyim al-Jawziyya, *Badāʾiʿ al-fawāʾid*, 1: 230f. Keeping in mind the following exegetical deliberation, it is highly likely that Ibn al-Qayyim also considers technical components of the classical Qurʾān science of *tajwīd*, i.e., the "rigorous system of rules that establish the proper vocalization of the Qurʾān, thereby determining its actual rhythm and sound," including the correct articulation of phonetic sounds. As such, *tajwīd* is part of the systems of Qurʾānic readings (*qirāʾāt*), recitations and vocal rendition (*tilāwat*); Anna M. Gade, "Recitation of the Qurʾān," in *EQ*, ed. Jane Dammen McAuliffe (Leiden: Brill, 2004), 4: 367f., 372–76. Elsewhere in his writing, Ibn al-Qayyim selectively refers to a Qurʾānic reading practiced by the people of Medina: Ibn Qayyim al-Jawziyya, *Badāʾiʿ al-fawāʾid*, 4:1619-1620.

22 Legal theorists tried to cope with the inherent ambiguity of plural pronouns (such as: *al-jamʿ*, *kull*, *man*, *mā*) in normative passages in the Qurʾān by accepting that these grammatical forms imply presumptive generality of reference. The great legal expert Muḥammad b. Idrīs ash-Shāfiʿī (d. 820/204) is considered to be the first scholar to employ the relevant terms of general (*ʿāmm*) and specific (*khāṣṣ*)

stands the actual speech and recognizes that meaning clearly (*man fahima ḥaqīqat al-kalām wa-kāna lahu dhawq ʿarafa hādhā wa-stabāna lahu*).[23]

Ibn al-Qayyim's linguistic explanation of *mā l-mawṣūla* includes an exegetical discussion on the six verses of sura *al-Kāfirūn*. This sura is a declaration made by the Prophet Muḥammad, while the audience is the infidels of Quraysh, which are his tribe. The following is a translation of the sura: "[1]Say (Prophet), 'Disbelievers: [2]I do not worship **what** you worship, [3]you do not worship **what** I worship, [4]I will never worship what you worship, [5]you will never worship what I worship: [6]you have your religion and I have mine.'" Ibn al-Qayyim explains that in verses 109:2–3 *mā* appears as such a relative pronoun: "[2]*lā aʿbudu* ***mā*** *taʿbudūna,* [3]*wa-lā antum ʿābidūna* ***mā*** *aʿbudu.*"[24] After listing four possibilities of exegesis based on traditional teachings (i.e. *tafsīr bi-l-maʾthūr*), Ibn al-Qayyim presents his interpretation as the fifth and preferable option: the use of *mā* in these verses is intended to describe the one being worshiped as worthy of reverence (*al-maʿbūd al-mawṣūf bi-kawnihi ahlan li-l-ʿibāda mustaḥiqqan lahā*). Had the verses stated the pronoun *man* instead of *mā* (as in a specific 'who' instead of general 'what'), it would signify a defining clause (*taʿrīf*); however, the verses emphasize the high merit of the object of worshipping, rather than defining its identity. Ibn al-Qayyim points to the equivalent structure of Q 4:3 "you may marry **whichever** (other) women seem good to you" (*fa-nkiḥū* ***mā*** *ṭāba lakum min an-nisāʾ*). In this case, *mā* opens an adjective clause (*waṣf*) intended to provide the motivating reason (*as-sabab ad-dāʿī*) and aim (*qaṣd*) of the marriage, which is its goodness (*aṭ-ṭīb*); hence, the intention in Q 4:3 is that one should marry women who are described as good. According to Arab grammarians, *mā* is commonly used in Arabic to indicate abstract nouns (the grammatical term in Arabic is *ṣifāt*, literally attributes). Ibn al-Qayyim adds that the use of *mā* to indicate abstract nouns is one of the most subtle ways of expression characterizing the Arabic Language (*min alṭaf masālik al-ʿArabiyya*).[25]

Ibn al-Qayyim's exegetical discussion continues as he didactically raises some ten additional grammatical benefits (*fawāʾid*) to be attained with relation to surat *al-Kāfirūn* and then meticulously expounds them one by one. Among the grammatical

as part of a coherent legal hermeneutics; Joseph E. Lowry, "al-ʿĀmm wa-l-khāṣṣ," in *EI*[3], ed. Kate Fleet et al., accessed on 22 September 2020, http://dx.doi.org/10.1163/1573-3912_ei3_SIM_0051.

23 Ibn Qayyim al-Jawziyya, *Badāʾiʿ al-fawāʾid*, 1: 232. In Ibn al-Qayyim's Ṣūfī compendium entitled *Madārij as-sālkīn*, he condemns the Ṣūfī scholars overestimating the epistemological value of such personal "spiritual tastes." Ibn al-Qayyim maintains that the Islamic scriptures of Revelation are the final judge (*ḥākim*) of any mystical knowledge, including the subjective spiritual taste, state, and ecstasy (*adh-dhawq wa-l-ḥāl wa-l-wajd*). Ibn Qayyim al-Jawziyya, *Ranks of the Divine Seekers: A Parallel English-Arabic Text*, trans. Ovamir Anjum (Leiden: Brill, 2020), 1: 61–63, 2: 126–29.

24 All of the Qurʾānic verses in this article are quoted from M.A.S. Abdel Haleem, *The Qurʾan: A new translation* (Oxford University Press, 2008). My additions or alterations are either written in bold script or given in brackets.

25 Ibn Qayyim al-Jawziyya, *Badāʾiʿ al-fawāʾid*, 1: 235f. The wording of this sentence is slightly different, and possibly clearer, in: *Tafsīr suwar al-Kāfirūn wal-muʿawwidhatayn*, 5.

issues Ibn al-Qayyim deliberates are the repetition of verbs in the sura and the meaning of their altering tenses (past, present or future), the form of complete negation uttered,[26] or the intention conveyed by the particular order of words in the verses.[27] In the course of this linguistic interpretation Ibn al-Qayyim repeatedly promises to reveal the hidden secrets (*asrār*) of surat *al-Kāfirūn*, which is one of the two *sūras* known to convey pure monotheistic belief (*iḥdā sūratay al-ikhlāṣ*; the second being sura 112).[28] He mentions that according to several prophetic traditions (*aḥadīth*), surat *al-Kāfirūn* alone is equal to one-quarter of the Qur'ān;[29] not everyone is able to comprehend it, but only those whom God has granted an apt understanding.[30] Here, Ibn al-Qayyim follows a line of argument already presented by Ibn Taymiyya,[31] claiming that these two *sūras* complement each other and together express a twofold meaning of monotheism (*tawḥīd*). Sura 112, *al-Ikhlāṣ* (Purity of Faith), positively expresses the knowledge and the dogmas of monotheism (*tawḥīd al-ʿilm wa-l-iʿtiqād*), namely describing the absolute transcendence of God using the attributes of perfection (*ṣifāt al-kamāl*).[32] Sura 109, *al-*

---

26 Ibn al-Qayyim applies the reasoning of sound symbolism also on the semantic distinction between the Arabic negators *lā* and *lan*: the extended, elongated sound of *lā* infers a long, contentious, and enduring negation, whereas the shorter sound of the word *lan* stands for a closer sense of immediate negation; Ibn Qayyim al-Jawziyya, *Badāʾiʿ al-fawāʾid*, 1: 166–69, 242.

27 Ibid., 236f.

28 The exact expression *sūratay al-ikhlāṣ* originates in a prophetic tradition (*ḥadīth*, pl. *aḥadīth*) of high level of authenticity (*ṣaḥīḥ*) transmitted by the *ṣaḥābī* Jābir b. ʿAbdallāh (d. 697/78) in relation to the Prophet's last pilgrimage (*ḥajj*): the Prophet incorporated *sūras* 109 and 112 in his two cycles of prayer during the circumambulation of the Kaʿba (*rakʿatay aṭ-ṭawāf*) which he performed. This *ḥadīth* is found in: Muḥammad b. ʿĪsā at-Tirmidhī, *al-Jāmiʿ al-kabīr – Sunan at-Tirmidhī*, ed. Bashār ʿAwwād Maʿrūf (Beirut: Dār al-Gharb al-Islāmī, 1998), 2: 213, no. 869. Several other *aḥadīth* also bring *sūras* 109 and 112 together as they depict the Prophet's prayers; they do not call them *sūratay al-ikhlāṣ*, but only mention the first verse of each sura. Interpreted in light of such *aḥadīth*, *sūras* 109 and 112 are often coupled in Qur'ānic exegetical tradition.

29 For example, in a *ḥadīth* transmitted by the prominent *ṣaḥābī* ʿAbdallāh b. ʿAbbās (d. 687/68) who reported that the Prophet said: "(sura 99, *az-Zalzala* (The Earthquake)) equals one-half of the Qur'ān, (sura 112, *al-Ikhlāṣ* (Purity of Faith)) equals one-third of the Qur'ān, and (sura 109, *al-Kāfirūn* (The Disbelievers)) equals one-quarter of the Qur'ān"; at-Tirmidhī, *al-Jāmiʿ al-kabīr*, 5: 166, no. 2894 (no. 2893, 2895 are additional versions of low level of authenticity narrated by other transmitters). Another *ḥadīth* mentions that each of these three *sūras* is equal to one-quarter of the Qur'ān. See: Aḥmad b. Ḥanbal, *Musnad al-Imām Aḥmad b. Ḥanbal*, ed. Shuʿayb al-Arnaʾūṭ et al. (Beirut: Muʾassasat ar-Risāla, 1997), 19: 472, no. 12488.

30 Ibn Qayyim al-Jawziyya, *Badāʾiʿ al-fawāʾid*, 1: 238, 241f., 246f.

31 Ibn Taymiyya's pertinent paragraphs of interpretation from his work *Bayān talbīs al-jahmiyya* will soon be elaborated on in the end of this section (III).

32 In the traditionalistic theological discourse, divine attributes and anthropomorphic descriptions of God were perceived as representations of God's perfection. This principle element was borrowed from the traditionalistic scholars with rationalistic tendencies such as Abū Muḥammad b. Qutayba (d. 889/276, a grammarian), and also appeared in the thought of Muḥammad b. Fūrak (d. 1015/406, an early Ashʿarī and one of the formalizers of the Ashʿarī thought). Ibn Taymiyya adopted the idea of divine perfection and incorporated it in his method of reading the anthropomorphisms found in the Islamic texts; Jon Hoover, "Ḥanbalī Theology," in *Oxford Handbook of Islamic Theology*, ed. Sabine Schmidtke (Oxford:

Kāfirūn negatively presents the aim and intention of monotheism (*tawḥīd al-qaṣd wa-l-irāda*), that is, worshiping God alone with no copartners (*shirk*). Ibn al-Qayyim asserts that in the *sunna* the Prophet himself is reported to have opened his prayer at daybreak (*sunnat al-fajr*) with a recitation of the two monotheistic *sūras*,[33] as well as to have closed his prayer with them at dusk after sunset (*sunnat al-maghrib*).[34] Another *ḥadīth* states the Prophet also used to recite these *sūras* when closing his final prayer at night before sleep, therefore the two monotheistic *sūras* concluded his daily activities.[35] As we can see, Ibn al-Qayyim intertwines pieces of Prophetic narrative from the Ḥadīth literature together with core doctrinal issues, and interioristic knowledge of secrets using Ṣūfī terminology,[36] into his primarily linguistic commentary of surat *al-Kāfirūn*.

Additional important aspects of Ibn al-Qayyim's exegetical writing unfold in the ending section of his *tafsīr*-unit on surat *al-Kāfirūn*, where he focuses on its last verse 109:6 "you have your religion and I (i.e. Muḥammad) have mine."[37] Ibn al-Qayyim presents two possible interpretations concerning the verse if understood as a confirmation (*iqrār*) of its content, only ultimately to categorically deny them both and offer his perspective instead. According to Ibn al-Qayyim, several scholars presumed (*ẓannū*) that Q 109:6 could be understood as an acknowledgement of the idolaters and their faith, since this acknowledgement is supposedly prescribed in this verse (*al-āya iqtaḍat at-taqrīr lahum ʿalā dīnihim*); these scholars thus believed that Q 109:6 was abrogated (*mansūkha*) by "the sword verse" (*āyat as-sayf*, 9:5).[38] Other scholars presumed that Q 109:6 merely referred to the People of the Book (*ahl al-kitāb*), whose faith was acknowledged in the Qur'ān, i.e. Jews, Christians and Zoroastrians. Following these scholars' observation, the verse thus carried a restricted specification (*makhṣūṣa*) of

Oxford University Press, 2016), 637–39; Livnat Holtzman, *Anthropomorphism in Islam: The Challenge of Traditionalism, 700–1550* (Edinburgh: Edinburgh University Press, 2018), 318–20.

33 "It was narrated from Abū Hurayra that the Messenger of Allāh recited in the two *rakʿa* of *fajr*: 'Say: O disbelievers' (Sura 109, *al-Kāfirūn* (The Disbelievers)) and 'Say: He is Allāh, (the) One' (sura 112, *al-Ikhlāṣ* (Purity of Faith))." Abū l-Ḥusayn Muslim b. al-Ḥajjāj, *English Translation of Ṣaḥīḥ Muslim*, ed. Ḥāfiẓ abū Ṭāhir Zubayr ʿAlī Zaʾī et al. (Riyadh: Darussalam Publications, 2007), 2: 239, (no. 1690) 98 – (726) (*Kitāb ṣalāt al-musāfirīn wa-qasrihā*).

34 "It was narrated from Ibn ʿUmar that the Messenger of Allāh recited in the two *rakʿa* before dawn and in the two *rakʿa* after dusk, more than twenty times or more than ten times: 'Say: O disbelievers' (sura 109, *al-Kāfirūn* (The Disbelievers)) and 'Say: He is Allāh, (the) One' (sura 112, *al-Ikhlāṣ* (Purity of Faith))." Ibn Ḥanbal, *Musnad*, 8: 381, no. 4763; 9: 501, no. 5691; 10: 32 no. 5742 (another version which states 25 or 35 times).

35 Ibn Qayyim al-Jawziyya, *Badāʾiʿ al-fawāʾid*, 1: 242–44.

36 See: Louis Gardet, "Ikhlāṣ," in *EI*[2], ed. Peri J. Bearman (Leiden: Brill, 1971), 3: 1059–160.

37 Although his initial count of linguistic benefits to be perused is only ten, Ibn al-Qayyim registers the following one as "the eleventh question" and mentions it is one of the most honorable and important issues he discusses; Ibn Qayyim al-Jawziyya, *Badāʾiʿ al-fawāʾid*, 1: 237, 247.

38 Q 9:5 "When the (four) forbidden months are over, wherever you encounter the idolaters, kill them, seize them, besiege them, wait for them at every lookout post; but if they repent, maintain the prayer, and pay the prescribed alms, let them go on their way, for God is most forgiving and merciful."

the intended disbelievers.[39] Ibn al-Qayyim firmly rejects both interpretations as downright false. According to his reading, Q 109:6 conveys a clearly defined meaning originating in the Preserved Tablet (*al-lawḥ al-maḥfūẓ*, namely: the heavenly source of all divine revelations and decrees), which universally rejects any faith of the disbelievers (*hiya muḥkama ʿumūmuhā naṣṣ maḥfūẓ*). Equally absurd would be to assume any abrogation (*naskh*) in this case, for the whole sura connotes pure monotheistic belief (*wa-hādhihi s-sūra akhlaṣat at-tawḥīd*). Both misinterpretations stem from the false presumption (*ẓann*) that the verse acknowledges another religion which is not Islam. Ibn al-Qayyim declares Q 109:6 instructs nothing but an absolute disassociation of God and His Messenger from the idolaters (*wa-innamā l-āya iqtadat al-barāʾa al-maḥḍa*); hence, this verse resolutely differentiates between the believers and the disbelievers, asserting God will cleanse his worshipers and lands (*āya qāʾima muḥkama thābita bayna al-muʾminīn wa-l-kāfirīn*).[40] Thus far, Ibn al-Qayyim mostly upholds the traditional interpretation of Q 109:6.[41] His next remark is likely aimed at Muslim rationalists: Ibn al-Qayyim considers the judgment of disassociation (*barāʾa*) in Q 109:6 to also be applicable to the struggle between the followers of the Prophetic path (*ahl sunnatihi* (i.e. *sunnat ar-rasūl*), namely the traditionalists), and the innovators (*ahl al-bidaʿ*) who contradict the Qurʾān and the Prophetic path (namely the rationalists). Had the successors of the Prophet said to those innovators "you have your religion and I have mine," that would not mean an acknowledgement of their unwanted innovation, but rather a call for distancing from that innovation while striving to refute it and battle the innovators in *jihād*, if possible.[42]

Ibn al-Qayyim's overall treatment of surat *al-Kāfirūn* is a brilliant illustration of his versatile exegetical proficiencies, as well as his deep familiarity with the various Islamic sciences. Relying primarily on the Islamic textual sources, Ibn al-Qayyim does not hesitate to employ personal reasoning – with a mystical twist – into the rich fabric of his linguistic commentary. In the undercurrents of his interpretation of surat *al-Kāfirūn*, the polemical tone known from his theological debates on the issue of divine attributes (*ṣifāt Allāh*) is highly conspicuous. This becomes evident as Ibn al-Qayyim opens the discussion with his conception of the grammatical function of *mā l-mawṣūla* as a description (*waṣf*), and then comments that verses 109:2–3 are meant to describe God as praiseworthy.[43] Particularly intriguing is a comment at the very end of Ibn al-Qayyim's

39 On the terms specific (*khāṣṣ*) vs. general (*ʿāmm*), see note 22 above.

40 Ibn Qayyim al-Jawziyya, *Badāʾiʿ al-fawāʾid*, 1: 247f.

41 Relevant here is the interpretation of the Qurʾānic term *barāʾa* as it appears in Q 9, see: Nadia Maria El-Cheikh, "Immunity," in *EQ*, ed. Jane Dammen McAuliffe, accessed on 06 September 2020, http://dx.doi.org/10.1163/1875-3922_q3_EQSIM_00216.

42 Ibn Qayyim al-Jawziyya, *Badāʾiʿ al-fawāʾid*, 1: 248f.

43 The polemic tone accompanying the theological discourse is also apparent in Ibn al-Qayyim's use of the derogatory names *jahmiyya* and *muʿattila* in his writing (see *Badāʾiʿ al-fawāʾid*, 1: 242). These labels are usually reserved for the Muʿtazila and his contemporary Ashʿarī opponents, respectively.

interpretation of surat *al-Kāfirūn*, where he declares that he wrote this commentary without seeking the help of any commentary (*min ghayr istiʿāna bi-t-tafsīr*). Ibn al-Qayyim discloses to his readers that the interpretation was handed down to him from what God had taught him and gave him as inspiration (*istimlāʾ mimmā ʿallamahu Allāh wa-alhamahu*), therefore it cannot be found anywhere else. In other words, the multi-layered interpretation presented above mostly represents Ibn al-Qayyim's independent understanding of surat *al-Kāfirūn*. This finding stands out given Ibn al-Qayyim's tendency to deliver his claims by the means of citing previous scholars in a persuading rearrangement of traditionalistic teachings.[44] In addition, since he mentions a divine inspiration (*ilhām*), Ibn al-Qayyim's authentic testimony of independent writing uses Ṣūfī vocabulary; this can explain the mystic attitude and reference to hidden matters throughout his commentary. Closing the discussion of surat *al-Kāfirūn*, Ibn al-Qayyim informs us that he commented on multiple parts of the Qurʾān in the same method of argumentation during his stay in Mecca (ca. 1332/733).[45] Locating these paragraphs of independent *tafsīr* is a worthy large-scale endeavor which will not be pursued here.

Since Ibn al-Qayyim's scholarship heavily relies on Ibn Taymiyya's, it is impossible to discuss the disciple's Qurʾānic hermeneutics without addressing his mentor's thought. Although Ibn Taymiyya did not take on the task of composing a full length Qurʾān commentary, his exegetical efforts alongside his works have been progressively explored in research.[46] In theory, Ibn Taymiyya restricted the permissible interpretation of the Qurʾān to that of the *salaf*, namely the first generations of the Muslim community and their early *tafsīr* material found in the Ḥadīth literature.[47] However, several studies have already shown Ibn Taymiyya displayed a far less radical, but rather a more pragmatic hermeneutical approach throughout his works; this ap-

Another example of combative language is found in Ibn al-Qayyim's ending comment where he compares the disbelievers of Q 109:6 to "the innovators" (*ahl al-bidaʿ*; ibid., 1: 249).

44 A case in point is Ibn al-Qayyim's systemization of the Taymiyyan epistemological and hermeneutical theories in his work *aṣ-Ṣawāʿiq*; see Miriam Ovadia, *Ibn Qayyim al-Jawziyya and the Divine Attributes: Rationalized Traditionalistic Theology* (Leiden: Brill, 2018), 75–81.

45 Ibn Qayyim al-Jawziyya, *Badāʾiʿ al-fawāʾid*, 1: 249.

46 A good example is a paper by Younus Y. Mirza, "Ibn Taymiyya as Exegete: Moses' Father-in-Law and the Messengers in Sūrat Yā Sīn," *JQS* 19, no. 1 (2017): 39–71. Another study discusses exegetical references made by Ibn Taymiyya on both Qurʾān and Ḥadīth: Youshaa Patel, "Whoever Imitates a People Becomes One of Them': A Ḥadīth and its Interpreters," *ILS* 25 (2018): 359–426. In Arabic, significant scholarly endeavors are seek to artificially assemble Ibn Taymiyya's exegetic references into a pseudo-*tafsīr*, such as: *Tafsīr shaykh al-islām Ibn Taymiyya: al-jāmiʿ li-kalām al-imām Ibn Taymiyya fī t-tafsīr*, eds. ʿĪyād b. al-Laṭīf b. Ibrāhīm al-Qaysī and Saʿd b. Fuwwāz aṣ-Ṣamīl, 7 vols. (Riyadh: Dār Ibn al-Jawzī, 2010). Jon Hoover's studies on Ibn Taymiyya's theological thought are also highly relevant: Jonn Hoover, *Ibn Taymiyya's Theodicy of Perpetual Optimism* (Leiden/Boston: Brill, 2007); Hoover, "Ḥanbalī Theology"; Jon Hoover, *Ibn Taymiyya* (London: Oneworld Academic, 2020).

47 The critical shift Ibn Taymiyya introduced in his theoretical treatise entitled *Muqaddima fī uṣūl at-tafsīr* (Introduction to the Foundations of Qurʾānic Interpretation) was the open dismissal of any other hermeneutical method apart from tradition-based exegesis. In modern Arabic and Islamic *tafsīr* histo-

proach was full of intricacies that were the product of reason-based deliberations.[48] Ibn Taymiyya positively advocated for a contextual understanding of the expressions found in the Qur'ān and Ḥadīth within their encompassing semantic context (*qarīna/ qarā'in ma'nawiyya*; contextual meaning) and in accord with the habitual conventions of the Arabic discourse (*'urf*), using linguistic and grammatical reasoning.[49] Ibn Taymiyya also integrated additional hermeneutical tools in his interpretation of the Qur'ān, such as Biblical traditions of Jewish or Christian origin which he reclaimed as part of the Islamic tradition.[50] Taymiyyan tradition-based contextual hermeneutics is the bedrock of Ibn al-Qayyim's approach to the Qur'ānic text.

Particularly relevant to Ibn al-Qayyim's reading of surat *al-Kāfirūn* presented above is Ibn Taymiyya's succinct comments on this sura in his substantial – yet thus far understudied – work *Bayān talbīs al-jahmiyya fī ta'sīs bida'ihim al-kalāmiyya* (Exposing the Deceit of the Heretics in their Establishment of *Kalām*ic Undesired-Innovations).[51] Ibn Taymiyya wrote *Bayān talbīs al-jahmiyya* as a direct response to Fakhr ad-Dīn ar-Rāzī's theological work *Asās at-taqdīs* which was dedicated entirely to the expanded statement on the *ta'wīl* rationalistic method, i.e. figurative interpretation of divine attributes and anthropomorphisms.[52] In *Bayān talbīs al-jahmiyya*, Ibn Taymiyya refers to *sūratay al-ikhlāṣ* as scriptural evidence when he discusses the Islamic consensus regarding the belief in an undividable and incomposite God (*ittifāq al-muslimīn 'alā nafy at-tajzi'a wa-l-inqisām wa-t-tarkīb bi-ma'nāhā l-lughawī 'an Allāh*).[53] In his view, surat *al-Kāfirūn* expresses the practical sense of monotheism as it emphasizes the bestowing of religious deeds on God alone (*tawḥīd al-'amal*), whereas surat *al-Ikhlāṣ* articulates

---

riography, this hermeneutical method of adherence to scriptural evidence from the Qur'ān and Ḥadīth in the course of interpretation has come to be known as *tafsīr bi-l-ma'thūr*; Walid A. Saleh, "Ibn Taymiyya and the Rise of Radical Hermeneutics: An Analysis of An Introduction to the Foundations of Qur'ānic Exegesis," in *Ibn Taymiyya and His Times*, eds. Yossef Rapoport and Shahab Ahmed (Karachi: Oxford University Press, 2010), 123–62; Hoover, *Ibn Taymiyya*, 111–15, 130.

48 Nadjet Zouggar, "Interprétation autorisée et interprétation proscrite: selon le 'Livre du rejet de la contradiction entre raison et Écriture' de Taqī al-Dīn Aḥmad b. Taymiyya," *AI* 44 (2010): 195–206; Wolfhart Heinrichs, "Contacts between Scriptural Hermeneutics and Literary Theory in Islam: The Case of Majāz," *ZGAIW* 7 (1991/92): 253–84.

49 Mohamed Mohamed Yunis Ali, *Medieval Islamic Pragmatics: Sunni Legal Theorists' Models of Textual Communication* (Richmond, Surrey: Routledge, 2000), 87–135; Binyamin Abrahamov, "Ibn Taymiyya on the Agreement of Reason with Tradition," *MW* 82, no. 3–4 (1992): 256–72.

50 Mirza, "Ibn Taymiyya as Exegete," 41–43, 54–61; Younus Y. Mirza, "The Disciples as Companions: Ibn Taymiyya's and Ibn al-Qayyim's Evaluation of the Transmission of the Bible," *Medieval Encounters* 24 (2018), 530–60.

51 The work is also known as *Naqḍ ta'sīs al-jahmiyya* (Criticizing the Foundations of the Jahmites); Ibn Taymiyya, *Bayān talbīs al-jahmiyya fī ta'sīs bida'ihim al-kalāmiyya*, eds. Yaḥya b. Muḥammad al-Hunaydī et al., 10 vols. (Medina: Majma' al-Malik Fahd li-Ṭibā'at al-Muṣḥaf ash-Sharīf, 2005).

52 Fakhr ad-Dīn ar-Rāzī, *Asās at-taqdīs*, ed. Aḥmad Ḥijāzī as-Saqā (Cairo: Maktabat al-Kuliyyāt al-Azhariyya, 1986); Hoover, "Ḥanbalī Theology," 634; Holtzman, *Anthropomorphism in Islam*, 317.

53 Ibn Taymiyya, *Bayān talbīs al-jahmiyya*, 3: 129.

monotheistic intention in its verbal declaration of belief in the one and only God (*tawḥīd al-ʿilm/al-irādā*). In this manner, Ibn Taymiyya sustains that God's singular divinity means that He is worthy of acts of worship and invocations in prayer (*wa-l-ilāhiyya tataḍammanu istiḥqāqahu li-l-ʿibāda wa-d-duʿāʾ*).[54] In light of this proclamation, it becomes evident that Ibn al-Qayyim's understanding of the "two monotheistic *sūras*" in his interpretation of surat *al-Kāfirūn* in *Badāʾiʿ al-fawāʾid* reflects his understanding of Ibn Taymiyya's theological teachings on divine attributes and is not entirely independent. A similar observation can be made with regard to Ibn al-Qayyim's mystical attitude in interpreting the sura. Ibn al-Qayyim probably drew his inspiration from Ibn Taymiyya, because Ibn Taymiyya corroborated his perception of purity of intention when worshipping God alone by referring to the view of "the majority of Ṣūfī scholars" (*ikhlāṣ ad-dīn li-Allāh bi-l-qaṣd wa-l-irāda wa-huwa alladhī yatakallamu bihi mashāyīkh at-taṣawwuf ghāliban*).[55] Notwithstanding, in *Badāʾiʿ al-fawāʾid* Ibn al-Qayyim provides a significant expansion of his mentor's comments on *sūratay al-ikhlāṣ* in a tidy and developed exegetical discussion on surat *al-Kāfirūn*, which emphasizes a linguistic approach. Moreover, Ibn al-Qayyim's appropriated interpretation of surat *al-Kāfirūn* conveys an intensified Ṣūfī inclination which deliberately accentuates the place of secrets and hidden matters. In this respect, Ibn al-Qayyim's exegetical discussion above bears witness to his own perception of the Taymiyyan hermeneutic approach.

4 Rationalized-Traditionalistic Hermeneutics

Highly developed discussions on the correct manner of comprehending the Islamic texts can be found in one of Ibn al-Qayyim's major and mature theological works entitled *aṣ-Ṣawāʿiq al-mursala ʿalā l-jahmiyya wa-l-muʿaṭṭila* (The Unleashed Thunderbolts Directed against the Heretics and the Deniers (of the Divine Attributes)).[56] In this voluminous work, Ibn al-Qayyim systemized his mentor's theories of epistemology and hermeneutics on the much-disputed theological issue of divine attributes and anthropomorphisms in the Qurʾān and Ḥadīth. The essence of *aṣ-Ṣawāʿiq* is an attack on the later-Ashʿarīs for deploying the *majāz* (metaphor) device and practicing their version of *taʾwīl* i.e., figuratively interpreting problematic divine attributes which appear in the textual sources as metaphors rather than understanding them literally and with no inquiry into their mo-

54 Ibid, 141f.

55 This statement appears in a different work by Ibn Taymiyya, in which he deliberates *sūratay al-ikhlāṣ*. See: Taqī d-Dīn Aḥmad b. Taymiyya, *Amrāḍ al-qulūb wa-shifāʾuhā, yalīhā at-tuḥfa al-irāquyya*, no editor mentioned (Cairo: al-Maṭbaʿa as-Salafiyya wa-Maktabatuhā, 1981³), 61, https://archive.org/details/waq58161/page/n60/mode/2up.

56 Ibn Qayyim al-Jawziyya, *aṣ-Ṣawāʿiq al-mursala ʿalā l-jahmiyya wa-l-muʿaṭṭila*, ed. ʿAlī b. Muḥammad ad-Dakhīl Allāh, 4 vols. (Riyadh: Dār al-ʿAṣīma, 1998³).

dality (*bi-lā kayfa*).[57] Negating the Ashʿarī view on the very existence of metaphors in the Arabic language, Ibn al-Qayyim rigorously defends Ibn Taymiyya's stance of closely adhering to the literal and apparent expressions in the Qurʾān and their full acceptance at face value (*ʿalā ẓāhirih*), and only as ontological realities (*ḥaqāʾiq*, sing. *ḥaqīqa*) when describing God.[58] Ibn al-Qayyim stresses the Taymiyyan element of context-dependent meaning (*al-maʿnā* (. . .) *yakūnu maʿahu qarāʾin taḥtaffu bihī*) which is ingrained in the speakers' habitual use of language (*ʿurf ahl al-lugha*). According to Ibn al-Qayyim, human language originated in Revelation (*tawqīfiyya*) and was pre-imposed (or *mawḍūʿa*, assigned) by God, while the human creation was necessarily granted an intuitive knowledge of it (as a part of the idea of *fiṭra*, i.e. the natural disposition of Muslim believers); thus, one should follow the customary use of grammar and semantics.[59] In *aṣ-Ṣawāʿiq* Ibn al-Qayyim sustains this contextual hermeneutics by asserting theological-linguistic reasoning pertaining to the pre-imposed nature of the human language.[60] Once again, Ibn al-Qayyim's hermeneutic approach in *aṣ-Ṣawāʿiq* was formulated in thematic relation to the inter-traditionalistic theological clash over divine attributes.

The verses of divine attributes (*ayāt aṣ-ṣifāt*) were traditionally considered among the ambiguous verses in the Qurʾān (*mutashābihāt*) whose meaning is known to God alone and hence are not to be interpreted.[61] However, in the introduction of *aṣ-Ṣawāʿiq* Ibn al-Qayyim argues that the verses of divine attributes (*āyāt aṣ-ṣifāt*) and the corre-

57 Binyamin Abrahamov, "The 'Bi-lā Kayfa' Doctrine and Its Foundations in Islamic Theology," *AR* 42, no. 1–3 (November 1995): 366–70, 378; Binyamin Abrahamov, "Scripturalist and Traditionalist Theology," in *Oxford Handbook of Islamic Theology*, ed. Sabine Schmidtke (Oxford: Oxford University Press, 2016), 263–79; Livnat Holtzman, "Anthropomorphism," in *EI³*, ed. Kate Fleet et al. (Leiden: Brill, 2011), 1: 46–55.

58 Ibn al-Qayyim relies on his mentor and claims that later-Ashʿarī hermeneutics is rooted in Muʿtazilī ultra-rationalist theories. More specifically, Ibn al-Qayyim denounces the Muʿtazilī conventionalist linguistic theory (*iṣṭilāḥ*), which hypothesized that language and meanings were established by humankind and hence the choice of names linked to meanings is basically arbitrary. This premise led them to conceive the *majāz* as a hermeneutical device to metaphorically interpret problematic divine attributes in the textual sources, such as anthropomorphic descriptions of God. Mustafa Shah, "The Philological Endeavors of the Early Arabic Linguists: Theological Implications of the *tawqīf–iṣṭilāḥ* Antithesis and the *majāz* Controversy (Parts I and II)," *JQS* 1, no. 1 (1999): 27–46, and 2, no. 1 (2000): 44–66.

59 Bernard G. Weiss, "Waḍʿ al-Luġa," in *Encyclopedia of Arabic Language and Linguistics Online*, eds. Lutz Edzard and Rudolf de Jong, accessed March 2016, https://referenceworks.brillonline.com/entries/encyclopedia-of-arabic-language-and-linguistics/wad-al-luga-EALL_SIM_0141?s.num=1; Bernard G. Weiss, "Medieval Muslim Discussions of the Origin of Language," *ZDMG* 124, no. 1 (1974): 33–41; Kees Versteegh, "Linguistic Attitudes and the Origin of Speech in the Arab World," in *Understanding Arabic: Essays in Contemporary Arabic Linguistics in Honor of El-Said Badawi*, ed. Alaa Elgibali (Cairo: American University in Cairo Press, 1996), 27f.; Mustafa Shah, "Classical Islamic Discourse on the Origins of Language: Cultural Memory and the Defense of Orthodoxy," *Numen* (Brill) 58, no. 2 (2011): 315f.

60 Fuller investigation of Ibn al-Qayyim's hermeneutics in the work *aṣ-Ṣawāʿiq* appears in my doctoral project: Ovadia, *Ibn Qayyim al-Jawziyya and the Divine Attributes*, 218–35.

61 Leah Kinberg, "*Muḥkamāt* and *Mutashābihāt* (Koran 3/7): Implication of a Koranic Pair of Terms in Medieval Exegesis," *AR* 35, no. 2 (July 1988): 156f.

sponding traditions (*aḥādīth aṣ-ṣifāt*) are clarified by God in the Qur'ān (and the Prophet in Ḥadīth) beyond any doubt (*bayān shāfī lā yaqaʿu fīhi lubsun wa-lā ishkāl*), which runs counter to the conventional exegesis. Those believers whose hearts God opens to receive these verses "with kindling light" (*wa-man sharaḥa Allāh lahā ṣadra wa-nawwara lahā qalba*) know that they convey clearer content than many – although certainly not all – of the legal verses (*āyāt al-aḥkām*).[62] In other words, only a selected few (*al-khāṣṣa*) can fully grasp the legal verses, whereas both the enlightened elite and lay audience (*al-khāṣṣ wa-l-ʿāmm*) are able to understand the verses of divine attributes (*āyāt aṣ-ṣifāt*). Ibn al-Qayyim is careful in restricting his last statement: he explains that he refers to understanding *āyāt aṣ-ṣifāt* in principle (*aṣl al-maʿnā*), but not understanding their very nature and modality (*lā fahm al-kawn wa-l-kayfiyya*), in accordance with the traditionalistic formula of *bi-lā kayfa*.[63] For example, God clearly specified the time in which one must finish eating during the fasting days of Ramadan in the legal verse Q 2:187 "eat and drink until the white thread becomes distinct from the black, **when it is dawn** (*min al-fajr*)." God also left no room for doubt in the preceding verse Q 2:186 which contains a divine attribute: "(Prophet), if My servants ask you about Me, I am **near** (*qarīb*). (namely:) I respond to those who call Me." Ibn al-Qayyim asserts that the meaning of the divine attribute "near" here is perfectly clear, and therefore none of the Prophet's companions found it vague. In effect, Ibn al-Qayyim advocates for interpreting the Qur'ān by means of the Qur'ān itself. On the other hand, several of the legal verses convey a general edict, thus one must consult the Ḥadīth in order to understand them. For example, regarding the obligatory pilgrimage to Mecca which is mentioned in verse Q 22:29 "and circle around the Ancient House," the mentioned circumambulation (*ṭawāf*) is general (*mujmal*) while the Ḥadīth specifies that one must circle around the Kaʿba seven times.[64] Ibn al-Qayyim demonstrates here an interpretation of the Qur'ān using the Ḥadīth. Conversely, the divine attributes in the Qur'ān and their related *aḥadīth* are not general and therefore do not need an interpretive aid outside of the Qur'ān. The Ḥadīth literature may supply additions or elaborate further on certain divine attributes, but they are understandable on their own within the Qur'ān, unlike the verses of practical law.[65]

Ibn al-Qayyim posits a potential counter-argument to his approach, to which he provides a suitable response. According to the conventional dichotomous perception, the legal verses (*āyāt al-aḥkām*) are definite in meaning (*muḥkamāt*) while verses of

62 Ibn Qayyim al-Jawziyya, *aṣ-Ṣawāʿiq al-mursala*, 1: 210; Ibn Qayyim al-Jawziyya, *Iʿlām al-muwaqqiʿīn ʿan rabb al-ʿālamīn*, ed. Mashhūr b. Ḥasan Āl Sulaymān abū ʿUbayda (Riyadh: Dār Ibn al-Jawzī, 2002), 2: 91.

63 See footnote 57.

64 "Jābir (b. ʿAbdallāh) said: We did not intend anything other than *ḥajj*, and we were not thinking of *ʿumrah*. When we came to the Kaʿbah with him, he touched the corner then trotted (*ramala*) for three circuits, and walked normally (*mashā*) for four (. . .)"; Muslim b. al-Ḥajjāj, *English Translation of Ṣaḥīḥ Muslim*, 3: 345f. (no. 2950), 147 – (1218) (chapter 19: The Ḥajj of the Prophet).

65 Ibn Qayyim al-Jawziyya, *aṣ-Ṣawāʿiq al-mursala*, 1: 211f.

divine attributes (*āyāt aṣ-ṣifāt*) are ambiguous (*mutashābihāt*). How, then – an opponent may argue – is it possible for the ambiguous to be more lucid than the definite? Ibn al-Qayyim answers that two kinds of references to ambiguous and definite meanings exist in the Qur'ān (*at-tashābuh wa-l-iḥkām naw'ān*); one is general (*'āmm*) whereas the other is specific (*khaṣṣ*). According to the general meaning, the scripture as a whole is either *mutashābih* or *muḥkam*. In this case, the meaning of *mutashābih* is not ambiguous but rather a text whose parts are consistent and resemble each other. Two examples of the *mutashābih/muḥkam* description concerning the entire Qur'ān are Q 39:23 "God has sent down the most beautiful of all teachings: a Scripture that is consistent and draws comparisons (*kitāban mutashābihan*)," Q 11:1 "(This is) a Scripture whose verses are perfected, then set out clearly (*kitābun uḥkimat āyātuhu*)." According to the specific meaning, particular verses are either *mutashābih* (ambiguous) or *muḥkam* (definite). Specific reference to verses of either ambiguous or definite meanings appears in Q 3:7 "it is He who has sent this Scripture down to you (Prophet). Some of its verses are definite in meaning (*muḥkamāt*) – these are the cornerstone of the Scripture – and others are ambiguous (*mutashābihāt*)." In the case of general references to the Qur'ān as a whole, verses are consistent in their meanings throughout the entire text, be they verses of divine attribute or legal verses. However, in the case of specific references to verses of either ambiguous or definite meanings, we are dealing with a relative matter (*amr nisbī*); the verses of divine attributes and the legal verses are ambiguous only in certain instances, and how this is determined depends on the person reading the text. Not only does Ibn al-Qayyim interpret the verses cited, he also distinguishes between two separate kinds of general vs. specific[66] by pointing out possible implications of each kind in a way similar to the logical form of exclusive disjunction (i.e.: either α or β, but not both; in Arabic: *taqsīm* or *qisma*).[67] Well aware of the multitude of scholarly disputes concerning *muḥkamāt* vs. *mutashābihāt*, Ibn al-Qayyim additionally emphasizes that not a single member of the *ṣaḥāba* considered the ambiguous verses in the Qur'ān to consist of verses of divine attributes; furthermore, the information that was ascribed to them indicates the contrary. Ibn al-Qayyim posits a few rhetorical questions: Since there was a consensus among the *ṣaḥāba* regarding the meaning of *āyāt aṣ-ṣifāt*, how is it possible to consider these verses as "ambiguous"? Furthermore, how can the legal verses be taken as "definite in meaning," when they stirred controversies and disagreements? Ibn al-Qayyim concludes that the hypothetical claim with which he opened his discussion came first from scholars of the generation succeeding the *ṣaḥāba*.[68]

Drawing on the hermeneutics of Ibn Taymiyya, Ibn al-Qayyim upheld the notion that an adequate interpretation is an explanatory one (*ta'wīl at-tafsīr*). This interpreta-

66 This pair of technical terms is additionally surveyed regarding Qur'ān exegesis in: Joseph E. Lowry, "al-'Āmm wa-l-khāṣṣ".

67 Josef van Ess, "The Logical Structure of Islamic Theology," in *Classical Islamic Culture*, ed. Gustav E. von Grunebaum (Wiesbaden: Harrassowitz, 1970), 40.

68 Ibn Qayyim al-Jawziyya, *aṣ-Ṣawā'iq al-mursala*, 1: 212–14.

tion is permissible and applies to both *muḥkamāt* and *mutashābihāt*. Therefore, interpretation that distorts the intended meaning of the text (*ta'wīl at-taḥrīf*) is categorically prohibited. The hermeneutical discussion depicted thus far represents the manner in which Ibn al-Qayyim employs Qur'ānic verses to support his view and simultaneously offers his interpretation thereof. He exhibits cases of Qur'ānic interpretation using the Qur'ān itself or additional *aḥadīth*, whilst actively proclaiming that he prefers the interpretations of the *ṣaḥāba*. However, Ibn al-Qayyim makes use of the traditionalistic formula *bi-lā kayfa* and refrains from interpreting the modality of divine attributes in the texts. Another prominent characteristic of Ibn al-Qayyim's discussion here is his deployment of speculative arguments which may come from his opponents, as well as his effort to prove them wrong. All of these aspects are intertwined in *aṣ-Ṣawā'iq* into a well-developed scholarly discourse expressing a highly rationalized version of traditionalistic hermeneutics which befits the discourse of *Kalām* (i.e., speculative theology).

5 Conclusions

Much of Ibn al-Qayyim's scholarship was shaped by the theological controversies that stood at the heart of the scholarly agenda in the vehement setting of Mamluk Damascus. As previously noted, Ibn al-Qayyim was persecuted by the Ashʿarī religious establishment in the city since the day his mentor died in prison. The theological issue that stimulated a large part of the political dispute surrounded the divine attributes. Under these circumstances of political inferiority, Ibn al-Qayyim endeavored to propagate Taymiyyan hermeneutics. Each of the samples of commentary discussed above (of surat *al-Kāfirūn* in *Badā'iʿ al-fawā'id*, as well as the examples from *aṣ-Ṣawā'iq*) epitomize Ibn al-Qayyim's didactic and organized authorship. However, each commentary sample simultaneously highlights different aspects of Ibn al-Qayyim's hermeneutical approach; whereas the interpretation of surat *al-Kāfirūn* highlights the place of hidden knowledge and mysticism, the deliberations in *aṣ-Ṣawā'iq* feature a rationalized-traditionalistic discourse. Such discrepancies can reflect the different stages of Ibn al-Qayyim's writing career, *aṣ-Ṣawā'iq* being a later and more mature work compared to *Badā'iʿ al-fawā'id*.[69] This variability may also result from Ibn al-Qayyim's awareness to the particular audience he intended to address. The interpretation of surat *al-Kāfirūn* is shrouded in mysticism and incorporates elements of popular belief to tug at the heartstrings of the laypeople, while the *aṣ-Ṣawā'iq* is a theological monograph that was written to counter Ibn al-Qayyim's ideological rivals and peers amongst the Damascene scholarly elite. In any case, Ibn al-Qayyim's

69 For a chronology of Ibn al-Qayyim's works, see: Livnat Holtzman, "Ibn Qayyim al-Jawziyya (1292–1350)," in *Essays in Arabic Literary Biography*, eds. Roger Allen and Joseph E. Lowry (Wiesbaden: Harrassowitz, 2009), 201–22.

use of philology, linguistics, and Arabic grammar dominates his exegetical endeavors.[70] Inspired by the Taymiyyan model of Qurʾānic interpretation, Ibn al-Qayyim's engagement with the Qurʾānic text in his oeuvre contains a considerable use of personal reasoning, while still firmly grounded in Islamic traditionalism.

6 Bibliography

6.1 Sources

Abdel Haleem, M.A.S. *The Qurʾan: A New Translation*. Oxford: Oxford University Press, 2008.

Ibn Ḥanbal, Aḥmad. *Musnad al-Imām Aḥmad b. Ḥanbal*. Ed. Shuʿayb al-Arnaʾūṭ et al., 50 vols. Beirut: Muʾassasat ar-Risāla, 1997.

Ibn al-Mawṣilī, Shams ad-Dīn Abū ʿAbdallāh Muḥammad b. Muḥammad. *Mukhtaṣar aṣ-Ṣawāʿiq al-mursala ʿalā l-jahmiyya wa-l-muʿaṭṭila*, originally written by Ibn Qayyim al-Jawziyya. Ed. Al-Ḥasan b. ʿAbd ar-Raḥmān al-ʿAlawī, 4 vols. Riyadh: Aḍwāʾ as-Salaf, 2004.

Ibn Qayyim al-Jawziyya. *At-Tafsīr al-qayyim li-l-imām Ibn al-Qayyim*, complied by Muḥammad Uways an-Nadawī. Ed. Muḥammad Ḥāmid al-Fiqī. Beirut: Dār al-Kutub al-ʿIlmiyya, 1948.

Ibn Qayyim al-Jawziyya. *Tafsīr suwar al-Kāfirūn wa-l-muʿawwidhatayn*. Ed. Muḥammad Ḥāmid al-Fiqī. Cairo: Maṭbaʿat as-Sunna al-Muḥammadiyya, 1949.

Ibn Qayyim al-Jawziyya. *Tafsīr sūrat al-Fātiḥa*. Ed. Muḥammad Ḥāmid al-Fiqī. Cairo: Dār al-Jihād, ca. 1979.

Ibn Qayyim al-Jawziyya. *Tafsīr al-muʿawwidhatayn*. Ed. Sayyid Ibrāhīm. Cairo: Dār al-Ḥadīth, 1989.

Ibn Qayyim al-Jawziyya. *Badāʾiʿ at-tafsīr: al-jāmiʿ li-tafsīr al-imām Ibn Qayyim al-Jawziyya*. Ed. Yusayrī s-Sayyid Muḥammad, 5 vols. Riyadh: Dār Ibn al-Jawzī, 1993.

Ibn Qayyim al-Jawziyya. *Aḍ-Ḍawʾ al-munīr ʿalā t-tafsīr – min kutub al-imām al-muḥaddith al-mufassir al-faqīh . . . al-maʿrūf bi-Ibn Qayyim al-Jawziyya*. Ed. ʿAlī l-Ḥamad al-Muḥammad aṣ-Ṣāliḥī, 6 vols. Dakhna/Riyadh: Muʾassasat an-Nūr in collaboration with Maktabat Dār as-Salām, 1994.

Ibn Qayyim al-Jawziyya. *Aṣ-Ṣawāʿiq al-mursala ʿalā l-jahmiyya wa-l-muʿaṭṭila*. Ed. ʿAlī b. Muḥammad ad-Dakhīl Allāh, 4 vols. Riyadh: Dār al-ʿAṣīma, 1998[3].

Ibn Qayyim al-Jawziyya. *Iʿlām al-muwaqqiʿīn ʿan rabb al-ʿālamīn*. Ed. Mashhūr b. Ḥasan Āl Sulaymān Abū ʿUbayda, 7 vols. Riyadh: Dār Ibn al-Jawzī, 2002.

Ibn Qayyim al-Jawziyya. *Badāʾiʿ al-fawāʾid*. Ed. ʿAlī b. Muḥammad al-ʿImrān and Bakr b. ʿAbdallāh abū Zayd, 5 vols. Mecca: Dār ʿĀlam al-Fawāʾid, 2003.

Ibn Qayyim al-Jawziyya. *Al-Badāʾiʿ fī ʿulūm al-Qurʾān*. Ed. Yusayrī s-Sayyid Muḥammad, Beirut: Dār al-Maʿrifa, 2003.

Ibn Qayyim al-Jawziyya. *Badāʾiʿ at-tafsīr*. Ed. Yusayrī s-Sayyid Muḥammad, revised by Ṣāliḥ Aḥmad ash-Shāmī. 3 vols. Riyadh: Dār Ibn al-Jawzī, 2005.

Ibn Qayyim al-Jawziyya. *Al-Intifāʿ bi-l-Qurʾān li-Ibn Qayyim al-Jawziyya*. Ed. Jamīl Ibrāhīm Ḥabīb. Beirut: Dār al-Arqam, 2008.

Ibn Qayyim al-Jawziyya. *At-Tibyān fī l-īmān bi-l-Qurʾān*. Ed. ʿAbdallāh b. Salām al-Baṭāṭī and Bakr b. ʿAbdallāh abū Zayd. Mecca: ʿAlām al-Fawāʾid, 2008.

Ibn Qayyim al-Jawziyya. *Maṣābiḥ at-tafāsīr al-Qurʾāniyya: al-jāmiʿ li-tafsīr Ibn Qayyim al-Jawziyya*. Ed. ʿAbd ar-Raḥmān al-Qammash, 22 vols. 2016. https://ketabonline.com/en/books/104346/read?page=1.

70 See also: Abdessamad Belhaj, "Ibn Qayyim al-Ǧawziyyah et sa contribution à la rhétorique arabe," in *Essays in the Legal and Theological Thought of Ibn Qayyim al-Ǧawziyyah*, eds. Caterina Bori and Livnat Holtzman (Rome: Istituto per l'Oriente C.A. Nallino, 2010), 103–12.

Ibn Qayyim al-Jawziyya. *Ranks of the Divine Seekers: A Parallel English-Arabic Text*, trans. Ovamir Anjum. 4 vols. Leiden: Brill, 2020.
Ibn Taymiyya, Taqī d-Dīn Aḥmad. *Muqaddima fī uṣūl at-tafsīr*, no editor mentioned. Beirut: Dār Maktabat al-Ḥayāt, 1980.
Ibn Taymiyya, Taqī d-Dīn Aḥmad. *Amrāḍ al-qulūb wa-shifā'uhā, yalīhā at-tuḥfa al-irāquyya*, no editor mentioned. Cairo: al-Maṭba'a as-Salafiyya wa-Maktabatuhā, 1981³. https://archive.org/details/waq58161/page/n60/mode/2up.
Ibn Taymiyya, Taqī d-Dīn Aḥmad. *Bayān talbīs al-jahmiyya fī ta'sīs bida'ihim al-kalāmiyya*. Eds. Yaḥya b. Muḥammad al-Hunaydī et al., 10 vols. Medina: Majma' al-Malik Fahd li-Ṭibā'at al-Muṣḥaf ash-Sharīf, 2005.
Ibn Taymiyya, Taqī d-Dīn Aḥmad. *Tafsīr shaykh al-islām Ibn Taymiyya: al-jāmi' li-kalām al-imām Ibn Taymiyya fī t-tafsīr*. Eds. 'Īyād b. al-Laṭīf b. Ibrāhīm al-Qaysī and Sa'd b. Fuwwāz aṣ-Ṣamīl, 7 vols. Riyadh: Dār Ibn al-Jawzī, 2010.
Muslim b. al-Ḥajjāj, Abū l-Ḥusayn. *English Translation of Ṣaḥīḥ Muslim*. Ed. Ḥāfiẓ abū Ṭāhir Zubayr 'Alī Za'ī et al., 7 vols. Riyadh: Darussalam Publications, 2007.
Ar-Rāzī, Fakhr ad-Dīn. *Asās at-taqdīs*. Ed. Aḥmad Ḥijāzī as-Saqā. Cairo: Maktabat al-Kuliyyāt al-Azhariyya, 1986.
At-Tirmidhī, Muḥammad b. 'Īsā. *Al-Jāmi' al-kabīr – Sunan at-Tirmidhī*. Ed. Bashār 'Awwād Ma'rūf, 6 vols. Beirut: Dār al-Gharb al-Islāmī, 1998.

6.2 Secondary Literature

Abdul-Raof, Hussein. *Schools of Qur'anic Exegesis: Genesis and Development*. New York: Routledge, 2010.
Abrahamov, Binyamin. "Ibn Taymiyya on the Agreement of Reason with Tradition." *MW* 82, no. 3–4 (1992): 256–72.
Abrahamov, Binyamin. "The 'Bi-lā Kayfa' Doctrine and Its Foundations in Islamic Theology." *AR* 42, no. 1–3 (November 1995): 366–78.
Abrahamov, Binyamin. *Islamic Theology: Traditionalism and Rationalism*. Edinburgh: Edinburgh University Press, 1998.
Abrahamov, Binyamin. "Scripturalist and Traditionalist Theology." In *Oxford Handbook of Islamic Theology*, edited by Sabine Schmidtke. Oxford: Oxford University Press, 2016: 263–79.
Abū Zayd, Bakr 'Abdallāh. *At-Taqrīb li-'ulūm Ibn al-Qayyim*. Riyadh: Dār al-'Aṣīma, 1996².
Abū Zayd, Bakr 'Abdallāh. *Ibn Qayyim al-Jawziyya: ḥayātuhu, athāruhu, mawāriduhu*. Riyadh: Dār al-'Āṣima li-n-Nashr wa-t-Tawzī', 2002².
Ali, Mohamed Mohamed Yunis. *Medieval Islamic Pragmatics: Sunni Legal Theorists' Models of Textual Communication*. Richmond, Surrey: Routledge, 2000.
Anjum, Ovamir. *Politics, Law and Community in Islamic Thought: The Taymiyyan Moment*. Cambridge/New York: Cambridge University Press, 2012.
Belhaj, Abdessamad. "Ibn Qayyim al-Ǧawziyyah et sa contribution à la rhétorique arabe." In *Essays in the Legal and Theological Thought of Ibn Qayyim al-Ǧawziyyah*, edited by Caterina Bori and Livnat Holtzman. Rome, Istituto per l'Oriente C.A. Nallino, 2010: 103–12.
Bori, Caterina, and Livnat Holtzman. "A Scholar in the Shadow." In *A Scholar in the Shadow: Essays in the Legal and Theological Thought of Ibn Qayyim al-Ǧawziyyah*, edited by Caterina Bori and Livnat Holtzman. Rome: Instituto per l'Oriente C.A. Nallino, 2010: 13–44.
Ad-Dusrī, Muḥammad, and Muḥammad b. 'Abdallāh al-Qaḥṭānī. *Ikhtiyārāt Ibn al-Qayyim wa-tarjīḥātuhu fī t-tafsīr*. 2 vols. Riyadh: Jāmi'at al-Mālik Su'ūd, 2018.

El-Cheikh, Nadia Maria. “Immunity.” In *EQ*, edited by Jane Dammen McAuliffe. Brill, 2001–2006. Accessed on 06 September 2020. http://dx.doi.org/10.1163/1875-3922_q3_EQSIM_00216.

Gade, Anna M. “Recitation of the Qur’ān.” In *EQ*, edited by Jane Dammen McAuliffe. 6 vols. Leiden: Brill, vol. 4, 2004: 367–85. http://dx.doi.org/10.1163/1875-3922_q3_EQCOM_00168.

Gardet, Louis. “Ikhlāṣ.” In *EI²*, edited by Peri J. Bearman et al. 12 vols. Leiden: Brill, vol. 3, 1971: 1059–160. http://dx.doi.org/10.1163/1573-3912_islam_SIM_3512.

Gilliot, Claude. “Exegesis of the Qur’ān: Classical and Medieval.” In *EQ*, edited by Jane Dammen McAuliffe. 6 vols. Leiden: Brill, vol. 2, 2002: 99–124.

Heinrichs, Wolfhart. “Contacts between Scriptural Hermeneutics and Literary Theory in Islam: The Case of Majāz.” *ZGAIW* 7 (1991/92): 253–84.

Holtzman, Livnat. “Ibn Qayyim al-Jawziyya (1292–1350).” In *Essays in Arabic Literary Biography*, edited by Roger Allen and Joseph E. Lowry (Wiesbaden: Harrassowitz, 2009): 201–22.

Holtzman, Livnat. “Anthropomorphism.” In *EI³*, edited by Kate Fleet et al. 5 vols. Leiden: Brill, vol. 1, 2011: 46–55.

Holtzman, Livnat. *Anthropomorphism in Islam: The Challenge of Traditionalism, 700–1550*. Edinburgh: Edinburgh University Press, 2018.

Holtzman, Livnat. “Ibn Qayyim al-Jawziyya.” In *EI³*, forthcoming.

Hoover, Jon. *Ibn Taymiyya’s Theodicy of Perpetual Optimism*. Leiden/Boston: Brill, 2007.

Hoover, Jon. “Ḥanbalī Theology,” in *Oxford Handbook of Islamic Theology*, edited by Sabine Schmidtke. Oxford: Oxford University Press, 2016: 625–46.

Hoover, Jon. *Ibn Taymiyya*. London: Oneworld Academic, 2020.

Kinberg, Leah. “*Muḥkamāt* and *Mutashābihāt* (Koran 3/7): Implication of a Koranic Pair of Terms in Medieval Exegesis.” *AR* 35, no. 2 (July 1988): 143–72.

Krawietz, Birgit. “Ibn Qayyim al-Jawzīyah: His Life and Works.” *MSR* 10, no. 2 (2006): 19–64.

Lāshīn, ʿAbd al-Fattāḥ. *Ibn Qayyim al-Jawziyya wa-ḥissuhu al-balāghī fī tafsīr al-Qur’ān*. Beirut: Dār ar-Rā’id al-ʿArabī, 1982.

Lev, Yaacov. “Symbiotic Relations: Ulama and the Mamluk Sultans.” *MSR* 13, no. 1 (2009): 1–26.

Lowry, Joseph E. “al-ʿĀmm wa-l-khāṣṣ.” In *EI³*, edited by Kate Fleet et al. Brill, 2007–. Article published online 2007. Accessed on 22 September 2020. http://dx.doi.org/10.1163/1573-3912_ei3_SIM_0051.

Al-Matroudi, Abdul Hakim I. *The Hanbali School of Law and Ibn Taymiyyah: Conflict or Conciliation*. London: Routledge, 2006.

Mirza, Younus Y. “Ibn Taymiyya as Exegete: Moses’ Father-in-Law and the Messengers in Sūrat Yā Sīn.” *JQS* 19, no. 1 (2017): 39–71.

Mirza, Younus Y. “The Disciples as Companions: Ibn Taymiyya’s and Ibn al-Qayyim’s Evaluation of the Transmission of the Bible.” *Medieval Encounters* 24 (2018), 530–60.

Al-Mutawallī, Ṣabrī. *Manhaj ahl as-sunna fī tafsīr al-Qur’ān al-karīm: Dirāsa mawḍūʿiyya li-juhūd Ibn al-Qayyim at-tafsīriyya*. Cairo: Maktabat Zahrā’ ash-Sharq, 2002.

Ovadia, Miriam. *Ibn Qayyim al-Jawziyya and the Divine Attributes: Rationalized Traditionalistic Theology*. Leiden: Brill, 2018.

Patel, Youshaa. “Whoever Imitates a People Becomes One of Them’: A Ḥadīth and its Interpreters.” *ILS* 25 (2018): 359–426.

Perho, Irmeli. “Man Chooses his Destiny: Ibn Qayyim al-Jawziyya’s Views on Predestination.” *Islam and Christian-Muslim Relations* 12, no. 1 (January 2001): 61–70.

Sadan, Arik. *The Subjunctive Mood in Arabic Grammatical Thought*. Leiden: Brill, 2012.

Saleh, Walid A. “Ibn Taymiyya and the Rise of Radical Hermeneutics: An Analysis of An Introduction to the Foundations of Qur’ānic Exegesis.” In *Ibn Taymiyya and His Times*, edited by Yossef Rapoport and Shahab Ahmed. Karachi: Oxford University Press, 2010: 123–62.

Saleh, Walid A. “Preliminary Remarks on the Historiography of *tafsīr* in Arabic: A History of the Book Approach.” *JQS* 12 (2010): 6–40.

Shah, Mustafa. "The Philological Endeavours of the Early Arabic Linguists: Theological Implications of the *tawqīf-iṣṭilāḥ* Antithesis and the *majāz* Controversy (Parts I and II)." *JQS* 1, no. 1 (1999): 27–46, and 2, no. 1 (2000): 44–66

Shah, Mustafa. "Classical Islamic Discourse on the Origins of Language: Cultural Memory and the Defense of Orthodoxy." *Numen* (Brill) 58, no. 2 (2011): 314–33. doi: 10.1163/156852711X562335.

Van Ess, Josef. "The Logical Structure of Islamic Theology." In *Classical Islamic Culture*, edited by Gustav E. von Grunebaum. Wiesbaden: Harrassowitz, 1970: 21–50.

Versteegh, Kees. "Linguistic Attitudes and the Origin of Speech in the Arab World." in: *Understanding Arabic: Essays in Contemporary Arabic Linguistics in Honor of El-Said Badawi*, edited by Alaa Elgibali. Cairo: American University in Cairo Press, 1996: 15–31.

Versteegh, Kees. "Sound Symbolism." In *Encyclopedia of Arabic Language and Linguistics Online*, edited by Lutz Edzard and Rudolf de Jong. Brill, 2005–2009. Article published online 2011. Accessed on 06 September 2020. http://dx.doi.org/10.1163/1570-6699_eall_EALL_COM_0315.

Weiss, Bernard G. "Medieval Muslim Discussions of the Origin of Language." *ZDMG* 124, no. 1 (1974): 33–41.

Weiss, Bernard G. "Waḍʿ al-Luġa." In *Encyclopedia of Arabic Language and Linguistics Online*, edited by Lutz Edzard and Rudolf de Jong. Brill, 2005–2009. Article published online 2011. Accessed March 2016. https://referenceworks.brillonline.com/entries/encyclopedia-of-arabic-language-and-linguistics/wad-al-luga-EALL_SIM_0141?s.num=1.

William Wright. *A Grammar of the Arabic Language*, trans. Carl Paul Caspari and edited with numerous additions and corrections by W. Wright. 2 vols. Beirut: Librairie du Liban, vol. 2, 1996[3].

Zouggar, Nadjet. "Interprétation autorisée et interprétation proscrite: selon le 'Livre du rejet de la contradiction entre raison et Écriture' de Taqī al-Dīn Aḥmad b. Taymiyya." *AI* 44 (2010): 195–206.

Ibn Kathīr

Mohammed Nekroumi

1 An Introduction to Ibn Kathīr's Qur'ānic Exegesis

Ismāʿīl b. ʿUmar b. Kathīr's Qur'ān commentary, written in the fourteenth century, is considered one of the most essential works in this field, focusing on the transmission of the Prophetic tradition (*ʿilm riwāyat al-ḥadīth*). Ibn Kathīr was a hadith scholar and critic, which is why his exegesis is mainly based on the hadith narratives concerning the verse in question. In reference to Ibn Kathīr's exegesis, Jalāl ad-Dīn as-Suyūṭī (d. 1505/911) writes that "the exegesis he has conducted is unlike anyone else's."[1] Shams ad-Dīn adh-Dhahabī (d. 1347–48/748), in turn, states that "Ibn Kathīr was highly knowledgeable, a fact witnessed by other scholars, especially in hadith studies, exegesis and history (. . .) In general (. . .) Ibn Kathīr's knowledge is for anyone who reads his exegesis or history, which are both best of what has been written for people."[2]

Ibn Kathīr's method of interpreting the Qur'ān consists of naming the verse, summarizing its main idea, and then explaining it using other verses from the Qur'ān, narratives or sayings from the sunna, or statements of the Prophet's companions or their followers. In addition, he includes the cases relevant to jurisprudence (*fiqh* or *aḥkām*) that the verse might entail. Thus, he gathers evidence related to the Qur'ān and the sunna to demonstrate what he considers to be right in those cases.

Ibn Kathīr's commentary encompasses all Qur'ānic verses, starting with sura 1, *al-Fātiḥa* (The Opening), and ending with sura 114, *an-Nās* (Mankind). Ibn Kathīr divides each sura into a group of relevant verses, then explains it in the way described above. In addition, Ibn Kathīr provides an introduction to his exegetical method, in which he discusses matters related to the interpretation of the Qur'ān, stating that this process of contemplating and explaining Qur'ānic verses is a mandatory task for scholars. As he writes, "It is the scholar's duty to uncover the meanings of the Word of God, to explain it, extract its subjects, learn all of it, and teach it."[3] Moreover, his attitude towards *isrāʾīliyyāt* can be divided into three sections:

> (t)he first one consists of that which we know to be authentic thanks to the witnesses we have that confirm it. The second section consists of that which we know its falsehood thanks to witnesses we have that disclaim it. The third kind belongs to neither of these groups because they

1 Jalāl ad-Dīn as-Suyūṭī, *Dhayl Ṭabaqāt al-ḥuffāẓ li-dh-Dhahabī*, ed. Zakariyā ʿUmayrāt (Beirut: Dār al-Kutub al-ʿIlmiyya, n.d.), 239.

2 Ibid.

3 Ismāʿīl b. ʿUmar b. Kathīr, *Tafsīr al-Qur'ān al-ʿaẓīm*, ed. Sāmī b. Muḥammad Salāma (Medina: Dār Ṭayba li-n-Nashr wa-t-Tauzīʿ, 1999), 1: 6.

https://doi.org/10.1515/9783111320083-011

are unsaid. Hence, we neither believe nor deny them, and they are thus acceptable to narrate but are mostly of not benefit in religious matters.[4]

Ibn Kathīr's introduction states that taking the point of view of a follower (*tābiʿ*) does not necessarily validate an argument for future generations. Nevertheless, if all the followers agree on one view, then it undoubtedly binds the next generation. In the event of disagreement, one should refer to the Qurʾān, the sunna, the standard Arabic language, or the sayings of the companions. Drawing on ʿAbd Allāh b. ʿAbbās's (d. 688/86) saying, "Whoever uses rational discretion (to explain) the Qurʾān, or says what he is ignorant of, let him take his place in Hell,"[5] Ibn Kathīr states that explaining the Qurʾān using one's rational discretion (*ar-raʾy*) is forbidden. What makes Ibn Kathīr's interpretation of the Qurʾān distinctive is the way in which he comments on and criticizes the stories it contains. He clearly states whether a hadith is true or false and discusses scholars' opinions on it, sometimes even criticizing other exegetes' arguments.

His commentary mentions around 160 sources. Some of them have been printed, some remain in the form of manuscripts, whereas others are now missing. These sources can be divided into two groups. The first group is the original sources of exegesis. Although Ibn Kathīr relies on Ibn Jarīr aṭ-Ṭabarī's (d. 923/310) commentary more than any other book, he also criticizes it. In addition, he uses a number of narratives by Abū Ḥātim ar-Rāzī (d. 890/277), generally criticizing his texts. He also refers to Abū ʿAbd Allāh al-Qurṭubī (d. 1272/671), Fakhr ad-Dīn ar-Rāzī (d. 1209/605), and Maḥmūd b. ʿUmar az-Zamakhsharī (d. 1144/538) in the first volume of his work. Finally, he relies on al-Qurṭubī when discussing matters of jurisprudence, while drawing on ar-Rāzī to address logical matters, which were apparently the object of discussion.[6]

The second group consists of his diverse works on the sunna and includes distinct works of hadith scholarship: *Musnads*, the six books, *Muʿjams* (Abū Muḥammad al-Baghāwī (d. 1122/516), Abū l-Qāsim aṭ-Ṭabarānī (d. 971/360)), and history books (*Tārīkh Baghdad, Tārīkh Dimashq, Tārīkh Ibn al-Athīr*). The fact that he favored *Musnad Aḥmad* over other hadith books can be inferred from his use of eleven hadiths from Aḥmad's *Musnad* to explain one argument.[7]

4 Ibid., 9.

5 Ibid., 10.

6 Maṭar az-Zahrānī, *Ibn Kathīr al-mufassir* (Mekka: Kulliyyat ash-Sharīʿa wa-d-Dirāsāt l-Islāmiyya, 1981), 112ff.

7 Ibid., 122ff.

2 Ibn Kathīr's Life, Context and Major Works

Ibn Kathīr, whose full name was 'Imād ad-Dīn Abū l-Fidā' Ismā'īl b. 'Umar b. Ḍaw' b. Kathīr b. Dar', was born in Majdil, a village in Buṣrā,[8] in either 1299/700 or 1300/701.[9] He was a hadith scholar, historian, exegete, and *faqīh* who descended from the Qurashī dynasty and was of Basran origins, although he grew up and studied in Damascus, where he affiliated himself with the Shāfi'ī school of *fiqh*.[10]

Ibn Kathīr grew up in a learned household. His father was a preacher in the mosque as well as a *faqīh* and scholar. He wrote about language and Arabic grammar, and memorized Arabic poetry.[11] In 1306/707, four years after his father's death, Ibn Kathīr moved with his family to Damascus. By the age of ten, he had memorized the whole Qur'ān and had learned its different ways of recitation (*qirā'āt*). He then focused on hadith sciences, listening to *Saḥīḥ Muslim* being read by Najm ad-Dīn b. al-'Asqalānī.[12]

Ibn Kathīr was concerned with not only reading the hadith texts but also hearing them being narrated. Furthermore, he was very knowledgeable in the fields of jurisprudence, hadith, exegesis, history, grammar, and mathematics.[13] One of his most important teachers was chief *Qāḍī* (magistrate) Kamāl ad-Dīn Abū Muḥammad b. Shuhba's. In addition, Abū l-Ḥajjāj Yūsuf al-Mazī, his father-in-law, and Taqī d-Dīn b. Taymiyya played a key role in his education.[14] He became a well-known scholar and, as his reputation grew, students gathered around him to benefit from his knowledge. They included Abū l-Maḥāsin ad-Dimashqī, known as al-Ḥusaynī, Badr ad-Dīn az-Zarkashī, Zayn ad-Dīn al-'Iraqī, and Aḥmad b. Ḥajjī d-Dimashqī.[15] Ibn Kathīr died in Sha'bān in 1372/774.[16]

---

8 Buṣrā is an ancient town in the south of modern Syria; see Abū Khalīl Shawqī, *Aṭlās al-ḥadīth an-nabawī min al-kutub aṣ-ṣiḥāḥ as-sitta* (Damascus: Dār al-Fikr, 2003), 68.

9 Shams ad-Dīn adh-Dhahabī, *Tadhkirat al-huffāẓ*, ed. Zakariyā 'Umayrāt (Beirut: Dār al-Kutub al-'Ilmiyya, 1998), 4: 201.

10 Shams ad-Dīn adh-Dhahabī, *al-Mu'jam al-mukhtaṣṣ bi-l-muḥaddithīn li-l-imām adh-Dhahabī*, ed. Muḥammad al-Ḥabīb al-Hīla (Taif: Maktabat aṣ-Ṣiddīq, 1988), 74; Shams ad-Dīn Abū l-Maḥāsin al-Ḥusaynī, *Dhayl Tadhkirat al-ḥuffāẓ*, ed. n.n. (Beirut: Dār al-Kutub al-'Ilmiyya, 1998), 1: 83; Ibn Ḥajar al-'Asqalānī, *ad-Durar al-kāmina fī a'yān al-mi'a ath-thāmina*, ed. Muḥammad 'Abd al-Ma'īd Ḍān (Haydarabad: Majlis Dā'irat al-Ma'ārif al-'Uthmāniyya, 1972), 445; as-Suyūṭī, *Dhayl Ṭabaqāt al-ḥuffāẓ li-dh-Dhahabī*, 238.

11 Ismā'īl b. 'Umar b. Kathīr, *al-Bidāya wa-n-nihāya*, ed. 'Alī Shīrī (Beirut: Iḥyā' at-Turāth al-'Arabī, 1988), 14: 36.

12 Ibid., 172.

13 Ibn al-'Imād al-Ḥanbalī, *Shadharāt adh-dhahab fī akhbār man dhahab*, ed. Maḥmūd al-Arnā'ūṭ and 'Abd al-Qādir al-Arnā'ūṭ (Damascus/Beirut: Dār Ibn Kathīr, 1986), 1: 68; Aḥmad Shākir, *'Umdat at-tafāsīr 'an al-ḥāfiẓ Ibn Kathīr "Mukhtaṣar Tafsīr Ibn Kathīr" ash-Shaykh* (Alexandria: Dār al-Wafā', 2005), 25.

14 Ibn Ḥajar, *ad-Durar al-kāmina*, 445; Shams ad-Dīn ad-Dāwūdī, *Ṭabaqāt al-mufassirīn*, ed. 'Alī Muḥammad 'Umar (Beirut: Dār al-Kutub al-'Ilmiyya, 1983), 1: 111.

15 Sulaymān b. Ibrāhīm al-Lāḥim, *Manhaj Ibn Kathīr fī t-tafsīr* (Riyad: Dār al-Muslim, 1999), 30.

16 Ibn Ḥajar, *ad-Durar al-kāmina*, 446.

Ibn Kathīr spent most of his life in Damascus, the urban center of the Levant. At that time, the Levant and Egypt were ruled by the Mamluk Sultanate, which had its headquarters in Egypt and its second capital in Damascus.

The Mamluks were initially slaves of the Ayyubids, often assigned military or administrative positions in service of the Ayyubids. As the Mamluks' power grew, they conspired against the Ayyubids and eventually took over Egypt and the Levant, which they ruled for two centuries.[17] This era was characterized by instability, coups, and betrayal. At that time, the Crusades and Tatars also posed a threat to the Islamic world.[18]

In contrast to the political situation, scientific life flourished during this era, which manifested in the scientific schools established by the Mamluks, which included:

- Al-Madrasa aẓ-Ẓāhiriyya, which was founded by aẓ-Ẓāhir Baybars in 1263/662[19] and was home to a library containing ancient resources from all fields of science. Scholars at the Madrasa aẓ-Ẓāhiriyya offered lectures on the Ḥanafī and Shāfiʿī legal schools as well as in hadith and *qirāʾāt*;[20];
- Al-Madrasa al-Manṣūriyya, which was founded by Sultan al-Manṣūr Qalāwūn.[21] It offered schooling on the four Islamic schools of jurisprudence, as well as hadith, exegesis, and medicine;[22] and
- the Mosque-Madrasa of Sultan Ḥasan,[23] which he started building in 1357/758. Architecturally, this was one of the greatest schools from that period. It offered schooling on the four schools of Islamic jurisprudence.

This era was also characterized by the emergence of numerous scholars in the Islamic world, particularly in Egypt and the Levant. Moreover, Cairo and Damascus became scholarly hubs, drawing in academics from different disciplines, especially after the onset of Mongol rule in Damascus. Ibn Kathīr was a contemporary of many scholars who contributed to the scientific renaissance of that era,[24] including Taqī d-Dīn Abū l-Fatḥ Muḥammad b. ʿAlī Wahb, known as Ibn Daqīq al-ʿĪd, the Grand *Qāḍī*, the Shaykh

---

17 The Mamluk era lasted from 1250/648 until 1525/932, see Maḥmūd Shākir, *at-Tārīkh al-islāmī* (Beirut: al-Maktab al-Islāmī, 2000), 7: 35.

18 Ibn Kathīr, *al-Bidāya wa-n-nihāya*, 13: 261.

19 Jamāl ad-Dīn Abū l-Maḥāsin b. Taghrī Bardī, *an-Nujūm az-zāhira fī mulūk Miṣr wa-l-Qāhira*, ed. Ṭāhā ʿAbd ar-Raʾūf Saʿd and Muṣṭafā l-Huwārī (Cairo: Maktabat al-Qāhira, 1983), 7: 158.

20 Jalāl ad-Dīn as-Suyūṭī, *Ḥusn al-muḥāḍara fī tārīkh Miṣr wa-l-Qāhira*, ed. Muḥammad Abū l-Faḍl Ibrāhīm (Cairo: al-Kutub al-ʿArabiyya ʿĪsā l-Bābī l-Ḥalabī, 1976), 2: 228.

21 Al-Ḥusayn ʿUmar b. Ḥabīb, *Tadhkirat an-nabīh fī ayyām al-Manṣūr wa-banīh*, ed. Muḥammad Amīn (Cairo: Wizārat ath-Thaqāfa bi-Miṣr, Markaz Taḥqīq at-Turāth, n.d.).

22 As-Suyūṭī, *Ḥusn al-muḥāḍara fī tārīkh Miṣr wa-l-Qāhira*, 2: 228.

23 Ibid., 232. The one mentioned here is an-Nāṣir Ḥasan b. Muḥammad b. Qalāwūn (d. 1361/762); see ʿUmar b. Ḥabīb, *Tadhkirat an-nabīh*, 30: 240.

24 Az-Zuḥaylī, *Ibn Kathīr ad-Dimashqī*, 24.

of Egypt, who died in 1305/705;[25] Sharaf ad-Dīn ʿAbd al-Muʾmin b. Khalaf Abū Muḥammad ad-Dumyāṭī sh-Shāfiʿī, one of the four great Imams;[26] and Shams ad-Dīn Muḥammad b. Aḥmad b. ʿUthmān adh-Dhahabī, the author of a great number of valuable works. In addition to the learned milieu in which he grew up, this era of great intellectual activity undoubtedly had a strong influence on Ibn Kathīr's development as a scholar.

Ibn Kathīr holds an outstanding position among the scholars of the Ḥijaz due to his important scientific contributions as a teacher, Hadith scholar, Imam, exegete and jurist. Because of his skillful issuing of *fatwās*, some kings and princes consulted him as their advisor.[27] His writings include *Tafsīr al-Qurʾān al-ʿaẓīm*, the focus of this article; *Faḍāʾil al-Qurʾān al-karīm*; *Ikhtiṣār ʿulūm al-ḥadīth*, in which he summarized the book *ʿUlūm al-ḥadīth* by Ibn aṣ-Ṣalāḥ ash-Shahrazūrī; *al-Bidāya wa-n-nihāya*, presenting the history of humanity from the beginning of creation up to Ibn Kathīr's time; *ʿUmar b. ʿAbd al-ʿAzīz*, a biography of the caliph; *al-Fuṣūl fī ikhtiṣār sīrat ar-rasūl*; *al-Ijtihād fī ṭalab al-jihād*; *Mawlid rasūl Allah*, in which he mentions the narratives relevant to the birth of the Prophet Muḥammad; *Aḥādīth at-tawḥīd wa-r-radd ʿalā ahl ash-shirk*; and *Takhrīj aḥādīth mukhtaṣar Ibn Ḥājib fī usūl al-fiqh*.[28]

3 An Analysis of Ibn Kathīr's Method of Interpretation

3.1 The Theological Background

There are different views regarding the nature of Ibn Kathīr's theological background. Some studies claim that Ibn Kathīr was Ashʿarī, while others argue that he was Salafī. The Ashʿarites see him as Ashʿarī for three main reasons: First, Ibn Kathīr himself stated that his doctrine was Ashʿarite. In his biography of Ibrāhīm, the son of Ibn Qayyim al-Jawziyya, Ibn Ḥajar al-ʿAsqalānī writes, "A conflict occurred between him (Ibrāhīm) and ʿImād ad-Dīn b. Kathīr concerning how to teach people. Ibn Kathīr said, 'You hate me because I am Ashʿarī.' The other replied, 'Even if your hair grew down to your feet, nobody would believe you are Ashʿarī or that your teacher is Ibn Taymiyya.'"[29] Second, He was the Grand Imam of the Dār al-Hadith, a position only Ashʿarites were allowed to

25 Shams ad-Dīn adh-Dhahabī, *Muʿjam ash-shuyūkh al-kabīr*, ed. Muḥammad al-Ḥabīb al-Hīla (Taif: Maktabat aṣ-Ṣiddīq, 1998), 2: 249; Ibn Ḥajar, *ad-Durar al-kāmina*, 5: 384.

26 Adh-Dhahabī, *Muʿjam ash-shuyūkh*, 2: 249; Ibn Ḥajar, *ad-Durar al-kāmina*, 3: 221.

27 Muḥammad az-Zuḥaylī, *Ibn Kathīr ad-Dimashqī l-ḥāfiẓ al-mufassir al-muʾarrikh al-faqīh* (Damascus: Dār al-Qalam, 1995), 95–109.

28 Az-Zahrānī, *Ibn Kathīr al-mufassir*, 58–71; az-Zuḥaylī, *Ibn Kathīr ad-Dimashqī*, 150.

29 Ibn Ḥajar, *ad-Durar al-kāmina*, 1: 65.

hold.[30] Third, in his commentary, he denies anthropomorphism, as one can see in his exegesis of the verse "He ascended (*istawā*) on the Throne" (Q 7:54) as well as in other clear examples proving his Sunnī Ashʿarism.[31]

In contrast, Salafists see him as a follower of their school for three main reasons: First, they argue that Ibn Kathīr explicitly declared that his doctrine was Salafi. When explaining the verse, "He ascended (*istawā*) on the Throne," he states that there were several conflicting opinions about its meaning:

> We, however, follow the path that our righteous predecessors (*as-salaf aṣ-ṣāliḥ*), such as Mālik, al-Awzāʿī, ath-Thawrī, al-Layth b. Saʿd, ash-Shāfiʿī, Aḥmad, Isḥāq b. Rāhawayh, and the rest of the scholars of Islam, both in past and present times, took in this regard. Surely, we accept the apparent meaning of "*istawā*," without discussing its true essence, equating it (with the attributes of the creation), or altering or denying it (in any way or form). We also believe that the idea held by some who equate God with the creation is to be rejected, for nothing is similar to God: "There is nothing like Him, and He is the All-Hearer, the All-Seer (Q 42:11)." Indeed, we assert and affirm the statements of Imams, such as Nuʿaym b. Ḥammād al-Khuzāʿī, al-Imām al-Bukhārī's teacher, who said, "Whoever likens God with His creation will have committed blasphemy (*kufr*). Whoever denies what God has described Himself with, will have committed blasphemy (*kufr*)." Certainly, there is no resemblance (of God to the creation) in what God and His Messenger have described Him with. Whoever attests to God the attributes that the plain *āya* and authentic hadiths have mentioned, in the manner that suits God's majesty, all the while rejecting any shortcomings on His part, will have taken the path of guidance.[32]

The author of the book *Ibn Kathīr bayna at-taʾwīl wa-t-tafwīḍ* comments on this passage, remarking, "This statement alone is sufficient for the definition of Ibn Kathīr's doctrine. Do you not see that he says '*athbata*' (attests) and not '*fawwaḍa*' (relegates), and that he says, 'in the manner that suits God's majesty,' which means that he attested the attribute and delegated the manner to God Almighty."[33] Second, in many examples, Ibn Kathīr explains verses containing the attributes of God without esoterically interpreting or delegating them,[34] such as:

> They did not make a just estimate of God such as is due to Him. And on the Day of Resurrection, the whole of the earth will be grasped by His Hand and the heavens will be rolled up in His Right Hand. Glorified is He, and High is He above all that they associate as partners with Him! (Q 39:67)."[35]

He explains, "how it and other verses like it, according to the school (*madhhab*) of *as-salaf*, are to be interpreted and accepted without any attempts to twist or change the

30 Ḥamd as-Sinān and Fawzī l-ʿAnjarī, *Ahl as-sunna al-ashāʿira: shahādat ʿulamāʾ al-umma wa-adillatuhum* (Kuweit: Dār aḍ-Ḍiyāʾ, 2010), 249.

31 Ibid., 250.

32 Ibn Kathīr, *Tafsīr*, 3: 383.

33 ʿAbd al-Ākhar al-Ghanīmī, *ʿAqīdat al-ḥāfiẓ Ibn Kathīr bayna at-tafwīḍ wa-t-taʾwīl* (Dammam: Dār al-Akhillāʾ, 1994).

34 Ibn Kathīr, *Tafsīr*, 3: 72.

35 Ibid., 7: 101.

meaning."[36] Third, he was Ibn Taymiyya's special student.[37] As Ibn Ḥajar writes, "He learned from Ibn Taymiyya and he was charmed by his love and afflicted because of him."[38] According to Taqī d-Dīn b. Qāḍī Shuhba, Ibn Kathīr defended and followed Ibn Taymiyya in many views. "He repeated Ibn Taymiyya's *fatwā* concerning divorce, which caused him a lot of harm."[39]

However, Khālid al-Ḥusayn 'Abd ar-Raḥmān Ḥamdān, presents the following argument:

> Concerning Ibn Kathīr's attitude, it is not possible to conclude that he held one view about esoteric interpretation or delegation. That is to say, if we exhibit the interpretation of qualities of the Lord Almighty, we notice that Ibn Kathīr adopts the Salafis' method when it concerns delegation and that he follows the Khalaf (Ash'arites) in their esoteric interpretation, as if he does not mind using interpretations (. . .). What makes us think that we are not wrong is his interpretation of God's mockery of blasphemers as ridiculing them, which is an interpretation made before by Abū Ja'far aṭ-Ṭabarī. (. . .) What is the meaning of his interpretation of God's accompaniment (presence), and what is the meaning of how he interprets the statement "We too will forget you" (Q 20:126), by saying "We will treat you as we would treat one who is forgetful"? And many other examples can be stated. Parallel to that, he followed the path of the *Salaf* regarding ascendance (*al-istiwā'*), God's utterances and speech (*al-kalām*), arrival, love (. . .).[40]

Ibn Kathīr is a *salafī*, although his use of the *khalaf* methods in many issues demonstrates that the controversy between the two schools (*khalaf* and *salaf*) has no negative impact on Islamic theology.

Based on previous remarks, we can clearly see the difference between Ibn Kathīr and his teacher, Ibn Taymiyya, as Ibn Kathīr made use of both esoteric interpretation and delegation, whereas Ibn Taymiyya used only delegation, completely excluding esoteric interpretation.[41] For instance, as Ibn Taymiyya writes,

> The total statement about this whole matter is that God is described the way he described Himself, or as the Prophet described Him, and as the first followers described Him without surpassing the Qur'ān or hadith. Moreover, as-Salaf's method is to describe God the way He described Himself and the Prophet without twisting (*taḥrīf*) or denying His attributes (*ta'tīl*) or using descriptive designations of God's attributes (*takyīf*) or likening Him (*tamthīl*).[42]

It is worth mentioning that Ibn Kathīr discussed some matters of faith in the same way as the *ahl al-athar* (a movement of *hadith* scholars) and the Salafis, such as the

36 Ibid.

37 Al-Ghanīmī, *'Aqīdat al-ḥāfiẓ Ibn Kathīr bayna at-tafwīḍ wa-t-ta'wīl*, 65.

38 Ibn Ḥajar, *ad-Durar al-kāmina*, 1: 445.

39 Taqī d-Dīn b. Qāḍī Shuhba, *Ṭabaqāt ash-Shāfi'iyya*, ed. 'Abd al-'Alīm Khān (Beirut: 'Ālam al-Kutub, 1986), 3: 86.

40 Khālid Ḥusayn 'Abd ar-Raḥīm Ḥamdān, *al-Imām Ibn Kathīr wa-manhajuhu fī dirāsat qaḍāyā al-'aqīda* (Omdurman: Jāmi'at al-Qur'ān al-Karīm wa-l-'Ulūm al-Islāmiyya, 1996), 123.

41 Ibid., 124.

42 Taqī d-Dīn b. Taymiyya, *al-Fatwā l-Ḥamawiyya al-kubrā*, ed. Ḥamad b. 'Abd al-Muḥsin at-Tuwayjirī (Riyad: Dār aṣ-Ṣamī'ī, 1998), 271.

ebb and flow of faith, whether faith is a saying and a deed, invalidations of *tawḥīd*, God's attributes, and proving that there is no god but God. In addition, he tends to argue with atheists (*dahriyya*), pagans, Muʿtazilites, Shiits (*rawāfiḍ*), and Khawārij in his exegesis to defend the doctrines of *sunna wa-l-jamāʿa*.

3.2 Ibn Kathīr's Hadith- and Jurisprudence-Critical Stance and Its Influence

Ibn Kathīr was a great scholar and distinguished hadith critic, which is discernible from his work in the field of hadith scholarship. He focused on discerning between false and authentic hadiths, criticized hadith scriptures, and, when a hadith contradicted the Qurʾān, an authentic hadith, or history, he condemned it as a false hadith.[43]

Indeed, Ibn Kathīr not only benefited from the hadith scholars who preceded him but also transcribed their narratives explaining the Qurʾān as well as their points of view, their ways of juridical deduction, and their verdicts on some hadiths and their narrators. In addition, he discussed the status of the hadith's chain of transmission and whether a hadith was weak (*ḍaʿīf*), transmitted (*mursal*), or sound (*ṣaḥīḥ*), and incorporated the opinions of scholars like Muḥammad b. ʿĪsā t-Tirmidhī, Muḥammad b. ʿAbd Allāh al-Ḥākim, Abū ʿUbayd al-Qāsim b. Sallām, Muḥammad b. Abī Shayba, and ʿAbd Allāh b. ʿAbd al-Barr on text-related matters, such as the question of which versions contained exaggerations or oddities.[44]

Although he greatly benefited from earlier exegetes such as aṭ-Ṭabarī, Abū ʿAbd Allāh al-Qurṭubī, and Ibn Abī Ḥātim ar-Rāzī, read their works, and presented their views, he adopted only that which suited his exegetical method. The way in which he retraces the thought of these *tafsīr* scholars exemplifies the accuracy of his scientific approach. Indeed, he relied on legitimate arguments to dispute unsubstantiated statements and usually justified the reasons for which he denied the statement in question. Ultimately, he decided what statement is more likely to be true based on the Qurʾān and the sunna.[45] His criticism was objective and fair, as he never used his critique to target or insult scholars, instead analyzing their ideas using logic alone.[46]

The majority of Ibn Kathīr's teachers were Shāfiʿī. Thus, Ibn Kathīr followed the Shāfiʿī school of jurisprudence that dominated in both Egypt and the Levant. At a very young age, he memorized Shafiʿī books like Abū Isḥāq ash-Shirāzī's *at-Tanbīh fī l-fiqh*.

43 Sālim al-Ghalīlāt Aḥlām and al-Quḍāt Sharaf, "Naqd al-matn ʿinda Ibn Kathīr fī tafsīrihi," *Majallat al-Jāmiʿa li-d-Dirāsāt al-Islāmiyya* 26, no. 2 (2018): 339–70.

44 Az-Zahrānī, *Ibn Kathīr al-mufassir*, 383–401; Ibn Ibrāhīm, *Manhaj Ibn Kathīr fī t-tafsīr*, 241.

45 Aḥmad b. ʿUmar as-Sayyid, *Taʿqībāt al-imām Ibn Kathīr ʿalā man ṣabaqahu min al-mufassirīn min khilāl kitābihi Tafsīr al-Qurʾān al-ʿaẓīm* (Mekka: Jāmiʿat Umm al-Qurā: Kulliyyat Uṣūl ad-Dīn wa-d-Daʿwa, 2010), 116–20.

46 Az-Zahrānī, *Ibn Kathīr al-mufassir*, 402.

He practiced this school's method throughout his life and wrote a book on Shafiʿī legal scholarship entitled *Tabaqāt al-fuqahāʾ ash-Shāfiʿiyīn*. This jurisprudential background had a significant impact on Ibn Kathīr's interpretive method, as he implicitly defended the Shafiʿī school in his exegesis. As a scholar, however, he did not blindly support Shāfiʿī statements, for he was a thinker who relied primarily on the Qurʾān and the sunna. He thus disagreed with ash-Shāfiʿī on many occasions, for example, regarding his support of Aḥmad b. Ḥanbal and Mālik b. Anas in the matter of the ritual cleansing (*masḥ*) of the head in ablution (*wuḍūʾ*).[47]

3.3 Ibn Kathīr's Principles of Interpretation

Ibn Kathīr applies the method of interpretation through transmitted earlier interpretations (*at-tafsīr bi-l-maʾthūr*),[48] which he clearly states in his introductory work:

> The best way to explain the Qurʾān is using the Qurʾān itself because what was mentioned briefly in part is detailed in another part. However, if you still cannot find the meaning then look in the sunna, as it explains and clarifies the Qurʾān's text (. . .). Otherwise, we use the companions' statements because they witnessed the events and situations related to them, and this fact in particular made them more knowledgeable than others in addition to their being righteous.[49]

These are the main principles that Ibn Kathīr followed, the details of which will be explained in the following sections.

3.3.1 Interpreting the Qurʾān Using the Qurʾān Itself

Using the Qurʾān itself to interpret the Qurʾān was always Ibn Kathīr's methodological approach. His book is best known for its citations of verses with a corresponding meaning,[50]

47 Ibid., 54f.

48 "The interpretation is divided according to the method applied by the scholars in two parts: first, the method of received interpretation (*at-tafsīr al-maʾthūr*), interpreting the Qurʾān using the Qurʾān itself and details of some of its verses, and what was reported from the Messenger, and what was reported from the Companions, may God be pleased with them, and what is transmitted from the followers, clarifying the statements and the intention of the Almighty God from the texts of his noble book. The second way is *at-tafsīr bi-r-raʾy*, which means the interpretation of the Qurʾān by *ijtihād* based on the words of the Arabs, the meaning of the words, the circumstances of revelation (*asbāb an-nuzūl*), abrogating and abrogated verses (*an-nāsikh wa-l-mansūkh*), the knowledge of the purposes of the sharia and the rest of the tools that the interpreter needs, see Jalāl ad-Dīn as-Suyūṭī, *al-Itqān fī ʿulūm al-Qurʾān*, ed. Muḥammad abū l-Faḍl Ibrāhīm (Cairo: al-Hayʾa al-Miṣriyya al-ʿĀmma li-l-Kitāb, 1974), 4: 221; Muḥammad az-Zurqānī, *Manāhil al-ʿirfān fī ʿulūm al-Qurʾān* (Cairo: Maṭbaʿat ʿIsā l-Bābī l-Ḥalabī, 1943), 2: 12; Muḥammad Ḥusayn adh-Dhahabī, *DhTM* (Cairo: Maktabat Waḥba, 1977), 112; az-Zuḥaylī, *Ibn Kathīr ad-Dimashqī*, 211.

49 Ibn Kathīr, *Tafsīr*, 1: 8 (*muqaddima*).

50 *DhTM*, 175.

a method largely unused by other scholars before Ibn Kathīr. Additionally, Ibn Kathīr compiled verses that are similar in terms of their construction, that deal with a particular subject, and that lead to a possible meaning or suggest it. The fact that Ibn Kathīr paid great attention to this methodological principle is evident from his use of specific verses that resemble the verse being explained in only some aspects.[51] For example, in the verse, "And indeed, We created man from sounding clay of altered black smooth mud" (Q 15:26), after citing what preceding exegetes have said about the meaning of *aṣ-ṣalṣāl* (clay), he writes that "it seems that this is similar to what God says: 'He created man (Ādam) from sounding clay like the clay of pottery' (Q 55:14)." Hence, interpreting the verse by using another one is helpful.[52] Sulaymān Allāḥim elaborates on the different types of this method of interpretation, including the interpretation of an abridged verse using one or several more detailed verses, elucidating a collective verse (mujmal) with an exemplified one, reconciling the unconditional (*muṭlaq*) with the restricted (*muqayyad*), specifying the general, clarifying the ambiguous, and indicating the abrogation of one verse by another, as well as confirming the harmony of the Qur'ān and using the recitation forms for additional elaboration.[53]

3.3.2 Interpreting the Qur'ān using the Sunna

Ibn Kathīr stated that, due to its position in Islamic theology, he relied mainly on the sunna to interpret the Qur'ān. These are the main aspects of interpreting the Qur'ān using the sunna in Ibn Kathīr's work:[54]

- Clarifying a collective verse: some verses seem to be unclear and in need of clarification using the sunna, which usually provides detailed information. For instance, concerning aṣ-Ṣafā and al-Marwa and the interpretation of the doctrines related to *al-Hajj* (pilgrimage), he says, "Verily! Aṣ-Ṣafā and al-Marwa (two mountains in Mecca) are among the Symbols of God" (Q 2:158).[55] This interpretation is based on the practices of the Prophet Muḥammad concerning the pilgrimage.
- Clarifying ambiguities: The verse, "eat and drink until the white thread (light) of dawn appears to you distinct from the black thread (darkness of night)" (Q 2:187),[56] is an example of this form, as Ibn Kathīr interpreted it according to the Prophet's

51 Ibn Ibrāhīm, *Manhaj Ibn Kathīr fī t-tafsīr*, 185f.
52 Ibn Kathīr, *Tafsīr*, 4: 458.
53 Ibn Ibrāhīm, *Manhaj Ibn Kathīr fī t-tafsīr*, 188–221.
54 Ibn Ibrāhīm, *Manhaj Ibn Kathīr fī t-tafsīr*, 229–33.
55 Ibn Kathīr, *Tafsīr*, 1: 340.
56 Ibid., 378.

statement, which was recounted by ʿAdī b. Ḥātim: "It represents the white light that appears in early morning and the darkness that appears at night."[57]

- Specifying generalizations in the Qurʾān using the sunna: For instance, the verse, "Cut off the hand of the thief, male or female" (Q 2:38)[58] is specified in the following statement ascribed to the prophet by ʿĀʾisha, "The hand of the thief should not be cut off (if he stole) a quarter dinar or less."[59]
- Restricting some verses of the Qurʾān using the sunna: The verse cited above does not specify which hand. However, according to the Prophet's application of this verse, i.e., the order to "Cut off her hand," "her right hand was cut off."[60]

3.3.3 Interpreting the Qurʾān Using the Narratives Transmitted by the Companions and their Followers

As mentioned above, Ibn Kathīr explains that he uses the narratives of the companions and their followers when interpreting the Qurʾān. Nevertheless, he does not recite the Salaf's sayings without analyzing them, instead choosing the narratives that seem most plausible or more likely to be true. This distinguishes his work from that of the scholars who only compiled narratives without evaluating and discerning between the true and false ones.[61]

3.3.4 Interpreting the Qurʾān by Relying on Classical Arabic

Ibn Kathīr mentions his reliance on Classical Arabic for interpreting the Qurʾān when talking about the sayings of the followers:

> If they (the followers) agree about something, then there is no question that it is authentic. If they do not agree with one another, one side does not have authority over the other. It goes without saying that their understanding has no authority over later generations, which forces us to rely on the language of the Qurʾān and the sunna or what we can understand based on Classical Arabic and, in particular, the companions' language.[62]

Thus, he draws on both poetry and prose to reveal a verse's meaning. Moreover, Ibn Kathīr emphasizes the use of Classical Arabic in his introduction by mentioning and

---

57 Muslim b. al-Hajjāj, *Ṣaḥīḥ Muslim*, ed. Muḥammad Fūʾād ʿAbd al-Bāqī (Beirut: Iḥyāʾ at-Turāth al-ʿArabī, 1991), 2: 766f. (Nr. 1090).

58 Ibn Kathīr, *Tafsīr*, 3: 98.

59 Muslim, *Saḥīḥ*, 2: 1312 (Nr. 1684).

60 Aḥmad b. Ḥanbal, *al-Musnad*, ed. Shuʿayb al-Arnāʾūṭ et al. (Beirut: Muʾassasat ar-Risāla, 2001), 11: 238 (Nr. 6657).

61 Ibid., 304.

62 Ibn Kathīr, *Tafsīr*, 1: 11.

applying it when interpreting the Qur'ān according to the previously stated order.[63] In his interpretation of the verse *"wa-qad balaghtu min al-kibari 'itiyyan"* ("and I have reached the extreme old age," Q 19:8), for instance, he relies on Classical Arabic, writing that ancient Arabs say *"'atā, ya'tū, 'itiyan, and 'utuwwa, 'asā, ya'sū, and 'usuwwan"* when a stick desiccates.[64] Based on this description of the word's root, he concludes that the meaning is "growing extremely in age."

3.3.5 Interpreting the Qur'ān using *isrā'īliyyāt*

Isrā'īliyyāt refers to the tales that exegetes drew on from the *ahl al-kitāb* (Jews and Christians) to explain the Qur'ān. Regardless of whether the narrator was a Jew or Christian, these tales were considered by Ibn Kathīr parasites by books about interpretation because of the exaggeration and distortion that were said to characterize them.[65] As already mentioned above concerning his methodological view of *isrā'īliyyāt*, Ibn Kathīr believed that

> (t)he first one consists of that which we know to be authentic thanks to the witnesses we have that confirm it. The second section consists of that which we know its falsehood thanks to witnesses we have that disclaim it. The third kind belongs to neither of these groups because they are unsaid. Hence, we neither believe nor deny them, and they are thus acceptable to narrate but are mostly of no benefit in religious matters.[66]

He thus ignores the group of *isrā'īliyyāt* that contradict the Qur'ān and the sunna, although some other exegetes mention them. When Ibn Kathīr mentions these *isrā'īliyyāt*, he either criticizes them or does not comment on them at all.[67] Aḥmad Shākir argues that Ibn Kathīr's lack of criticism concerning the citation of these *isrā'īliyyāt* means that what we do not understand from the word of God should not be proven using what we neither accept nor deny. Indeed, the prophet allowed us to speak about them without having to believe or deny what we hear about them. "Does that not mean that placing the *isrā'īliyyāt* alongside the word of God in order to interpret it is nothing but a confirmation of its truth? May God forgive!"[68]

63 Ibn Ibrāhīm, *Manhaj Ibn Kathīr fī t-tafsīr*, 334–63.

64 Ibn Kathīr, *Tafsīr*, 5: 190.

65 *DhTM*, 1: 130; Ramzī Na'nā'a, *al-Isrā'īliyyāt wa-athāruhā fī kutub at-tafsīr* (Damascus: Dār al-Qalam, 1970), 74.

66 Ibn Kathīr, *Tafsīr*, 1: 10 (*muqaddima*).

67 Muhammad 'Abd Allāh al-Fāliḥ, *Ḥayāt Ibn Kathīr wa-kitābuhu Tafsīr al-Qur'ān al-'aẓīm* (Irbid: Dār al-Bayān, 2004), 107.

68 Shākir, *'Umdat at-tafāsīr*, 8: 135.

3.3.6 Clarifying the Meanings of the Verses Using the Saying of the Scholars of Islamic Jurisprudence

Adh-Dhahabī states that Ibn Kathīr became involved in jurisprudential discussions by citing the sayings of jurisprudence scholars and their arguments when explaining verses related to practical issues.[69] Nonetheless, unlike other exegetes, he limited the use of these scholars' views. This has been emphasized by Maṭar az-Zuhranī, who mentions that Ibn Kathīr used a moderate method when addressing jurisprudential issues and that he was not overly prescriptive in citing their views and differences. However, because of the benefit of these views for the interpretation of the Qur'ān, he did mention them concisely.[70]

4 The Historical Importance of Ibn Kathīr's Exegesis for the Past and Present

Various features distinguish Ibn Kathīr's exegesis from that of others, which is why it gained the approval and blessings of Ibn Kathīr's contemporaries. Moreover, the subsequent influence of his work made it a primary source for later scholars in this field. However, it is the impact that his work has had as an important source for most Muslims that constitutes its historical relevance.

4.1 Contemporary Approval of Ibn Kathīr's Work

Ibn Kathīr's scholarly peers studied his work as soon as it was published and noticed its unique features, which differentiated it from preceding works. This made his work famous among the masses and scholars, who praised and approved it. One of his peers, adh-Dhahabī, claimed that Ibn Kathīr's exegesis was one of his greatest works.[71] In another context, he wrote, "Indeed, this exegesis is the best *at-Tafsīr bi-l-ma'thūr* ever written."[72] Jamāl ad-Dīn b. Taghrī Bardī was another of Ibn Kathīr's contemporaries who mentioned his skill as an exegete.[73] His peers were not the only ones who praised him. He was later lauded by scholars such as Jalāl ad-Dīn as-Suyūṭī, who said that "the exege-

69 *DhTM*, 1: 176.
70 Az-Zahrānī, *Ibn Kathīr al-mufassir*, 8: 76.
71 Shams ad-Dīn adh-Dhahabī, *Siyar a'lām an-nubalā'* (Cairo: Dār al-Kutub al-'Ilmiyya, n.d.), 1: 29.
72 *DhTM*, 1: 176.
73 Jamāl ad-Dīn b. Taghrī Bardī, *al-Manhal aṣ-ṣāfī l-mustawfī ba'd al-wāfī*, ed. Muḥammad Amīn (Cairo: al-Hay'a al-Miṣriyya, 1984), 2: 414.

sis he has done is unlike anyone else's,"[74] and Ibn al-ʿImād al-Ḥanbalī.[75] Aḥmad ash-Shawkānī said, "he is one of the greatest exegetes, if not the greatest,"[76] while Shaykh Muḥammad az-Zurqānī claimed that his "exegesis is one of the most correct interpretations (*at-tafsīr bi l-ma'thūr*) ever received, if not the most correct one."[77] Moreover, as Shaykh Aḥmad Shākir wrote, "the exegesis of al-Ḥāfiẓ Ibn Kathīr could be the best one we have ever seen in terms of quality and punctuality, unless the work of Abū Jaʿfar aṭ-Ṭabarī is the best. We cannot even compare them with other works we have because they are so much better than the other ones."[78] The historical significance and preservation of Ibn Kathīr's work has been ensured by the dedication of scholars from his time to the present age, who have diligently studied and summarized it. Muḥammad b. al-Buʿalī (known as Ibn al-Junāniya) and Ibn Saʿd al-Kāzurūnī were the first to summarize Ibn Kathīr's book. Further attempts from the present age include Aḥmad Shākir's book, *ʿUmdat at-Tafsīr ʿan al-Ḥāfiẓ Ibn Kathīr*, Muḥammad Karīm's *Mukhtaṣar Tafsīr Ibn Kathīr,* Muḥammad ʿAlī ṣ-Ṣābūnī's work *Mukhtaṣar Tafsīr Ibn Kathīr*, and Muḥammad Nasīb ar-Rifāʿī's book *Taysīr al-ʿalī l-qadīr li-ikhtiṣār Tafsīr Ibn Kathīr.*[79]

4.2 Ibn Kathīr's Influence on Later Audiences

Ibn Kathīr's exegesis has been an important source for Salafi exegetes, who have drawn from, relied on, and greatly benefitted from it. Qur'ānic interpretations that do not in some way refer to Ibn Kathīr's view are rare. The most famous figures to have been influenced by Ibn Kathīr's work are Abū s-Suʿūd al-ʿImādī (d. 1574/982), the author of *Irshād al-ʿaql as-salīm ilā mazāyā al-Qur'ān al-karīm*; Muḥammad ash-Shawkānī (d. 1834/1250), the author of *Fatḥ al-qadīr al-jāmiʿ bayna fannay ar-riwāya wa-d-dirāya min ʿilm at-tafsīr*; Shihāb ad-Dīn al-Alūsī (d. 1854/1270), the author of *Rūḥ al-maʿānī fī tafsīr al-Qur'ān al-ʿaẓīm wa-s-sabʿ al-mathānī*; and Jamāl al-Dīn al-Qāsimī (d. 1904/1322), the author of *Maḥāsin at-ta'wīl.*

Twentieth-century scholars who have used this work in their interpretations include Muḥammad Rashīd Riḍā (d. 1935/1354), the author of *Tafsīr al-manār*; Sayyid Quṭb (d. 1965/1385), the author of *Fī ẓilāl al-Qur'ān*; and Muḥammad Amīn ash-Shinqīṭī (d. 1973/1393), the author of the book *Aḍwā' al-bayān fī īdāḥ al-Qur'ān bi-l-Qur'ān.*

74 As-Suyūṭī, *Dhayl Ṭabaqāt al-ḥuffāẓ*, 239.
75 Ibn ʿImād al-Ḥanbalī, *Shadharāt adh-dhahab*, 1: 67.
76 Aḥmad ash-Shawkānī, *al-Badr aṭ-ṭāliʿ bi-maḥāsin min baʿd al-qarn as-sābiʿ* (Beirut: Dār al-Kitāb al-Islāmī, 2012), 1: 153.
77 Az-Zurqānī, *Manāhil al-ʿirfān fī ʿulūm al-Qur'ān*, 2: 30.
78 Shākir, *ʿUmdat at-tafāsīr*, 1: 9.
79 Az-Zuḥaylī, *Ibn Kathīr ad-Dimashqī*, 228.

5 Conclusion

Ibn Kathīr's exegetical method is considered an example of the *atharī* interpretive tradition. This is primarily visible in his methodological principle of explaining the Qur'ān using the Qur'ān itself. He prioritized the sayings of the Prophet, followed by the sayings of the Companions and the Followers. Finally, he relied considerably on the Arabic language. According to Ibn Kathīr, these form the key reference points for interpretation, and they exclude any other point of reference. Thus, other points of references can only be used by advocates of rational discretion (*ahl ar-ra'y*).

Ibn Kathīr was a hadith scholar predominately influenced by the methods of the hadith sciences in his interpretation of the Qur'ān. He criticized certain narratives, either the text itself, its chain of transmission (*sanad*), or both. He also provided commentary on the narratives he used to elucidate their meanings. The critique of the narratives presented in his exegesis serves as a model for other scholars. The features of his exegesis include his brief and clear citations of relevant jurisprudential issues and his exact, scholarly methodology when dealing with the *isrā'īliyyāt*. Many exegetes and scholars have benefited from Ibn Kathīr's works, his exegesis, his method and pure doctrine, both in the past and present. The historical importance of Ibn Kathīr's work is reflected in the fact that it has become a central point of reference in Qur'ān interpretation for both laymen and experts alike.

6 Bibliography

6.1 Sources

Al-ʿAsqalānī, Ibn Ḥajar. *An-Nukat ʿalā kitāb Ibn aṣ-Ṣalāḥ*. Ed. Majdī Fatḥī s-Sayyid and Muṣṭafā Shatāt. Cairo: Dār al-Kutub al-ʿIlmiyya, n.d.

Ad-Dāwūdī, Shams ad-Dīn. *Ṭabaqāt al-mufassirīn*. Ed. ʿAlī Muḥammad ʿUmar, 2 vols. Beirut: Dār al-Kutub al-ʿIlmiyya, 1983.

Adh-Dhahabī, Muḥammad Ḥusayn. *DhTM*. Cairo: Maktabat Waḥba, 1977.

Adh-Dhahabī, Shams ad-Dīn. *Al-Muʿjam al-mukhtaṣṣ bi-l-muḥaddithīn li-imām adh-Dhahabī*. Ed. Muḥammad al-Ḥabīb al-Hīla. Taif: Maktabat aṣ-Ṣiddīq 1988.

Adh-Dhahabī, Shams ad-Dīn. *Muʿjam ash-shuyūkh al-kabīr*. Ed. Muḥammad al-Ḥabīb al-Hīla, 2 vols. Taif: Maktabat aṣ-Ṣiddīq, 1998.

Adh-Dhahabī, Shams ad-Dīn. *Tadhkirat al-ḥuffāẓ*. Ed. Zakariyā ʿUmayrāt, 5 vols. Beirut: Dār al-Kutub al-ʿIlmiyya, 1998.

Adh-Dhahabī, Shams ad-Dīn. *Siyar aʿlām an-nubalāʾ*. Cairo: Dār al-Kutub al-ʿIlmiyya, n.d.

Al-Ḥanbalī, Ibn al-ʿImād. *Shadharāt adh-dhahab fī akhbār man dhahab*. Ed. Maḥmūd al-Arnāʾūṭ and ʿAbd al-Qādir al-Arnāʾūṭ, 10 vols. Damascus/Beirut: Dār Ibn Kathīr, 1986.

Al-Ḥusaynī, Shams ad-Dīn Abū l-Maḥāsin. *Dhayl Tadhkirat al-ḥuffāẓ*. Ed. n.n. Beirut: Dār al-Kutub al-ʿIlmiyya, 1998.

Ibn Ḥabib, al-Ḥusayn ʿUmar. *Tadhkirat an-nabīh fī ayyām al-Manṣūr wa-banīh*. Ed. Muḥammad Amīn. Cairo: Wizārat ath-Thaqāfa bi-Miṣr, Markaz Taḥqīq at-Turāth, n.d.

Ibn Ḥanbal, Aḥmad. *Al-Musnad*. Ed. Shuʿayb al-Arnāʾūṭ et al., 45 vols. Beirut: Muʾassasat ar-Risāla, 2001.
Ibn Kathīr, Ismāʿīl b. ʿUmar. *Al-Bidāya wa-n-nihāya*. Ed. ʿAlī Shīrī, 14 vols. Beirut: Iḥyāʾ at-Turāth al-ʿArabī, 1988.
Ibn Kathīr, Ismāʿīl b. ʿUmar. *Tafsīr al-Qurʾān al-ʿaẓīm*. Ed. Muḥammad Ḥusayn Shams ad-Dīn, 9 vols. Beirut: Dār al-Kutub al-ʿIlmiyya, 1998.
Ibn Kathīr, Ismāʿīl b. ʿUmar. *Tafsīr al-Qurʾān al-ʿaẓīm*. Ed. Sāmī b. Muḥammad Salāma, 8 vols. Medina: Dār Ṭayba li-n-Nashr wa-t-Tauzīʿ, 1999.
Ibn Taghrī Bardī, Jamāl ad-Dīn. *An-Nujūm az-zāhira fī mulūk Miṣr wa-l-Qāhira*. Ed. Ṭāhā ʿAbd ar-Raʾūf Saʿd and Muṣṭafā l-Huwārī, 16 vols. Cairo: Maktabat al-Qāhira, 1983.
Ibn Taghrī Bardī, Jamāl ad-Dīn. *Al-Manhal aṣ-ṣāfī l-mustawfī baʿd al-wāfī*. Ed. Muḥammad Amīn, 13 vols. Cairo: al-Hayʾa al-Miṣriyya, 1984.
Ibn Taymiyya, Taqī d-Dīn. *Al-Fatwā l-Ḥamawiyya al-kubrā*. Ed. Ḥamad b. ʿAbd al-Muḥsin at-Tuwayjirī. Riyad: Dār aṣ-Ṣamīʿī, 1998.
Ibn Qāḍī Shuhba, Taqī d-Dīn. *Ṭabaqāt ash-Shāfiʿiyya*. Ed. ʿAbd al-ʿAlīm Khān, 4 vols. Beirut: ʿĀlam al-Kutub, 1986.
Muslim b. al-Hajjāj. *Ṣaḥīḥ Muslim*. Ed. Muḥammad Fūʾād ʿAbd al-Bāqī, 5 vols. Beirut: Iḥyāʾ at-Turāth al-ʿArabī, 1991.
As-Suyūṭī, Jalāl ad-Dīn. *Al-Itqān fī ʿulūm al-Qurʾān*. Ed. Muḥammad abū l-Faḍl Ibrāhīm, 4 vols. Cairo: al-Hayʾa al-Miṣriyya al-ʿĀmma li-l-Kitāb, 1974.
As-Suyūṭī, Jalāl ad-Dīn. *Ḥusn al-muḥāḍara fī tārīkh Miṣr wa-l-Qāhira*. Ed. Muḥammad abū l-Faḍl Ibrāhīm, 2 vols. Cairo: al-Kutub al-ʿArabiyya ʿĪsā l-Bābī l-Ḥalabī, 1976.
As-Suyūṭī; Jalāl ad-Dīn. *Dhayl Ṭabaqāt al-ḥuffāẓ li-dh-Dhahabī*. Ed. Zakariyā ʿUmayrāt. Beirut: Dār al-Kutub al-ʿIlmiyya, n.d.

6.2 Secondary Literature

Aḥlām, Sālim al-Ghalīlāt, and Sharaf al-Quḍāt. “Naqd al-matn ʿinda Ibn Kathīr fī tafsīrihi.” *Majallat al-Jāmiʿa li-Dirāsāt al-Islāmiyya* 26, no. 2 (2018): 339–70.
Al-ʿAsqalānī, Ibn Ḥajar. *Ad-Durar al-kāmina fī aʿyān al-miʾa ath-thāmina*. Ed. Muḥammad ʿAbd al-Maʿīd Ḍān. Haydarabad: Majlis Dāʾirat al-Maʿārif al-ʿUthmāniyya, 1972.
Al-Fāliḥ, Muḥammad ʿAbd Allāh. *Ḥayāt Ibn Kathīr wa-kitābuhu Tafsīr al-Qurʾān al-ʿaẓīm*. Irbid: Dār al-Bayān, 2004.
Al-Ghanīmī, ʿAbd al-Ākhar. *ʿAqīdat al-ḥāfiẓ Ibn Kathīr bayna at-tafwīḍ wa-t-taʾwīl*. Dammam: Dār al-Akhillāʾ, 1994.
Ḥamdān, Khālid Ḥusayn ʿAbd ar-Raḥīm. *Al-Imām Ibn Kathīr wa-manhajuhu fī dirāsat qaḍāyā al-ʿaqīda*. Omdurman: Jāmiʿat al-Qurʾān al-Karīm wa-l-ʿUlūm al-Islāmiyya, 1996.
Al-Lāḥim, Sulaymān b. Ibrāhīm. *Manhaj Ibn Kathīr fī t-tafsīr*. Riyad: Dār al-Muslim, 1999.
Maḥmūd, Manīʿ b. ʿAbd al-Ḥalīm. *Manāhij al-mufassirīn*. Cairo: al-Kitāb al-Miṣrī/Beirut: al-Kitāb al-Lubnānī, 2000.
Naʿnāʿa, Ramzī. *Al-Isrāʾiliyyāt wa-athāruhā fī kutub at-tafsīr*. Damascus: Dār al-Qalam, 1970.
As-Sayyid, Aḥmad b. ʿUmar. *Taʿqībāt al-imām Ibn Kathīr ʿalā man ṣabaqahu min al-mufassirīn min khilāl kitābihi Tafsīr al-Qurʾān al-ʿaẓīm*. Mekka: Jāmi ʿat ʾUmm al-Qurā: Kulliyyat Uṣūl ad-Dīn wa-d-Daʿwa, 2010.
Shākir, Aḥmad. *ʿUmdat at-tafāsīr ʿan al-ḥāfiẓ Ibn Kathīr “Mukhtaṣar Tafsīr Ibn Kathīr”: ash-Shaykh*. Alexandria: Dār al-Wafāʾ, 2005.
Shākir, Maḥmūd. *At-Tārīkh al-islāmī*. 22 vols. Beirut: al-Maktab al-Islāmī, 2000.
Ash-Shawkānī, Aḥmad. *Al-Badr aṭ-ṭāliʿ bi-maḥāsin min baʿd al-qarn as-sābiʿ*. 2 vols. Beirut: Dār al-Kitāb al-Islāmī, 2012.
Shawqī, Abū Khalīl. *Aṭlās al-ḥadīth an-nabawī min al-kutub aṣ-ṣiḥāḥ as-sitta*. Damascus: Dār al-Fikr, 2003.

As-Sinān, Ḥamd and Fawzī al-ʿAnjarī. *Ahl as-sunna al-ashāʿira: shahādat ʿulamāʾ al-umma wa-adillatuhum*. Kuweit: Dār aḍ-Ḍiyāʾ, 2010.

Yāqūt, Abū ʿAbd Allāh al-Ḥamawī. *Muʿjam al-udabāʾ*. Ed. Iḥsān ʿAbbās, 3 vols. Beirut: Dār al-Gharb al-Islāmī, 1993.

Az-Zahrānī, Maṭar. *Ibn Kathīr al-mufassir*. Mekka: Kulliyyat ash-Sharīʿa wa-d-Dirāsā l-Islāmiyya, 1981.

Az-Zuḥaylī, Muḥammad. *Ibn Kathīr ad-Dimashqī l-ḥāfiẓ al-mufassir al-muʾarrikh al-faqīh*. Damascus: Dār al-Qalam, 1995.

Az-Zurqānī, Muḥammad. *Manāhil al-ʿirfān fī ʿulūm al-Qurʾān*. 2 vols. Cairo: Maṭbaʿat ʿIsā l-Bābī l-Ḥalabī, 1943.

Sayyid Ḥaydar al-Āmulī

Mathieu Terrier

1 Introduction

Bahā' ad-Dīn Ḥaydar b. ʿAlī b. Ḥaydar al-ʿAlawī l-Ḥusaynī l-Āmulī, known as Sayyid Ḥaydar Āmulī (b. 1320/719, d. after 1385–86/787), holds an important place in the history of Islamic thought as the first to have combined ancient Shiite esotericism, Sufi mysticism, and philosophy.[1] He was also the first Shiite Imāmī to compose a personal commentary on the Qur'ān, or rather, a treatise on Qur'ānic hermeneutics, after the "Occultation" of the twelfth Imam (940–41/329).

2 His Life and Works

2.1 The Life of Sayyid Ḥaydar Āmulī

Ḥaydar Āmulī belonged to a large family of *sayyids* (descendants of the Prophet Muḥammad) from Āmul, in the Iranian province of Ṭabaristān. His family had already converted to Twelver Shiism long before it was established as the state religion under the Safavids (1501–1722/906–1135). Most of the information available on Ḥaydar Āmulī's life come from two autobiographical accounts, the first of which was composed in 1375–76/777[2] and the second in 1380–1/782.[3]

According to these accounts, Ḥaydar Āmulī first spent twenty years studying Imāmī religious sciences in Āmul, then in Khurasān and Iṣfahān. After that, he went back to Āmul and became the minister of the Prince of Ṭabaristān, Fakhr ad-Dawla b. Yazdigard (r. 1333–1349/734–750). Around the age of thirty, he suddenly relinquished his worldly career, donned the Sufi cloak (*khirqa*), and set out on a journey to the Shiite shrines of Iraq and to Mecca. In Isfahan, he received a second *khirqa* from the hand of

1 For a general presentation, see Henry Corbin, *En islam iranien: Aspects spirituels et philosophiques* (Paris: Gallimard, 1971–72), 3: 149–213; Kāmil Muṣṭafā ash-Shaybī, *aṣ-Ṣila bayna at-taṣawwuf wa-t-tashayyuʿ* (Beirut: al-Jamal, 2011), 2: 104–15.

2 Sayyid Ḥaydar Āmulī, *Tafsīr al-Muḥīṭ al-aʿẓam wa-l-baḥr al-khiḍamm fī taʾwīl kitāb Allāh al-ʿazīz al-muḥkam*, ed. Muḥsin al-Mūsawī at-Tabrīzī (Qum: Nūr ʿalā Nūr, 1996) (henceforth abbreviated as MAa), 1: 526–31; ed. n.n. (Beirut: Dār al-Maḥajja al-Bayḍā, 2012) (henceforth abbreviated as MAb), 202–4.

3 Sayyid Ḥaydar Āmulī, *Le texte des textes: commentaire des "Fosûs al-hikam" d'Ibn ʿArabi*, ed. Henry Corbin and Osman Yahya (Tehran/Paris: Département d'iranologie de l'institut franco-iranien de recherche/Librairie d'Amérique et d'Orient Adrien-Maisonneuve, 1975) (henceforth abbreviated as NNa), 535–37; Sayyid Ḥaydar Āmulī, *Naṣṣ an-nuṣūṣ fī sharḥ al-Fuṣūṣ*, ed. Muḥsin Bīdārfar (Qum: Bīdār, 2015) (henceforth abbreviated. as NNb), 2: 851–53.

https://doi.org/10.1515/9783111320083-012

a Sufi master called Nūr ad-Dīn at-Ṭihrānī al-Iṣfahānī. This *khirqa* was reputed to have been transmitted by the Prophet, the first eight Shiite Imams, and Sufi masters such as Abū l-Qāsim al-Junayd (d. 911/298).[4] After a long period in Najaf, near the shrine of the first Shiite Imam ʿAlī b. Abī Ṭālib (d. 661/40), Āmulī established himself in Baghdad, where he studied theology (*kalām*) under Fakhr al-Muḥaqqiqīn al-Ḥillī (d.1370/771), the son of the Shiite scholar al-ʿAllāma al-Ḥillī (d. 1326/726), before returning to Najaf, where he spent the rest of his life. He died sometime after 1385/787.

2.2 His Works

According to Ḥaydar Āmulī himself, he composed around forty works in Persian and Arabic, most of which have been lost. His earliest surviving work is *Anwār al-ḥaqīqa wa-aṭwār aṭ-ṭarīqa wa-asrār ash-sharīʿa*,[5] written before 1351–52/752 and reproduced in the *Tafsīr al-Muḥīṭ al-aʿẓam* discussed below. *Jāmiʿ al-asrār wa-manbaʿ al-anwār*, achieved around 1351–52/752, is an essay that harmonizes Imāmī Shiism with Sufism. The *Risālat al-Wujūd fī maʿrifat al-maʿbūd*, composed in 1359/760, is a treatise on ontology comparing the views of theologians, philosophers and mystics. The *Risālat Naqd an-nuqūd fī maʿrifat al-wujūd* is a summary of the previous work. *Risālat Rafʿ al-khilāf wa-l-munāzaʿa*, composed in 1367–68/769, is a polemic epistle seeking to justify the attitude of Imam ʿAlī during the reigns of the three first caliphs. *Tafsīr al-Muḥīṭ al-aʿẓam wa-l-baḥr al-khiḍamm fī taʾwīl kitāb Allāh al-ʿazīz al-muḥkam* was completed in 1375–76/777. Finally, *Naṣṣ an-nuṣūṣ fī sharḥ al-Fuṣūṣ*, authored in 1380–1/782, is a Shiite commentary on Muḥyī d-Dīn b. ʿArabī's (d. 1240/633) *Fuṣūṣ al-ḥikam*.

2.3 *Al-Muḥīṭ al-aʿẓam*

Projected as a monumental Qurʾānic commentary, the book remained unfinished or mostly lost. It was conceived in seven parts, the first containing seven prolegomena, following the model of the Qurʾānic commentary of the Sufi Najm ad-Dīn Dāyih ar-Rāzī (d. 1256/654), *Baḥr al-ḥaqāʾiq wa-manbaʿ ad-daqāʾiq*.[6] It is based on a hadith attributed to the Prophet: "The Qurʾān has both an exoteric (*ẓahr*) and an esoteric (*baṭn*) meaning; the esoteric meaning itself has an esoteric meaning, and so on until there are seven esoteric meanings."[7] As it stands, it consists of six introductions (*muqaddamāt*) and a commentary on the first sura, *al-Fātiḥa* (The Opening). It reproduces nine chapters of Ibn ʿArabī's *al-Futūḥāt al-makkiyya* in full (Chapters 6, 7, 9, 11, 60, 61, 63,

4 MAa, 1: 534f.; MAb, 206.

5 For complete references, see the bibliography at the end of this article.

6 NNa, 12; NNb, 1: 16; MAa, 1: 202; MAb, 21.

7 MAa, 1: 203; MAb, 21.

64, and 65), which mainly concern sainthood (*walāya*) and eschatological events. Rather than a *tafsīr* in the strict sense, the work is a treatise on Qur'ānic hermeneutics from the standpoint of Imāmī tradition, mysticism, and philosophy altogether.

3 Āmulī's Views on Hermeneutics

3.1 Qur'ānic Hermeneutics in the Shiite Tradition

According to the early doctrine of Twelver Shiism, the Imams are the only men able and designed to interpret the Qur'ān. This is by virtue of their impeccability and infallibility (*ʿiṣma*), and because ʿAlī, the first Imam, received the whole Book – its letter (*tanzīl*) and interpretation (*ta'wīl*) – directly from the Prophet. According to their own reading and exegesis of Q 3:7,

> (i)t is He who sent down upon thee the Book, wherein are verses clear that are the Essence of the Book, and others ambiguous. As for those in whose hearts is swerving, they follow the ambiguous part, desiring dissension, and desiring its interpretation; and none knows its interpretation, save only God and those firmly rooted in knowledge. They say, "We believe in it; all is from our Lord"; yet none remembers, but men possessed of minds.[8]

The Imams are "those firmly rooted in knowledge" (*ar-rāsikhūn fī l-ʿilm*), sharing with God the science of hermeneutics. Consequently, the first Shiite *tafsīr*s composed before and just after the "great Occultation" of the twelfth Imam in 940–1/349, like those of ʿAlī b. Ibrāhīm al-Qummī (d. ca 919/307), and Muḥammad b. Masʿūd al-ʿAyyāshī (d. ca 932/320), were mere compilations of Imāmī exegetic hadiths.[9]

3.2 The Legacy of Hermeneutics

In the first introduction to his commentary, Ḥaydar Āmulī deals with the definition of and justification for Qur'ānic hermeneutics (*ta'wīl*), which he first of all considers to be obligatory for those who are qualified and designated for it, i.e., "those firmly rooted in knowledge."[10] This duty is ascertained by tradition (*naql*), reason (*ʿaql*), and spiritual unveiling (*kashf*). According to an argument of Muʿtazilī origin, the permanence of the Book and the existence of a divinely inspired interpreter are required as

8 Arberry translation slightly modified. Notably, this verse is read in Sunni Islam as follows: "and none knows its interpretation, save only God. Those firmly rooted in knowledge say, 'We believe in it (. . .).'"
9 See Meir M. Bar-Asher, *Scripture and Exegesis in Early Imami Shi'ism* (Leiden: Brill, 1999).
10 MAa, 1: 205; MAb, 22.

it is necessary for God to dispense His grace (*luṭf*).[11] The return of the Mahdī, the twelfth Imam, at the End of Time will be the advent of the integral *taʾwīl*, the unveiling of the whole hidden meaning of the Book. This is the meaning of the Major Resurrection (*al-qiyāma al-kubrā*) according to Q 7:52–53:

> And We have brought to them a Book that We have well distinguished, resting on knowledge, a guidance and a mercy unto a people that believe. Do they look for aught else but its interpretation? The day its interpretation comes, those who before forgot it shall say, "Indeed, our Lord's Messengers came with the truth . . ."[12]

However, it appears that, for our author, the need to understand the Book can no longer wait until that promised day. Thus, Āmulī gives a broader interpretation of "those firmly rooted in knowledge" including, after the Imams, the "masters of Divine Unity" (*arbāb at-tawḥīd*), to whom he naturally intends to belong: "We say that 'those firmly rooted in knowledge,' absolutely speaking, are first the prophets, then the messengers, then the saints (*awliyāʾ*), then the Imams, then the learned heirs (of the latter) called 'the masters of Divine Unity.'"[13] These learned men have as their patron Salman the Persian, about whom Prophet Muḥammad said: "Salman is among the People of my House (*ahl baytī*)."[14] Reference is made to them in statements by the Prophet such as "The learned are the heirs of the prophets" and "The learned of my community are like the prophets of the sons of Israel."[15]

They are endowed with inspiration (*ilhām*) and spiritual unveiling (*kashf*) according to Q 50:22 – "Thou wast heedless of this; therefore, We have now removed from thee thy covering, and so thy sight today is piercing." According to Q 18:65, they have also received "bestowed knowledge" (*ʿilm ladunī*) from God without any human intermediary: "Then they found one of Our servants unto whom We had given mercy from Us, and We had taught him knowledge proceeding from Us (*min ladunnā*)." Here, Āmulī quotes an epistle of Shiite tendency attributed to Abū Ḥāmid al-Ghazālī (d. 1111/505), entitled *ar-Risāla al-laduniyya*:

> Since revelation has been interrupted, since the door of messengerhood has been closed, men are exempt from messengers (. . .), for the proof has been made explicit and the religion perfect, as God said: "Today I have perfected your religion for you" (Q 5:3) (. . .). But the door of inspiration does not close, nor does the assistance of the Universal Soul stop, as long as the soul remains in need of confirmation, renewal, and reminder. As men are exempt from messengerhood, while they still need reminding and awakening (. . .), God has closed the door of revelation and has

11 MAa, 1: 357–61; MAb, 103–5.
12 MAa, 1: 458; MAb, 164.
13 MAa, 1: 447; MAb, 157.
14 MAa, 1: 450, 452; MAb, 159f.
15 MAa, 1: 514; MAb, 193.

opened the door of inspiration by His Mercy (. . .), to make known that "God is All-gentle to His servants, providing for whomsoever He will" (Q 42:19).[16]

This knowledge is a legacy of the Imams, especially the first one:

(ʿAlī's) knowledge was of two kinds. The first was received from God by way of bestowal and inspiration, referred to as the bestowed science (*al-ʿilm al-ladunī*) (. . .). This knowledge is exclusive to the saints legatees (the Imams), and their learned heirs, and it is known that (ʿAlī) is the prince, the center, and guide of these sciences. The second kind was received by instruction from the Prophet and attachment to him (. . .).[17]

The masters of Divine Unity, as the inheritors of the Imams, are the gatherers of all exoteric and esoteric sciences:

The knowledge is of three kinds: there is the shell (*qishr*), the kernel (*lubb*), and the kernel of kernels (*lubb al-lubb*). All belong to them (the Imams), especially the knowledge that is the kernel of kernels, defined as the bestowed, revealed knowledge, proceeding from the fountain of God's Unity and from the source of His Separation. The shell is the exoteric sciences, such as solely traditional insights (*al-manqūlāt*). The kernel is the esoteric sciences, such as solely rational insights (*al-maʿqūlāt*). The kernel of kernels is the esoteric sciences of the esoteric, such as solely bestowed and unveiled insights (*al-laduniyyāt wa-l-kashfiyyāt*). (. . .) Since the shell, the kernel, and the kernel of kernels belong to them and return to their presence, the rootedness in knowledge (*ar-rusūkh fī l-ʿilm*) is their own, the (Qurʾānic) hermeneutics is their own (. . .); it can only be learned from them and their successors from among the masters of Divine Unity.[18]

3.3 A Revealed Hermeneutics?

Āmulī not only counts himself among these "masters of Divine Unity"[19] but also claims to have a "sensory and suprasensory" (*ṣūriyya wa-maʿnawiyya*) filiation with the Imams, the first by virtue of his lineage, which goes back to the fourth Imam ʿAlī b. al-Ḥusayn (d. ca 714/95), the second by way of his theological training under Shiite scholars as well as his mystical initiation under Sufi shaykhs.[20]

Moreover, while commenting on Ibn ʿArabī's claim that he received the *Fuṣūṣ al-ḥikam* from Prophet Muḥammad in a dream,[21] Āmulī states in his *Naṣṣ an-nuṣūṣ* that two books belong to the Prophet: the Qurʾān, which was revealed to him (*an-nāzil ʿalayhi*), and the *Fuṣūṣ al-ḥikam*, which proceeded from him (*aṣ-ṣādir minhu*). Equally,

16 MAa, 1: 477f.; MAb, 174–78; Abū Ḥāmid al-Ghazālī, "ar-Risāla al-laduniyya," in *Majmūʿa rasāʾil al-imām al-Ghazālī*, ed. n.n. (Beirut: Dār al-Kutub al-ʿIlmiyya, 2011), 3: 69f.

17 MAa, 1: 479f.; MAb, 178.

18 MAa, 1: 516f.; MAb, 194f.

19 MAa, 1: 478f.; MAb, 177f.

20 MAa, 1: 527–36; MAb, 202–7.

21 Muḥyī ad-Dīn b. ʿArabī, *Fuṣūṣ al-ḥikam*, ed. Abū l-ʿAlā ʿAfīfī (Beirut: Dār al-Kitāb al-ʿArabī, 2002), 47f.

there are two books belonging to Ibn ʿArabī: the *Fuṣūṣ al-ḥikam*, which was sent to him (*al-wāṣil ilayhi*), and the *Futūḥāt al-makkiyya*, which proceeded from him. Finally, two books belong to Āmulī himself: *al-Muḥīṭ al-aʿẓam*, which "overflowed" onto him (*al-fāʾiḍ ʿalaynā*), and *Naṣṣ an-nuṣūṣ*, which proceeded from him. Thus, the *Muḥīṭ al-aʿẓam* is to him what the *Fuṣūṣ al-ḥikam* was to Ibn ʿArabī and what the Qurʾān was to the Prophet Muḥammad.[22] This amazing claim lends his commentary the same status as that of the Imams' hermeneutics, that is, as a continuation of the Qurʾānic revelation.

3.4 The Three Books of Qurʾān, Universe, and Soul

Āmulī's conception of hermeneutics is based on an original interpretation of the "Book" (*kitāb*) as evoked many times in the Qurʾān. He states that the Divine Book not only refers to the Qurʾān itself but also to "the Book of horizons" (*al-kitāb al-āfāqī*), i.e., the universe, and "the Book of souls" (*al-kitāb al-anfusī*), i.e., man, according to Q 41:53: "We shall show them Our signs in the horizons and in themselves (literally 'in their souls' (*fī anfusihim*)), until it is clear to them that it is the truth." Thus, Introductions 2 to 5 respectively deal with the correspondence between the analytical (*tafṣīlī*) book of universe, the synthetic (*ijmālī*) book of soul, and the gathering (*jamʿī*) book of the Qurʾān (2); the correspondence between the letters of these three Books (3); the correspondence between the words of these three Books (4); and the correspondence between the verses or signs (*āyāt*) of these three books (5).

> Know that the Book of God is not limited to the Qurʾān, the Torah, the Gospel, and other heavenly Books; that His verses are not limited to the verses of the Qurʾān or other heavenly Books; nor His words to their words, nor His letters to their letters. But the world called "the horizons" is a whole Book of God with its verses, words and letters; it is the great divine Book. Similarly, the man designated as "the souls" is a divine synthetic book containing His verses, words, and letters; it is the small divine Book. The first (the world) is called "the great man," the second "the little man," according to their (the philosophers') saying: "The world is a great man (*insān kabīr*), and man is a little world (*ʿālam ṣaghīr*)."[23]

Thus, the task of hermeneutics is not limited to the Qurʾān but extends to the cosmos and the human soul. Hermeneutics, according to the masters of the esoteric (*arbāb al-bāṭin*), is tantamount to the knowledge of God (*al-ḥaqq*), the world, and mankind through the knowledge of the Qurʾān, correlated with the Book of horizons, which is the world in its detail, and with the book of souls, which is man as a whole.[24]

Consequently, the Book whose letters and words are said to be infinite in number in Q 18:109 – "Say: 'If the sea were ink for the Words of my Lord, the sea would be

22 NNa, 148f.; NNb, 216f.
23 MAa, 2: 15; MAb, 230.
24 MAa, 1: 240f.; MAb, 36.

spent before the Words of my Lord are spent, though We brought replenishment the like of it'" – and Q 31:27 – "Though all the trees in the earth were pens, and the sea-seven seas after it to replenish it, yet would the Words of God not be spent. God is All-mighty, All-wise" – is neither the Qur'ān nor any other revealed Book, as the exegetes (*al-mufassirūn*) say, because the letters and words of the Qur'ān are spent on the sensory level (*ṣūrī*), although they are unspent on the suprasensory one (*maʿnawī*). However, this Book is the world or the human soul whose words – that is, the spiritual and corporeal realities – are unspent and in infinite number.[25]

3.5 A Breakthrough Gesture

Āmulī is thus the author of a real breakthrough in the history of Shiite hermeneutics, first of all, because he carries out *ta'wīl* in the absence of the Imam and in his place. Consequently, while traditional Shiite *tafsīr* is essentially composed of traditions attributed to the Imams, Āmulī systematically associates with these traditions insights borrowed from Neoplatonic philosophy and Ibn ʿArabī's mysticism. And while traditional Shiite *tafsīr* is often limited to "personalized commentary"[26] focused on the historical personalities of the Prophet's familiars (*ahl al-bayt*) and their enemies, Āmulī proposes an anthropocentric commentary focusing on man as a microcosm and/or a macrocosm. Thus, in the analysis of the first verse of the Qur'ān:

> Just as "In the name of God, the Merciful, the Compassionate" (Q 1:1) is a locus of manifestation of the supreme Name (*al-ism al-aʿẓam*) and of the supreme Word of God (*kalimat Allāh al-ʿulyā*), man is the locus of manifestation of the supreme Name and the supreme Word of God, for man is to the universe ("the horizons") what "In the name of God, the Merciful, the All-Merciful" is to the Qur'ān, because he gathers and unites in himself all divine Names and Attributes.[27]

Moreover, the dualistic vision of early Shiism that sets the forces of Knowledge and Good in opposition to those of Ignorance and Evil[28] only takes shape from a spiritual and ahistorical perspective that is ultimately absorbed within the doctrine of the Unity of Existence (*waḥdat al-wujūd*). Āmulī substitutes the traditional opposition between the Imams and their enemies with that between the people of the exoteric (*ahl aẓ-ẓāhir*) and the people of the esoteric (*ahl al-bāṭin*) in terms of their respective knowledge of the Qur'ān and of the universe. For the knowledge of the former goes

25 MAa, 2: 16f., 447–49; MAb, 230f., 430f.
26 See Mohammad Ali Amir-Moezzi, *The Silent Qur'an and the Speaking Qur'an. Scriptural Sources of Islam Between History and Fervor*, trans. Eric Ormsby (Columbia: Columbia University Press, 2015), 201, chapter 3.
27 MAa, 2: 457; MAb, 436.
28 See Mohammad Ali Amir-Moezzi Christian and Jambet, *What is Shiʿi Islam? An Introduction*, trans. Kenneth Casler and Eric Ormsby (London/New York: Routledge, 2018), 16–18.

no further than the official religious sciences and the contemplation of the *mulk*, the manifest sensible world, whereas that of the latter reaches the hermeneutics (*ta'wīl*) of the Qur'ān and the contemplation of the *malakūt*, the invisible world of intellects, souls, and spirits.[29] As Āmulī states,

> Whoever knows the two Books as he should, rises from the word to its meanings, from the *mulk* to the *malakūt* and from the *malakūt* to the *jabarūt* (i.e. the realm of God's essence). Thus, he knows by direct contemplation (*shuhūd*) that all spiritual and corporeal beings are signs of God, from which he derives His Essence, His Attributes, and His Words, according to Q 41:53: "We shall show them Our signs in the horizons and in themselves" (. . .). Those who are deprived of this knowledge are designated by God in this way: "They know an outward part of the present life, but of the Hereafter they are heedless" (Q 30:7) (. . .), and they are blamed in many verses such as: And recite to them the tiding of him to whom We gave Our signs, but he cast them off; and Satan followed after him, and he became one of the perverts. And had We willed, We would have raised him up thereby; but he inclined toward the earth and followed his lust. So, the likeness of him is as the likeness of a dog (. . .) (Q 7:175–176).[30]

4 The Triad of Law, Path, and Truth

Āmulī's hermeneutics is broadly structured by the conceptual triad of the Law (*sharīʿa*), the Path (*ṭarīqa*), and the Truth (or Reality per se) (*ḥaqīqa*), referring simultaneously to different modes of knowledge, levels of reality, and classes of God's men. This triad is the subject of the sixth introduction, the longest in the book,[31] taken up in full from Āmulī's early work Anwār al-ḥaqīqa wa-aṭwār aṭ-ṭarīqa wa-asrār ash-sharīʿa. It consists of an analysis conducted from these three successive standpoints of the five principles of Shiite Islam: Divine Unity, Divine Justice, Prophecy, Imamate, and Resurrection (*maʿād*);[32] and of its five branches: prayer, fasting, almsgiving, pilgrimage, and fighting (*jihād*).[33]

It is based on a prophetic hadith appreciated by Sufi milieus in particular: "The Law is my speeches, the Path is my actions, and the Truth is my (spiritual) states."[34] As Āmulī remarks, "The Law means the assent to the words of the prophets by heart, as well as the action taken in conformity with them. The Path means the realization

29 On these notions, see Mathieu Terrier, "Sufi Hierarchies of the Worlds or Degrees of Existence: *Mulk, Malakūt, Jabarūt*, and Related Concepts," in *Handbook of Sufi Studies*, vol. 2, *Sufi Cosmology*, ed. Alexander Knysh and Christian Lange (Leiden: Brill, 2023), 277–98.

30 MAa, 2: 513–15; MAb, 461f.

31 MAa, 3: 5; MAb, 491.

32 MAa, 3: 177; MAb, 538.

33 MAa, 3: 7; MAb, 678.

34 MAa, 3: 16f.; MAb, 492.

of their deeds and characters, as well as their fulfilment by virtue. The Truth means the contemplation of their spiritual states and ranks through unveiling."[35] The harmonization of these three aspects of religion aims to resolve the conflict between Shiite lawyers and theologians (*fuqahāʾ*), Sufis, and philosophers (*ʿurafāʾ, ḥukamāʾ*).[36] Āmulī's statements on the subject would eventually be taken up by thinkers concerned with the same conciliation, like Ibn Abī Jumhūr al-Aḥsāʾī (d. after 1499/904) and Quṭb ad-Dīn Ashkiwarī (d. between 1677/1088 and 1684/1095).[37] For example,

> Know that Law, Path, and Truth are synonyms designating the same reality, that of the Muḥammadian Revelation, viewed from different standpoints and at different levels. There is, on examination, no contradiction between these standpoints and levels. The Revelation is like the complete almond, which contains the shell, the kernel, and the kernel of kernels; the shell may be likened to the exoteric Law, the kernel to the esoteric Path, the kernel of kernels to the esoteric Truth of the esoteric (Path), and the almond is the union of all.[38]

And further: "The shell is all exoteric knowledge protecting from corruption the esoteric one which is its kernel, just as the Law protects the Path, and the Path protects the Truth."[39]

Āmulī considers these three ranks as deriving from the principles of messengerhood (*risāla*), prophethood (*nubuwwa*), and sainthood (*walāya*) respectively. The Law is required by messengerhood, the Path by prophethood, and the Truth by sainthood.[40] This distinction corresponds to the three categories of God's men, namely the messengers or prophets with Law (*rusul*), the prophets without Law (*anbiyāʾ*), and the saints or the friends of God (*awliyāʾ*). For the Shiites, the term *walī* (*awliyāʾ* in the plural) refers solely to the Imams, and for the Sufis, to all their spiritual masters. Āmulī seems to be playing on this ambiguity in order to bring Shiites and Sufis into agreement.

This triadic vision commands a philosophy of the history of Revelation:

> Know that if Law, Path, and Truth are three (perspectives on) the same reality, the Truth is higher than the Path, and the Path higher than the Law, and so are their respective adepts. (. . .) Since the perfection attributed to Moses was the universal perfection of the species, he inclined to the perfection of the lowest part of man, which is the body; the Torah is therefore full of measures beneficial to material life. Since Jesus was more perfect than (Moses), his perfection concerned the noblest part of man, which is the soul; the Gospel is therefore full of measures beneficial to the resurrection. And since Muḥammad was more perfect than the universal perfec-

35 MAa, 3: 20; MAb, 493.

36 MAa, 3: 5–9; MAb, 491.

37 See Mathieu Terrier, "The Defence of Sufism among Twelver Shiʿi Scholars of Early Modern and Modern Times: Topics and Arguments," In *Shiʿi Islam and Sufism: Classical Views and Modern Perspectives*, ed. Denis Hermann and Mathieu Terrier (London: I.B. Tauris, 2020), 27–63.

38 MAa, 1: 227f.; MAb, 29.

39 MAa, 1: 517; MAb, 195.

40 MAa, 3: 21–24; MAb, 494.

> tion of the species, his perfection concerned both parts of man at the same time, (and the Qur'ān therefore contains both types of measures).[41]

Muslim lawyers and exoteric scholars are thus "the people of the Law," tantamount to that of Moses; the philosophers and masters of rational knowledge (*arbāb al-maʿqūl*) are the people of the Path, tantamount to that of Jesus; and the Gnostics (*ʿurafā'*) are the people of the Truth, tantamount to that of Muḥammad. The latter rank stands above the other two. In the Qur'ān, they are associated at one point with God and angels – "There is no god but He: that is the witness of God, the angels, and men endued with knowledge, standing firm on justice" (Q 3:18) – and at another with God alone – "and none knows its interpretation, save only God and those firmly rooted in knowledge" (Q 3:7).[42]

5 Shiite Esoteric Topics

Despite his innovative approach toward hermeneutics, Āmulī supports the fundamental positions of original Shiite esotericism with both traditional and rational arguments.

5.1 The Imams in the Qur'ān

In line with the Shiite exegetic tradition, Āmulī states that many verses in the Qur'ān were revealed about the Imams in particular. He thus cites the following Imāmī tradition from the most ancient, preserved Shiite source, the book attributed to Sulaym b. Qays al-Hilālī (d. ca 695–96/76)[43]:

> We are those to whom God has commanded men to obey by saying, "O believers, obey God, and obey the Messenger and those in authority among you" (Q 4:59); we are those whom God ordered to love and cherish by saying, "Say: 'I do not ask of you a wage for this, except love for the kinsfolk'" (Q 42:23); (. . .) we are the inheritors about whom He said, "Those are the inheritors who shall inherit Paradise therein dwelling forever" (Q 23:10–11); we are the well-guided and the guides about whom He said, "who give ear to the Word and follow the fairest of it. Those are they whom God has guided; those – they are men possessed of minds" (Q 39:18).[44]

ʿAlī is personally referred to in Q 5:55, among other verses: "Your friend is only God, and His Messenger, and the believers who perform the prayer and pay the alms, and

41 MAa, 3: 82; MAb, 509.

42 MAa, 3: 83; MAb, 509. Q 3:18 is given in the translation of Abdullah Yusuf Ali, *The Holy Qur'an* (Medina: King Fahd Holy Qur-ān Printing Complex, 1410), 144, slightly modified.

43 On this book, see Amir-Moezzi, *The Silent Qur'an*, chap. 1.

44 MAa, 1: 440; MAb, 154.

bow them down."[45] The five "people of the clock" (*ahl al-kisāʾ*), i.e., Muḥammad, ʿAlī, Fāṭima, al-Ḥasan, and al-Ḥusayn, are referred to in Q 7:33 – "People of the House (*ahl al-bayt*), God only desires to put away from you abomination and to cleanse you" – and in the verse of "the mutual cursing" (*al-mubāhala*), Q 3:61 – "And whoso disputes with thee concerning him, after the knowledge that has come to thee, say: 'Come now, let us call our sons and your sons, our wives and your wives, our selves and your selves, then let us humbly pray and so lay God's curse upon the ones who lie.'" "Our sons" refers to al-Ḥasan and al-Ḥusayn, "our wives" to Fāṭima, and "our selves" to Muḥammad and ʿAlī.[46] The manifestation of the twelfth Imam, the Mahdī, is referred to in Q 7:52–53, as previously mentioned. Verses like Q 28:5 – "Yet We desired to be gracious to those that were abased in the land, and to make them leaders, and to make them the inheritors" – and Q 24:55 – "God has promised those of you who believe and do righteous deeds that He will surely make you successors in the land, even as He made those who were before them successors, and that He will surely establish their religion for them that He has approved for them, and will give them in exchange, after their fear, security . . ." – were revealed generally for the *ahl al-bayt* and for the Mahdī in particular.[47]

5.2 Prophetology and Imamology

Āmulī also strongly supports the tenet according to which the ending of the cycle of prophecy with Muḥammad corresponds to the beginning of the last cycle of sainthood or Imamate (*walāya*) with ʿAlī. Following the Imāmī tradition, he asserts that each of the six prophets with a Law (Adam, Noah, Abraham, Moses, Jesus, Muḥammad) has been followed by twelve Imams,[48] and that the number and the names of Muḥammad's Imams were announced by Muḥammad himself in numerous hadiths.[49] The last Imam must therefore remain hidden but alive until the end of time, which is also demonstrated by reason.[50]

In line with early Shiite esotericism, Āmulī supports both the primacy of sainthood or Imamate over prophethood – the former being the esoteric dimension of the latter[51] – and the superiority of the Prophet over the Saint or the Imam, that is, on the exoteric level:

45 MAa, 1: 445; MAb, 156.
46 MAa, 1: 503–5; MAb, 190f.
47 MAa, 1: 461; MAb, 166.
48 MAa, 1: 537–39; MAb, 208f.
49 MAa, 1: 555–58; MAb, 220f.
50 MAa, 1: 429; MAb, 150.
51 MAa, 3: 275; MAb, 566.

> Know that in every messenger, prophethood precedes messengerhood, and that in every prophet, sainthood precedes prophethood. This is because prophethood can only be fulfilled after sainthood, just as the existence of messengerhood can only be conceived after that of prophethood. Thus, every messenger is a prophet, but not every prophet is a messenger, just as every prophet is a saint, but not every saint is a prophet. Yet sainthood is greater than prophethood, and prophethood is greater than messengerhood, but this is true by considering the three in the same individual, not in three separate individuals. Sainthood is greater than prophethood, but the prophet is greater than the saint. Likewise, prophethood is greater than messengerhood, but the messenger is greater than the prophet. This is because the prophet holds both prophethood and sainthood when the saint holds only sainthood.[52]

6 Akbarian Topics

Two major topics of Āmulī's hermeneutics are borrowed from the mystical philosophy of Ibn ʿArabī (*ash-shaykh al-akbar*): the concept of God's self-disclosure (*tajallī*) through the Divine Names (*al-asmāʾ al-ilāhiyya*); and the famous and controversial doctrine – in both Sunni and Shiite Islam – of the Unity of Existence (*waḥdat al-wujūd*).

6.1 The Theophany of Divine Names

According to this doctrine, all existent beings are determined and hierarchized by the Divine Names – as revealed in the Qurʾān – of which they are loci of manifestation. These Divine Names are of three kinds: those of God's Essence (*adh-dhāt*), those of His Attributes (*aṣ-ṣifāt*), and those of His Acts (*al-afʿāl*). The Names of God's Attributes and Acts are also divided into Names of Beauty and Subtlety (*al-jamāliyya wa-l-luṭfiyya*) and Names of Majesty and Domination (*al-jalāliyya wa-l-qahriyya*).

The existent beings that manifest the Names of Essence are mankind, the first Intellect, the prophets, the Imams, and the perfect men. The existent beings that manifest the Names of Beauty and Attributes of Subtlety are the celestial angels and the higher immaterial beings such as intellects and souls. The existent beings that manifest the Names of Majesty and Attributes of Domination are the demons, bad djinns, Pharaohs, and Nemruds.[53] Man is the only creature that manifests the Names of Beauty and Majesty at once, therefore man is said to have been created by God with His two hands: "Said He, 'Iblis, what prevented thee to bow thyself before that I created with My own hands?'"(Q 38:75).[54]

52 MAa, 5: 385f.; MAb, 988.
53 MAa, 1: 336f.; MAb, 89f.
54 MAa, 6: 34; MAb, 1014.

This theory is illustrated by the "diagram of the Divine Names," in which the latter are matched with levels of reality and divine knowledge. At the center is the absolutely one, pure, and absolute Divine Essence. Around it, the first circle is that of the Names of Essence, designating the world of Jabarūt and corresponding to the revelation (*waḥy*) that descended upon the prophets and messengers, which is the principle of the Qur'ān and the other Divine Books. It counts thirty Divine Names, including God (*Allāh*), the Living (*al-ḥayy*), the Manifest (*aẓ-ẓāhir*), the Hidden (*al-bāṭin*), the First (*al-awwal*), the Last (*al-ākhir*), the Real (*al-ḥaqq*), and the Light (*an-nūr*). The second circle is that of the Names of Divine Attributes, designating the world of Malakūt and corresponding to the inspiration (*ilhām*) bestowed on the saints (*awliyā'*) and the legatees (of the prophets) (*awṣiyā'*) – i.e., the Imams –, which is the principle of the hadith *qudsī* and the Imāmī traditions of the same nature. It counts twenty-two Names including the Dominator (*al-qahhār*), the Compassionate (*ar-raḥmān*), the Merciful (*ar-raḥīm*), the Knower (*al-'alīm*), the Powerful (*al-qādir*), and the Wise (*al-ḥakīm*). The third circle is that of the Names of Divine Acts, designating the world of Mulk, i.e., the macrocosm, corresponding to the unveiling (*kashf*) of the legatees and their successors among the learned men, which is the principle of the prophetic hadith and the equivalent Imāmī traditions. It counts thirty-eight Names, including the Creator (*al-khāliq*), the Provider (*al-rāziq*), the Author (*al-bārī'*), the Shaper (*al-muṣawwir*), the Life-Giver (*al-muḥyī*), and the Death-Giver (*al-mumīt*). Finally, at the four outer corners of the diagram are the four Names of Q 57:3: the First, the Last, the Apparent, and the Hidden.[55] Two other diagrams represent the "world of the horizons" and the "world of the souls," described in the terms of philosophy and natural sciences as the loci of manifestation of the Divine Names.[56] The Divine Names revealed in the Qur'ān are therefore the key to the knowledge of God, the world, and mankind.

6.2 The Unity of Existence

The theory of the Unity of Existence is developed by Āmulī in the interpretation of many Qur'ānic verses, for example, in relation to Q 41:53, "We shall show them Our signs in the horizons and in themselves":

> Know that the meaning of His word is to address His pure servants in this way: "We will perfect the eye of your insight by the light of Our providence (*'ināya*) and guidance (*hidāya*), so that you may acquire the ability to know Our cosmic and psychic signs, the ability to contemplate Us directly in each of these signs, and clearly see that there is nothing in existence but Us, Our Names, Our Attributes, and Our Acts. For that which is other than Us is nothing but pure non-existence and nothingness. This is why the Gnostic says: 'There is nothing in existence but God, His Names, His Attributes, and His Acts. All is He, through Him, from Him, and to Him,' and We said about

55 MAa, 1: 337–40; MAb, 91f.
56 MAa, 1: 550–54; MAb, 216–19.

> Ourselves: 'All things perish, except His Face. His is the Judgment, and unto Him you shall be returned.' (Q 28:88)" (. . .). "The Face" is the expression of His existence, His Essence, and His Reality; therefore, His Word means that everything perishes and disappears except His existence, His Essence, and His Reality.[57]

This doctrine inspires Āmulī to embrace a broad interpretation of Q 24:35:

> God is the Light of the heavens and the earth; the likeness of His Light is as a niche, in which there is a lamp (the lamp in a glass, the glass as it were a glittering star), kindled from a Blessed Tree, an olive that is neither of the East nor of the West, whose oil wellnigh would shine, even if no fire touched it; Light upon Light. God guides to His Light whom He will. And God strikes similitudes for men, and God has knowledge of everything:
>
> Light is real and divine existence. The heavens, the earth, and everything in between are His higher and lower loci of manifestation (. . .). Thus, the niche is the world of bodies and bodily things; the glass is the world of spirits and spiritual beings; the lamp is the world of Intellects and separate essences. The resemblance is based on the fact that the divine Lights, rising from the East of absolute and real Existence in order to shine on the temples of the existing and created beings (. . .), shine first on the world of separate Intellects and essences which are like the lamp (. . .), then on the world of spirits which are like the glass (. . .), then on the world of bodies which are like the niche (. . .).The lamp is lit at the tree of absolute Existence by which all defined and relative existence is illuminated (. . .). This tree is said to be neither Eastern nor Western because the true East refers to the world of spirits and spiritual beings (. . .), while the true West refers to the world of bodies and bodily things (. . .). Now the absolute Existence which is true light comes neither from the mere world of spirits nor from the mere world of bodies. (. . .) This is the meaning of His word "Light upon Light": the light of the Real upon the light of the created, or the light of the Essence upon the light of the inner vision, or the light of the universal Intellect upon the light of the partial intellect.[58]

The same interpretative grid is applied to the psychic world or the human microcosm:

> In the same way, the body and senses of the 'little man' are correspond to the niche, his heart to the glass, his spirit to the lamp, and his totality to the tree (. . .). Thus, each organ and each faculty of man corresponds to a branch and a leaf of the cosmic tree. Through this correspondence, one understands the meaning of "We shall show them Our signs in the horizons and in themselves" (Q 41:53).[59]

Following Ibn ʿArabī, Āmulī distinguishes between two forms of the affirmation of Divine Unity (*tawḥīd*), which is the first creed of Islam: the affirmation of theological unity (*at-tawḥīd al-ulūhī*) and the affirmation of ontological unity (*at-tawḥīd al-wujūdī*):

57 MAa, 2: 522; MAb, 466.
58 MAa, 2: 534–40; MAb, 473–75.
59 MAa, 2: 546; MAb, 479.

> The manifestation of all prophets and messengers aimed to call men to affirm theological unity and to break out of open associationism. The manifestation of all saints and Imams aimed to call men to affirm ontological unity and to break out of hidden associationism. Whoever turns from the determined deity to the absolute Deity, from the worship of the created to the worship of the Creator, and utters the formula of theological unity, "There is no god save God," has broken out of open associationism and enters the Law as a Muslim, believer, and Unitarian (*muwaḥḥid*) on the exoteric level. (. . .) Whoever turns from determined existence toward absolute Existence, from the contemplation of the possible being to that of the necessary one, and utters the formula of ontological uniqueness, "There is no existent save God," has broken out of hidden associationism and enters the Truth as a Unitarian, a Gnostic (*ʿārif*), and a man of realization (*muḥaqqiq*) on the esoteric level.[60]

Notably, the spiritual rank of the Imam is here clearly above that of the Prophet. And incidentally, the Imams are credited with the theory of the Unity of Existence, which is far from obvious in their hadiths.

Āmulī finally distinguishes between three modes of affirming Divine Unity: the Unity of Acts (*at-tawḥīd al-fiʿlī*), the Unity of Attributes (*at-tawḥīd al-waṣfī*), and the Unity of Existence and Essence (*at-tawḥīd al-wujūdī adh-dhātī*). The first mode entails disregarding all agents and acts until the true and unique Agent is grasped, which is the origin of all acts, and thereby perfecting one's intellect (*ʿaql*). The second mode involves disregarding all attributes and subjects until grasping the true and unique subject, which is the source of all attributes and subjects, and thereby perfecting one's insight (*baṣīra*). The third mode requires disregarding any essence and existence until the absolute and pure existence is grasped that is the Existence-giver of all existents, the pure and exclusive Essence, the cradle of all essences, and thereby perfecting one's spiritual vision (*shuhūd*). The one who affirms Divine Unity through this final mode is a perfect Gnostic, a man of perfection and realization (*mukammil muḥaqqiq*).[61]

In his commentary on Q 1:2, "Praise belongs to God" (*al-ḥamdu li-llāh*), Āmulī also writes,

> "Praise," at the level of the affirmation of the Unity of God's Essence, means the affirmation of the single existence of God, the Real, and the absolute negation of any other's existence, for he who affirms the existence of any other in addition to God's associates something else with His Unity. (. . .) He who remains blind to the fact that any existence other than His is merely a metaphorical and abstract existence, devoid of extramental reality, and that His existence is the only essential and real one, such that in reality there is no extramental existence but His, does not really praise God. True praise is based on the true affirmation of God's Unity, and whoever does not affirm the ontological Unity of God cannot in any way realize His true praise.[62]

60 MAa, 3: 190f.; MAb, 543.

61 MAa, 3: 215f.; MAb, 550f.

62 MAa, 6: 16f.; MAb, 1005–6.

7 Philosophical Topics

Although Āmulī incorporates various classical themes from *falsafa* and Greek philosophy, particularly Neoplatonism, he offers a critique of philosophy and rational knowledge that must itself be viewed as philosophical.

7.1 The Legacy of Neoplatonism

Āmulī broadly takes up the Neoplatonist theory of the procession of Intellect, Soul, and Nature, from the absolute One. He thus applies a philosophical interpretative framework to the Qur'ānic verses, referring to the Book, the Calamus, and the Tablet, as in Q 13:39 – "God blots out, and He establishes whatsoever He will; and with Him is the Essence of the Book" – and Q 17:58 – "that is in the Book inscribed":

> By these two (Books), God means the first Intellect and the universal Soul, which are the two forms, comprehensive and detailed, of the Book. The first Intellect is the comprehensive, universal Book in which all things are confirmed synthetically. The universal Soul is the universal detailed Book in which all things are confirmed analytically. (. . .) (According to another interpretation), the universal Soul is referred to as "the guarded Tablet" (Q 85:21–22) and the first Intellect as "the Pen" (Q 68:1). Indeed, the first Intellect is similar to a pen as it emits the sciences and essential realities onto the universal Soul, the universal Soul is similar to a tablet due to its receptivity to these insights, and "Nūn. By the Pen, and what they inscribe" (Q 68:1) is an allusion to both, because "Nūn" refers to the universal Soul and "the Pen" to the first Intellect.[63]

In the commentary on the opening verse of the Qur'ān, "In the name of God, the Merciful, the Compassionate," Āmulī sets out another diagram representing the universal worlds in which the nineteen letters of the verse correspond to the nineteen degrees of the hierarchy of beings according to "the wise (philosophers) among the masters of rational knowledge" (*al-ḥukamā' min arbāb al-ma'qūl*): 1 (*bā'*): the world of Order (*al-amr*), the level of the absolute Unity of existence; 2 (*sīn*): the world of Intellect, the first existing world resulting from God's Order without intermediary; 3 (*mīm*): the world of Soul, the first existing world deriving from the Order through the intermediary of Intellect, referred to in the Qur'ān as "the Tablet," "the Throne," and "Nūn"; 4 (*alif*): the world of Nature, the deputy (*khalīfa*) of the universal Soul; 5 (*lām*): the outermost sphere, sometimes called (in the Qur'ān) "the Throne"; 6 (*lām*): the sphere of the Zodiac; 7 (*hā*): Saturn; 8 (*alif*): Jupiter; 9 (*lām*): Mars; 10 (*rā'*): the Sun; 11 (*ḥā'*): Venus; 12 (*mīm*): Mercury; 13 (*nūn*): the Moon; 14 (*alif*): the prime matter (*hayūlā*); 15 (*lām*): Fire; 16 (*rā'*): Air; 17 (*ḥā'*): Water; 18 (*yā'*): Earth; 19 (*mīm*): the three begotten

63 MAa, 2: 21f.; MAb, 231f.

species (minerals, plants, animals).[64] In this way, the philosophical sciences appear to be necessary for deciphering the Book of God.

7.2 The Harmonization between Revelation and Reason

In order to establish harmony between reason (*ʿaql*) and revelation (*sharʿ*), Āmulī takes up an argument made by Ashʿarī theologian Abū l-Qāsim ar-Rāghib al-Iṣfahānī (d. 1108–9/502) that a *faylasūf* would not have denied:

> Know that reason can only be guided by revelation, and that revelation can only be understood by reason. Reason is like a foundation and revelation like a building: a foundation is useless if there is no building, a building does not stand if there is no foundation. Alternatively, reason is like sight and revelation like a ray of light: sight is useless if there is no ray of light coming from outside, the ray is useless if there is no sight. This is why God says: "There has come to you from God a light, and a Book Manifest, whereby God guides whosoever follows His good pleasure in the ways of peace, and brings them forth from the shadows into the light by His leave" (Q 5:15-16). Alternatively, reason is like a lamp and revelation like the oil that feeds it: the lamp does not work if there is no oil, the oil does not shine if there is no lamp. Of this, God has warned by His word: "God is the Light of the heavens and the earth . . ." until: "Light upon light!" (Q 24:35), that is, the light of revelation upon the light of reason, (the former) illuminating only through (the latter). Alternatively, revelation is a reason from the outside and reason is a revelation from the inside: both help each other, or rather, are united. Because revelation is a reason from the outside, God has denied unbelievers the name of reason in other places in the Qurʾān, such as: "deaf, dumb, blind – they do not understand" (Q 2:171). Because reason is a revelation from the outside, God described reason as follows: "(So set thy face to the religion, a man of pure faith) – God's original upon which He originated mankind. There is no changing God's creation. That is the right religion; but most men know it not" (Q 30:30), where He called reason "religion". And because the two are united, He said: "Light upon light!," that is, the light of reason and the light of revelation; then He said, "God guides to His Light whom He will," making the two one light.[65]

7.3 Criticism of Philosophy

Elsewhere, Āmulī voices harsh criticism of *falsafa* in the name of Revelation (*sharʿ*) and gnosis (*ʿirfān*) or (alternatively) spiritual unveiling (*kashf*). Thus, in his analysis of *jihād* from the standpoint of the people of the Truth, he states,

> For them, (. . .) *jihād* means a struggle against theoretical reason in order to repel its sophisms and doubts. This is because theoretical reason always remains in determination and definition,

64 MAa, 5: 207–11; MAb, 899–902.

65 MAa, 3: 94f.; MAb, 514f.; Abū l-Qāsim ar-Rāghib al-Iṣfahānī, *Tafṣīl an-nashʾatayn fī taḥṣīl as-saʿādatayn*, ed. Aḥmad Nājī and Muḥammad Amīn Khānjī (Cairo: al-Maṭbaʿa al-Ḥamīdiyya al-Miṣriyya, 1905), 67–69; resumed in Abū Ḥāmid al-Ghazālī, *Maʿārij al-quds fī madārij maʿrifat an-nafs*, ed. n.n. (Beirut: Dār al-Āfāq al-Jadīda, 1981), 57–59.

> whereas the object sought always exists in absoluteness and detachment, as required by love and spiritual tasting (*al-ʿishq wa-dh-dhawq*). What does reason know about love? (. . .) Spiritual realization (*taḥqīq*) shows that reason is to love, its knowledge, its revelations, and contemplations, what fantasy (*wahm*) is to reason, its knowledge and perceptions. (. . .) Hence the opposition between the insights provided by reason and demonstration on the one hand, and those gained from inspired preaching and spiritual tastings on the other. (. . .) Hence the sophisms of philosophers denying the bodily resurrection and God's knowledge of temporal individuals (. . .) or professing that the world is eternal (. . .) according to the sole judgment of their weak reasons, powerless to grasp the secrets of Revelation (. . .). That is why the Unitarians are always fighting against philosophers and Brahmins in all parts of the world with the suprasensory sword. Such is the *jihād* of the people of the Truth, led against the masters of reasoning in order to eliminate their sophisms and to push back their doubts, so that they may finally return from the way of theoretical reason to the way of true tasting and divine love, respectively referred to as revelation (*waḥy*) and inspiration (*ilhām*).[66]

8 Conclusion

In *al-Muḥīṭ al-aʿẓam*, Ḥaydar Āmulī aims at nothing less than reviving the Imāmī hermeneutics of the Qurʾān by establishing that he, as a Gnostic or a "master of Divine Unity," is the Imam's spiritual heir. He does so by combining the original Imāmī tradition with the theoretical mysticism of Ibn ʿArabī and Neoplatonic philosophy. While supporting the idea that one cannot understand the Book of God without studying "the great book of the world" and "the small book of the soul" in parallel, he paved the way to a spiritual, universal, and always renewed reading of the sacred book of Islam. Despite the loss of most of his commentary, this was an unprecedented essay heralding in seventeenth-century Shiite philosophy and its hermeneutical output. Thus, while Ḥaydar Āmulī's endeavor certainly constitutes a departure from the original Imāmite tradition, it may have contributed to restoring to Imāmī Shiism its spiritual and hermeneutical dimension.

9 Bibliography

9.1 Sources

Āmulī, Sayyid Ḥaydar. *Le texte des textes: commentaire des "Fosûs al-hikam" d'Ibn ʿArabi*. Ed. H. Corbin and O. Yahya. Tehran/Paris: Département d'iranologie de l'institut franco-iranien de recherche/Librairie d'Amérique et d'Orient Adrien-Maisonneuve, 1975.

66 MAa, 4: 300–2; MAb, 776.

Āmulī, Sayyid Ḥaydar. *La philosophie shi'ite*. Ed. Henry Corbin and Osman Yahya, 2 vols. (vol. 1: *Jāmiʿ al-asrār wa-manbaʿ al-anwār*, vol. 2: *Risālat Naqd an-nuqūd fī maʿrifat al-wujūd*). Tehran: Departement d'Iranologie de l'Institut Franco-Iranien de Recherche, 1968; 1989[2].

Āmulī, Sayyid Ḥaydar. *Tafsīr al-Muḥīṭ al-aʿẓam wa-l-baḥr al-khiḍam fī taʾwīl kitāb Allāh al-ʿazīz al-muḥkam*. Ed. Muḥsin al-Mūsawī at-Tabrīzī, 7 vols. Qum: Nūr ʿalā Nūr, 1996; Ed. n.n., Beirut: Dār al-Maḥajja al-Bayḍā, 2012.

Āmulī, Sayyid Ḥaydar. *Naṣṣ an-nuṣūṣ fī sharḥ al-Fuṣūṣ*. Ed. Muḥsin Bīdārfar, 3 vols. Qum: Bīdār, 2015.

Āmulī, Sayyid Ḥaydar. *Anwār al-ḥaqīqa wa-aṭwār aṭ-ṭarīqa wa-asrār ash-sharīʿa*. Ed. Muḥsin al-Mūsawī at-Tabrīzī. Qum: Nūr ʿalā Nūr, 2015–16.

Āmulī, Sayyid Ḥaydar. *Risālat Rafʿ al-khilāf wa-l-munāzaʿa*. Ed. Ḥusayn Kalbāsī Ashtarī. Tehran: Muʾassasa-yi Pajūhashī Ḥikmat wa Falsafa, 2017.

Āmulī, Sayyid Ḥaydar. *Risālat al-wujūd fī maʿrifat al-maʿbūd*. Ed. Mahdī Ḥājiyān. Tehran: Madrasa-yi ʿāli-yi Shahīd-i Moṭahharī, 2020.

Al-Ghazālī, Abū Ḥāmid. *Maʿārij al-quds fī madārij maʿrifat an-nafs*. Ed. n.n. Beirut: Dār al-Āfāq al-Jadīda, 1981.

Al-Ghazālī, Abū Ḥāmid. "Ar-Risāla al-laduniyya." In *Majmūʿa rasāʾil al-imām al-Ghazālī*. 7 parts. Beirut: Dār al-Kutub al-ʿIlmiyya, part 3, 2011: 57–74.

Ibn ʿArabī, Muḥyī d-Dīn. *Fuṣūṣ al-ḥikam*. Ed. Abū l-ʿAlā ʿAfīfī. Beirut: Dār al-Kitāb al-ʿArabī, 2002.

Ar-Rāghib al-Iṣfahānī, Abū l-Qāsim. *Tafṣīl an-nashʾatayn fī taḥṣīl as-saʿādatayn*. Ed. Aḥmad Nājī and Muḥammad Amīn Khānjī. Cairo: al-Maṭbaʿa al-Ḥamīdiyya al-Miṣriyya, 1905.

The Koran Interpreted, trans. Arthur J. Arberry. Oxford: University Press, 1964 (first ed. 2 vols. London: Allen and Unwin, 1955).

9.2 Secondary Literature

Amir-Moezzi, Mohammad Ali. *The Silent Qur'an and the Speaking Qur'an: Scriptural Sources of Islam Between History and Fervor*, trans. Eric Ormsby. Columbia: Columbia University Press, 2015.

Amir-Moezzi, Mohammad Ali, and Christian Jambet. *What is Shi'i Islam? An Introduction*, trans. Kenneth Casler and Eric Ormsby. London/New York: Routledge, 2018.

Bar-Asher, Meir M. *Scripture and Exegesis in Early Imami Shi'ism*. Leiden: Brill, 1999.

Corbin, Henry. *En islam iranien: Aspects spirituels et philosophiques*. 4 vols. Paris: Gallimard, 1971–72.

Ash-Shaybī, Kāmil Muṣṭafā. *Aṣ-Ṣila bayna at-taṣawwuf wa-t-tashayyuʿ*. 2 vols. Beirut: al-Jamal, 2011.

Terrier, Mathieu. "The Defence of Sufism among Twelver Shiʿi Scholars of Early Modern and Modern Times: Topics and Arguments." In *Shiʿi Islam and Sufism: Classical Views and Modern Perspectives*, edited by Denis Hermann and Mathieu Terrier. London: I.B. Tauris, 2020: 27–63.

Terrier, Mathieu. "Sufi Hierarchies of the Worlds or Degrees of Existence: *Mulk*, *Malakūt*, *Jabarūt*, and Related Concepts." In *Sufi Cosmology*. Vol. 2, *Handbook of Sufi Studies*, edited by Alexander Knysh and Christian Lange 2 vols. Leiden: Brill, 2023: 277–98.

Ibrāhīm al-Biqāʿī

Roy Michael McCoy III

"Now this is an astonishing book, distinguished in its art, unlike anything I have seen from those who preceded me, and beyond the reach of any keen mind to surpass."
Ibrāhīm b. ʿUmar b. Ḥasan al-Biqāʿī (d. 1480/885), *Naẓm ad-durar fī tanāsub al-āyāt wa-s-suwar*, 1:2.

1 Introduction

Al-Biqāʿī's notable attributes did not include humility. He was a learned scholar of the fifteenth century, but a tendency towards self-aggrandizement coupled with an antagonistic attitude towards his critics over the course of his career, would ultimately leave him on the outskirts of the scholarly world, both figuratively and literally. He died ostracized and alone just beyond the walls of his Damascene community three years after the completion of his magnum opus, *Naẓm ad-durar fī tanāsub al-āyāt wa-s-suwar* (The String of Pearls: On the Interrelatedness of the Verses and Chapters).[1] Under Circassian Mamluk rule (1382–1517/782–923), eleven sultans reigned during the span of al-Biqāʿī's life, from al-Malik an-Nāṣir Faraj (r. 1399–1412/801–815), who was the eleven-year-old son of al-Malik aẓ-Ẓāhir Barqūq (d. 1399/801) when he acceded to the throne, to the sultanate of Qāytbāy (r. 1468–1496/872–901).[2] This final age of Mamluk prestige provoked less amicable interactions between scholars that made the political climate more critical towards al-Biqāʿī and his inclusion of the Arabic Bible within his Qurʾānic interpretations.

To demonstrate why his *tafsīr* was beyond reproach, al-Biqāʿī employed a prevalent literary motif of his era — the "dreams and visions" trope. Dreams and visions

1 Al-Biqāʿī, *Naẓm ad-durar fī tanāsub al-āyāt wa-s-suwar*, ed. ʿAbd ar-Razzāq Ghālib al-Mahdī, 8 vols. (Beirut: Dār al-Kutub al-ʿIlmiyya, 2011); al-Biqāʿī, *Naẓm ad-durar fī tanāsub al-āyāt wa-s-suwar*, 22 vols. (Hyderabad: Dār al-Maʿārif al-ʿUmāniyya, 1976–1982) (all references are to this Hyderabad edition).
2 Li Guo, "Al-Biqāʿī's Chronicle: A Fifteenth-Century Learned Man's Reflections on His Time and World" in *The Historiography of Islamic Egypt (C. 950–1800)*, ed. Hugh Kennedy (Leiden: Brill, 2001), 121–48, here 121–23; see also Jo Van Steenbergen, *A History of the Islamic World, 600–1800: Empire, Dynastic Formations, and Heterogeneities in Pre-Modern Islamic West-Asia* (London: Routledge, 2020); Jean-Claude Garcin, "The Regime of the Circassian Mamlūks," in *CHE*, vol. 1, *Islamic Egypt, 640–1517*, ed. Carl F. Petry (Cambridge: Cambridge University Press, 1998), 291–92. Barqūq is credited with establishing the Circassian dynasty (ca. 1382/783) following the Baḥrī Mamlūks (r. 1250–1390/647–792). For details on this transition, see Linda S. Northrup, "The Baḥrī Mamlūk sultanate, 1250–1390," in *CHE*, vol. 1, ed. Petry, 287–89.

https://doi.org/10.1515/9783111320083-013

were used to validate the author's work through the retelling of an extraordinary experience or event that aided the author in some way.[3]

One of the initial topics al-Biqāʿī explores in his prolegomena is a dream that serves as proof of the divine support he received in the process of writing *Naẓm ad-durar*. Al-Biqāʿī comments that at the age of ten, he saw the *rūḥ al-qudus* (holy spirit), Gabriel, "the one who brought down this spirit," that is, the Qurʾān; and he saw the Prophet Muḥammad, "the one whom this spirit descended upon." Al-Biqāʿī continues to describe their appearance as men riding on horses towards the east. He then concludes, "God aided me through blessings from (Gabriel and Muḥammad) in interpreting (the Qurʾān) and compiling it through a spirit from Him, as attested by those who have read (my book) and contemplated it."[4] An episode is also documented in al-Biqāʿī's historical/autobiographical work, *Taʾrīkh al-Biqāʿī*, regarding a dream he had on the day of Sultan Jaqmaq's (d. 1453/857) death, which subsequently repeated itself upon the demise of Sultan Īnāl (d. 1461/865).[5] In his mind, the dream and the portion of the Qurʾān that al-Biqāʿī lectured on become linked to these events – here his reality becomes intertwined with his anagogical reading of the Qurʾān.

To clarify, the scriptural hermeneutics of the Cairene polymath were not only influenced by his historical context but also driven by his sense of nostalgia for scholars of the past. Al-Biqāʿī's capacity for learning from earlier centuries is linked to four primary influences on his hermeneutics: Abū l-Faḍl al-Maghribī l-Mashdāllī (d. 1460/864), Ibn az-Zubayr al-Gharnāṭī (d. 1308/708), Abū l-Ḥasan al-Ḥarrālī (d. 1241/638) and Abū Maḥmūd b. ʿAbd ar-Raḥmān al-Aṣbahānī (d. 1349/749).[6] It is clear that the foundational principle used by al-Biqāʿī to expand on quranic "connections" came from his time studying with al-Mashdāllī:

> The general guideline for elucidating the coherence among the verses throughout the Qurʾān is to ascertain the aim *(gharaḍ)* for which a particular sūra has been crafted. Subsequently, one must identify the prerequisite elements needed to fulfill that aim, and evaluate the extent to which these elements have fulfilled their purpose. It is also crucial to understand how the dis-

3 On these phenomena as a means of validating knowledge transmission, see Jonathan Berkey, *Popular Preaching and Religious Authority in the Medieval Islamic Near East* (Seattle: University of Washington Press, 2001), 81.

4 Al-Biqāʿī, *Naẓm ad-durar*, 1: 4f.

5 Al-Biqāʿī, *Iẓhār al-ʿaṣr li-asrār ahl al-ʿaṣr: Taʾrīkh al-Biqāʿī*, ed. Muḥammad Sālim b. Shadīd al-ʿAwfī (Riyadh: np, 1992), 305–7; cf. Guo, "Al-Biqāʿī Chronicle," 133–35.

6 Walid A. Saleh, *In Defense of the Bible: A Critical Edition and an Introduction to al-Biqāʿī's Bible Treatise* (Leiden: Brill, 2008), 14–18. Al-Biqāʿī mentions at the opening of his commentary that he will reference the work of Ibn az-Zubayr at the beginning of each sūra because of its importance to him (Al-Biqāʿī, *Naẓm ad-durar*, 1:9). Saleh also notes that al-Aṣbahānī is found in Carl Brockelmann's *Geschichte der arabischen Litteratur* under the name "al-Iṣfahānī" (Brockelmann, *GAL* (Leiden: Brill, 1936), 2: 110–11) for reasons unknown.

> course following these prerequisites guides the audience's anticipation of what comes next, whether it pertains to divine commandments or other necessary topics that complement the sūra's intended purpose. The sūra's discourse should, therefore, adhere to the principles of rhetoric, ensuring that the audience is not unduly burdened in discerning the sūra's aim. This fundamental principle unravels the web of connections within the Qur'ān's various sections. By following this principle, you will gain insights into how the Qur'ān's composition was achieved between each verse in every sūra.[7]

Al-Mashdāllī's description of the process for interpreting the Qur'ān constitutes the essence of al-Biqāʿī's hermeneutic, *ʿilm munāsabāt al-Qur'ān* (the science of connections of the Qur'ān). Applying *munāsabāt* to one's reading of the Qur'ān, according to al-Biqāʿī, is contingent upon one's knowledge of the global objective (*maqṣūd*) of the particular chapter in which connections are being studied.[8] This provides an understanding of what all of the sentences in the chapter mean, and that is why one's knowledge of the global objective is extremely precious (*ghāyat an-nafāsa*).[9] While the work of both al-Mashdāllī and Ibn az-Zubayr, as well as al-Aṣbahānī, were formative for al-Biqāʿī, they fail to be called to mind as regularly in *Naẓm ad-durar* as that of al-Ḥarrālī; one cannot go more than three pages in al-Biqāʿī's *tafsīr* without seeing a reference to the Moroccan mystic. Al-Biqāʿī comments, "I have benefited in (*Naẓm ad-durar*) a great deal with a particular *tafsīr*," that is, al-Ḥarrālī's treatise *Miftāḥ al-bāb al-muqfal li-fahm al-Qur'ān al-munzal* (*The Key to the Locked Door for Understanding the Revealed Qur'ān*).[10] It is al-Ḥarrālī's anagogical method of interpretation that al-Biqāʿī imbibes *in toto* and uses to delve even deeper into *ʿilm munāsabāt al-Qur'ān*. Thus, of the four scholars, the remainder of this paper will address the connection between al-Ḥarrālī and al-Biqāʿī at different points for his scriptural hermeneutics.[11]

7 Saleh, *In Defense of the Bible*, 15; al-Biqāʿī, *Naẓm ad-durar*, 1: 18; see Shams ad-Dīn as-Sakhāwī, *aḍ-Ḍaw' al-lāmiʿ li-ahl al-qarn at-tāsiʿ* (Cairo: n.p., 1934), 9: 185. Al-Biqāʿī quotes the same text from al-Mashdāllī which as-Sakhāwī uses in his biography of several scholars.

8 Al-Biqāʿī uses "global objective" (*maqṣūd*) here to denote a thematic approach focused on common leitmotifs in each sūra (Abdul-Raof's "Consonance Approach"; see idem *Theological Approaches to Qur'anic Exegesis: A practical comparative-contrastive analysis* (London: Routledge, 2012), 94–99), which forms the link between all of the "connections" (*munāsabāt*) across the Qur'ānic corpus.

9 Al-Biqāʿī, *Naẓm ad-durar*, 1: 6.

10 Much of al-Ḥarrālī's *Miftāḥ* is found in quoted form in al-Biqāʿī's *Naẓm ad-durar*. However, these parts were extracted and published by Maḥmādī b. ʿAbd as-Salām Khayyāṭī, *Turāth Abī l-Ḥasan al-Marrākushī fī t-tafsīr* (Casablanca: Matbaʿat an-Najāḥ al-Jadīda, 1997).

11 Even in Saleh's "Hermeneutic Biography," found in *In Defense of the Bible*, the connection between al-Ḥarrālī and al-Biqāʿī remains elusive in comparison to the other three influential scholars.

2 Al-Biqāʿī's Hermeneutics of *ʿilm munāsabāt al-Qurʾān*

The centrality of *ʿilm al-munāsabāt* for al-Biqāʿī's Qurʾānic exegesis can be summed up in one statement from the prolegomena to his *tafsīr*: "(T)he more hidden a connection is, the more powerful it is once you see it" (*wa-ʿalā qadr ghumūḍ tilka al-munāsabāt yakūn wuḍūḥuhā baʿda inkishāfi-hā*).[12] For al-Biqāʿī then, it is through the science of *munāsabāt* that one can figure out why the text appears in particular forms rather than others. He defines the principle of *munāsabāt* in general terms first:

Whether in the Qurʾān or in (something) other than the Qurʾān, *munāsabāt* is a science through which is known the subtle reasons behind orderings (*tartīb*). What is the thing the subject is studying? You apply this knowledge to the parts in which you want to understand the connections.[13]

That is to say, one applies *munāsabāt* in the study of any field of knowledge to determine whether in that branch of learning.

The fruit of *munāsabāt*, continues al-Biqāʿī, is "to be aware of the position that this part you are studying deserves by virtue of what is behind it and what is in front of it, according to the similarities between them, which amounts to a collection of distinguishing attributes."[14] Therefore, according to al-Biqāʿī the science of Qurʾānic *munāsabāt* is expressed as "the science through which you know the subtle reasoning behind the ordering of the Qurʾān's parts, and it is the secret of rhetoric (*balāgha*) because it leads to really understanding how these meanings suit what the circumstances need."[15] An awareness of these connections directly informs al-Biqāʿī's interpretation of the Qurʾānic text thus allowing for an intratextual hermeneutic.[16] It is by virtue of *munāsabāt* that parts of the Qurʾān are understood to connect with each other and then are used to inform the interpretation of both. What is remarkable about this methodological approach is that on the one hand, it allows for *tafsīr al-Qurʾān bi-l-Qurʾān* (the interpreta-

12 Al-Biqāʿī, *Naẓm ad-durar*, 1: 14.

13 Ibid, 1: 14.

14 Ibid., 1: 5. He further elaborates this point in similar language after beginning his exegesis of *sūrat al-Fātiḥa* (The Opening, Q 1: 17–18); see also Saleh, *In Defense of the Bible*, 15.

15 Al-Biqāʿī, *Naẓm ad-durar*, 1: 5f.; for a definition of *balāgha* and its function in Islamic literature, see A. Schaade and G.E. von Grunebaum, "Bālagha," in *EI*², ed. P. Bearman et al., accessed November 20, 2023, http://dx.doi.org/10.1163/1573-3912_islam_SIM_1123; for a detailed study on Arabic rhetoric and the three "sciences" that comprise *balāgha* (*ʿilm al-maʿān*, *ʿilm al-bayān* and *ʿilm al-badīʿ*), see Hussein Abdul-Raof, *Arabic Rhetoric: A Pragmatic Analysis* (London: Routledge, 2006).

16 The kind of methodology suggested here by "*intra*textual hermeneutic" derives its meaning from the interpretation of textual units that occur within a single text, rather than texts *ab extra* (e.g., *inter*-textuality) (Roland Meynet, *Treatise on Biblical Rhetoric*, trans. Leo Arnold, Rubianto Solichin, and Llane B. Briese (Leiden: Brill, 2012), 233, 235–37 *passim*); for a detailed study on the usage of intratextual and intertextual exegesis in Islamic sources, see Michel Cuypers, *The Composition of the Qurʾan: Rhetorical Analysis*, trans. Jerry Ryan (London: Bloomsbury Academic, 2015), 155–76.

tion of the Qurʾān with/through the Qurʾān) and, on the other, it encompasses *tafsīr al-Qurʾān bi-l-Kitāb* (the interpretation of the Qurʾān with/through the Bible).[17]

Al-Biqāʿī then places his hermeneutical theory within the broader context of Qurʾānic exegesis by relating the knowledge of *munāsabāt* to the general knowledge of exegesis, which, he states, is like the relationship between the science of rhetoric and the science of grammar.[18] In other words, it is through understanding rhetoric that we can find out why the speaker chooses a particular grammatical structure over others. Therefore, through *munāsabāt* the scholar can figure out why the Qurʾānic text appears in particular forms (e.g., sūra placement) over others. This is made explicit in his comment that,

> (T)hrough (*munāsabāt*) it will become clear also that there is no such thing as a complete pause in the book of God, not even at the end of *sūrat an-Nās*; that is not a stop either. Rather, it connects with the very first chapter (*sūrat al-Fātiḥa*) — the beginning of the Qurʾān — just like it connects with the sūra right in front of it (*sūrat al-Falaq*); in fact, its connection is even stronger with the *Fātiḥa*.[19]

Al-Biqāʿī makes a similar connection between the introductions to sūras *al-Baqara* and *Āl ʿImrān*. Here al-Biqāʿī focuses on the "sending down" of revelation, or the Scripture, in the opening lines of Q 2:4: *wa-lladhīna yuʾminūn bi-mā unzila ilayka wa-mā unzila min qablika wa-bi-l-ākhirati hum yūqinūn* ("And who believe in what has been sent down to you and what was sent down before you and are sure of the world to come"); al-Biqāʿī then links this with the introduction in Q 3:3, *nazzala ʿalayka l-kitāba bi-l-ḥaqqi muṣaddiqan li-mā bayna yadayhi* ("He revealed to you the Scripture in truth, confirming that which was before it"). Therefore, he argues, "what is strung together (*muntaẓim*) in the introduction to sūra *Āl ʿImrān* is something that parallels what is strung together as the introduction of *sūrat al-Baqara*"; the opening of sūra *Āl ʿImrān* "gradually descends to the *aṣl* (root) of the sending-down of the revealed scripture."[20]

Further, the "sending down" commented on at Q 3:3 is connected with Q 3:7: *huwa lladhī anzala ʿalayka l-kitāba* ("It is He Who sent down the Scripture upon you"); "sent down" is a "common Qurʾānic locution for its revelation and connotes a transi-

---

17 Here "Bible" denotes the general Qurʾānic bifurcation of the "previous scriptures" into the *Tawrāt* and the *Injīl*. Thus, the hermeneutical approach becomes *inter*textual here by comprising "all those texts belonging to biblical books to which a particular text is related by way of quotation, reference or allusion" (Meynet, *Treatise on Biblical Rhetoric*, 21; cf. Cuypers, *The Composition of the Qurʾan*, 156). For a full analysis of al-Biqāʿī's biblical scholarship, see R. Michael McCoy III, *Tafsīr al-Qurʾān bi-l-Kitāb: Reading the Arabic Bible in the Tafsīrs of Ibn Barraǧān (d. 536/1141) and al-Biqāʿī (d. 885/1480)* (Leiden: Brill, 2021).

18 Al-Biqāʿī, *Naẓm ad-durar*, 1: 6.

19 Ibid., 14.

20 Ibid., 4: 223f.

tive transmission from a celestial archetype."[21] This locution occurs twice in the introduction to *sūrat Āl ʿImrān* (Q 3:3), first with reference to the Qurʾān, and second to the Torah and the Gospel. Following al-Ḥarrālī's exegesis of Q 3:7, al-Biqāʿī interprets this text as the verse of worship from generation to generation (*āy at-taʿabbud min khalq li-l-khalq*), whose ruling (*ḥukm*) does not change among any of the three books.[22] The closeness of al-Biqāʿī's orientation to that of al-Ḥarrālī can be observed in what Faris Casewit has described as the "principle of moral symmetry," captured succinctly in al-Ḥarrālī's own words:

> Among that which has prevented (*ṣadda*) the majority of this community (*akthar hādhihi l-ʾumma*) from fully apprehending (*tafahhum*) the Qurʾān, is their supposing (*ẓannahum*) that what it contains in the way of stories of the ancients (*qiṣaṣ al-awwalīn*), and of reports of those who were recompensed and those who were punished (*akhbār al-muthābīn wa-l-muʿāqabīn*), from among the members of all religions – that the purpose (of all that) is solely to inform and recount (*al-ikhbār wa-l-qaṣaṣ faqaṭ*). Nay! It is not so! Its intended purpose (*maqṣūduhu*) is crossing-over (*al-iʿtibār*) and warning (*at-tanbīh*) by way of observing its recurrence in this community (*bi-mushāhadat mutakarririhi fī hādhihi l-ʾumma*), (wherein are found) the counterparts of all those human collectivities, all those postures, and all those historical signs (*al-āthār*).[23]

The Qurʾān is not to be understood, according to al-Ḥarrālī, as a static text but rather a means for every community to live out the Qurʾān's maxims.

3 Al-Biqāʿī's Hermeneutics of Q 3:7

Al-Biqāʿī approaches access to God through the *muḥkamāt* (clear/firm verses) and the *mutashābihāt* (self-similar verses) derived from his interpretation of Q 3:7 in a distinctive way. Commenting on Q 3:7b, he states that the *mutashābihāt* lead astray the ones in whose hearts is deviation (*zayghun*) and who seek mischief (*fitna*). Through stopping (*waqfa*) at the *mutashābihāt*, however, those who are firm in knowledge (*rāsikhūn*) actually open the door for God to impart knowledge on the *mutashābih* verses.[24] That is

21 Jane Dammen McAuliffe, "The Qurʾānic Context of Muslim Biblical Scholarship," *Islam and Christian-Muslim Relations* 7, no. 2 (1996): 142. Q 2:4 opens with the same locution as well: "And who believe in what has been sent down to you (Muḥammad) and what was sent down before you, and are sure of the world to come."

22 Al-Biqāʿī, *Naẓm ad-durar*, 4: 224f.

23 Faris Casewit, "'Jewish' Muslims and 'Christian' Muslims: Ḥarrālī's (d. 638/1241) Reflexive Transposition of the Qurʾān's Polemic Against the Religious Other," (unpublished paper, Microsoft Word file, accessed August 7, 2020), 7.

24 At this point in the verse, how one chooses to read the text significantly effects the interpretation. The meaning of this part of the verse hinges on how to understand the conjunction *wāw* between "God" and the *rāsikhūn fī l-ʿilmi* ("those who are well-grounded in knowledge") in the words *wa-mā yaʿlamu taʾwīlahū ʾillā Allāh wa-r-rāsikhūna fī l-ʿilmi yaqūlūna āmannā bihi*. Either the *wāw* is a letter indicating the beginning of a new sentence (*wāw al-istiʾnāf*) – this would signify that God alone knows

to say, the *mutashābih* verses exist to test those who are firm in knowledge. The test is to stop at the *mutashābihāt* and focus, rather, on practicing what is enjoined in the *muḥkam*, which ultimately leads to a realization of the *mutashābihāt*. Here again, the influence of al-Ḥarrālī on al-Biqāʿī's thinking is clear. Al-Ḥarrālī explains:

> These verses (*muḥkamāt*) are those whose rulings have been strengthened, just like a rope is fortified when someone uses a bit (*ḥakama*), in other words, a bridle, to control something they fear may escape their grasp. Just as firm (*muḥkam*) verses serve to rein in (*tuḥkima*) the soul to prevent it from going astray. They also keep the soul from breaking loose, like a horse that breaks free from its owner and gains the upper hand, and instead, they guide and discipline the soul to its rightful state.[25]

The *muḥkam* verses are used to keep the soul from breaking loose and running wild with the *mutashabihāt*; they keep the believer bridled lest they go astray. Al-Biqāʿī embraces this ideal and explains the interaction between *muḥkam* and *mutashābih* using a common motif in Sufi writings[26] that plays on the duality of veiling (*as-satr wa-l-ikhfāʾ*) and the unveiling (*kashf*):

> Hence, the command of God and His manifestations can only truly be attained through (divine) will from Him. A transformation occurs within the servant that lifts the veils of darkness and veils of light (*hujub aẓ-ẓulmāniyya wa-n-nūrāniyya*), a point at which all the scholars halted. However, within the *mutashābih* revelation, there exists only one of two approaches: one is the approach of adhering to the limits of faith, tailored for those who possess true knowledge and who immerse themselves in humility, lowliness, God-fearing-ness, and righteousness. The Prophet has commanded individuals to follow these qualities until they reach a stage where God's love embraces them.

the meaning of the *mutashābihāt* – or it is a conjunctive particle (*wāw al-ʿatf*), implying that the *rāsikhūn* do as well (see Ismail Albayrak, "The Notions of Muḥkam and Mutašābih in the Commentary of Elmalılı Muḥammad Ḥamdi Yazır," *JQS* 5, no. 1 (2003): 27; for a comparison of the variants, see Arthur Jeffery, *Materials for the History of the Text of the Qurʾān* (Leiden: Brill, 1937); this variant is observed in three of the fifteen non-canonical, primary "old codices": ʿAbd Allāh b. Masʿūd, Ubay b. Kaʿb and ʿAbd Allāh b. ʿAbbās. For a table of the canonical variants on Q 3:7, see Aḥmad Mukhtār ʿUmar and ʿAbd al-ʿĀl Sālim Mukram, *Muʿjam al-qirāʾāt al-Qurʾāniyya*, 8 vols. (Kuwait: Dhāt as-Salāsil, 1982–1985), 2: 7; for a comparative study on the ten readings, see Christopher Melchert, "The Relation of the Ten Readings to One Another," *JQS* 10, no. 2 (2008): 73–87; for a detailed analysis of variants in early Qurʾānic manuscripts, see Yasin Dutton, "Some Notes on the British Library's 'Oldest Qurʾān Manuscript' (Or. 2165)," *JQS* 6, no. 1 (2004): 43–71; Dutton, "An Early *Muṣḥaf* According to the Reading of Ibn ʿĀmir," *JQS* 3, no. 1 (2001): 71–89; Stefan Wild, "Why self-referentiality?" in *Self-Referentiality in the Qurʾān*, ed. Stefan Wild (Wiesbaden: Harrassowitz-Verlag, 2006), 20–21). Al-Biqāʿī has little to say about the differences of opinion here; he accepts the reading that God alone knows the interpretation: "Therefore it is not possible for someone to know the interpretation (*taʾwīl*) of everything except God, Most High," (al-Biqāʿī, *Naẓm ad-durar*, 4: 247).

25 Al-Biqāʿī, *Naẓm ad-durar*, 4: 224.

26 Nasr Abu Zayd, "Towards understanding the Qurʾān's worldview," in *New Perspectives on the Qurʾān: The Qurʾān in Its Historical Context 2*, ed. Gabriel Said Reynolds (London: Routledge, 2011), 62.

> At that juncture, when you merely follow and practice without delving into the *mutashābihāt*, God's love will be showered upon you as a result of your dedication to the *muḥkam*. God will then remove your incapacity to pause at the *mutashābihāt* and free you from the veil of light, making nothing of subtlety (*daqīq*) difficult for you to comprehend. It is through God's love for you that you will be free from fault due to hidden aspects.[27]

Therefore, it is the practice of *muḥkam* that allows God to remove the veils (*ḥujub*) so that what was once *daqīq* will become clear.

Then al-Biqāʿī interprets the correct posture for interpreting God's revelation at the end of Q 3:7: *wa-r-rāsikhūna fī l-ʿilmi yaqūlūna āmannā bihi kullun min ʿindi rabbinā wa-mā yadhdhakkaru illā ūlū l-albābi* ("Those who are well-grounded in knowledge say, 'We believe in it. All is from our Lord.' Only men of understanding are reminded"). To be "well-grounded in knowledge" is the key for comprehending the relationship between the clear and self-similar verses within al-Biqāʿī's hermeneutical paradigm. The *rāsikhūna fī l-ʿilm*, he states, are those people who are versed in the great points of knowledge, who have not been carried off in the least by imaginations (*takhyīl*) about God.[28]

Al-Biqāʿī conceives of these "imaginations" as excessive contemplation on the anthropomorphic verses (*āyāt aṣ-ṣifāt*).[29] That is to say, those verses in which God communicates *vis-à-vis* descriptions of himself that are "perceived by any of the senses"[30] of his creation, such as things based on the creation in terms of the form, like the eye, the hand, the foot and the face, everything connected with the human form. Likewise, the attributes of the soul are counted among the *āyāt aṣ-ṣifāt*, which consist of knowledge, wisdom, anger and pleasure, for example. All of these are *mutashābihāt* verses in al-Biqāʿī's estimation.[31]

There is a divine purpose for these anthropomorphisms, however, that relates back to the clear verses, as explained by al-Biqāʿī. It is because the intellect of creation cannot actually grasp God and "their eyes cannot really look to Him" that he has introduced himself to them with "similitudes from themselves (i.e., anthropomorphisms),

27 Al-Biqāʿī, *Naẓm ad-durar*, 4: 244f.

28 Ibid., 4: 244.

29 Here Michel Lagarde's treatment on anthropomorphic expressions and the *mutashābihāt* in the *tafsīr* tradition is helpful ("De l'ambiguïté (mutašābih) dans le Coran: tentatives d'explication des exégètes musulmans," *Quaderni di studi arabi* 3 (1985): 56–62).

30 Edward W. Lane, *An Arabic-English Lexicon* (Edinburgh: Williams and Norgate, 1863)., *ḥ-s-s*, 2: 564.; here the Arabic reads, *maḥsūs*, or "palpable," the sense of which connotes something much stronger than just "perceiving" with one's senses.

31 Al-Biqāʿī, *Naẓm ad-durar*, 4: 243. The use of *mutashābih* as a category for interpreting anthropomorphisms in the Qurʾān was first adopted by the Muʿtazilites as a key theological hermeneutic (Abu Zayd, "Towards understanding the Qurʾān's worldview," 55–58; Abu Zayd, *The Rational Trend in Exegesis: A Study of the Concept of Metaphor in the Muʿtazilites' Thought* (Beirut: al-Markaz ath-Thaqāfī l-ʿArabī, 1982); ʿAbd al-Jabbār al-Asadābādī, *Mutashābih al-Qurʾān*, ed. ʿAdnān Muḥammad Marzūq (Cairo: Dār at-Turāth, 1966)).

so that the *muḥkam* verses are for practice and the *mutashābih* verses are to show people their inability to grasp (Him)." Therefore, he concludes, the fundamental reason behind revealing the *muḥkamāt* and the *mutashābihāt* is for worship:

> The purpose of revealing the *mutashābihāt* verses is to make (God) known to (His creation) through a test for them, without dwelling excessively on the *mutashābihāt*, and instead focusing on worship. Hence, the advantage of elucidating the mutually resembling verses lies in the practice of the *muḥkam* and then pausing at the *mutashābih*, enabling the test from two perspectives: one concerning practical application when approaching the *muḥkamāt*, and the other related to halting when encountering the *mutashābihāt*.[32]

Probing into the intricacies of what is considered *mutashābih* is deemed an innovation (*bidʿa*) and is cautioned against, as it has the potential to misguide those "in whose hearts is deviation." Conversely, abstaining from delving into the *mutashābihāt* is viewed as adhering to the sunna, representing correctness and propriety. Al-Biqāʿī's decision to delve into praxis as an aspect of the *muḥkam* aligns him with like-minded exegetes, such as those observed in Abū Jaʿfar b. Jarīr aṭ-Ṭabarī's treatment of the verse. However, it illustrates that while al-Biqāʿī is in harmony with the interpretive tradition on this matter, his *muḥkam*/*mutashābih* formula is distinct in the sense that both components function in tandem rather than as separate approaches to the text of the Qurʾān.

Continuing with the latter part of the verse, al-Biqāʿī employs al-Ḥarrālī's interpretation to elucidate the term *kull* ("all") in *kullun min ʿindi rabbinā* ("all is from our Lord"). This elucidation highlights how both *muḥkam* and *mutashābih* stem from a shared origin. Al-Ḥarrālī explains:

> The term *kull* gains definiteness through the encompassment (*bi-taʿrīf al-iḥāṭa*) of all its constituents, a concept acknowledged by grammarians as a form of definiteness, except for some who do not include it. It represents one of the most comprehensive forms of definiteness, as its essence lies in specification, whether through observation or intellectual understanding. It serves as an indication of the comprehensive nature of what He has revealed (obscurely (*ibhāmihi*)). Thus, *kull* is the common origin to which both the *mutashābih* and the *muḥkam* refer, as they both trace back to the same point where these two branches diverge. This is because every distinct reality always comes back from a single point of meeting. When they express their belief by saying "we believe in it," they refer to this point of meeting (of the *muḥkam* and *mutashābih*) in the encompassment of the Book before any separation occurs.[33]

Al-Ḥarrālī's notion of *muḥkam* and *mutashābih* originating from "one origin" is examined by al-Biqāʿī through the analogy of a "root" (*umm*) and the branches of a tree. This analogy aligns his anagogical interpretation closely with that of al-Ḥarrālī:

32 Al-Biqāʿī, *Naẓm ad-durar*, 4: 243.
33 Al-Biqāʿī, *Naẓm ad-durar*, 4: 248.

> Regarding the task of making the verses clear (*bayān*), it is imperative to note that they possess the utmost clarity, as do the rulings within their internal coherence. Some verses are interconnected with others to the extent that they appear as a single entity. When it comes to tracing the *mutashābihāt* verses back (to the Root), this is externally straightforward – for those with firmly established faith, right intentions, and extensive knowledge. As a consequence, the entirety (of meanings) merges into a singular entity, because (God) employs a singular term (i.e., *al-umm*) to refer to a plural (i.e., *al-āyāt*).[34]

With this understanding of *al-umm* (the Root) and its relationship to the clear and the self-similar verses, al-Biqāʿī's method of connections becomes the hermeneutical thread that binds together revelation for all communities. This concept closely mirrors the interpretation of the early Qurʾān exegete Muqātil b. Sulaymān al-Balkhī (d. 767/150). He, too, envisions *umm al-kitāb* and the *muḥkamāt* as a unified entity that encompasses the earlier revelations of the *Tawrāt* and *Injīl.*[35] For both Muqātil and al-Biqāʿī, there exists a shared revelatory core among the three books, and the *muḥkamāt* are intended for all of the children of Adam.[36]

Furthermore, al-Biqāʿī's perspective emphasizes that God's division of what is revealed into clear and self-similar verses, in his view, serves to envision them as a coherent whole, as exemplified by the reference to "*al-kitāb*" in Q 3:7. He sees all three books—Qurʾān, Torah, and Gospel—as harmoniously agreeing on the concept of *mutashābihāt,* which form an integral part of this comprehensive revelation. The link between these three texts lies in their expression of clear and self-similar verses collectively. As al-Biqāʿī directs his attention to these three scriptures, the significance of his Qurʾānic hermeneutics in interpreting the Torah and the Gospel becomes increasingly evident.

According to al-Biqāʿī, Q 3:7 elucidates that, just as God renders specific actions "*mutashābih*" – for instance, the creation of Jesus without male sperm and the miraculous deeds attributed to him – He also employs words that are "*mutashābih,*" such as "*bunuwwa*" (sonship) and "*ubuwwa*" (fatherhood), as al-Biqāʿī will further elaborate in his discussion on the four canonical Gospels.[37]

Al-Ḥarrālī's engagement with the Gospels also reflects deeper coalescence with his own hermeneutic of transposition (*inṭibāq*) as he develops his concept for the "wisdom of the intellect" (*al-ḥikma al-ʿaqliyya*). For al-Ḥarrālī, the wisdom of the intellect is focused on:

---

34 Ibid., 4: 225.

35 Muqātil b. Sulaymān, *Tafsīr Muqātil b. Sulaymān*, ed. ʿAbd Allah Maḥmūd Shiḥāta (Beirut: al-Hayʾa al-Miṣriyya al-ʿĀmma li-l-Kitāb, 1979), 1: 263–64.

36 Jane Dammen McAuliffe, "Text and Textuality: Q 3:7 as a Point of Intersection," in *Literary Structures of Religious Meaning in the Quran*, ed. Issa J. Boullata (London: Routledge, 2000), 60; cf. Muqātil b. Sulaymān, *Tafsīr al-khams miʾat āya min al-Qurʾān*, ed. I. Goldfeld (Shfaram: Dār al-Mashriq, 1980), 275.

37 Ibid., 4: 226f.

> (a) specific facet of reality corresponds to a realm where the objects of cognition are directly observable and readily understandable (*al-mashhūd al-maʿlūm*). This particular level of reality is referred to as (*ʿālam al-mulk*), signifying the earthly realm. However, there is also a wisdom associated with a reality beyond the corporeal, a concealed and unseen world. Al-Ḥarrālī designates this level of reality as *ʿālam al-malakūt*, denoting the celestial realm.[38]

Within the framework of dualistic realms, earthly and celestial, al-Ḥarrālī introduces a teaching attributed to Jesus. He quotes Jesus's words, which appear to reference the Gospel of John (3:3): "None shall attain the celestial kingdom but those who are born twice." However, the first birth, in al-Ḥarrālī's interpretation, does not refer to a biological birth, as in the context of John. Instead, it symbolizes the acquisition of intellectual wisdom – the initial detachment from the physical, earthly realm. The second birth signifies the subsequent immersion into the celestial realm, transcending the boundaries of rational intellect. Al-Ḥarrālī's use of Jesus serves to convey a feature of his hermeneutics, emphasizing an anagogical connection between his interpretation of the Qurʾān, drawing on Q 2:269 and the Gospel text.

Similarly, when al-Biqāʿī begins to interpret the Qurʾān and Gospels together, he appeals to the figurative interpretation of Jesus's words in order to return what is self-similar (*mutashābih*) back to what it is clear (*muḥkam*), an integration of *ʿilm munāsabāt al-Qurʾān* (science of connections of the Qurʾān) with biblical exegesis. When considering the concept of "fatherhood" and its meaning in the Gospels, al-Biqāʿī dedicates more time to explaining this concept than the term "sonship" at Q 3:7. He advocates for the use of the term "Father" in reference to God in various places, starting with Luke 11:1–4, which features the Lord's Prayer: "Oh Lord, teach us to pray as John taught his disciples. So he said to them, 'If you pray, then say: Our Father who is in heaven (. . .).'"

Al-Biqāʿī harks back to this text to emphasize that Jesus instructed his followers to address God as "Father." Consequently, this description does not necessarily indicate a unique relationship between Jesus and God.[39] He then elaborates on the dynamic between Jesus as a son and God as his father:

> If (the Christians) refer the phrases "father and son" back to this *muḥkam* (clear) text and the likes of it, and there are many in all of the Gospels, they would know without any doubt that its meaning is that God dealt with (Jesus) as a father deals with his son; namely, nurturing, protecting, aiding, venerating and honoring, just like they are forced to interpret his saying in what I mentioned earlier, "Our Father Who is in Heaven."[40]

---

38 Faris Casewit, "The Sayings of Jesus (ʿIsá) in Ḥarrālī's (d. 1241) Exegetical Treatise *Saʿd al-wāʿī wa-uns al-qārī*," (paper presented at the International Qurʾānic Studies Association, Boston, 2017, Microsoft Word file, accessed by permission July 27, 2020), 6.; this part of the treatise is also found in Paris, BnF MS Ar. 1398, ff. 188r–v, henceforth referred to as *Saʿd al-wāʿī*.

39 Al-Biqāʿī, *Naẓm ad-durar*, 4: 230f.

40 Ibid., 4: 231.

Al-Biqāʿī provides various passages to support this interpretation of *ubuwwa* in the Gospels, stating: "Let your light shine forth in front of people that they may see your good deeds and they may glorify your Father who is in the heavens" (Matthew 5:16); "Be good to those who hate you and pray for those who expel you and let you down in order that you may be sons to your Father who is in heaven, because he is the one Who shines the sun on the good and the evil, and He gives rain to the saints and the oppressors" (Matthew 5:44–45); "Pray to your Father secretly, and your Father sees the secret and He will give to you openly" (Matthew 6:6). Al-Biqāʿī conveys the idea that "they should venerate (God) more than they venerate their (earthly) fathers in order that He may care for them more than fathers care for their children."[41]

4 Concluding Remarks

Through *ʿilm munāsabāt al-Qurʾān* and the *muḥkam/mutashābih* paradigm, al-Biqāʿī developed a method of scriptural interpretation that extended beyond the text of the Qurʾān to encompass the previous scriptures of the *ahl al-kitāb* (People of Scripture). By incorporating the "Torah and Gospel" into his hermeneutical framework as legitimate forms of revelation, al-Biqāʿī explicitly acknowledges within the genre of tafsīr that "what has come in (the Qurʾān), has not been changed in what the Christians have with them (i.e., the Gospels)"; that is to say, the *kitāb* the Christians possess "right now at the end of this ninth century" contains "from the *muḥkamāt* what suffices in returning the self-similar (*mutashābih*) (verses) to it."[42] In essence, he utilizes the *muḥkam* and *mutashābih* formula as hermeneutical categories, applying them to the Gospels, to interpret the *mutashābihāt* in the light of the *muḥkam* verses. For al-Biqāʿī and, to some extent, al-Ḥarrālī, God has done in the Qurʾān what He has done with the *Tawrāt* and the *Injīl* for those communities of a shared revelatory experience, which is separating the verses into categories of *muḥkam* and *mutashābih* in order to test the people of faith so that those who are well-grounded in knowledge and have certitude will know that their *kitāb* is from Him.[43]

41 A similar use of *muḥkam* and *mutashābih* has been observed in Saadiah Gaon's *Tafsīr* on the Hebrew Bible (David M. Freidenreich, "The Use of Islamic Sources in Saadia Gaon's Tafsīr of the Torah," *JQR* 93, no. 3–4 (2003): 353–95, esp. 360).

42 Al-Biqāʿī, *Naẓm ad-durar*, 4:227–28.

43 Ibid., 4:226.

5 Bibliography

5.1 Sources

Al-Biqāʿī, Abū l-Ḥasan. *Iẓhār al-ʿaṣr li-asrār ahl al-ʿaṣr: Taʾrīkh al-Biqāʿī*. Ed. Muḥammad Sālim b. Shadīd al-ʿAwfī. Riyadh: n.p., 1992.
Al-Biqāʿī, Abū l-Ḥasan. *Naẓm ad-durar fī tanāsub al-āyāt wa-s-suwar*. 22 Vols. Hyderabad: Dār al-Maʿārif al-ʿUthmāniyya, 1969–84.
Al-Biqāʿī, Abū l-Ḥasan. *Naẓm ad-durar fī tanāsub al-āyāt wa-s-suwar*. Ed. ʿAbd al-Razzāq Ghālib al-Mahdī. 8 Vols. Beirut: Dār al-Kutub al-ʿIlmiyya, 2011.
Ibn Sulaymān, Muqātil. *Tafsīr al-khams miʾat āya min al-Qurʾān*. Ed. I. Goldfeld. Shfaram: Dār al-Mashriq, 1980.
Ibn Sulaymān, Muqātil. *Tafsīr Muqātil b. Sulaymān*. Edited by ʿAbd Allah Maḥmūd Shiḥāta. Beirut: al-Hayʾa al-Miṣriyya al-ʿĀmma li-l-Kitāb, 1979.
Khayyāṭī, Maḥmādī b. ʿAbd as-Salām. *Turāth Abī l-Ḥasan al-Marrākushī fī t-tafsīr*. Casablanca: Matbaʿat an-Najāḥ al-Jadīda, 1997.
As-Sakhāwī, Muḥammad b. ʿAbd al-Raḥmān. *Aḍ-Ḍawʾ al-lāmīʿ li-ahl al-qarn at-tāsiʿ*. 12 Vols. Cairo: Maktabat al-Qudsī, 1934–36.
At-Ṭabarī, Abū Jaʿfar b. *Jarīr*. *Jāmiʿ al-bayān ʿan taʾwīl āy al-Qurʾān*. Ed. ʿAbd Allāh b. ʿAbd al-Muḥsin Turkī. Riyadh: Dār ʿĀlam al-Kutub, 2003.

5.2 Secondary Literature

Abdul-Raof, Hussein. *Arabic Rhetoric: A Pragmatic Analysis*. London: Routledge, 2006.
Abdul-Raof, Hussein. *Theological Approaches to Qurʾanic Exegesis: A practical comparative-contrastive analysis*. London: Routledge, 2012.
Abu Zayd, Nasr. *The Rational Trend in Exegesis: A Study of the Concept of Metaphor in the Muʿtazilites' Thought*. Beirut: al-Markaz ath-Thaqāfī l-ʿArabī, 1982.
Abu Zayd, Nasr. "Towards understanding the Qurʾān's worldview." In *New Perspectives on the Qurʾān: The Qurʾān in Its Historical Context 2*, edited by Gabriel Said Reynolds. London: Routledge, 2011.
Al-Asadābādī, ʿAbd al-Jabbār. *Mutashābih al-Qurʾān*. Ed. ʿAdnān Muḥammad Marzūq. Cairo: Dār at-Turāth, 1966.
Berkey, Jonathan. *Popular Preaching and Religious Authority in the Medieval Islamic Near East*. Seattle: University of Washington Press, 2001.
Casewit, Faris. "The Sayings of Jesus (ʿIsá) in Ḥarrālī's (d. 1241) Exegetical Treatise *Saʿd al-wāʿī wa-uns al-qārī*." Paper presented at the International Qurʾānic Studies Association, Boston, 2017. Microsoft Word file, accessed by permission July 27, 2020.
Cuypers, Michel. *The Composition of the Qurʾan: Rhetorical Analysis*, trans. Jerry Ryan. London: Bloomsbury Academic, 2015.
Freidenreich, David M. "The Use of Islamic Sources in Saadia Gaon's Tafsīr of the Torah." In *The Jewish Quarterly Review* 93, no. 3–4 (2003): 353–95.
Garcin, Jean-Claude. "The Regime of the Circassian Mamlūks." In *The Cambridge History of Egypt, vol. 1: Islamic Egypt, 640*–1517, edited by Carl F. Petry. Cambridge: Cambridge University Press, 1998: 291f.
Guo, Li. "al-Biqāʿī's Chronicle: A Fifteenth-Century Learned Man's Reflections on His Time and World." In *The Historiography of Islamic Egypt (c. 950/1800)*, edited by Hugh Kennedy. Leiden: Brill, 2001: 121–48

Lagarde, Michel. “De l’ambiguïté (mutašābih) dans le Coran: tentatives d’explication des exégètes musulmans.” *Quaderni di studi arabi* 3 (1985): 56–62.

Lane, Edward W. *An Arabic-English Lexicon*. 8 vols. Edinburgh: Williams and Norgate, 1863.

McAuliffe, Jane Dammen. “The Qur’ānic Context of Muslim Biblical Scholarship.” *Islam and Christian-Muslim Relations* 7, no. 2 (1996): 141–58.

McAuliffe, Jane Dammen. “Text and Textuality: Q 3:7 as a Point of Intersection.” In *Literary Structures of Religious Meaning in the Quran*, edited by Issa J. Boullata. London: Routledge, 2000: 141–58.

Meynet, Ronald. *Treatise on Biblical Rhetoric*, trans. Leo Arnold, Rubianto Solichin, and Llane B. Briese. Leiden: Brill, 2012.

Northrup, Linda S. “The Baḥrī Mamlūk sultanate, 1250–1390.” In *The Cambridge History of Egypt, vol. 1: Islamic Egypt, 640*–1517, edited by Carl F. Petry. Cambridge: Cambridge University Press, 1998: 287–89

Schaade, A., and G.E. von Grunebaum. “Bālagha.” In *EI*^{2}, edited by P. Bearman et al. Brill, 1954–2005. Article published online 2012. Accessed on November 20, 2023. http://dx.doi.org/10.1163/1573-3912_islam_SIM_1123.

Saleh, Walid A. *In Defense of the Bible: A Critical Edition and an Introduction to al-Biqāʿī’s Bible Treatise*. Leiden: Brill, 2008.

Van Steenbergen, Jo. *A History of the Islamic World, 600–1800: Empire, Dynastic Formations, and Heterogeneities in Pre-Modern Islamic West-Asia*. London: Routledge, 2020.

Jalāl ad-Dīn as-Suyūṭī

Nimet Seker

1 Introduction

Even during his lifetime in the Mamluk era, Jalāl ad-Dīn as-Suyūṭī (d. 1505/911) gained fame as a scholar and author of numerous significant writings. His full name was Abū l-Faḍl ʿAbd ar-Raḥmān b. Abī Bakr b. Muḥammad Jalāl ad-Dīn al-Khuḍayrī s-Suyūṭī. With an extraordinarily comprehensive body of work covering all branches of Islamic sciences, and particularly valued for its specific systematic approach, richness in detail, and tradition-oriented character, as-Suyūṭī is considered today as one of the most prolific polymaths among Sunni scholars. Touching upon nearly every topic in premodern Islamic scholarship, as-Suyūṭī was especially productive in the disciplines of exegesis, hadith sciences, jurisprudence, as well as Arabic language and literature. His most renowned and arguably most crucial work, the Qurʾānic encyclopedia titled *al-Itqān fī ʿulūm al-Qurʾān*, continues to be regarded as the most comprehensive and influential source for the field of Qurʾānic sciences (*ʿulūm al-Qurʾān*). Despite the enduring significance attributed to as-Suyūṭī's work, it remains relatively underexplored in relation to its importance.

2 Biography

The name as-Suyūṭī is derived from the town of Asyut (Asyūṭ) in Upper Egypt, which is also the namesake of the present-day Asyut Governorate. As-Suyūṭī's ancestors, before settling in Asyut, had lived in the Baghdad neighborhood of al-Khuḍayriyya. As-Suyūṭī himself attributes the nisba al-Khuḍayrī in his name to this Baghdad quarter. As-Suyūṭī hailed from a family of Sufis and scholars; the tomb of his Sufi grandfather in Asyut remains a Sufi shrine to this day.[1] His father, Kamāl ad-Dīn, initially served as a deputy judge (*qāḍī*) in Asyut before moving to Cairo for further studies, where he counted Ibn Ḥajar al-ʿAsqalānī (d. 1449/852) among his teachers. In Cairo, Kamāl ad-Dīn worked as a preacher, teacher of Shāfiʿī law, and deputy judge.[2]

As-Suyūṭī was born on October 3, 1445/Rajab 1, 849 in Cairo. In his early years, he accompanied his father to the Shaykhūniyya across from the Shaykū Mosque, where

1 Halit Özkan, "Süyûtî," in *TDVIA*, ed. İslâm Araştırmaları Merkezi (Ankara: TDV, 2010), 38: 188.
2 Elizabeth Sartain, *Jalāl al-dīn al-Suyūṭī: Biography and background* (Cambridge: Cambridge University Press, 1975), 1: 21f. This book includes a translation and analysis of as-Suyūṭī's autobiography *at-Taḥadduth bi-niʿmat Allāh.*

https://doi.org/10.1515/9783111320083-014

his father also taught.[3] His studies included, in addition to a foundational education in the Qur'ān, Arabic language and grammar, a particular focus on the study of Shāfiʿī legal works. His other areas of study encompassed *ʿilm al-adab*, rhetoric, and Hadith, with a specific emphasis on hadith evaluation (*dirāya*) as opposed to the compilation and transmission of hadiths (*riwāya*).[4]

He received his *ijāzāt* for the works of Ḥasan b. Jaʿfar al-Jamāʿīlī, Yaḥyā b. Sharaf an-Nawawī, and Nāṣir ad-Dīn al-Bayḍāwī from his teachers ʿAlam ad-Dīn al-Bulqīnī (d. 1463/868), Sharaf ad-Dīn al-Munāwāzī (d. 1467/871), ʿIzz ad-Dīn al-Kinānī l-Ḥanbalī (d. 1471/876), and Amīn ad-Dīn al-Aqṣarāʾī (d. 1476/880).[5] In the year 1462/866, at the age of 16, he obtained permission from his teacher al-Bulqīnī to teach Shāfiʿī law and provide legal opinions according to this legal school.[6] His other teachers included Sayf ad-Dīn Muḥammad b. Quṭlūbughā al-Ḥanafī (d. 1477/881), Taqī d-Dīn Aḥmad ash-Shummūnī (d. 1468/872), and Muḥyī d-Dīn al-Kāfiyajī (d. 1474/879), with whom he studied intensively for many years. From the latter, as-Suyūṭī received a comprehensive *ijāzā* at the age of nineteen, allowing him to teach all the works he had studied with al-Kāfiyajī, including his own writings.[7]

As-Suyūṭī mentions in his autobiographical work *at-Taḥadduth bi-niʿmat Allāh* that Ibn Ḥajar al-ʿAsqalānī was also among his teachers. Given that as-Suyūṭī was born in the year 1445/849 and Ibn Ḥajar died in 1449/852, as-Suyūṭī must have encountered him at the tender age of three years and five months.[8] Elizabeth Sartain notes that most of as-Suyūṭī's teachers taught at the Shaykhūniyya and therefore speculates that as-Suyūṭī benefited from his father's connections as a teacher.

After his studies, he held teaching positions at the Shaykhūniyya and the Ibn Ṭūlūn Mosque. For approximately twelve years, he was the responsible shaykh at the *khānqāh* al-Baybarsiyya.[9]

Through the broad reception and circulation of his writings, al-Suyūṭī gained recognition both within and outside Egypt from a young age. His fame and his contentious personality led him to engage in scholarly debates that sometimes spanned over several years. These debates often revolved around questions of legal interpretation but could also involve dogmatic themes or issues regarding the legitimacy of certain religious positions or opinions.[10] These disputes were carried out in writing through the exchange of legal opinions (*fatāwā*). He authored individual treatises on many is-

3 Ibid., 26.

4 Ibid., 30f.

5 Ibid., 27.

6 Ibid., 28.

7 Ibid., 29.

8 Ibid., 26.

9 Éric Geoffroy, "al-Suyūṭī," in *EI*[2], ed. P. Bearman et al., accessed December 7, 2023, <http://dx.doi.org/10.1163/1573-3912_islam_COM_1130>.

10 For further insights into the various issues as-Suyūṭī grappled with throughout his life, see Takao Ito, "Al-Suyūṭī and Problems of the Waqf," in *Al-Suyūṭī, a Polymath of the Mamlūk Period*, ed. A. Ghersetti (Leiden: Brill, 2017), 49–53; Sartain, *Jalāl al-Dīn al-Suyūṭī*, 53–60.

sues, serving the purpose of argumentative rejection or deconstruction of opposing views. These writings not only provide insights into the positions as-Suyūṭī took in the scholarly debates of his time but also offer a glimpse into his personality and attitude displayed in his interactions with colleagues. In summary, we encounter a scholar of deep religiosity, great confidence in his judgment, and, above all, significant self-assurance regarding his standing in Islamic scholarship, for which he faced accusations of arrogance and haughtiness from his opponents. As-Suyūṭī is also known for his love of debate and his sharp-tongued expression in these written disputes.

As-Suyūṭī discussed various issues, such as whether logic is legally forbidden (*ḥarām*) (which, according to him, it was).[11] He also participated in the significant debate on whether the Sufi poet ʿUmar b. al-Fāriḍ[12] (d. 1235/632) engaged in *ḥulūl* and *ittiḥād*, i.e., the idea of divine incarnation in the created or the unity of God with humans or His creation. Another debate concerned the poems of the mystic Muḥyī d-Dīn b. al-ʿArabī (d. 1240/638). As-Suyūṭī defended both against charges of heresy.[13]

From the sources available to us, it is evident that these scholarly disputes not only dealt with professional matters but sometimes escalated into personal conflicts, leading to mutual insults.[14] As-Suyūṭī often accused his antagonists of ignorance. This must be understood against the backdrop of as-Suyūṭī declaring himself a *mujtahid* and *mujaddid* of his age,[15] an assessment vehemently rejected by some of his contemporaries.[16] As-Suyūṭī, in turn, engaged in personal confrontations with his critics, such as the Hanafi jurist Abū l-Wafāʾ Burhān ad-Dīn Ibrāhīm b. ʿAbd ar-Raḥmān b. Muḥammad b. al-Karakī[17] (d. 1516/922), a student of Ibn Ḥajar and al-Bulqīnī, and Shams ad-Dīn as-Sakhāwī[18] (d. 1497/902), also a student of Ibn Ḥajar and al-Kāfiyajī, through treatises targeting each scholar and their arguments.[19] As evident from the biographies of many scholars with whom as-Suyūṭī disagreed, they were sometimes contemporaries, former classmates, and scholars from as-Suyūṭī's immediate circle in Cairo. For example, as-Suyūṭī responded to Ibn al-Karakī's criticism of his self-proclaimed claim to be a *mujaddid* with the specially written work *Ṭarz al-imāma fī l-farq bayn al-maqāma wa-l-qumāma*.[20]

---

11 Sartain, *Jalāl al-dīn al-Suyūṭī*, 53.

12 Th. Emil Homerin, "Ibn al-Fāriḍ," in *EI*[3], ed. Kate Fleet et al., accessed November 22, 2023, <http://dx.doi.org/10.1163/1573-3912_ei3_COM_30774>.

13 Sartain, *Jalāl al-dīn al-Suyūṭī*, 54.

14 Ibid., 53.

15 See Özkan, "Süyûtî," 189; Sartain, *Jalāl al-dīn al-Suyūṭī*, 62–65, 69–71. As-Suyūṭī extensively justifies in his autobiography, *at-Taḥadduth bi-niʿmat Allāh*, why he fulfilled all the conditions for the status of a *mujtahid*.

16 Sartain, *Jalāl al-dīn al-Suyūṭī*, 59, 71.

17 Kemal Yıldız, "İbnü'l-Kerekî," in *TDVIA*, ed. İslâm Araştırmaları Merkezi (Ankara: TDV, 2000), 21: 111f.

18 Cengiz Tomar, "Sehâvî," in *TDVIA*, ed. İslâm Araştırmaları Merkezi (Ankara: TDV, 2009), 36: 313–16.

19 Sartain, *Jalāl al-dīn al-Suyūṭī*, 60.

20 Ibid., 77–79.

In the last years before his death, as-Suyūṭī lived in seclusion on the island of Rawḍa, dedicating himself exclusively to the editing of his writings. He passed away there in the year 1505/911.

3 As-Suyūṭī's Works on Qur'ānic Exegesis and Qur'ānic Sciences

In total, there are thirty-six individual writings by as-Suyūṭī, including the *Itqān*, that address questions of Qur'ānic exegesis and sciences.[21] Many of these individual books served as the basis for specific chapters in the *Itqān*. The following are explanations of some writings for which at least one edition is known.

3.1 Exegesis

3.1.1 *Ad-Durr al-manthūr fī t-tafsīr bi-l-ma'thūr*

Among tradition-based Qur'ānic commentaries, this work is one of the most extensive. In this commentary, the traditions are not listed with their chains of transmission (*asānīd*) but rather with references to the collections from which they are drawn.[22]

The structure of *ad-Durr* is based on a classical verse-by-verse commentary, commencing with the first verse of Sura 1, *al-Fātiḥa* and concluding with the last verse of Sura 114, *an-Nās*. In addition, as-Suyūṭī provides exegetical traditions in a separate chapter on the two Suras, *al-Khal'* and *al-Ḥafd* (*dhikr mā warada fī sūrat al-khal' wa-sūrat al-ḥafd*),[23] commonly known as the *qunūt* Suras, the texts of a type of supplication prayer, which are said to have been transmitted in the Codex of Ubayy b. Ka'b.[24] By doing so, as-Suyūṭī also incorporates texts from the Qur'ān that did not find their way into the canonical *muṣḥaf*, which is highly unusual for a Qur'ānic commentary. A

21 M. Suat Mertoğlu, "Süyûtî/Kur'an İlimleri," in *TDVIA*, ed. İslâm Araştırmaları Merkezi (Ankara: TDV, 2010), 38: 198–201, lists thirty-six writings, whereas Geoffry acknowledges only twenty works related to "ḳur'ānic sciences," see Geoffroy, "al-Suyūṭī."

22 Editions: Ed. Tehran: al-Maṭba'a al-Islāmiyya, 1957; Ed. Beirut: Dār al-Fikr, 1983; Ed. Shaykh Najdat Najī (Beirut: Dār Iḥyā' at-Turāth al-'Arabī, 2001); Ed. 'Abd al-Muḥsin at-Turkī (Cairo: Markaz Hijr li-l-Buḥūth wa-d-Dirāsāt al-'Arabiyya wa-l-Islāmiyya, 2003).

23 Jalāl ad-Dīn as-Suyūṭī, *ad-Durr al-manthūr fī tafsīr bi-l-ma'thūr*, ed. 'Abd Allāh b. 'Abd al-Muḥsin at-Turkī (Cairo: Markaz Hijr li-l-Buḥūth wa-d-Dirāsāt al-'Arabiyya wa-l-Islāmiyya, 2003), 15: 810–16.

24 Sean Anthony, "Two 'Lost' Suras of the Qur'ān: Surat al-Khal' and Surat al-Ḥafd between Textual and Ritual Canon (1st–3rd/7th–9th centuries)," *JSAI* 46 (2019): 67–112.

final supplement to the commentary is the chapter on the supplication at the conclusion of the Qur'ān (*dhikr du'ā' khatm al-qur'ān*).[25] This, too, is an appendix typically absent from verse-by-verse commentaries. With this, as-Suyūṭī expands the boundaries of what constitutes the Qur'ānic text, including additional texts, specifically prayer texts, into his *Tafsīr*.

In the main part of *ad-Durr*, as-Suyūṭī confines himself to a meticulous selection of traditions[26] for each chapter and verse. Some estimates suggest that as-Suyūṭī cites a total of approximately 37,691 traditions in this Qur'ānic commentary.[27]

In each chapter, the author begins with introductory traditions about the respective chapter as a whole. This includes information about the place of revelation – or the phase of revelation (Meccan or Medinan)[28] – and the occasions of revelation (*asbāb an-nuzūl*). It may also encompass traditions about the occasion of recitation of the chapter or the specific verse. Following these are the actual traditions explaining the verses. However, what is highly noticeable is as-Suyūṭī's abstention from any personal commentary. In a few instances, he limits himself to evaluating the reliability of the cited tradition, expressed in brief remarks about sound (*ṣaḥīḥ*) or weak (*ḍa'īf*) isnāds. Similarly, he does not analyze the presented traditions, as, for example, aṭ-Ṭabarī does in his *Jāmi' al-bayān*. The personal opinions of the scholar as-Suyūṭī thus completely disappear behind the extensive list of works and names of scholars from the early generations. His own opinion and views are not found at the detailed level of his commentary but only in its overarching methodological orientation, clearly attributable to his preference for working based on transmitted knowledge. This preference is ultimately reflected, although not in identical form, in his magnum opus *al-Itqān*.

Another noticeable difference between *ad-Durr al-manthūr* and any other transmission-based Qur'ān commentary is the radical omission of all *isnāds*. Instead, as-Suyūṭī provides for each tradition an indication of the written source in which the respective report is listed. This represents a radical shift in the prevailing *isnād* paradigm in the science of hadith. No longer is the *isnād* the guarantor of reliable (oral) transmission; rather, it is the written source that has already verified the *isnād* and included it in the respective collection. As-Suyūṭī justifies the omission of *isnāds* in *ad-Durr* by citing the readership's need for a shorter exegetical work.

25 As-Suyūṭī, *ad-Durr al-manthūr*, ed. at-Turkī, 15: 817f.

26 This becomes evident when comparing traditions related to specific verses across various works by as-Suyūṭī. It is apparent that he does not list all traditions in all works but rather adopts a selective approach.

27 Shabir Ally, "The Culmination of Tradition-based *Tafsīr*: The Qur'ān Exegesis *al-Durr al-manthūr* of as-Suyūṭī (d. 911/1505)" (PhD diss., University of Toronto, 2012), 56.

28 In doing so, as-Suyūṭī cites multiple traditions, which may also contain contradictory information. However, he does not take a position on the question of which revelational context (Meccan or Medinan) is more likely.

In the remarkably brief introduction (*muqaddima*) to *ad-Durr*, the author explains the genesis of the commentary and thereby provides insight into the method he chose: Initially, he wrote a comprehensive Qur'ān commentary titled *Tarjumān al-Qur'ān*. This commentary included traditions from the Prophet and his companions and spanned multiple volumes (*mujalladāt*). In this commentary, he meticulously listed all the *isnāds*, from the origin of each tradition, that is, from the Prophet or one of his companions, to the books from which he extracted the reports. However, his contemporaries showed little interest in the full detail of the *isnāds*, prompting him to create a work that was shortened by excluding the *isnāds*.[29]

Unlike many other commentaries that cover the entire Qur'ān, as-Suyūṭī does not preface this work with an extensive prologue. Typically, the *muqaddima* is the place where programmatic, hermeneutical, and methodical considerations about the respective work are articulated. Detailed declarations of intent from the otherwise loquacious as-Suyūṭī, as seen in the elaborate introduction of the *Itqān*, are absent in the introduction to *ad-Durr al-manthūr*. In a sense, the introduction is as uncommented as the entire work itself.

Overall, *ad-Durr* thus resembles more of a collection of hadiths than a Qur'ānic commentary.[30] It takes on the character of a Qur'ān-related compilation of traditions from a broad spectrum of hadith-based sources, all while following the arrangement of the canonical Qur'ānic text.

With this approach, as-Suyūṭī theoretically follows the ideal "radical hermeneutics"[31] of Ibn Taymiyya.[32] According to this perspective, the Qur'ān is first to be interpreted using itself (*tafsīr al-Qur'ān bi-l-Qur'ān*), followed by the incorporation of the exegetical knowledge of the Prophet, his companions (*ṣaḥāba*), and the subsequent generation (*tābi'ūn*). Ibn Taymiyya's ideal hermeneutics assume absolute consistency in the exegesis of the Prophet and the first generations. He argues that the exegetical knowledge of the companions and successors is based on the knowledge transmitted to them by the Prophet. Ibn Taymiyya equates this knowledge with divine knowledge, asserting that all prophetic knowledge is inspired by God. This idea elevates the tradition-based exegesis to a throne of impeccability, simultaneously providing an additional foundation for the *isnād* paradigm.[33]

---

29 As-Suyūṭī, *ad-Durr al-manthūr*, ed. at-Turkī, 1: 3f.

30 Ally, "The Culmination of Tradion-based *Tafsīr*," 50.

31 Walid A. Saleh, "Ibn Taymiyya and the Rise of Radical Hermeneutics. An Analysis of 'An Introduction to the Foundations of Qur'ānic Exegesis'," in *Ibn Taymiyya and His Times*, ed. Yossef Rapoport and Shahab Ahmed (Oxford/New York: Oxford University Press, 2010), 123–62.

32 Abū l-'Abbās Tāqī d-Dīn b. Taymiyya, *Muqaddima fī uṣūl at-tafsīr*, ed. 'Adnān Zarzūr (Beirut: Dār al-Qur'ān al-Karīm, 1972).

33 Nimet Seker, *Der Koran als Rede und Text. Hermeneutik sunnitischer Koranexegeten zwischen Textkohärenz und Offenbarungskontext* (Berlin: EB Dr. Brandt, 2019), 160–70, 184–88.

In the debate over the legitimacy of independent reasoning (*ijtihād*) and the subjective opinion of a scholar (*raʾy*) in exegesis, Ibn Taymiyya takes an extreme position. In a legitimate exegesis, according to Ibn Taymiyya, *raʾy* has no place, and considering *raʾy* leads to incorrect, erroneous, and even sectarian interpretations. The ambiguity and diversity of opinions in the classical Tafsīr tradition, particularly, are problematic for Ibn Taymiyya. He suggests that this issue can be avoided by restricting exegesis to transmitted knowledge, and *raʾy* should not deviate from the provided framework of meaning in the traditions. Alongside Ibn Kathīr and Ibn Abī Ḥātim ar-Rāzī (d. 938/327), it was primarily as-Suyūṭī who contributed to this form of commentary.

However, behind the formal, transmission-based structure of *ad-Durr* lies a significant refutation of Ibn Taymiyya's broader assumptions by as-Suyūṭī. As-Suyūṭī demonstrates that the presumed unanimity and lack of contradiction in the exegesis of the first generations are not accurate. He achieves this by highlighting the diverse nature of the Qurʾānic text due to the circulation of phonetic variations in the modes of reading (*qirāʾāt*) among the companions.[34] According to as-Suyūṭī, this varied interpretive foundation resulted in the emergence and circulation of divergent interpretations among the Prophet's companions:

> The Companions were not all elucidating the same text. The Qur'ān was available to them in various readings. Hence their exegeses were varied, one from another, at their very cores. Not only is the exegesis of the Qur'ān polyvalent; the text of the Qur'ān itself is polyvalent.[35]

3.1.2 *Qaṭf al-azhār fī kashf al-asrār* (*Asrār at-tanzīl*)

This work delves into the rhetorical devices of the Qurʾān, the harmonious connections between verses and suras (*munāsabāt*), as well as the richness of meaning derived from differences in recitation. These themes are intricately linked to the inimitability of the Qurʾān (*iʿjāz al-Qurʾān*). The commentary concludes with Q 9:92.[36]

3.1.3 *Al-Iklīl fī istinbāṭ at-tanzīl*

This work is a commentary on the derived legal rulings (*aḥkām*) from Qurʾānic verses, incorporating theological questions from *kalām* as well.[37]

---

34 Ally, *The Culmination of Tradion-based* Tafsīr, 262–311.

35 Ibid., 316f.

36 Ed. Aḥmad al-Ḥamādī (Doha: Wizārat al-Awqāf wa-sh-Shuʾūn al-Islāmiyya, 1994).

37 Ed. Sayf ad-Dīn ʿAbd al-Qādir al-Kātib (Beirut: Dār al-Kutub al-ʿIlmiyya, 1985).

3.1.4 *Takmilat tafsīr al-Jalāl al-Maḥallī* (*Tafsīr al-Jalālayn*)

This work represents as-Suyūṭī's completion of the Qur'ānic commentary by his teacher Jalāl ad-Dīn al-Maḥallī, from which the name *Tafsīr al-Jalālayn* is derived. Due to its popularity, it continues to circulate widely. The first prints of this commentary appeared in 1796/1211 in New Delhi. Subsequently, it has been printed on the margins of the Qur'ānic text or alongside other Qur'ānic commentaries, often as part of supercommentaries (*ḥawāshī*).[38]

3.1.5 *Nawāhid al-abkār wa-shawārid al-afkār*

This work serves as a supercommentary on the Qur'ānic commentary *Anwār at-tanzīl wa-asrār at-ta'wīl* by al-Bayḍāwī, extending up to Sura 11, *Hūd*.[39]

3.2 Qur'ānic Sciences (*'ulūm al-Qur'ān*)

3.2.1 *Kitāb at-taḥbīr fī 'ilm at-tafsīr*

See below 4.3.

3.2.2 *Ṭabaqāt al-mufassirīn*

With this collection of biographies, as-Suyūṭī presented the first work of its kind, featuring the biographies of 136 Qur'ān commentators.[40]

---

38 Ali Akpınar, "Tefsîrü'l-Celâleyn," in *TDVIA*, ed. İslâm Araştırmaları Merkezi (Ankara: TDV, 2011), 40: 294f. Modern editions (Selection): Ed. Beirut: Dār Ibn Kathīr, 2008; Ed. Cairo: Dār al-Ḥadīth, 2001; Ed. Istanbul: Dār al-Kutub al-'Arabiyya, 2019.

39 Editions of the *tafsīr* on Sura 2, *al-Baqara* (The Cow) to Sura 9, *at-Tawba* (Repentance): Ed. Aḥmad Ḥāj Muḥammad 'Uthmān, *(Q 2:1–2:20)*; Ed. Muḥammad Kamāl 'Alī, *(Q 2:21–3:112)*; Ed. I'dād aṭ-Ṭālib, *(Q 3:113–9:48)* (Mekka: Jāmi'at Umm al-Qurā, 1424/25).

40 Editions: Ed. Albert Meursinge and Hendrik Weyers Engelinus (Leiden: Leiden University, 1839); Ed. 'Alī Muḥammad 'Umar (Cairo: Maktaba Wahba, 1976).

3.2.3 *Mufḥimāt al-aqrān fī mubhamāt al-Qur'ān*

This is a commentary that explains ambiguous references in the Qur'ān (*mubhamāt*) concerning addressed individuals, times, and places.[41]

3.2.4 *Marāṣid al-maṭāli' fī tanāsub al-maqāṭi' wa-l-maṭāli'*

In this book, as-Suyūṭī discusses the connections between the beginnings and endings of chapters (*fawātiḥ as-suwar, khawātim as-suwar*). In the introduction to this brief treatise, the author also refers to the *Itqān*.[42]

3.3 Individual Disciplines of Qur'ānic Sciences

3.3.1 *Tanāsuq ad-durar fī tanāsub as-suwar*

This work delves into the relationships between the suras while referencing their arrangement in the *muṣḥaf*.[43] It is largely based on the corresponding chapter of *Qaṭf al-azhār*.

3.3.2 *Lubāb an-nuqūl fī asbāb an-nuzūl*

The book stands among the most comprehensive sources on the occasions of revelation, presenting them in relation to 950 Qur'ānic verses. It also incorporates traditions that, according to the methodology of hadith evaluation, are deemed less reliable.[44]

3.3.3 Additional works

Additional works by as-Suyūṭī that have influenced his Qur'ānic encyclopedia include:
- *Mu'tarak al-aqrān fī i'jāz al-Qur'ān* (A treatise on the miraculous nature of the Qur'ān and the resulting polysemy)
- *al-Muhadhdhab fī mā waqa'a fī l-Qur'ān min mu'arrab* (An index of non-Arabic loanwords in the Qur'ān)
- *al-Mutawakkilī* (A summary of *al-Muhadhdhab*)

41 Ed. Ṭāhā 'Abd ar-Ra'ūf Sa'd (Cairo: al-Maktabat al-'Arabiyya li-t-Turāth, 1992).
42 Ed. 'Abd al-Muḥsin b. 'Abd al-'Azīz al-'Askar (Riyadh: Maktabat Dār al-Minhāj, 2005).
43 Ed. 'Abd al-Qādir Aḥmad 'Aṭā (Beirut: Dār al-Kutub al-'Ilmiyya, 1985).
44 Ed. 'Abd ar-Razzāq al-Mahdī (Beirut: Dār al-Kitāb al-'Arabī, 2005).

4 Textual Genesis of the *Itqān*

The most extensive Qur'ānic encyclopedia, *al-Itqān*, which continues to set the standard in this genre, is based on more than 352 different written sources from various disciplines that as-Suyūṭī systematically incorporated into his work.[45] Another count indicates that as-Suyūṭī cites from 450 different titles authored by 412 scholars.[46] Of these sources, 205 are already found in az-Zarkashī's *al-Burhān*.[47] Additionally, as-Suyūṭī integrates other sources whose titles and authors he does not mention, using specific expressions such as *qīla* ("it was said"), *qāla baʿḍuhum* ("some of them said"), and *qāla ghayruhu* ("someone else said").[48]

Consistent with his compilatory approach and preference for working with hadiths and the science of transmission, the most frequently cited authors are those who have presented relevant works in the field. Thus, as-Suyūṭī most frequently cites Abū Bakr b. al-Anbārī (d. 940/328), Ibn Abī Ḥātim ar-Razī (d. 938/327), Ibn al-Jazarī (d. 1429/833), Abū Jaʿfar aṭ-Ṭabarī (d. 923/310), Abū l-Qāsim az-Zarkashī, Abū Bakr al-Bayhaqī (d. 1066/458), and Muḥammad b. Ismāʿīl al-Bukhārī (d. 870/256) – in the listed order. Among the most frequently cited scholars whose exegetical method is less influenced by tradition and more by Arabic philology and linguistics are Maḥmūd b. ʿUmar az-Zamakhsharī (d. 1144/538), ar-Rāghib al-Iṣfahānī (d. first quarter of the eleventh century), and Fakhr ad-Dīn ar-Rāzī (d. 1210/606).[49] Among all the sources consulted by as-Suyūṭī, there might be some that are no longer available to us today.

4.1 Al-Kāfiyajī's *Kitāb at-taysīr fī qawāʿid ʿilm at-tafsīr*

Muḥyī d-Dīn al-Kāfiyajī (modern Turkish "el-Kâfiyecî"),[50] born in 1386/788 in Anatolian Bergama (Pergamon), studied on extensive journeys in Anatolia and Persia, Damascus, Mecca, and Medina. He spent several years in Jerusalem and eventually settled in Cairo, where he spent most of his life. As an Ottoman scholar, he taught Hanafi law, Arabic language and grammar, and the entire classical theological curriculum in Cairo. His specialties also included philosophy, logic, and astronomy. As-Suyūṭī spent intense hours studying with him and seemed to respect him greatly. Al-Kāfiyajī is

45 Olaf Tyllack, "Die Quellen des Itqān von Suyūṭī" (PhD diss., LMU München, 1952).

46 Kenneth E. Nolin, "The 'Itqān' and its Sources: A Study of al-Itqān fī ʿulūm al-Qur'ān by Jalāl al-Dīn al-Suyūṭī with special reference to al-Burhān fī ʿulūm al-Qur'ān by Badr al-Dīn al-Zarkashī" (PhD diss., Hartford Seminary Foundation, 1968), 17.

47 İbrahim Hakkı İmamoğlu, "Celâleddîn es-Süyûtî ve Kur'an İlimlerindeki Yeri. El-Itkân Örneği" (PhD diss., University of Ankara, 2011), 274–89.

48 Nolin, "The 'Itqān' and its Sources," 17.

49 Ibid., 187f.

50 Hasan Gökbulut, "Kâfiyeci," in *TDVIA*, ed. İslâm Araştırmaları Merkezi (Ankara: TDV, 2001), 24: 154f.

known for twenty works, among which as-Suyūṭī refers to the book *Kitāb at-taysīr fī qawāʿid ʿilm at-tafsīr*[51] in *al-Itqān*. This is likely one of his most cited works, as many manuscripts of it exist. As-Suyūṭī briefly references this work in the introduction of the *Itqān*, quoting his teacher, and quickly concludes that this work "did not quench his thirst" and did not satisfy him.[52] However, explicitly mentioning *Kitāb at-taysīr* in the list of sources quoted in his introduction effectively amounts to an acknowledgment and highlighting of this short work.

The concise book *at-Taysīr* stands out from other sources of *Itqān* in that it is less classified into the genre of *ʿulūm al-Qurʾān*. It deals with principles of Qurʾānic interpretation (*uṣūl at-tafsīr*) in two main chapters and a concluding section, following the discussion of *qawāʿid* (rules, methods) of Qurʾānic exegesis.[53] In contrast, many compilations of Qurʾānic sciences include the terms *ʿilm* or *ʿulūm* in their titles. The latter suggests a different guiding claim by the authors, who wanted to present encyclopedic collected knowledge with these works. The claim of writings with the term *qawāʿid* in the title discusses questions of methodology in a narrower sense. For this reason, despite the substantive points of overlaps, these sources can be classified into two groups: on the one hand, the encyclopedic, sources on the sciences of the Qurʾān (the *ʿulūm al-Qurʾān* literature), and on the other hand, methodological works (the *uṣūl at-tafsīr* literature).

Al-Kāfiyajī takes up an existing tradition of *uṣūl at-tafsīr* (principles of exegesis) with his work: Already al-Ḥārith al-Muḥāsibī (d. 857/243) had presented such a work with *Fahm al-Qurʾān*. Other important writings of this genre include *Qānūn at-taʾwīl* (*al-Qānūn al-kullī fī t-taʾwīl*) by Abū Ḥāmid al-Ghazālī (d. 1111/505), *Muqaddimat jāmiʿ at-tafāsīr* by ar-Rāghib al-Iṣfahānī, *al-Iksīr fī qawāʿid ʿilm at-tafsīr* by Najm ad-Dīn aṭ-Ṭūfī (d. 1316/716), and finally *Muqaddima fī uṣūl at-tafsīr* by Ibn Taymiyya (d. 1328/728). Most of these writings have strongly influenced the *Itqān* in terms of content and methodology.

Many of these writings discuss linguistic aspects from rhetoric with its subdisciplines (*ʿilm al-balāgha, badīʿ, bayān*, and *maʿānī*). In al-Kāfiyajī's work, the role of logic is particularly emphasized. His writing reveals the influence of Māturīdī's systematization of the hermeneutical terms *tafsīr* and *taʾwīl*.[54] Overall, al-Kāfiyajī, like many other authors of *uṣūl at-tafsīr* works, places the question of legitimate exegesis

---

51 Muḥyī d-Dīn Abū ʿAbd Allāh al-Kāfiyajī, *Kitābu't-Taysīr Fī Ḳavāidi ʿilmi't-Tafsīr*, ed. İsmail Cerrahoğlu (Ankara: Ankara Üniversitesi İlâhiyat Fakültesi Yayınları, 1974).

52 Jalāl ad-Dīn as-Suyūṭī, *al-Itqān fī ʿulūm al-Qurʾān*, ed. Markaz ad-Dirāsāt al-Qurʾāniyya (Madina: al-Maktabat al-ʿArabiyya as-Suʿūdiyya, 2005), 1: 5.

53 Fatma Yalnız, "Ulûmu'l-Kur'ân Külliyatı Yazımı Konusunda Süyûtî'nin, Bulkînî'nin ve Kafiyeci'nin Birbirlerinde Ne Ölçüde Etkilendiklerinin Tespiti," *Gümüşhane Üniversitesi İlahiyat Fakültesi Dergisi* 11, no. 21 (2022): 216, 235.

54 Nimet Seker, *Der Koran als Rede und Text. Hermeneutik sunnitischer Koranexegeten zwischen Textkohärenz und Offenbarungskontext* (Berlin: EB Dr. Brandt, 2019), 170–78.

and its limits at the center of his reflections. He discusses *raʾy maḥmūd* (permissible subjective opinion) and *raʾy madhmūm* (impermissible subjective opinion), connecting them with aspects of the theory of *ijtihād* and the question of whether and how scholars can grasp the intention of God (*iṣābat al-mujtahid*).[55] It is essential to note that al-Kāfiyajī was strongly influenced by the linguistic-theoretical *uṣūl al-fiqh* methodology. His approach includes knowledge of various disciplines, a total of fifteen, such as lexicography, etymology, morphology, syntax, the science of the readings (*al-qirāʾāt*) of the Qurʾān, the occasions of revelation (*asbāb an-nuzūl*), the sunna of the Prophet, legal interpretation and its sources (*uṣūl al-fiqh*), and more, as necessary tools for successful exegesis.[56] These are elements that run through the hermeneutical literature and find its reflection in chapters such as *shurūṭ al-mufassir wa-ādābuhu* (The conditions required for the commentator and his rules of conduct) in the *Itqān*. The *Kitāb at-taysīr* belonged to the works whose authors were still alive when as-Suyūṭī was working on his Qurʾānic encyclopedia, meaning that this source was not merely a written book for him.

4.2 Al-Bulqīnī's *Mawāqiʿ al-ʿulūm* and as-Suyūṭī's *Kitāb at-taḥbīr fī ʿilm at-tafsīr*

When examining the long list of writings that as-Suyūṭī claims to have used as sources for his Qurʾānic encyclopedia, a central source of the *Itqān* seems to be missing: the work *Mawāqiʿ al-ʿulūm min mawāqiʿ an-nujūm* by al-Qāḍī Jalāl ad-Dīn ʿAbd ar-Raḥmān b. ʿUmar al-Bulqīnī[57] (d. 1421/824), the brother of his teacher ʿAlam ad-Dīn. Jalāl ad-Dīn al-Bulqīnī was a Shafiʿī jurist, hadith scholar, and, from 1402/804 until his death with interruptions, the chief judge (*qāḍī l-quḍāt*) of the Mamluk Empire. Additionally, he taught subjects such as exegesis, law, and hadith at various madrasas. He authored several works, including the Qurʾānic commentary *Nahr al-ḥayāt* and the *ʿulūm al-Qurʾān* work *Mawāqiʿ al-ʿulūm*. This work was the model and inspiration for as-Suyūṭī's first *ʿulūm al-Qurʾān* work, *Kitāb at-Taḥbīr fī ʿilm at-tafsīr*.[58] As-Suyūṭī describes it in the introduction of his encyclopedia, quoting from al-Bulqīnī's work. The quote also makes it clear that as-Suyūṭī was not the inventor of the genre *ʿulūm al-Qurʾān* but that, along with az-Zarkashī's *al-Burhān*, at least al-Bulqīnī's *Mawāqiʿ al-ʿulūm* was known to the compiler. As-Suyūṭī especially praises the innovative charac-

55 Ibid., 174–76.

56 Al-Kāfiyajī, *Kitābu't-Taysīr*, 10–12.

57 M. Yaşar Kandemir, "Bulkînî, Abdurrahman b. Ömer," in *TDVIA*, ed. İslâm Araştırmaları Merkezi (Ankara: TDV, 1992), 6: 409f.; H.A.R. Gibb, "al-Bulḳīnī," in *EI*[2], ed. P. Bearman et al., accessed December 7, 2023, http://dx.doi.org/10.1163/1573-3912_islam_SIM_1523.

58 Also known as "at-Taḥbīr fī ʿulūm at-tafsīr."

ter of al-Bulqīnī's *Mawāqi'* and particularly lauds the reordering of knowledge by him.[59] However, as-Suyūṭī does not introduce this source in the introduction as part of his *Itqān* bibliography and refers to al-Bulqīnī only a total of seventeen times throughout the *Itqān*, thereby strongly marginalizing al-Bulqīnī's contribution to the creation of his work.[60] Perhaps as-Suyūṭī focused more on the textual structure of the Mawāqi' and sees a minimal contribution of this source to the substantive part of the *Itqān*.

Al-Bulqīnī's *Mawāqi' al-'ulūm* seems to have been of greater importance for the precursor *at-Taḥbīr*. This is evident from the chapter division and thematic arrangement of the two works. It is clear that as-Suyūṭī was influenced by the arrangement and chapter titles of al-Bulqīnī's *Mawāqi'* in the composition of *at-Taḥbīr*. Overall, as-Suyūṭī's thematic structure is identical to al-Bulqīnī's work, with the difference that as-Suyūṭī devotes a separate chapter to individual sub-disciplines. For example, al-Bulqīnī has a chapter titled *al-firāshī* ("revelations in bed"), which in *at-Taḥbīr* becomes two separate chapters titled *al-firāshī* and *an-nawmī*, and in *Itqān*, they are combined into one chapter titled *al-firāshī wa-n-nawmī* ("revelations in bed and during sleep").[61] Thus, from the fifty-two chapter headings in *Mawāqi'*, 102 chapters have been created in *at-Taḥbīr*.[62] Regarding the overall thematic spectrum of *'ulūm al-Qur'ān*, it seems that al-Bulqīnī, in his much more compact work, already laid the structural foundation for the selection and arrangement of the topics treated. As-Suyūṭī oriented himself to this structure, expanded it, and collected additional knowledge on each topic. The arrangement and ordering of chapters in al-Bulqīnī's work are not random. It follows a specific order guided by the idea that the Qur'ānic sciences consist of six meta-disciplines, which al-Bulqīnī summarizes himself. These meta-disciplines include 1. The temporal and local contexts of the Qur'ānic revelations and the reasons for their descent, 2. The sciences of Qur'ānic recitation along with their transmission, 3. The practical application of recitation, 4. Qur'ānic linguistics, 5. The meanings of Qur'ānic verses with legal contents, and 6. The science of wording (*alfāẓ*) and meaning (*ma'ānī*), including disciplines like rhetoric.[63] Both of as-Suyūṭī's *'ulūm al-Qur'ān* works follow this meta-thematic chapter arrangement, whose cornerstone was laid not by as-Suyūṭī but by al-Bulqīnī. All other Qur'ānic scientific disciplines found in the *Itqān* but not in *Mawāqi' al-'ulūm* probably go back to the influence of az-Zarkashī's *al-Burhān*.

---

59 As-Suyūṭī, *al-Itqān*, 1: 5f.

60 Stephen R. Burge, "Evidence of Self-Editing in al-Suyūṭī's Taḥbīr and Itqān: A Comparison of His Chapters on Asbāb al-Nuzūl," in *Al-Suyūṭī, a Polymath of the Mamlūk Period*, ed. Antonella Ghersetti (Leiden: Brill, 2017), 164f.

61 For a comparison of chapter headings in *Mawāqi'*, *at-Taḥbīr*, and *al-Itqān*, see Yalnız, "Süyûtî, Bulkînî ve Kâfeyicî," 227–29; for a comparison of chapter headings in *Mawāqi'* and *at-Taḥbīr*, see İmamoğlu, "el-Itkân," 227–30.

62 İmamoğlu, "el-Itkân," 227.

63 Jalāl ad-Dīn al-Bulqīnī, *Mawāqi' al-'ulūm fī mawāqi' an-nujūm*, ed. Anwar M. al-Mursī Khaṭṭāb (Tanta: Dār aṣ-Ṣaḥāba li-t-Turāth, 2007), 28f.; as-Suyūṭī, *al-Itqān*, 1: 6f.

Moreover, there is also an influence of al-Bulqīnī's primary interest in writing *Mawāqiʿ al-ʿulūm* as a systematic handbook for exegetical sciences on as-Suyūṭī's systematics, both in *at-Taḥbīr* and later in *Itqān*. In his introduction, al-Bulqīnī justifies the need for a work similar to those for hadith evaluation and narration, equipped with an appropriate structure and systematic collection.[64] But unlike the large, encyclopedic work that as-Suyūṭī would later compose, *Mawāqiʿ al-ʿulūm* is intended to serve as a textbook in studying.

According to his compilatory method, which especially considers transmitted knowledge in the form of reports, as-Suyūṭī also includes transmissions taken from the *Mawāqiʿ* in *at-Taḥbīr*. However, in terms of both transmitted reports and the citation of al-Bulqīnī's opinions, as-Suyūṭī has been selective.[65] A significant comparison between the various *ʿulūm al-Qurʾān* works that as-Suyūṭī himself mentions should therefore include the little-known work of al-Bulqīnī, at least as much as as-Suyūṭī's *at-Taḥbīr* and az-Zarkashī's *al-Burhān*, if not more.

4.3 As-Suyūṭī's *Kitāb at-Taḥbīr fī ʿilm at-tafsīr*

As-Suyūṭī's general working method can be reconstructed based on his handling of textual sources and references he used to write a new book. In general, in addition to a compilatory approach, often focused on the transmission, collection, and citation of Qurʾānic knowledge and exegesis from previous generations, he also followed a strongly editorial approach. The goal of *al-Itqān* might, on the one hand, lie in a comprehensive, systematic collection of all the knowledge available to as-Suyūṭī. However, the compiler pursued an independent approach that went beyond mere collection and arrangement, relying much more on an editorial and selective hand. Through this approach, preferences were drawn, emphases were made, and irrelevant information was excluded. We can confidently assume that in this editorial process, as-Suyūṭī was guided by his own formed scholarly opinion. Ultimately, this means that his often unexpressed personal opinion can be found between the lines. Reading between the lines is only possible through a comparative reading of as-Suyūṭī's own writings and his source materials.

As-Suyūṭī's editorial activity encompasses not only the processing of external sources but also his own writings, such as his earlier *ʿulūm al-Qurʾān* work *at-Taḥbīr fī ʿilm at-tafsīr*. The forms of processing this source were diverse: at times, as-Suyūṭī adopts entire passages verbatim, but he also shortens, revises, and supplements them. In doing so, specific transmissions as well as discussions of particular verses may be subject to cuts and additions.[66] As-Suyūṭī has, in places, significantly revised his text,

64 Yalnız, "Süyûtî, Bulkînî ve Kâfeyicî," 227; al-Bulqīnī, *Mawāqiʿ*, 28; as-Suyūṭī, *al-Itqān*, 1: 6.
65 Yalnız, "Süyûtî, Bulkînî ve Kâfeyicî," 230.
66 Burge, "Evidence of Self-Editing," 147–55.

for instance, in the chapter on *asbāb an-nuzūl* (the occasions of revelation).[67] He has rearranged certain parts, at times entirely restructured them, and substantially expanded them with new information from other sources. Thus, *al-Itqān* appears as a completely independent work in terms of content. While it was generally aligned with the *at-Taḥbīr* template, it was shaped into a distinct, separate work through substantial "self-editing"[68] and considerable additions. Throughout this process, as-Suyūṭī not only copies his own text but also, in some instances, shortens and expands upon it. This may also involve omitting quotes from other sources, such as al-Wāḥidī's *Asbāb nuzūl al-Qur'ān*, found in *at-Taḥbīr*.[69]

The *'ulūm al-Qur'ān* framework created by al-Bulqīnī was thus expanded and supplemented by as-Suyūṭī in *at-Taḥbīr* and then extensively enriched, revised, and filled with comprehensive information by the same author in *al-Itqān*.[70] A notable highlight in a comparison between *at-Taḥbīr* and *al-Itqān* is the inclusion of methodological discussions in *al-Itqān*. For instance, it addresses handling exegetical problems like resolving diverse and contradictory transmission materials related to the *asbāb an-nuzūl*-material.[71] In contrast to *at-Taḥbīr*, as-Suyūṭī presents not only a greater wealth of detail on Qur'ānic knowledge in the *Itqān* but he also reflects on his hermeneutical profile. This is particularly evident in methodological discussions, as seen in one of the concluding chapters of the *Itqān* on the necessary tools for effective exegesis (The conditions required for the commentator and his rules of conduct, *shurūṭ al-mufassir wa-ādābuhu*).

4.4 Az-Zarkashī's *al-Burhān fī 'ulūm al-Qur'ān*

A consideration of *al-Itqān* is not possible without at least mentioning the second most important classical Qur'ānic encyclopedia. Together with the *Itqān*, this work forms the culmination and pinnacle of the classical discipline of *'ulūm al-Qur'ān*. Both sources are the most important reference works in this field to date and have achieved quasi-canonical status. *Al-Burhān* is one of the works to which as-Suyūṭī attributes a significant influence on the creation of the *Itqān*. In his introduction, he elevates it to his most important source. He adopted the content of al-Burhān and reorganized it, meaning he selected sections from the chapters of Burhān and integrated them into the chapters of *al-Itqān*. Furthermore, he rearranged the chapters and enriched the knowledge con-

67 Ibid., 158f.
68 Ibid.
69 Ibid., 151.
70 For a comparison of chapter headings in *at-Taḥbīr* and *al-Itqān*, see Burge, ibid., 167–72.
71 Burge, "Evidence of Self-Editing," 158; Nimet Seker, "As-Suyūṭīs Diskussion der Mehrfachherabsendung am Beispiel von Vers 33:35," in *Sure Al-Aḥzāb 33*, ed. Ayşe Başol (Berlin: EB Dr. Brandt, 2019), 103–38.

tained in al-Burhān with additional details.[72] However, as-Suyūṭī also emphasizes the differences: he reordered the chapters in the *Itqān* differently and separated and added many topics from *al-Burhān*.[73]

The influence and processing of *al-Burhān* on *al-Itqān* deserves its own study.[74] Even though this work is less extensive in content than the *Itqān*, it is much more marked by the personal touch and the personal judgment of the author compared to as-Suyūṭī's Qur'ānic encyclopedia.

An important programmatic commonality, which also conditioned the creation of both Qur'ānic encyclopedias, is the need for a work that is systematically similar and summarizes various disciplines, as had already been done in the past in the science of hadith.[75] A similar need in Qur'ānic sciences had already been addressed by al-Bulqīnī in *Mawāqiʿ al-ʿulūm*.[76] This expressed desire by all three scholars provides insight into the underlying understanding of the *ʿulūm al-Qur'ān* genre: it should not only serve to collect existing exegetical knowledge in all its possible aspects and discuss it in an organized manner by topics and problems but also provide a systematic overview of methods of correct exegesis. In the background are methodological questions of Qur'ānic exegesis, which elsewhere in classical as well as contemporary literature are referred to as *uṣūl at-tafsīr* (Principles of Qur'ānic Exegesis).

The intensive engagement with the *ʿulūm al-ḥadīth* and close contact with the work and achievements of significant hadith scholars from the Mamluk era, such as Ibn Ḥajar al-ʿAsqalānī and Ibn aṣ-Ṣalāḥ (died 1245/643), may have contributed to recognizing and naming such a desideratum in Qur'ānic studies. Additionally, there might have been a need to systematize, sort, and organize the accumulated knowledge in Qur'ānic sciences over centuries.

Another interpretive aspect that connects both scholars is their affiliation with the Shāfiʿī legal school. In their respective Qur'ānic encyclopedias, this is evident in the use of the systematics, terminology, and general interpretation doctrine of *uṣūl al-fiqh*. The influence of *uṣūl al-fiqh* literature is already found in al-Bulqīnī's *Mawāqiʿ* and in as-Suyūṭī's *at-Taḥbīr*. As-Suyūṭī had already used the works of ash-Shāfiʿī,[77] *al-Jāmiʿ li-aḥkām al-Qur'ān* by al-Qurṭubī,[78] and *Majmūʿ sharḥ al-muhadhdhab* by an-Nawawī.[79]

72 As-Suyūṭī, *al-Itqān*, 1: 14f.

73 Ibid.

74 The first comparative studies of these two works are presented in Nolin, "The 'Itqān' and its Sources," and Harun Bekiroğlu, *Tefsir Metodolojisi Açısından El-Burhân ve El-Itkân* (Ankara: Araştırma Yayınları, 2013).

75 Badr ad-Dīn Muḥammad az-Zarkashī, *al-Burhān fī ʿulūm al-Qur'ān*, ed. Muḥammad Abū Faḍl Ibrāhīm (Cairo: Maktabat Dār at-Turāth, 1983), 1: 9; as-Suyūṭī, *al-Itqān*, 1: 12.

76 Al-Bulqīnī, *Mawāqiʿ*, 28.

77 As-Suyūṭī, *at-Taḥbīr fī ʿilm at-tafsīr*, ed. Fatḥī ʿAbd al-Qādir Farīd (Riyadh: Dār al-ʿUlūm, 1981), 236, 323.

78 Ibid., 305.

79 Ibid., 146, 436.

As-Suyūṭī utilizes the content from *al-Burhān* expansively for his Qurʾānic encyclopedia. One of the methods as-Suyūṭī used to include az-Zarkashī's work in the *Itqān* was quoting. However, it is not always clear where the quote begins and ends, and where as-Suyūṭī continues with his own thoughts. Sometimes, he quotes from the *Burhān* without specifying it as a source (a method he uses for many other sources as well).

Another form is the creation of new chapters from topics that az-Zarkashī lists under sections with the names *fāʾida* or *faṣl*. For example, as-Suyūṭī formed a separately expanded chapter on the topic *mā takarrara nuzūluhu*[80] from the section (*faṣl*) *fī mā nazala mukarraran*[81] in az-Zarkashī, and from the section (*faṣl*) *taqaddum nuzūl al-āya ʿalā ḥukmihi*,[82] he developed a detailed chapter with the name *mā taʾakhkhara ḥukmuhu ʿan nuzūlihi wa-mā taʾakharra nuzūluhu ʿalā ḥukmihi.*[83] In this way, as-Suyūṭī generated sixteen new chapters for his own collection.

As-Suyūṭī's compilatory and amalgamative approach to his sources is also evident in the chapter *munāsabāt bayn al-āyāt wa-s-suwar*. Here, he orients himself to sources that az-Zarkashī had already mentioned, such as *al-Burhān fī tartīb suwar al-Qurʾān* or *al-Burhān fī munāsabāt as-suwar* by Abū Jaʿfar b. az-Zubayr al-Gharnāṭī (d. 1308/708).[84] However, he also draws inspiration from additional works which played no role for az-Zarkashī, including as-Suyūṭī's own writings on the topic, titled *Asrār at-tanzīl*, also known as *Qaṭf al-azhār fī kashf al-asrār*, where he addresses the connections (*munāsabāt*, *tanāsub*) between the verses and chapters of the Qurʾān. In addition, he processed in this chapter his own work on the same topic, *Tanāsuk ad-durar fī tanāsub as-suwar*. Another source for this chapter was al-Biqāʿī's extensive exegetical work *Naẓm ad-durar fī tanāsub al-āyāt wa-s-suwar*. In total, as-Suyūṭī cites over 200 different sources, including those quoted by az-Zarkashī.[85]

> The *Itqān* is, therefore, the culmination of a chain of four works: al-Bulqīnī's *Mawāqiʿ al-ʿulūm*, al-Suyūṭī's *Taḥbīr*, al-Zarkashī's *Burhān* (although written before both the *Mawāqiʿ al-ʿulūm* and the *Taḥbīr*), and al-Suyūṭī's *Itqān*.[86]

In summary, it can be noted that as-Suyūṭī's approach to *al-Burhān* is characterized by an adoption and expansion of its content, a more detailed treatment of topics, and the addition of further chapters and sub-disciplines in the Qurʾānic sciences.

---

80 As-Suyūṭī, *al-Itqān*, 1: 234–37.

81 Az-Zarkashī, *al-Burhān*, 1: 29–32.

82 Ibid.,1: 32f.

83 As-Suyūṭī, *al-Itqān*, 1: 238–43.

84 Francisco Vidal-Castro, "Ibn al-Zubayr al-Gharnāṭī," in *EI*^{3}, ed. Kate Fleet et al., accessed December 5, 2023, http://dx.doi.org/10.1163/1573-3912_ei3_COM_32316.

85 İmamoğlu, "el-Itkân," 274–89.

86 Burge, "Evidence of Self-Editing," 144.

4.5 Contents of the *Itqān*

4.5.1 Introduction (*muqaddima*) of the *Itqān*

The *Itqān* consists of an introduction by the author (*muqaddima*), followed by eighty thematic chapters (*naw'*, *anwā'*). The work concludes with some closing sentences praising and thanking God, stating that he completed the manuscript in the month of Dhū l-Qa'da in the year 883/ January 1497.[87]

From the introduction, not only the textual genesis and the history of the predecessors of the *Itqān* become apparent, but also the motivation for creating this extensive work in the first place: The *Itqān* is intended to serve as a comprehensive manual, described by as-Suyūṭī as an urgent desideratum in the field of Qur'ānic sciences. No work written up to that point had been able to fulfill this task: as-Suyūṭī mentions various predecessors and highlights certain deficiencies in all sources. Al-Kāfiyajī's *at-Taysīr* was too short and not comprehensive enough. Al-Bulqīnī's *Mawāqi'*, while very good in its systematic approach, was also too brief and did not contain all the knowledge. As-Suyūṭī's own predecessor work *at-Taḥbīr* was already an advancement by bringing together the two disciplines of Qur'ānic exegesis and Hadith, as evidenced by the greater incorporation of transmitted knowledge in the *Taḥbīr*. However, it still did not present comprehensive knowledge. This became evident only with the discovery of the *Burhān* by as-Suyūṭī, whose contents he editorially synthesized with hundreds of other sources in the *Itqān*. As-Suyūṭī's reference to these four predecessor sources serves the purpose of tracing the genesis of the *Itqān* and demonstrating the epistemological basis and sources on which the foundation of the *al-Itqān* was built. At the same time, he aims to highlight the unique significance of the *Itqān* as the culmination and pinnacle of Qur'ānic sciences.

The second half of the introduction of the *Itqān* consists of an organized list of primary sources that have been incorporated into the *Itqān*. The list includes transmission-based works from the genres of Qur'ānic commentary, collections of hadith, and their commentaries, works on the readings and phonetic variations of the Qur'ān, and the history of the Qur'ānic text.[88] Furthermore, he enumerates a series of works related to Qur'ānic linguistics, including lexicography and semantics of the Qur'ān.[89] Following that is a list of works on the exegesis of legal verses in the Qur'ān[90] and an extensive list of works on the inimitability of the Qur'ān and the resulting "arts of rhetoric" (*funūn al-balāgha*).[91]

87 As-Suyūṭī, *al-Itqān*, 6: 2454–57.
88 Ibid., 1: 19–23.
89 Ibid., 1: 23–26.
90 Ibid., 1: 27–29.
91 Ibid., 1: 29–35.

In addition to a series of "general" works (*al-kutub al-jāmiʿa*), which do not fit into a specific genre, such as Ibn al-Jawzī's (d. 1350/751) *Badaʾiʿ al-fawāʾid*, as-Suyūṭī also includes a series of non-tradition-based commentary works (*tafāsīr ghayr al-muḥaddithīn*). As research has shown, it cannot be inferred from the main part of the *Itqān* that all the sources mentioned in the introduction are quoted. Instead, as-Suyūṭī used hundreds of additional sources in the *Itqān*.[92]

4.5.2 Arrangement and Contents of Chapters

The arrangement of chapters in the *Itqān* follows the logic of thematically grouping multiple chapters under the meta-themes of Qur'ānic sciences. As-Suyūṭī did not explicitly mark these meta-themes, and the chapters 1–80 follow one another without further sub-division. The sub-division into meta-structures below is based on my assessment.

A closer look reveals that the first sixteen chapters focus on (I.) the temporal and local contexts of the Qur'ān's revelation and its reasons for descent. Within this first thematic group, as-Suyūṭī starts with the macro-contexts of Mecca and Medina, presenting more specialized knowledge about the contexts of revelation in the subsequent chapters by collecting information on the time, place, and situation in which the Prophet received revelations. These are followed by chapters providing additional information about the contexts of revelation. All chapters are related to the situation, moment, time, and place in which the Prophet found himself when receiving revelations. The first sixteen chapters deal with the following topics:

I. Temporal and Local Contexts of the Revelation of the Qur'ān and Its Reasons for Descent

Ch. 1 Revelations in Mecca and Medina (*al-makkī wa-l-madanī*)
Ch. 2 Revelations in Residence and during Travel (*al-ḥaḍarī wa-s-safarī*)
Ch. 3 Revelations during the Day and at Night (*an-nahārī wa-l-laylī*)
Ch. 4 Revelations in Summer and Winter (*aṣ-ṣayfī wa-sh-shitāʾī*)
Ch. 5 Revelations in Bed and during Sleep (*al-firāshī wa-n-nawmī*)
Ch. 6 Revelations on Earth and in the Sky (*al-arḍī wa-s-samāʾī*)
Ch. 7 The First Revelation (*awwal mā nazal*)
Ch. 8 The Last Revelation (*ākhir mā nazal*)
Ch. 9 Occasions of Revelation (*sabab an-nuzūl*)
Ch. 10 Revelations according to the Utterances of Some Companions of the Prophet (*ʿalā lisān baʿḍ aṣ-ṣaḥāba*)
Ch. 11 Repeated Descents (*mā takarrara nuzūluhu*)

92 Nolin, The 'Itqān' and its Sources, 17.

Ch. 12 Descents, whose decree has been delayed, and descents that have been revealed after their decree has taken effect (*mā taʾakhkhara ḥukmuhu ʿan nuzūlihi wa-mā taʾakhkhara nuzūluhu ʿan ḥukmihi*)[93]
Ch. 13 Individual and Collective Descents (*mā nazala mufarraqan wa-mā nazala jamʿan*)
Ch. 14 Descents when the Prophet was accompanied or alone (*mā nazala mushayyaʿan wa-mā nazala mufradan*)
Ch. 15 Descents on Some Prophets (*ʿalā baʿḍ al-anbiyāʾ*) and Descents on No One before the Prophet Muḥammad
Ch. 16 The Modalities of Descent (*fī kayfiyyati inzālihi*)

The arrangement of chapters in the *Itqān* begins with knowledge about the fundamental question of the revelation of the Qurʾān itself, its contextual references, and also addresses methodological and sometimes contradictory questions within this complex of topics. For example, it explores how to deal with repeated revelations or delayed revelations. These methodological questions arise from the multitude and contradictions of the traditions that as-Suyūṭī presents. The material he presents is not just a pure collection and sorting of knowledge but also touches on technical questions related to handling competing material and its exegetical evaluation.

The next thematic block deals with (II.) the external form of the Qurʾānic text and its transmission. It includes information about the form of the Qurʾān, such as the names of the suras, the number of textual units of the Qurʾān like sura, verse, etc., as well as the transmission and memorization of the Qurʾān. The associated chapters cover the following topics:

II. The Formal Structure of the Qurʾān and the Transmission of its Text
Ch. 17 The Names of the Qurʾān and the Names of the Suras
Ch. 18 The Collection and Arrangement of the Text
Ch. 19 The Number of Chapters, Verses, Words, and Letters in the Qurʾān
Ch. 20 The Memorizers (*ḥuffāẓ*) and Transmitters (*ruwāt*) of the Qurʾān
Ch. 21 The Elevated (*ʿālī*) and Lower (*nāzil*) Chains of Transmission of the Qurʾān
Ch. 22 to Ch. 28 Knowledge about the Frequently Transmitted (*al-mutawātir*), the Well-Known (*al-mashhūr*), the Singularly Transmitted (*al-āḥād*), the Irregularly Transmitted (*ash-shādhdh*), the Doubtfully Transmitted (*al-mawḍūʿ*), and the Interpolated (*al-mudraj*)

The chapters on the external form and formal transmission are followed by eight chapters on (III.) Qurʾānic recitation and the knowledge of its practical implementa-

93 This means that the decree was in effect before the Prophet received the corresponding Qurʾānic verse and, conversely, that a verse was revealed while the corresponding decree had not yet taken effect. See Seker, "As-Suyūṭī's Diskussion" for more details.

tion. This section delves into specific knowledge about the phonetic pronunciation of the Qurʾān, with a particular focus on the rules of Qurʾānic recitation.

III. Sciences of Qurʾānic Recitation and Their Practical Implementation

Ch. 28 Knowledge of Pause (*al-waqf*) and Resumption (*al-ibtidāʾ*)

Ch. 29 Knowledge of Verses that are formally connected but semantically separated

Ch. 30 Pronunciation of *imāla*[94] and *fatḥa*

Ch. 31 Assimilation (*idghām*), disclosure (*iẓhār*), concealment (*ikhfāʾ*), and conversion (*iqlāb*)

Ch. 32 Lengthening (*al-madd*) and Shortening (*al-qaṣr*)

Ch. 33 The Softening of the *hamza*

Ch. 34 Instruction in the Qurʾān and Methods of Memorization

Ch. 35 Etiquette of Recitation and Etiquette of the Listener

After covering the foundations of revelation and its genesis, as well as the formal structure and recitation of the Qurʾān, as-Suyūṭī proceeds with an extensive list of chapters that deal with the classification of Qurʾānic language. This part of the *Itqān* can be considered the philological centerpiece of the work. It begins with (IV.) an overview of sciences dealing with wording (*lafẓ*). Following that are (V.) classifications of Qurʾānic language according to the methods of norm derivation (*istinbāṭ al-ḥukm*) in legal theory (*uṣūl al-fiqh*) and from the perspective of rhetoric (*ʿilm al-balāgha*) and its sub-disciplines. The transition from chapters applying linguistic classifications from *uṣūl al-fiqh* to chapters drawing knowledge from rhetoric appears to be seamless.

IV. Qurʾānic Linguistics and Sciences Dealing with the Wording (*lafẓ*)

Ch. 36 Unusual Expressions (*gharīb*) in the Qurʾān

Ch. 37 Terms that do not correspond to the *ḥijāzī* dialect

Ch. 38 Non-Arabic Terms in the Qurʾān

Ch. 39 Polysemy and Synonymy (*al-wujūh wa-n-naẓāʾir*)

Ch. 40 Tools Needed by the Exegete

Ch. 41 Syntax Analysis (*iʿrāb*)[95]

Ch. 42 Important Rules that the Exegete Must Know

V. Linguistic Classification According to *Uṣūl al-Fiqh* and Rhetoric (*ʿilm al-balāgha*)

Ch. 43 The Clear (*al-muḥkam*) and the Ambiguous (*al-mutashābih*)

Ch. 44 The Preceding (*muqaddam*) and the Subsequent (*muʾakhkhar*)

Ch. 45 The General (*ʿāmm*) and the Specific (*khāṣṣ*)

Ch. 46 The Concise (*mujmal*) and the Elaborate (*mubayyan*)

Ch. 47 The Abrogating (*nāsikh*) and the Abrogated (*mansūkh*)

94 Pronunciation of *fatḥa* near *kasra* and pronunciation of *alif* near *yā*.

95 As-Suyūṭī uses the term *iʿrāb* in this context with the meaning of syntax (*naḥw*). The meaning in the sense of syntax becomes evident from the content of the chapter.

Ch. 48 The Doubtful (*mushkil*), Disagreement (*ikhtilāf*), and Contradiction (*tanāquḍ*)
Ch. 49 The Indeterminate (*muṭlaq*) and the Specified (*muqayyad*)
Ch. 50 The Utterance (*manṭūq*) and the Understood (*mafhūm*)
Ch. 51 Forms of Address (*al-mukhāṭaba*)
Ch. 52 The Literal (*al-ḥaqīqa*) and the Figurative (*al-majāz*) Meaning
Ch. 53 Comparison (*tashbīh*) and Metaphor (*istiʿāra*)
Ch. 54 Metonymy (*al-kināya*) and Allusion (*at-taʿrīḍ*)
Ch. 55 Restriction (*al-ḥaṣr*) and Specification (*al-ikhtiṣāṣ*)
Ch. 56 Conciseness (*al-ījāz*) and Prolixity (*al-iṭnāb*)
Ch. 57 Descriptive Statement (*al-khabar*) and Prescriptive Statement (*al-inshāʾ*)
Ch. 58 Innovative Rhetorical Forms of the Qurʾān (*al-badāʾiʿ*)

Following the chapters devoted to linguistic exploration of the Qurʾān, a series of chapters ensue that deal with various forms of stylistics in the Qurʾān. As-Suyūṭī moves away from the level of wording (*lafẓ*) and turns to the level of meaning units spanning verses and chapters. This thematic unit begins with the harmonic connections between the beginning and end of verses and chapters, illustrating literary harmonies both within individual textual units and across textual units (for example, spanning multiple suras). Moreover, within this thematic block, additional subjects are explored, encompassing both substantive and formal dimensions of Qurʾānic ambiguity. Stylistic elements such as similes and oaths are also delved into, culminating in an examination of distinctive features of the Qurʾān, including its unique merits (*faḍāʾil*).

VI. Stylistic Features of the Qurʾān
Ch. 59 Conclusions of Verses (*fawāṣil al-āy*)
Ch. 60 The Opening of Chapters (*fawātiḥ as-suwar*)
Ch. 61 The Conclusions of Chapters (*khawātim as-suwar*)
Ch. 62 The Harmonic Connections between Verses and Chapters (*munāsabāt al-āyāt wa-s-suwar*)
Ch. 63 The Ambiguous Verses (*al-āyāt al-mutashābihāt*)
Ch. 64 The Inimitability of the Qurʾān (*iʿjāz al-Qurʾān*)
Ch. 65 Sciences Derived from the Qurʾān
Ch. 66 The Similes (*al-amthāl*) of the Qurʾān
Ch. 67 The Oaths (*al-aqsām*) of the Qurʾān
Ch. 68 The Controversies (*al-jadal*) of the Qurʾān
Ch. 69 Proper Names (*al-asmāʾ*), Kinship Names (*al-kunā*), and Nicknames (*al-alqāb*)
Ch. 70 Ambiguous References (*al-mubhamāt*)
Ch. 71 The Names of Individuals to Whom the Qurʾān was Revealed
Ch. 72 The Merits of the Qurʾān (*faḍāʾil*)
Ch. 73 The Excellent (*afḍal*) and Outstanding (*faḍīl*) in the Qurʾān
Ch. 74 The Unique Verses (*al-mufradāt*) of the Qurʾān
Ch. 75 The Specific Attributes (*khawāṣṣ*) of the Qurʾān

The *Itqān* concludes with five chapters on Qur'ānic Calligraphy, Qur'ānic Exegesis, and their Etiquette. In the final sections of the work, it addresses matters concerning the handling of the Qur'ān, following its earlier discussions on text genesis, the external structure of the text, and the linguistic expression of the text.

VII. Qur'ānic Calligraphy, Qur'ānic Exegesis, and their Etiquette
Ch. 76 The Artistry and Etiquette of Qur'ānic Calligraphy
Ch. 77 Knowledge of Exegesis (*at-tafsīr*) and Interpretation (*at-ta'wīl*), Expounding its Glory and Necessity
Ch. 78 The Conditions Required for the Commentator and His Rules of Conduct
Ch. 79 The Peculiarities (*al-gharā'ib*) of Exegesis
Ch. 80 Biographies of Qur'ānic Exegetes

The foundation for the aforementioned arrangement of chapters based on six to seven meta-themes was already laid out by al-Bulqīnī in his *Mawāqi'*. The organization of topics in *al-Itqān* begins with the theological foundations of the revelation of the Qur'ān. After delving into the circumstances and the local-temporal contexts in which the Qur'ān originated, the *Itqān* focuses on the external form of the Qur'ān and the transmission of its text. Following this, it delves into the sciences of the phonetic pronunciation and recitation of the Qur'ān, along with the transmission of the Qur'ānic text. After describing the genesis, form, and phonetic properties of the Qur'ān in these three major thematic blocks, as-Suyūṭī progresses to the philological core of the *Itqān*, which deals with the description and classification of the Qur'ānic language and its textual units. Here, as-Suyūṭī places special emphasis on established linguistic disciplines from the *uṣūl al-fiqh* and Arabic rhetoric. An additional thematic block addresses the stylistic peculiarities and unique features of the Qur'ān. In its concluding chapters, the *Itqān* deals with Qur'ānic calligraphy and Qur'ānic exegesis. The author completes this comprehensive Qur'ānic encyclopedia with a chapter on the biographies of Qur'ānic exegetes.

5 Plagiarism Accusations

One of the significant criticisms directed at as-Suyūṭī during his lifetime was the allegation that he often presented sources and ideas from other scholars as his own contributions in his works. As-Sakhāwī accused him of copying texts from the al-Maḥmūdiyya Library[96] in Cairo, slightly expanding on the copied content, and presenting it as his

96 Doris Behrens-Abouseif, in her work *The Book in Mamluk Egypt and Syria (1250–1517): Scribes, Libraries and Market* (Leiden, Boston: Brill, 2018), provides insights into the al-Maḥmūdiyya Library. Contemporary scholars described this library as the most important one with a comprehensive collection of books, see ibid., pages 25–28.

own writings without proper citation.[97] As-Sakhāwī also leveled a more fundamental accusation, claiming that as-Suyūṭī's knowledge did not come from personal instruction and oral transmission by an authorized teacher – the preferred form of religious knowledge transmission – but rather was extracted solely from books.[98]

In response to as-Sakhāwī's allegations, as-Suyūṭī countered by asserting that Sakhāwī, a hadith scholar, historian, and former student of Ibn Ḥajar al-ʿAsqalānī, himself copied from Ibn Ḥajar's writings.[99] However, the information that as-Suyūṭī utilized the al-Maḥmūdiyya Library for his source research comes from as-Suyūṭī himself. Nevertheless, he emphasized the anonymous authorship of the books he used from the Maḥmūdiyya.[100] The accusation of plagiarism was not limited to as-Suyūṭī but extended to other scholars of this era, such as al-Maqrīzī (d. 1422/845), and even as-Sakhāwī himself.[101]

However, the question of whether as-Suyūṭī was an eminent scholar, or an outstanding plagiarist is difficult to answer in the context of the prevalent writing practices of his time. Practices such as paraphrasing, citing in shortened form, elaborating on quotes in one's own words, and similar methods of dealing with external sources were common and accepted in as-Suyūṭī's era.[102] Furthermore, as-Suyūṭī conveys at least a thorough and conscious handling of the sources in the *Itqān*, listing them extensively in its introduction. Moreover, a modern understanding of authorship and intellectual property cannot be unequivocally projected onto the Mamluk era, as the concept of the author with original ideas emerged only in the modern period.

Lastly, one may question whether *al-Itqān* and the most important works of as-Suyūṭī, which have more of an encyclopedic character, are actually driven by the idea of sole authorship. It is possible that as-Suyūṭī saw himself more as a compiler, seeking to preserve existing knowledge in a comprehensive and systematic manner.

97 Ito, "Al-Suyūṭī and Problems of the Waqf," 49; Burge, "Evidence of Self-Editing," 163; Behrens-Abouseif, *The Book in Mamluk Egypt and Syria*, 25; Sartain, *Jalāl al-dīn al-Suyūṭī*, 74.

98 Behrens-Abouseif, *The Book in Mamluk Egypt and Syria*, 95.

99 Sartain, *Jalāl al-dīn al-Suyūṭī*, 75.

100 Behrens-Abouseif, *The Book in Mamluk Egypt and Syria*, 25.

101 Burge, "Evidence of Self-Editing," 163; Frédéric Bauden, "Maqriziana IX: Should al-Maqrīzī be thrown out with the bath water? The question of his plagiarism of al-Awḥadī's Khiṭaṭ and the documentary evidence," *MSR* 14 (2010): 159–232.

102 Sartain, *Jalāl al-dīn al-Suyūṭī*, 74.

6 Influence of the Hadith Sciences on as-Suyūṭī's Method and Approach

The influence of hadith methodology and methods of hadith evaluation is evident in the handling of *isnād* (chain of transmission) in all tradition-based sciences mentioned in the *Itqān*. This principle is already applied by az-Zarkāshī in *al-Burhān*. The application of hadith evaluation principles can be seen, for instance, in the chapters on Qur'ānic recitation, where terminology from the sciences of hadith (such as *mutawātir* (the frequently transmitted), *āḥād* (the singularly transmitted), *shādhdh* (the irregularly transmitted), *mawqūf* (the stopped), *marfūʿ* (the elevated), *maqṭūʿ* (the disrupted) etc.) and criteria for evaluating transmissions are utilized. These criteria are consistently applied throughout the *Itqān* in chapters dealing with transmitted knowledge, including discussions on the occasions of revelation. At times, as-Suyūṭī's discussions take on a very technical character, delving into solving complex questions, such as reconciling conflicting transmissions.[103]

As-Suyūṭī's preference for a tradition-based approach is also reflected in his commentary work, *ad-Durr al-manthūr*. In this work, traditions are presented largely without *isnād* but with an indication of the collection they are from. This approach mirrors the prevalent method during the Mamluk period, where citation from written sources took precedence, and the citation of knowledge acquired through personal instruction diminished in importance.[104] Unlike aṭ-Ṭabarī's *Jāmiʿ al-bayān*, as-Suyūṭī's approach in *ad-Durr al-manthūr* involves presenting exegetical transmissions without any additional commentary. Thus, this work stands out in the tradition of tradition-based exegesis (*tafsīr bi-l-maʾthūr* or *bi-r-riwāya*). Similar to the *Itqān*, this work also relies extensively on hadith-transmitting sources, with a focus on transmission-oriented exegetes such as Ibn Abī Ḥātim, aṭ-Ṭabarī, Ibn al-Mundhir an-Nīsābūrī (d. 930/318), and hadith compilation literature.[105]

Regarding *ad-Durr al-manthūr*, one might question the purpose of a Qurʾānic commentary that lacks the author's own exegesis. The specificity of this commentary may lie in its position between a hadith compilation and a Qurʾānic commentary. The deeper layers of meaning that as-Suyūṭī does not explicitly convey might have been evident to other readers who knew how to interpret them through contextualization and comparison with other commentaries. For this reason, *ad-Durr al-manthūr* might have been written for a specific audience with profound prior knowledge who could better contextualize as-Suyūṭī's selection of transmissions.

103 Seker, "As-Suyūṭīs Diskussion."

104 Stephen R. Burge, "Scattered pearls: Exploring al-Suyūṭī's hermeneutics and use of sources in *ad-Durr al-manthūr fī t-tafsīr bi-l-maʾthūr*," *JRAS* 24 (2014): 271.

105 Ibid., 256f.

7 Reception

Many of as-Suyūṭī's writings circulated during his lifetime and afterward, including the *Itqān*. The text was transmitted through handwritten copies across generations, with some manuscripts now stored in European universities, state libraries, and the Süleymaniye Library in Istanbul. The first prints of this work circulated in India from the late eighteenth century. Today, various editions of his writings exist (for a brief list, see the bibliography).

As-Suyūṭī's influence on Qur'ānic hermeneutical thinking and the genre of *'ulūm al-Qur'ān* is most significant. Although numerous works have been produced in the centuries after his death, some even surpassing it in scope, the *Itqān* remains the standard work in this genre. It continues to shape the field of Qur'ānic encyclopedias both in content and structure.[106]

In terms of reception history, clear traces can be found in the Ottoman era in Ṭashkubrīzādaḥ's (Ṭāshköprüzāde) (d. 1561/968) *Miftāḥ as-sa'āda* (*Mawẓu'āt al-'ulūm*). This bio-bibliographical encyclopedia, also classifying sciences, follows the chapter headings of the *Itqān* in the section on the branches of the science of interpretation. It adopts the names of individual Qur'ānic disciplines and replicates their order as found in the *Itqān*.[107]

From a later period, *az-Ziyāda wa-l-iḥsān fī 'ulūm al-Qur'ān* by the scholar and shaykh of the Qādirī Order, Jamāl ad-Dīn b. 'Aqīla al-Makkī (d. 1737/1150), is noteworthy. It was written with the intention of supplementing new knowledge to the *Itqān*. It includes a partial revision of the *Itqān*, removing certain sections and pointing out errors. Additionally, seventy-four new chapters were added. Several manuscripts exist in the Istanbul Süleymaniye Library, one of which is believed to be by the author himself.[108]

Also from the eighteenth century is the more well-known work *al-Fawz al-kabīr fī uṣūl at-tafsīr* by Shāh Walī Allāh ad-Dihlawī (d. 1762/1176). It covers a limited selection of topics and is classified more as an abridged work.

From the twentieth century comes the work *Manāhil al-'irfān fī 'ulūm al-Qur'ān* by the Azhari scholar Muḥammad 'Abd al-'Aẓīm az-Zurqānī (d. 1948/1367). It is one of the more systematic modern writings and was long used as a text and study book. One unique aspect of this work is the author's expansion of the field of Qur'ānic sciences by incorporating contemporary exegetical debates.

---

106 For further information s. the chapter on *Encyclopaedias of Qur'ānic Sciences* by the author in the first volume of the *Handbook*.

107 Aḥmad Ṭashkubrīzādah, *Miftāḥ as-sa'āda wa-misbāḥ as-siyyāda*, ed. 'Alī Daḥrūj (Beirut: Maktaba Lubnān Nāshirūn, 1998), 285–328.

108 Abdülhamit Birışık, "İbn Akîle," in *TDVIA*, ed. İslâm Araştırmaları Merkezi (Ankara: TDV, 1999), 19: 304–6.

As-Suyūṭī's writings continue to be influential, with authors like ʿAdnān Zarzūr (born 1939/1358 in Damascus) working to make the preserved knowledge of as-Suyūṭī and his predecessors accessible to contemporary readers even after centuries, by editing classical writings and authoring books[109] in English and Arabic on the Qur'ānic sciences (*ʿulūm al-Qur'ān*).

8 Bibliography

8.1 Sources

Al-Bulqīnī, Jalāl ad-Dīn. *Mawāqiʿ al-ʿulūm fī mawāqiʿ an-nujūm*. Ed. Anwar M. al-Mursī Khaṭṭāb. Tanta: Dār aṣ-Ṣaḥāba li-t-Turāth, 2007.

Ibn Taymiyya, Abū l-ʿAbbās Tāqī d-Dīn. *Muqaddima fī uṣūl at-tafsīr*. Ed. ʿAdnān Zarzūr. Beirut: Dār al-Qur'ān al-Karīm, 1972.

Al-Kāfiyajī, Muḥyī d-Dīn Abū ʿAbd Allāh. *Kitābu't-Taysīr Fī Ḳavāidi ʿilmi't-Tafsīr*. Ed. İsmail Cerrahoğlu. Ankara: Ankara Üniversitesi İlâhiyat Fakültesi Yayınları, 1974.

As-Suyūṭī, Jalāl ad-Dīn. *Al-Itqān fī ʿulūm al-Qur'ān*. Ed. Markaz ad-Dirāsāt al-Qur'āniyya. 6 vols. Madina: al-Maktabat al-ʿArabiyya as-Suʿūdiyya, 2005.

Editions (selection)

- Ed. Bashīr ad-Dīn Mawlavī, Sadīd ad-Dīn Khān Mawlavī, Nūr al-Ḥaqq Mawlavī, Calcutta: Alois Sprenger, 1857.
- Ed. Muḥammad Abū l-Faḍl Ibrāhīm. Cairo: Maktabat wa-Maṭbaʿat al-Mashhad al-Ḥusaynī, 1967.
- Ed. Cairo: Maktabat wa-Matbaʿat Muṣtafā al-Bābī l-Halabī wa-Awlāduhu, 1979.
- Ed. and commentary Muṣtafā Dīb al-Bughā. Damascus: Dār Ibn Kathīr, 1987.
- Ed. Bombay: Abnā' Mawlawī Muḥammad b. Ghulām as-Sūratī, 1978.
- Ed. Lahore: Suhayl Akīdimī, 1980.
- Ed. Beirut: Mu'assasat al-Kutub ath-Thaqāfiyya, 1996.
- Ed. Fawwāz Aḥmad Zamarlī. Beirut: Dār al-Kitāb al-ʿArabī, 1999.
- Ed. Beirut: Dār Ibn Ḥazm, 2022.
- Ed. Muḥammad Ṣādiq Darwīsh. Beirut: Dār Ibn Kathīr, 2023.

Translations

- *el-İtkān fī ulûmi'l-Kur'ân: Kur'an İlimleri Ansiklopedisi*, trans. Sakıp Yıldız, Hüseyin Avni Çelik. Istanbul: Madve, 1987.
- *The Perfect Guide to the Sciences of the Qur'ān*, trans. Hamid Algar, Michael Schub, Ayman Abdel Haleem. Reading, UK: Garnet, 2011.
- *Le parfait manuel des sciences coraniques al-Itqān fī ʿulūm al-Qur'ān de Ǧalāl ad-Dīn as-Suyūṭī: Présentation, Traduction et Annotation*, trans. Michel Lagarde, 2 vols. Leiden: Brill, 2018.

As-Suyūṭī, Jalāl ad-Dīn. *Ad-Durr al-manthūr fī tafsīr bi-l-ma'thūr*. Ed. n.p. Tehran: al-Maṭbaʿa al-Islāmiyya, 1957.

- Ed. Beirut: Dār al-Fikr, 1983.
- Ed. Shaykh Najdat Najī. Beirut: Dār Iḥyā' at-Turāth al-ʿArabī, 2001.
- Ed. ʿAbd al-Muḥsin at-Turkī. Cairo: Markaz Hijr li-l-Buḥūth wa-l-Dirāsāt al-ʿArabiyya wa-l-Islāmiyya, 2003.

---

109 E.g. ʿAdnān M. Zarzūr, *The Qur'an and Its Study: An In-depth Exploration of Islam's Sacred Scripture*, trans. and ed. M. A. Salahi (Markfield: The Islamic Foundation, 2020).

As-Suyūṭī, Jalāl ad-Dīn. *Qaṭf al-azhār fī kashf al-asrār (Asrār at-tanzīl)*. Ed. Aḥmad al-Ḥamādī. Doha: Wizārat al-Awqāf wa-sh-Shu'ūn al-Islāmiyya, 1994.

As-Suyūṭī, Jalāl ad-Dīn. *Al-Iklīl fī istinbāṭ at-tanzīl*. Ed. Sayf ad-Dīn ʿAbd al-Qādir al-Kātib. Beirut: Dār al-Kutub al-ʿIlmiyya, 1985.

As-Suyūṭī, Jalāl ad-Dīn. *Takmilat tafsīr al-Jalāl al-Maḥāllī (Tafsīr al-Jalālayn*. Ed. n.p. Beirut: Dār Ibn Kathīr, 2008.

- Ed. Cairo: Dār al-Ḥadīth, 2001.

- Ed. Istanbul: Dār al-Kutub al-ʿArabiyya, 2019.

As-Suyūṭī, Jalāl ad-Dīn. *Nawāhid al-abkār wa-shawārid al-afkār*. Ed. Aḥmad Ḥāj Muḥammad ʿUthmān, *(Q 2:1–2:20)*. Makka: Jāmiʿat Umm al-Qurʾā, 2004.

- Ed. Muḥammad Kamāl ʿAlī, *(Q 2:21–3:112)*. Makka: Jāmiʿat Umm al-Qurʾā, 2004.

- Ed. Iʿdād aṭ-Ṭālib, *(Q3:113–9:48)*. Makka: Jāmiʿat Umm al-Qurʾā, 2004.

As-Suyūṭī, Jalāl ad-Dīn. *Ṭabaqāt al-mufassirīn*, (Arabic-Latin). Ed. Albert Meursinge, Hendrik Engelinus Weyers. Leiden: Leiden University, 1839.

- Ed. ʿAlī Muḥamamd ʿUmar. 2 vols. Cairo: Maktaba Wahba, 1976.

As-Suyūṭī, Jalāl ad-Dīn. *Mufḥimāt al-aqrān fī mubhamāt al-Qurʾān*. Ed. Ṭāhā ʿAbd ar-Raʿūf Saʿd. Cairo: al-Maktabat al-Azhariyya li-t-Turāth, 1992.

As-Suyūṭī, Jalāl ad-Dīn. *Marāṣid al-maṭāliʿ fī tanāsub al-maqāṭiʿ wa-l-maṭāliʿ*. Ed. ʿAbd al-Muḥsin b. ʿAbd al-ʿAzīz al-ʿAskar. Riyadh: Maktabat Dār al-Minhāj, 2005.

As-Suyūṭī, Jalāl ad-Dīn. *Tanāsuq ad-durar fī tanāsub as-suwar*. Ed. ʿAbd al-Qādir Aḥmad ʿAṭā. Beirut: Dār al-Kutub al-ʿIlmiyya, 1985.

As-Suyūṭī, Jalāl ad-Dīn. *Lubāb an-nuqūl fī asbāb an-nuzūl*. Ed. ʿAbd ar-Razzāq al-Mahdī Beirut: Dār al-Kitāb al-ʿArabī, 2005.

As-Suyūṭī, Jalāl ad-Dīn. *At-Taḥbīr fī ʿilm at-tafsīr*. Ed. Fatḥī ʿAbd al-Qādir Farīd. Riyadh: Dār al-ʿUlūm, 1981.

Ṭashkubrīzādah, Aḥmad. *Miftāḥ as-saʿāda wa-misbāḥ as-siyyāda*. Ed. ʿAlī Daḥrūj. Beirut: Maktaba Lubnān Nāshirūn, 1998.

Az-Zarkashī, Badr ad-Dīn Muḥammad. *Al-Burhān fī ʿulūm al-Qurʾān*. Ed. Muḥammad Abū Faḍl Ibrāhīm. 4 vols. Cairo: Maktaba Dār at-Turāth, 1983.

8.2 Secondary Literature

Akpınar, Ali. "Tefsîrü'l-Celâleyn." In *TDVIA*, edited by İslâm Araştırmaları Merkezi, 44 vols. Ankara: TDV, vol. 40, 2011, 40: 294–95.

Ally, Shabir. "The Culmination of Tradion-based *Tafsīr*: The Qurʾān Exegesis *al-Durr al-manthūr* of as-Suyūṭī (d. 911/1505)." PhD diss., University of Toronto, 2012.

Anthony, Sean. "Two 'Lost' Suras of the Qurʾān: Surat al-Khalʿ and Surat al-Ḥafd between Textual and Ritual Canon (1st–3rd/7th–9th centuries)," *JSAI* 46 (2019): 67–112.

Bauden, Frédéric. "Maqriziana IX: Should al-Maqrīzī be thrown out with the bath water? The question of his plagiarism of al-Awḥadī's Khiṭaṭ and the documentary evidence." *MSR* 14 (2010): 159–232.

Behrens-Abouseif, Doris. *The Book in Mamluk Egypt and Syria (1250–1517): Scribes, Libraries and Market*. Leiden, Boston: Brill, 2018.

Bekiroğlu, Harun. *Tefsir Metodolojisi Açısından El-Burhân ve El-Itkân*. Ankara: Araştırma Yayınları, 2013.

Birışık, Abdülhamit. "İbn Akîle." In *TDVIA*, edited by İslâm Araştırmaları Merkezi, 44 vols. Ankara: TDV, vol. 19, 1999: 304–6.

Burge, Stephen R. "Evidence of Self-Editing in al-Suyūṭī's Taḥbīr and Itqān: A Comparison of His Chapters on Asbāb al-Nuzūl." In *Al-Suyūṭī, a Polymath of the Mamlūk Period*, edited by Antonella Ghersetti. Leiden: Brill, 2017: 143–81.

Burge, Stephen R. "Scattered pearls: Exploring al-Suyūṭī's hermeneutics and use of sources in *al-Durr al-manthūr fī l-tafsīr bi-l-ma'thūr*." *JRAS* 24 (2014): 251–96.

Geoffroy, Éric. "al-Suyūṭī." In EI^2, edited by P. Bearman et al. Brill, 1954–2005. Article published online 2012. Accessed on December 7, 2023. http://dx.doi.org/10.1163/1573-3912_islam_COM_1130.

Gibb, H.A.R. "al-Bulḳīnī." In EI^2, edited by P. Bearman et al. Brill, 1954–2005. Article published online 2012. Accessed on December 7, 2023. http://dx.doi.org/10.1163/1573-3912_islam_SIM_1523.

Gökbulut, Hasan. "Kâfiyeci." In *TDVIA*, edited by İslâm Araştırmaları Merkezi, 44 vols. Ankara: TDV, vol. 24, 2001: 154–55.

Homerin, Th. Emil. "Ibn al-Fāriḍ." In EI^3, edited by Kate Fleet et al. Brill, 2007–. Article published online 2016. Accessed on November 22, 2023. <http://dx.doi.org/10.1163/1573-3912_ei3_COM_30774>.

İmamoğlu, İbrahim Hakkı. "Celâleddîn es-Süyûtî ve Kur'an İlimlerindeki Yeri. El-Itkân Örneği." PhD diss., University of Ankara, 2011.

Ito, Takao. "Al-Suyūṭī and Problems of the waqf." In *Al-Suyūṭī, a Polymath of the Mamlūk Period*, edited by A. Ghersetti. Leiden: Brill, 2017: 47–63.

Kandemir, M. Yaşar. "Bulkînî, Abdurrahman b. Ömer." *TDVIA*, edited by İslâm Araştırmaları Merkezi, 44 vols. Ankara: TDV, vol. 6, 1992: 409–10.

Mertoğlu, M. Suat. "Süyûtî/Kur'an İlimleri." In *TDVIA*, edited by İslâm Araştırmaları Merkezi, 44 vols. Ankara: TDV, vol. 38, 2010: 198–201.

Nolin, Kenneth E. "The Itqān and its Sources: A Study of al-Itqān fī ʿulūm al-Qur'ān by Jalāl al-Dīn al-Suyūṭī with special reference to al-Burhān fī ʿulūm al-Qur'ān by Badr al-Dīn al-Zarkashī." PhD diss., Hartford Seminary Foundation, 1968.

Özkan, Halit. "Süyûtî." In *TDVIA*, edited by İslâm Araştırmaları Merkezi, 44 vols. Ankara: TDV, vol. 38, 2010: 188–98.

Saleh, Walid A. "Ibn Taymiyya and the Rise of Radical Hermeneutics. An Analysis of 'An Introduction to the Foundations of Qur'ānic Exegesis'." In *Ibn Taymiyya and His Times*, edited by Yossef Rapoport and Shahab Ahmed. Oxford/New York: Oxford University Press, 2010: 123–62

Sartain, Elizabeth. *Jalāl al-dīn al-Suyūṭī. I: Biography and background*, 2 vols. Cambridge: Cambridge University Press, 1975.

Seker, Nimet. *Der Koran als Rede und Text. Hermeneutik sunnitischer Koranexegeten zwischen Textkohärenz und Offenbarungskontext.* Berlin: EB Dr. Brandt, 2019.

Seker, Nimet. "As-Suyūṭīs Diskussion der Mehrfachherabsendung am Beispiel von Vers 33:35." In *Sure Al-Aḥzāb 33*, edited by Ayşe Başol. Berlin: EB Dr. Brandt, 2019: 103–38.

Tomar, Cengiz. "Sehâvî." In *TDVIA*, edited by İslâm Araştırmaları Merkezi, 44 vols. Ankara: TDV, vol. 36, 2009: 313–16.

Tyllack, Olaf. "Die Quellen des Itqān von Suyūṭī." Phd diss., LMU München, 1952.

Vidal-Castro, Francisco. "Ibn al-Zubayr al-Gharnāṭī." In EI^3, edited by Kate Fleet et al. Brill, 2007–. Article published online 2018. Accessed on December 5, 2023. http://dx.doi.org/10.1163/1573-3912_ei3_COM_32316.

Yalnız, Fatma. "Ulûmu'l-Kur'ân Külliyatı Yazımı Konusunda Süyûtî'nin, Bulkînî'nin ve Kafiyeci'nin Birbirlerinden Ne Ölçüde Etkilendiklerinin Tespiti." *Gümüşhane Üniversitesi İlahiyat Fakültesi Dergisi* 11, no. 21 (2022): 208–41.

Yıldız, Kemal. "İbnü'l-Kerekî." In *TDVIA*, edited by İslâm Araştırmaları Merkezi, 44 vols. Ankara: TDV, vol. 21, 2000: 111–12.

Zarzūr, ʿAdnān Muḥammad. *The Qur'ān and Its Study: An In-depth Exploration of Islam's Sacred Scripture*, trans. and edited by M. A. Salahi. Markfield: The Islamic Foundation, 2020.

Abū s-Su'ūd Efendi

M. Taha Boyalik

1 Introduction

Abū s-Su'ūd b. Muḥyī d-Dīn Meḥmed al-'Imādī (d. 1574/982) was a prominent scholar who lived in Istanbul in the sixteenth century. He is known as Abū s-Su'ūd Efendi. At the height of the Ottoman state's strength, he spent a long period serving as a professor (*mudarris*), judge (*qāḍī*), and chief judge (*qāḍī 'askar*), and then, as the head of religious and legal affairs (*shaykh al-Islām*), he carried out the state's judicial and fatwa affairs for twenty-nine years.

Although Abū s-Su'ūd Efendi is mostly known as a jurist (*faqīh*), the most famous of his works was his Qur'ānic exegesis, titled *Irshād al-'aql as-salīm ilā mazāyā l-Kitāb al-karīm* (The Guide for the Sound Mind to the Virtues of the Noble Book), which he wrote after he became *Shaykh al-Islām*. There are justified reasons for the universal reputation of this work: the *Shaykh al-Islām* very skillfully applies his vast knowledge of the fields of recitation (*qirā'a*), linguistics, rhetoric, *fiqh*, and theology in this commentary. This work is regarded as an authority, in which linguistics and rhetoric intersect with the interpretation of the Qur'ān.

This article will examine Abū s-Su'ūd Efendi's *Irshād* in the light of various aspects, in particular from the perspective of methodology. After providing information about the life of the commentator, *Irshād* will be introduced and analyzed in terms of its sources, content, and method.

2 Life and Works

Abū s-Su'ūd Efendi spent almost his entire life in Istanbul, the capital of the Ottoman Empire, in the sixteenth century. During his lifetime, four great Ottoman sultans ruled. Abū s-Su'ūd Efendi witnessed the twenty-one years of Bayezid II's reign (1481–1512/886–918) and the full reigns of Selim I (1512–1520/918–926), Suleiman the Magnificent (1520–1566/926–974), and Selim II (1566–1574/974–982). Under the rule of these mighty sultans, the borders of the Ottoman Empire stretched from the Adriatic coast to Yemen.[1]

Abū s-Su'ūd Efendi's father, Muḥyī d-Dīn Meḥmed (d. 1514/920), was a great scholar and Sufi. Muḥyī d-Dīn Meḥmed had established close relations with Sultan Bayezid II when the latter was still a prince in Amasya. When Bayezid became sultan,

1 Colin Imber, *Ebu's-Su'ud: The Islamic Legal Tradition* (Edinburgh: Edinburgh University Press, 1997), 3f.

https://doi.org/10.1515/9783111320083-015

Muḥyī d-Dīn Meḥmed settled in Istanbul and continued his studies and advisory activities in the lodge allocated to him by the sultan. Muḥyī d-Dīn Meḥmed was known as the "sultan's shaykh" due to his close relationship with the Sultan. It was said that prominent statesmen often visited his lodge.[2] Being the son of a famous scholar and Sufi with good relations to statesmen provided Abū s-Suʿūd with significant advantages in his scientific career.

According to the commonly held view, Abū s-Suʿūd Efendi was born on 17th Ṣafar 896/December 30, 1490.[3] His date of birth has also been given as 19th Ṣafar 896/January 1, 1491,[4] 19th Ṣafar 897/December 22, 1491,[5] and the year 898/1493.[6] Most of the sources report that he was born in Istanbul.[7] However, statements from his own *wakfiyya* suggest that he was born in the town of Iskilip, his father's hometown.[8] Iskilip, which is now a town in Çorum, Turkey, was part of Amasya during the Ottoman period. Abū s-Suʿūd Efendi must have been born there, before migrating to Istanbul with his father at an early age.

There is no certainty about Abū s-Suʿūd Efendi's full name. He only used "Abū s-Suʿūd" to refer to himself in his fatwas, letters, and *wakfiyyas*, and in the introductions to his works. Scholarly contemporaries also referred to him as only "Abū s-Suʿūd."[9] This makes it more likely that "Abū s-Suʿūd" is a name, not a tag. However, some sources mention the name Meḥmed,[10] while others state that his name was

2 Ahmad Efendi Tāshkubrīzādah, *ash-Shaqā'iq an-nuʿmāniyya fī ʿulamā' ad-dawla al-ʿuthmāniyya* (Beirut: Dār al-Kitāb al-ʿArabī, 1975), 206; Meḥmed al-Majdī, *Tarjama-yi Shaqā'iq an-Nuʿmāniyya. Hadā'iq ash-Shaqā'iq* (Istanbul: Dār at-Tibāʿa al-ʿĀmira, 1853), 349.

3 Nawʿīzādah ʿAṭā'ī, *Ḥadā'iq al-ḥaqā'iq fī takmilat ash-Shaqā'iq*, prepared by Suat Donuk, ed. Derya Örs (Istanbul: Türkiye Yazma Eserler Kurumu, 2017), 1: 640; Sulaymān Sāʿd ad-Dīn Mustaqimzādah, *Devḥat al-mashāyikh* (Istanbul: Çağrı Yayınları, 1978), 24; Fā'iq Rashād, *Aslāf* (Istanbul: Asır Kütüphanesi, 1311–12 (1895/6–1896/7), 1: 24.

4 ʿAbd al-Qādir al-ʿAydarūs, *an-Nūr as-sāfir ʿan akhbār al-qarn al-ʿāshir*, ed. Maḥmūd al-Arna'ūṭ et al. (Beirut: Dār Ṣādir, 2001), 319.

5 Mustaqimzādah, *Devḥat al-mashāyikh*, 24.

6 Tāshkubrīzādah, *ash-Shaqā'iq*, 440; ʿAlī b. Bālī (Manq ʿAlī), *al-ʿIqd al-manẓūm fī dhikr afāḍil ar-Rūm*, prepared by Suat Donuk, ed. Muhammed Hekimoğlu (Istanbul: Türkiye Yazma Eserler Kurumu Başkanlığı, 2018), 461.

7 Tāshkubrīzādah, *ash-Shaqā'iq*, 440; ʿAṭā'ī, *Ḥadā'iq al-ḥaqā'iq*, 1: 640; Awliya Çalabi, *Seyahatnâme* (Istanbul: Ikdâm Matbaası, 1928–38), 1: 402; Mustaqimzādah, *Devḥat al-mashāyikh*, 24.

8 Al-ʿAydarūs, *an-Nūr as-sāfir*, 319; M. Cavid Baysun, "Ebüssu'ūd Efendi," in *MEB Islâm Ansiklopedisi* (Istanbul: Milli Eğitim Bakanlığı, 1977), 4: 92.

9 Tāshkubrīzādah, *ash-Shaqā'iq*, 440; Ibn Bālī, *al-ʿIqd al-manẓūm*, 461; ʿAṭā'ī, *Ḥadā'iq al-ḥaqā'iq*, 1: 640.

10 ʿAbd al-Ḥayy b. Aḥmad b. al-ʿImād, *Shadharāt adh-dhahab fī akhbār man dhahab*, ed. ʿAbd al-Qādīr al-Arnā'ūṭ and Maḥmūd al-Arnā'ūṭ (Damascus/Beirut: Dār Ibn Kathīr, 1986), 10: 584; Sulaymān Sāʿd ad-Dīn Mustaqimzādah, *Majallat an-nisāb* (Ankara: Kültür Bakanlığı, 2000), 471b; Abū l-Hasanāt Muḥammad al-Laknawī, *al-Fawā'id al-bahiyya fī tarājim al-Ḥanafiyya*, ed. Muḥammad Badr ad-Dīn Abū Firās an-Naʿsānī (Cairo: Maṭbaʿat as-Saʿāda, 1324 (1908/9)), 82; *Ilmiye Salnāmesi* (Istanbul: Matbaa-i Āmire,

Aḥmed.[11] Among his many nicknames, such as "the second Abū Ḥanīfa,"[12] "*al-mu^c^al-lim ath-thānī*" (The Second Teacher),[13] and the "Sultan of the commentators,"[14] "Khoja Çelebi" is the most well known. Abū s-Su'ūd Efendi generally used the *nisba* "al-'Imadī" for himself. He also used the *nisba* "al-Iskilībī" at least once.[15]

Abū s-Su'ūd Efendi received his primary education from his father who taught him al-Jurjānī's (d. 1413/816) *Ḥashiyat at-tajrīd, sharh Miftāḥ al-'ulūm* and *Sharh al-Mawāqif.*[16] He stood out at a young age due to his wisdom and intelligence.[17]

In a graduation certificate that Abū s-Su'ūd Efendi wrote for his student, he stated that his teachers were his father, Muḥyī d-Dīn Meḥmed, as well as Muayyadzādeh 'Abd ar-Raḥmān (d. 1516/922) and Seydī Meḥmed (d. 1524–25/931).[18] After completing his basic education, Abū s-Su'ūd Efendi performed the duties of teacher, judge, chief judge, and *Shaykh al-Islām*. He started his first professorship at Isḥāk Pasha Madrasa in Inegol (1516–1520/922–926) at the age of twenty-six.[19] After a ten-month interruption, he worked as a *muderris* in Dāvud Pasha (1521/927) and Maḥmud Pasha Madrasas (1522–1525/928–931) in Istanbul, Koja Muṣṭafa Pasha Madrasa (1525/931) in Gebze, Sultāniya Madrasa (1526–1528/932–934) in Bursa, and Ṣaḥn Thamān Madrasa (1528–1533/934–939) in Istanbul.[20] He then became the judge of Bursa (1533/939) and, after just six months, was promoted to judge of Istanbul (1533–1537/940–944). He subsequently served as chief judge for eight years (1537–1545/944–952). Finally, he assumed the position of *Shaykh al-Islām* (1545–1574/952–982), which was the highest position of the *'il-miyya* class, and continued in that role for twenty-nine years until his death.[21]

According to the sources, Abū s-Su'ūd Efendi was dignified, decent, and humble. His oratory skills were of the highest level. He impressed his interlocutors when he spoke and was a master of argument. He is said to have beaten all scholars in the

---

1916), 376; Khayr ad-Dīn az-Ziriklī, *al-A'lām: Qāmūs tarājim li-ashhar ar-rijāl wa-n-nisā' min al-'arab wa-l-musta'ribīn wa-l-mustashriqīn* (Beirut: Dār al-'Ilm li-l-Malāyīn, 2002), 7: 59.

11 Ismā'īl Pasha al-Baghdādī, *Hadiyyat al-'ārifīn*, ed. Kilisli Rifat Bilge-Ibnülemin Mahmud Kemal (Ankara: Milli Eğitim Bakanlığı, 1951–55), 2: 253; Shams ad-Dīn Sāmī, *Qāmūs al-a'lām* (Ankara: Kaşgar Neşriyat, 1996), 1: 722; Rashād, *Aslāf*, 1: 24.

12 Hasan Çalabi Qınalızādah, *Tadhkirat ash-shu'arā'*, ed. Ibrahim Kutluk (Ankara: Türk Tarih Kurumu Yayınları, 1978), 1: 348; Rashād, *Aslāf*, 1: 24.

13 Mustaqimzādah, *Devḥat al-mashāyikh*, 26.

14 'Aṭā'ī, *Ḥadā'iq al-ḥaqā'iq*, 1: 613; Mustaqimzādah, *Devḥat al-mashāyikh*, 23.

15 Abū s-Su'ūd Efendi, *Biḍā'at al-qādī*, Süleymaniye Library, Hacı Mahmūd Efendi, no. 631, 30b.

16 Tāshkubrīzādah, *ash-Shaqā'iq*, 440; 'Aṭā'ī, *Ḥadā'iq al-ḥaqāiq*, 1: 640.

17 'Aṭā'ī, *Ḥadā'iq al-ḥaqā'iq*, 1: 640; Mustaqimzādah, *Devḥat al-mashāyikh*, 24.

18 Abū s-Su'ūd Efendi, *Icāzatnāmah*, Süleymaniye Library, Pertevniyal, no. 910, 40b.

19 'Aṭā'ī, *Ḥadā'iq al-ḥaqā'iq*, 1: 640f.

20 Ibid., 641.

21 'Aṭā'ī, *Ḥadā'iq al-ḥaqā'iq*, 1: 641f.; Maḥmūd b. Sulaymān al-Kafawī, *Katā'ib a'lām al-akhyār min fuqahā' madhab an-Nu'mān al-mukhtār*, ed. Saffet Köse et al. (Istanbul: Maktabat al-Irshād, 2017), 4: 461.

debates held in the presence of Suleiman the Magnificent.[22] He wrote fluently and clearly, and was a skilled poet. He is reported to have demonstrated the most outstanding eloquence in his poems.[23] He was familiar with the subtleties of Arabic, Turkish, and Persian, and wrote very successful poems in these three languages. He gave his fatwas in Turkish, Arabic, or Persian, depending on the language spoken by the person requesting the fatwa.[24] He dedicated himself to knowledge and worship throughout his life, without wasting even an hour.[25] His fatwas also reveal his witty, jovial personality.[26]

Abū s-Suʿūd Efendi developed good relations with the ruling class.[27] As a member of a respectable family, his superior intelligence, literary style, knowledge, and charismatic personality made it easy for him to establish contact with administrators. It has been said that he was much inclined toward rulers and governors, and he was careful to behave gently and politely toward them.[28] This must be one of the reasons for his constant promotions and professional stability.

Abū s-Suʿūd Efendi reached the highest level in the fields of rhetoric, *fiqh*, legal philosophy, and Qurʾānic exegesis.[29] He was one of the greatest authorities of his time in the field of *fiqh*. In *fiqh*, he followed the Ḥanafī school, which was widespread in his environment. In the field of theology, he followed the Māturīdīan school of the Ahl as-Sunna sect. It is reported that his most important achievement was incorporating administrative law (*qānūn*) into religious law (*sharīʿa*).[30] According to the view commonly held by his biographers, he was the greatest scholar of his age and has never been surpassed.[31] He wrote over fifty works, most of them in the fields of *fiqh* and *tafsīr*.[32] They include *Tafsīr Sūrat*

---

22 Al-ʿAydarūs, *an-Nūr as-sāfir*, 319.

23 Tāshkubrīzādah, *ash-Shaqāʾiq*, 362–64; Ibn Bālī, *al-ʿIqd al-manẓūm*, 473; ʿAṭāʾī, *Ḥadāʾiq al-ḥaqāʾiq*, 1: 647. Tāshkubrīzādah and Ibn Bālī presented his biography as well as a selection of his Arabic poems, see: Tāshkubrīzādah, *ash-Shaqāʾiq*, 445–53; Ibn Bālī, *al-ʿIqd al-manẓūm*, 475–507.

24 Ibn Bālī, *al-ʿIqd al-manẓūm*, 465.

25 Al-Kafawī, *Katāʾib*, 4: 463.

26 Abdullah Demir, *Devlet-i Aliyye'nin Büyük Hukukçusu Şeyhülislam Ebussuud Efendi* (Istanbul: Ötüken, 2006), 39.

27 Imber, *Ebu's-Suʿud*, 11.

28 Tāshkubrīzādah, *ash-Shaqāʾiq*, 445; Ibn Bālī, *al-ʿIqd al-manẓūm*, 474.

29 Al-Kafawī, *Katāʾib*, 3: 460f.

30 ʿAṭāʾī, *Ḥadāʾiq al-ḥaqāʾiq*, 1: 643f.; Kātib Çalabi, *Mīzān al-ḥaqq fī ikhtiyār al-aḥaqq* (Istanbul: Ali Riza Efendi Matbaası, 1286 (1870–71)), 126; *Ilmiye Salnāmesi*, 377; Michael M. Pixley, "Suleyman's Şeyhülislam," *TSAB* 6, no. 2 (1982): 12.

31 Tāshkubrīzādah, *ash-Shaqāʾiq*, 443; Ibn Bālī, *al-ʿIqd al-manẓūm*, 470f.; al-Kafawī, *Katāʾib*, 4: 458; Qınalızādah, *Tadhkirat ash-shuʿarāʾ*, 1: 348.

32 For a complete list and brief introduction to his books, see: Nihal Atsız, *Istanbul Kütüphanelerine Göre Ebussuud Bibliyografyası* (Istanbul: Milli Eğitim Basımevi, 1967), 5–61; Ahmet Akgündüz, "Ebüssuûd Efendi," in *TDVIA*, ed. İslâm Araştırmaları Merkezi (Ankara: TDV, 1994), 10: 370f.; Demir, *Şeyhülislam Ebussuud Efendi*, 42–52; Pehlül Düzenli, *Şeyhülislām Ebussuūd Efendi ve Fetvāları* (Istanbul: Osmanlı Araştırmaları Vakfı, 1994), 63–82; Pixley, "Suleyman's Şeyhülislam," 10–12. Halil Şimşek, "The Missing Link in the History of Quranic Commentary: The Ottoman Period and the Quranic Commen-

al-Furqān, Tafsīr Sūrat ash-Shuʿarāʾ, Tafsīr Sūrat al-Muʾminūn, Tafsīr Sūrat an-Nūr, Maʿāqid aṭ-ṭirāf fī awwal Sūrat al-Fatḥ min al-Kashshāf, as well as several *hāshiyas*: *Ḥāshiya ʿalā l-Hidāya, Ḥāshiya ʿalā Sharh al-Hidāya li-Akmal ad-Dīn, Maʿrūzāt-ı Abū s-Suʿūd Efendi, Kānunnāmahs, Ḥāshiya ʿalā t-Tawdīḥ,* and *Ḥāshiya ʿalā Awāʾil al-Anwār.*

Abū s-Suʿūd Efendi died in Istanbul on 5th Jumādā al-Awwal 982/August 23, 1574.[33]

3 Irshād al-ʿaql as-salīm

Irshād was written over a period of twenty years. It is highly probable that writing commenced in 1546/953, one year after Abū s-Suʿūd Efendi took office as *Shaykh al-Islām.*[34] As can be gleaned from the dates given at the end of the suras in the author's copy,[35] the section between sura 9 *at-Tawba* (Repentance), and sura 18, *al-Kahf* (The Cave), was interpreted first. The volume then turns to the beginning, interpreting the suras in the order that they appear in the Qurʾān. The work was completed on 3rd Rajab 973/January 24, 1566 according to the date given by the author at the end of the volume, which Abū s-Suʿūd Efendi dedicated to Suleiman the Magnificent.

Abū s-Suʿūd Efendi presents a vision of the Qurʾān in the introduction to *Irshād* and classifies the history of *tafsīr*. The following sections will first present the explanations he provides in the introduction, followed by a discussion of the commentary's content and method.

3.1 The Qurʾān and the History of *tafsīr* in the Introduction

Abū s-Suʿūd Efendi provides a remarkable description of the Qurʾān in the introduction. He states that "the ultimate goal in the writing of the book of the world and the great wisdom in the kneading of Adam's dough is to know and worship the supreme designer of the world. The goal of knowing the Creator and worshiping him can only be achieved by knowing the Qurʾān."[36]

The *Shaykh al-Islām* explains these statements as follows: God wrote the evidence of his power onto the pages of the book of the universe and sewed the banner of his

tary of Ebussuud/Abū al-Suʿūd al-ʿImādī (d. 1574 CE) *Irshād al-ʿaql al-salīm ilā mazāyā al-Kitāb al-Karīm*" (PhD diss., University of Toronto, 2018), 144–51.

33 ʿAṭāʾī, *Ḥadāʾiq al-ḥaqāʾiq*, 1: 642.

34 Ādem Yerinde, "Nuskhat al-muʿallif wa-aqdam nusakh li-tafsīr Abī s-Suʿud," *Yakın Doğu Üniversitesi Ilahiyat Fakültesi Dergisi* 1, no. 2 (2015): 203.

35 Bayezid Library, Merzifonlu Kara Mustafa Paşa, nos. 18712, 18713, 18714, 18715.

36 Abū s-Suʿūd Efendi, *Irshād al-ʿaql as-salīm ilā mazāyā al-Kitāb al-karīm*, ed. Mehmet Taha Boyalık, Ahmet Aytep, Ziyâüddin el-Kaliş, Muhammed İmâd el-Nabulsî, 9 vols (Istanbul: Nashriyyāt al-Vakf al-Diyāna al-Turkī, 2020), 1: 8.

unity onto the surfaces of objects and lands. He has made every particle, every piece of knowledge, and every line of genius and creativity a mirror and unquestionable proof of the observance of his beauty and the attributes of his perfection. However, it is beyond human power to read the book of the universe properly, to learn from it, to learn the secrets, signs and symbols of this book, and to grasp the intricacies of its evidence. For this reason, the Creator, who has absolute power, spoke by addressing human beings and supported the cosmic (*takwīnī*) verses with verbal verses. The addition of verbal verses to the cosmic verses removed the obstacles human beings faced in attaining the secret of *ʿubūdiyya* and *rubūbiyya*.[37]

According to Abū s-Suʿūd's explanations, the Divine Word, the Qurʾān, gives details of religious principles, explains the incomprehensible points in the *takwīnī* verses, removes obstacles to the holy realms, reveals the hidden potential in man, and equips him with the knowledge, wisdom, and principles that will lead him to happiness in the world and in the hereafter. In a literary style, the divine speech presents the subtleties of the theoretical and practical sciences and secret arts, the details of religious provisions, the essence of original and secondary evidence, the truth of things, the states of the angelic world, the rules to be obeyed, and things to avoid. However, this Divine Word, which pulls back the veils of existence, is also of a nature that will push the limits of human comprehension. It is very difficult to grasp its lofty meanings and impossible to exhaust its interpretations.[38]

Proper understanding of the Divine Word and the aforementioned qualities will, of course, require expertise. Abū s-Suʿūd Efendi spends the rest of the introduction explaining how scholars of the past have admirably done their part in this regard. In every century, prominent exegetes have interpreted the obscure and complicated parts of the Qurʾān.[39] According to the author, the history of *tafsīr* can be divided into two main parts, which correspond to the attitudes of scholars: the period of earlier scholars and the period of later scholars.

> Previous scholars who were analysts (*al-mutaqaddimūn al-muḥaqqiqūn*) revealed the meanings of the Qurʾān as it came from the Prophet – best greetings to Him – analyzed its structures, explained its intentions, and arranged its provisions. As for the later scholars who were investigative (*al-mutaʾakhkhirūn al-mudaqqiqūn*), they tried to reveal the miraculous aspects of the Qurʾān that left people incapable and its superior and privileged place among other great rabbinical books, in addition to what the previous ones had done. For this purpose, they have produced outstanding works full of bright and important arts (*funūn*) that contain valuable meanings that only distinguished eyes can see and fine distinctions that can be heard and captured by keen ears. Among them, *al-Kashshāf* and *Anwār at-tanzīl* stand out with their outstanding features. These two are pioneering works that have achieved great success. They are mirrors that clearly reveal the face of inimitability (*iʿjāz*). Their content is an exhibition of the finest virtues, while their lines are pearl neckbands and pure gold necklaces.[40]

---

37 Ibid., 1: 8f.
38 Ibid., 1: 9.
39 Ibid., 1: 9f.
40 Ibid., 1: 10f.

According to this text, what distinguishes the later period from the earlier period in *tafsīr* is that the latter demonstrates the miraculousness and the inimitability of the Qur'ān by revealing the precious meanings that only distinguished eyes can see and the fine distinctions that only sharp ears can hear. This was exemplified most notably in Abū l-Qāsim Maḥmūd b. 'Umar az-Zamakhsharī's (d. 1144/538) *al-Kashshāf* and Nāṣir ad-Dīn al-Bayḍāwī's (d. probably 1292/961) *Anwār at-tanzīl.*

Abū s-Su'ūd Efendi clearly states that his main sources are these two commentaries and bases his work on these constructive texts from the later period. He notes that he will follow the two commentaries methodically and reflect their subtle meanings in his own work in a beautiful style that is suitable for the inimitability of the Qur'ān. However, he notes that he will not be content with quotations from his forebears, and he is far from being modest in expressing that he will surpass his two predecessors in many ways. Abū s-Su'ūd Efendi promises that his commentary will be more successful than the other two in terms of style, content, and the application of methods, drawing on various sources and building on his own thoughts and findings.[41]

3.2 Content and Sources

The verses of the Qur'ān are studied in many ways in *Irshād.* Although the tradition-based approach is not particularly prominent, traditions are conveyed when necessary. It has been determined that there are 1,200 hadiths in *Irshād.*[42] The author took care to convey the traditions about the reasons for the *nuzūl.*[43] In the work, there are few narratives originating from Judeo-Christian traditions (*isrā'īliyyāt*) about miracles, stories, or historical events. *Isrā'īliyyāt* are rarely criticized.[44] Abū s-Su'ūd Efendi mentions different recitations (*qirā'āt*) in the verses, but he does not generally differentiate between authentic recitations and unacceptable (*ash-shādh*) ones.[45]

As in its two main sources, az-Zamakhsharī's *al-Kashshāf* and al-Bayḍāwī's *Anwār at-tanzīl, Irshād* is characterized by its distinctive linguistic and literary approach. Abū s-Su'ūd Efendi regularly examines the phonological features of words, their morphological structures, derivations, and lexical and contextual meanings. He analyzes the sentence structure within the framework of the theory of *al-'āmil* in the *nahw*, paying particular attention to the subtleties of meaning that arise from the selective elements of syntax.

41 Ibid.

42 Abdullah Aydemir, *Büyük Türk Bilgini Şeyhülislam Ebussuud Efendi ve Tefsirdeki Metodu* (Ankara: Diyanet Işleri Başkanlığı Yayınları, n.d.), 165.

43 For examples, see: ibid., 125–31.

44 Abū s-Su'ūd Efendi, *Irshād*, 1: 318f.; 7: 288.

45 Şimşek, "The Missing Link," 233–48.

Although Abū s-Suʿūd Efendi was an expert in the field of *fiqh*, he does not go into the details of *fiqh* issues and is content with pointing out the provisions in the verses,[46] often conveying the views of *fiqh* scholars. For example, while discussing whether the *Basmala* is included in sura 1, *al-Fātiḥa* (The Opening), he recounts the views of the imams of the four *fiqh* sects. Sometimes he gives the views of Mālik b. Anas (d. 795/179), Abū Ḥanīfa (d. 767/150), and Muḥammad b. Idrīs Shāfiʿī (d. 820/204),[47] sometimes only the views of Ḥanafīs and Shāfʿīs,[48] and sometimes just the views of Shāfiʿī.[49] As a Ḥanafī jurist, Abū s-Suʿūd Efendi favors the Ḥanafī sect when it comes to controversial *fiqh* issues. Ḥanafī views are generally adopted with the expression "for us" (*ʿindanā*).[50]

Detailed discussions of *kalām* and evidence of theological sects are not included in *Irshād* either. The theological interpretations presented are based on the Ahl as-Sunna and adhere closely to the literal meanings of the verses. The views of the Ahl as-Sunna are referred to by names such as *ahl as-sunna*[51] and *ahl al-ḥaqq*,[52] or the expression "for us" (*ʿindanā*)[53] is used. When there are contentious issues between Ashʿariyya and Māturīdiyya within the Ahl as-Sunna, the author prefers the views of the Māturīdiyya school.[54]

Abū s-Suʿūd Efendi makes use of many sources in his commentary. As Abū s-Suʿūd Efendi states in the introduction, although his main sources are az-Zamakhsharī's *al-Kashshāf* and al-Bayḍāwī's *Anwār at-tanzīl*, he aims to go beyond them. There are direct references to many commentators in the *Irshād*. The most famous of these are Muqātil b. Sulaymān (d. 767/150) (forty-one times), Abū Zakāriyyā l-Farrāʾ (d. 822/207) (forty-three references), Abū Isḥāq az-Zajjāj (d. 923/311) (thirty three references), Abū Jaʿfar aṭ-Ṭabarī (d. 923/310) (three references), Ibn Abī Ḥātim ar-Rāzī (d. 938/327) (one reference), Aḥmad b. Muḥammad ath-Thaʿlabī (d. 1035/427) (three references), ʿAlī b. Aḥmad al-Wāḥidī (d. 1076/468) (eighteen references), Abū Muḥammad al-Baghawī (d. 1122/516) (three references), Fakhr ad-Dīn ar-Rāzī (d. 1210/606) (two references), Abū l-Baqāʾ al-ʿUkbarī (d. 1219/616) (nine references), and Abū Ḥayyān al-Gharnāṭī (d. 1344/745) (two references).

Considering that az-Zamakhsharī is mentioned only twice in *Irshād*, and al-Bayḍāwī is not mentioned directly, it would be inaccurate to draw conclusions about Abū s-Suʿūd Efendi's sources based solely on direct references. In addition, many of

46 Abū s-Suʿūd Efendi, *Irshād*, 1: 470; 3: 458.
47 Ibid., 1: 342, 410, 439f., 464f., 468.
48 Ibid., 1: 467, 502, 505, 518, 532.
49 Ibid., 1: 408, 474, 519.
50 Ibid., 1: 410, 440, 465, 467f., 505, 512, 518, 528.
51 Ibid., 1: 150; 4: 61, 425, 498; 6: 94, 413, 587; 7: 546, 603; 8: 22.
52 Ibid., 5: 84.
53 Ibid., 1: 95; 6: 306, 313.
54 Ibid., 1: 77f.

the names mentioned above are not directly cited by their works; rather, their explanations are conveyed through *al-Kashshāf* and *Anwār at-tanzīl.*

The sources for many explanations are given in the postscripts to the author's copy of *Irshād.*[55] The commentary that the author most often points to in the postscripts is 'Umar b. 'Alī b. 'Ādil's (d. 1475/880) *al-Lubāb fī 'ulūm al-Kitāb.*[56] There are also references to the commentaries of ath-Tha'labī,[57] al-Wāḥidī,[58] Ibn 'Aṭiyya al-Andalusī (d. 1147/541),[59] Abū l-Baqā' al-'Uqbarī,[60] and Abū 'Abd Allāh al-Qurṭubī (d. 1273/671).[61] Abū s-Su'ūd Efendi also made use of commentary glosses (*ḥāshiya*). In the postscripts to the author's copy, there are references to the glosses of *al-Kashshāf* written by scholars such as Sa'd ad-Dīn at-Taftāzānī (d. 1390/792),[62] Sirāj ad-Dīn al-Qazwīnī (d. 1344/745),[63] Qutb ad-Dīn ar-Rāzī (d. 1365/766),[64] al-Jārbardī (d. 1346/746),[65] and Sharaf ad-Dīn aṭ-Ṭībī (d. 1343/743).[66]

3.3 Linguistic and Literary Approach

The Qur'ān was revealed in Arabic and contains verses on different subjects such as creed, worship, law, and morality. While each verse can be examined in terms of Arabic linguistics and rhetoric, insights from disciplines such as *fiqh* and *kalām* can only be seen in some verses. Abū s-Su'ūd Efendi interprets the suras and verses in the order that they appear in the Qur'ān. He explains the verses by analyzing them down to their most atomic elements. Thus, explanations at the lexicographical, morphological, syntactic, and pragmatic levels can be seen in the analysis of each verse. Theological, philosophical, legal, mystical, and moral issues are dealt with only when the content of the verse requires it. Controversial issues in disciplines such as *kalām* and *fiqh* do not come to the fore in every verse or even in every sura. Therefore, the insights and methods from the conventional disciplines do not constitute the main framework of *Irshād.* When the issues concerning these disciplines are in question, Abū s-Su'ūd Efendi does not get into detailed discussions. There are few exceptions to this.

55 All of the author's notes are given in footnotes in the edition I use.
56 For example, see: Abū s-Su'ūd Efendi, *Irshād,* 1: 197, 206, 515; 2: 349, 454, 539; 3: 109, 125, 312, 515.
57 For example, see: Ibid., 2: 239, 342, 539; 3: 33, 258, 410.
58 For example, see: Ibid., 1: 112, 292, 495; 2: 140, 197, 224.
59 For example, see: Ibid., 4: 280, 597; 5: 327; 6: 535.
60 For example, see: Ibid., 1: 107, 225, 278, 298, 301, 323, 524.
61 For example, see: Ibid., 1: 21; 2: 66, 73, 96, 454, 536.
62 For example, see: Ibid., 1: 165, 193, 258, 277; 2: 339.
63 For example, see: Ibid., 2: 385; 3: 225; 5: 177.
64 For example, see: Ibid., 4: 305.
65 For example, see: Ibid., 5: 231.
66 For example, see: Ibid., 8: 66.

One natural consequence of taking an atomistic approach to *tafsīr* is that each linguistic unit, from phoneme to sentence structure, becomes the focus of the commentator. This results in the prominence of the methods of linguistics and rhetoric in the *tafsīr* activity. In fact, linguistics and rhetoric dominate in *Irshād.* Abū s-Suʿūd Efendi examines every word in a verse in terms of its phonetic features, lexical meaning, derivation, morphological structure, syntactic function, and contextual meaning, insofar as it requires explanation. Keeping with the claim he makes in the introduction, he is rather skilled at detecting the subtle meanings that appear contextually in the Divine Word. Thus, as promised in the introduction, he "pulls back the veils that covered the inimitable nature of the Qur'ān."

Behind Abū s-Suʿūd Efendi's linguistics and rhetorically and stylistically centered approach, there are *al-Kashshāf* and *Anwār at-tanzīl.* The latter source is also written in direct contact with the former. There is almost a consensus that az-Zamakhsharī's most important achievement in *al-Kashshāf* is his systematic application of the theory of *naẓm*, which had been previously developed by ʿAbd al-Qāhir al-Jurjānī, to the Qur'ān.[67] This theoretical approach concentrates on every choice made in constructing a sentence, showing how the preferences of the speaker can alter the subtleties of meaning. According to al-Jurjanī and followers of his like az-Zamakhsharī, the Qur'ān is inimitable due to such subtleties of meaning. Az-Zamakhsharī considered determining these subtleties to be the primary duty of the commentator. Going in a similar direction, Abū s-Suʿūd Efendi makes the utmost efforts to determine every nuance of meaning in the discourse. During this investigation, he is not content with following his predecessors, but makes contributions in many areas by revealing his own preferences and new findings.[68]

His first interpretation of the *Basmala* occurs in the *Irshād.* Examining the beginning of this statement gives an impression of Abū s-Suʿūd Efendi's *tafsīr* method. Abū s-Suʿūd Efendi discusses the question of whether the *Basmala* can be counted as a verse in sura 1, *al-Fātiḥa* (The Opening), giving place to the traditions (*riwāya*) and the views of the imams of the four *fiqh* sects, before beginning his examination of the *Basmala* itself.[69] He discusses the issue of what kind of verse the *Basmala* is much more broadly than the two main sources. The *Basmala* is then mainly explained linguistically.

Abū s-Suʿūd Efendi first determines that there is an elliptical expression in the *Basmala.* The verb to which the letter *b* (بـ) in the phrase "*bismi llāh*" (بسم الله) is attached has been dropped from the sentence. The verb should be determined according to the action to be taken "in the name of Allāh." Abū s-Suʿūd Efendi states that it is more appropriate to use a more specific verb rather than a general verb such as "I begin." For example, if the thing to do after the *Basmala* is reading, the reader may

67 Mehmet Taha Boyalık, *el-Keşşâf Literatürü: Zemahşerî'nin Tefsir Klasiğinin Etki Tarihi* (Istanbul: Türkiye Diyanet Vakfı (ISAM) Yayınları, 2019), 31f.

68 For examples of this, see: ibid., 201f.

69 Abū s-Suʿūd Efendi, *Irshād,* 1: 19ff.

prefer the verb "I read" (*aqra'u/atlū*). The verb here is the governing element in the sentence structure. It is also crucial whether the verb is placed before or after the ruled element (the name of Allāh). Abū s-Suʿūd Efendi argues that the verb should be placed after the ruled element, not at the head where it is located in the Arabic standard sentence structure. Thus, the governed element (the name of Allāh) is emphasized and specialized at the same time. Accordingly, the meaning is: "(I read only) in the name of Allāh (not someone else)." The letter *b* in the expression "*bismi llāh*" can contextually mean "blessing" (*at-tabarruk*) or "request for help" (*al-istiʿāna*). According to the former, the meaning is: "(I read) in the name of Allāh (not in someone else's name, in the hope that my work will be blessed)." Going by the latter, the meaning is: "(I read) in the name of Allāh, (not in someone else's name, with the desire to make me successful in my work)."

Both meanings are possible grammatically, but Abū s-Suʿūd Efendi prefers the first. He even thinks that a request for help can be understood as a blessing request, if the second meaning is preferred. According to him, the preference of "*bismi llāh*" over "*bi-llāh*" in the Basmala is related to this issue. There are two reasons why the preposition *b* is adjacent to the word "the name" (*ism*), which serves as the complement of the proper noun (*Allāh*), not the proper noun itself. First, in this usage, the meaning of "oath" brought to mind by the expression "*bi-llāh*" has been left out. But the main reason is that, even when *b* means a request for help rather than a blessing, the request for help from the person and the request for help through the name of the person are different things. According to the former, God is required to give power to act, while the latter requires that the deed be carried out in accordance with Allāh's approval. The latter is intended here. If the word "name" was not mentioned, the first meaning of "asking for help" would be inferred from the guidance provided in the subsequent verses, which is not the intended meaning in this case. Abū s-Suʿūd Efendi also considers the objection that the word "the name" will not need to be mentioned when *b* means "blessing" rather than "request for help." According to him, the addition of "name" is necessary in any case, since, without this addition, the first unintended meaning will continue to be possible. When "the name" is mentioned, the meaning of "blessing" becomes clear without any confusion.

Since the syntactic functions of letters are not variable in Arabic, they are not subject to *iʿrāb* and are usually pronounced as in the accusative case (*al-manṣūb*). But the preposition *b* in the *Basmala* is pronounced as in the genitive case (*al-majrūr*). Abū s-Suʿūd Efendi concludes his explanations about *b* in the *Basmala* by explaining the reason for rendering *b* in the genitive case.[70] As can be seen, the interpretation of the *Basmala* begins with the smallest unit of meaning, *b*, and this meaningful letter is explained in terms of its relation to the verb, its grammatical function, its contextual meaning, and its phonetic features.

---

70 Ibid., 1: 23f.

The second smallest significant unit of the *Basmala* is the word *“ism”* (the name), and, in *Irshād*, this word is explained in terms of its phonetic properties, derivation, morphological structure, syntactic function, and contextual meaning. Abū s-Suʿūd Efendi notes that the letter *hamza* at the beginning was not originally part of the word, but was added to facilitate vocalization. He states that the original of the word is *“as-sumuw,”* meaning sublimity, and that the derivation is made from this root due to the glorified natüre of what is invoked. He also remarks that Kūfa grammarians believe that the root of the word is *“wa-sa-ma”* and conveys his objection to this view.

Finally, Abū s-Suʿūd Efendi mentions that, in the spelling of *“bismi llāh,”* the initial *hamza* is not included due to the many uses, and the letter *b* may have been extended in the writing instead. As can be seen, lexicographic, linguistic, and literary explanations are in the foreground in the explanation of the word “name.”

The word “Allāh” in the *Basmala* is explained in a similar way. Abū s-Suʿūd Efendi first states that the original form of the word “Allāh” is *“al-ilāh”* (The God) and explains how this word developed into “Allāh.” Then, he determines that the word *“al-ilāh”* is a common noun used for all kinds of deities, whether they are true or not, but that it is widely used for the true *al-ilāh* / true *al-ilāhs*. According to him, the word “Allāh” is a special name used only for God. Abū s-Suʿūd Efendi later discusses the derivation of this noun and the verb in its root extensively.

In this way, Abū s-Suʿūd Efendi goes to great lengths to avoid overlooking the grammatical and stylistic features of the verses. This is where *Irshād* is most successful. He elegantly detects the subtleties of selective applications in syntax such as *taʾkīd, taqdīm, taʾkhīr, ḥadhf, taʿrīf, tankīr, faṣl,* and *waṣl.*

4 Conclusion

Abū s-Suʿūd Efendi is one of the most important figures in the history of sixteenth-century Islamic scholarship. He served as a *mudarris, qāḍī, qāḍī ʿaskar*, and *Shaykh al-Islām* in the Ottoman Empire, studied traditional scholarship from the leading scholars of his time, and specialized in linguistics, rhetoric, *fiqh*, and *tafsīr*. His best-known work is his commentary *Irshād al-ʿaql as-salīm*, written over a period of twenty years. The main sources of the work are az-Zamakhsharī’s *al-Kashshāf* and al-Bayḍāwī’s *Anwār at-tanzīl*, aside from which many other commentaries and glosses have been used as well. Abū s-Suʿūd Efendi aimed to overcome its two main sources in the application of the methods adopted, and he did indeed achieve this goal.

In *Irshād*, the verses are interpreted in the order of their appearance in the Qurʾān, and each verse is analyzed in detail, starting with the smallest linguistic element. In this regard, linguistic and literary considerations take precedence, overshadowing theological, legal, moral, and mystical issues. There are no extensive discussions of these issues;

on the contrary, the statements are briefly interpreted by adopting one of the existing interpretative (*ta'wīl*) systems.

Abū s-Su'ūd Efendi took care not to miss any details in the verses, from phonemes to sentence structure and literary uses. The theory of "*naẓm*" put forward by 'Abd al-Qāhir al-Jurjanī in the determination of the subtleties of meaning was systematically applied by following az-Zamakhsharī and al-Bayḍāwī. Abū s-Su'ūd Efendi's work deserves to be seen as the pinnacle of commentaries written in the tradition of az-Zamakhsharī.

5 Bibliography

5.1 Sources

'Aṭā'ī, Naw'īzādah. *Ḥadā'iq al-ḥaqā'iq fī takmilat ash-Shaqā'iq*. Prepared by Suat Donuk, ed. Derya Örs, 2 vols. Istanbul: Türkiye Yazma Eserler Kurumu, 2017.

Al-'Aydarūs, 'Abd al-Qādir. *An-Nūr as-sāfir 'an akhbār al-qarn al-'āshir*. Ed. Maḥmūd al-Arna'ūṭ et al. Beirut: Dār aṣ-Ṣādir, 2001.

Al-Baghdādī, Ismā'īl Pasha. *Hadiyyat al-'ārifīn*. Ed. Kilisli Rifat Bilge-Ibnülemin Mahmud Kemal, 2 vols. Ankara: Milli Eğitim Bakanlığı, 1951–55.

Çalabi, Awliya. *Seyahatnâme*. 10 vols. Istanbul: Ikdâm Matbaası, 1928–38.

Efendi, Abū s-Su'ūd. *Biḍā'at al-qāḍī*. Süleymaniye Library, Haji Mahmūd Efendi, no. 631.

Efendi, Abū s-Su'ūd. *Icāzatnāmah*. Süleymaniye Library, Pertevniyal, no. 910.

Efendi, Abū s-Su'ūd. *Irshād al-'aql as-salīm ilā mazāyā al-Kitāb al-karīm*. Ed. Mehmet Taha Boyalık, Ahmet Aytep, Ziyâüddin el-Kaliş, Muhammed İmâd el-Nabulsî, 9 vols. Istanbul: Nashriyyāt al-Vakf al-Diyāna al-Turkī, 2020.

Efendi, Abū s-Su'ūd. *Irshād al-'aql as-salīm ilā mazāyā al-Kitāb al-karīm*. Bayezid Library, Merzifonlu Kara Mustafa Paşa, nos. 18712, 18713, 18714, 18715.

Al-Ghazzī, Najm ad-Dīn. *Al-Kawākib as-sā'ira bi-a'yān al-mi'a al-'āshira*. Ed. Khalīl al-Manṣūr, 3 vols. Beirut: Dār al-Kutub al-'Ilmiyya, 1997.

Ibn Bālī, 'Alī (Manq 'Alī). *Al-'Iqd al-manẓūm fī dhikr afāḍil ar-Rūm*. Prepared by Suat Donuk, ed. Muhammed Hekimoğlu. Istanbul: Türkiye Yazma Eserler Kurumu Başkanlığı, 2018.

Ibn al-'Imād, 'Abd al-Ḥayy b. Aḥmad. *Shadharāt adh-dhahab fī akhbār man dhahab*. Ed. 'Abd al-Qadīr al-Arnā'ūṭ and Maḥmūd al-Arnā'ūṭ, 11 vols. Damascus/Beirut: Dār Ibn Kathīr, 1986.

Ilmiyye Salnāmesi. Istanbul: Matbaa-i Āmire, 1916.

Kafawī, Maḥmūd b. Sulaymān. *Katā'ib a'lām al-akhyār min fuqahā' madhab an-Nu'mān al-mukhtār*. Ed. Saffet Köse et al., 9 vols. Istanbul: Maktabat al-Irshād, 2017.

Kātib Çalabi. *Mīzān al-ḥaqq fī ikhtiyār al-ḥaqq*. Istanbul: Ali Riza Efendi Matbaası, 1870–71.

Kātib Çalabi. *Kashf aẓ-ẓunūn*. Ed. Şerefeddin Yaltkaya and Kilisli Rifat Bilge, 2 vols. Istanbul: Milli Eğitim Basımevi, 1941–43.

Al-Laknawī, Muḥammad 'Abd al-Ḥayy. *Al-Fawā'id al-bahiyya fī tarājim al-Ḥanafiyya*. Ed. Muḥammad Badr ad-Dīn Abū Firās an-Na'sānī. Cairo: Maṭba'at as-Sa'āda, 1908–9.

Majdī, Meḥmed. *Tarjama-yi Shaqā'iq an-nu'māniyya: Hadā'iq ash-Shaqā'iq*. Istanbul: Dār aṭ-Ṭibā'a al-'Āmira, 1853.

Mustaqimzādah, Sulaymān Sā'd ad-Dīn. *Devḥat al-mashāyikh*. Istanbul: Çağrı Yayınları, 1978.

Mustaqimzādah, Sulaymān Sā'd ad-Dīn. *Majallat an-nisāb*. Ankara: Kültür Bakanlığı, 2000.

Paçawī, Ibrāhīm. *Tārīkh-i Paçawī*. 2 vols. Istanbul: Matbaʿa-i ʿĀmira, 1865/6–67/8).
Qinalizādah, Ḥasan Çalabi. *Tadhkirat ash-shuʿarāʾ*. Ed. Ibrahim Kutluk, 2 vols. Ankara: Türk Tarih Kurumu Yayınları, 1978.
Rashād, Fāʾiq. *Aslāf*. 2 vols. Istanbul: Asır Kütüphanesi, 1895/6–1896/7.
Sāmī, Shams ad-Dīn. *Qāmūs al-aʿlām*. 5 vols. Ankara: Kaşgar Neşriyat, 1996.
Tāshkubrīzādah, Ahmad Efendi. *Ash-Shaqāʾiq an-nuʿmāniyya fī ʿulamāʾ ad-dawla al-ʿuthmāniyya*. Beirut: Dār al-Kitāb al-ʿArabī, 1975.
Az-Ziriklī, Khayr ad-Dīn. *Al-Aʿlām: Qāmūs tarājim li-ashhar ar-rijāl wa-n-nisāʾ min al-ʿarab wa-l-mustaʿribīn wa-l-mustashriqīn*. 8 vols. Beirut: Dār al-ʿIlm li-l-Malāyīn, 2002.

5.2 Secondary Literature

Akgündüz, Ahmet. “Ebüssuūd Efendi.” In *TDVIA*, edited by İslâm Araştırmaları Merkezi, 44 vols. Ankara: Türkiye Diyanet Vakfı Yayınları, vol. 10, 1994: 365–71.
Ateş, Süleyman. “Ebu's-suūd Efendi.” *Istanbul Üniversitesi Ilahiyat Fakültesi Dergisi* 1 (1999): 39–77.
Atsız, Nihal. *Istanbul Kütüphanelerine Göre Ebussuud Bibliyografyası*. Istanbul: Milli Eğitim Basımevi, 1967.
Aydemir, Abdullah. *Büyük Türk Bilgini Şeyhülislam Ebussuud Efendi ve Tefsirdeki Metodu*. Ankara: Diyanet Işleri Başkanlığı Yayınları, n.d.
Baysun, M. Cavid. “Ebüssu'ūd Efendi.” In *MEB Islām Ansiklopedisi*. 15 vols. Istanbul: Milli Eğitim Bakanlığı, vol. 4, 1977: 92–100.
Boyalik, Mehmet Taha. *El-Keşşāf Literatürü: Zemahşerī'nin Tefsir Klasiğinin Etki Tarihi*. Istanbul: Türkiye Diyanet Vakfı (ISAM) Yayınları, 2019.
Demir, Abdullah. *Devlet-i Aliyye'nin Büyük Hukukçusu Şeyhülislam Ebussuud Efendi*. Istanbul: Ötüken, 2006.
Düzenli, Pehlül. *Şeyhülislām Abū s-Suʿūd Efendi ve Fetvāları*. Istanbul: Osmanlı Araştırmaları Vakfı, 1994.
Imber, Colin. *Ebu's-Suʿud: The Islamic Legal Tradition*. Edinburgh: Edinburgh University Press, 1997.
Pixley, Michael M. “Suleyman's Şeyhülislam.” *TSAB* 6, no. 2 (1982): 7–16.
Şimşek, Halil. “The Missing Link in the History of Quranic Commentary: The Ottoman Period and the Quranic Commentary of Ebussuud/Abū al-Suʿūd al-ʿImādī (d. 1574 CE) *Irshād al-ʿaql al-salīm ilā mazāyā al-Kitāb al-Karīm*.” PhD diss., University of Toronto, 2018.
Yerinde, Ādem. “Nuskhat al-muʾallif wa-aqdam nusakh li-tafsīr Abī s-Suʿūd.” *Yakın Doğu Üniversitesi Ilahiyat Fakültesi Dergisi* 1, no. 2 (2015): 189–236.

Mullā Ṣadrā

Mukhtar H. Ali

1 Introduction

Ṣadr ad-Dīn Muḥammad b. Ibrāhīm ash-Shīrāzī (1571–1640/980–1050), popularly known as Mullā Ṣadrā, is considered the culminating figure of Islamic philosophy in Persia.[1] Born into an aristocratic family in Shīrāz, he was given the best education from an early age in the city of his birth. He then traveled to Isfahan to study under two of the greatest scholars of his time, Mīr Muḥammad Bāqir Dāmād Astarābādī (d. 1631/1041),[2] with whom he studied philosophy and theology, and the enigmatic sage and jurist, Shaykh Bahā' ad-Dīn 'Āmilī known as Shaykh Bahā'ī (d. 1620–21/1030)[3] with whom he studied transmitted sciences and Qur'ānic exegesis. Having mastered both intellectual and transmitted sciences, he retreated to Kahak, a village outside of Qum, Iran, where he composed his magnum opus, *al-Ḥikma al-muta'āliya fi l-asfār al-'aqliyya al-arba'a* (Transcendent Philosophy on the Four Intellectual Journeys), popularly known as *al-Asfār al-arba'a* (The Four Journeys). He composed more than fifty works, but it was in the *Asfār* that he formulated a comprehensive philosophical system that integrated reason, revelation, and mystical experience. Mullā Ṣadrā's school of "Transcendent Philosophy" (*al-Ḥikma al-muta'āliya*) synthesized several major streams of Islamic thought, including scholastic theology (*kalām*), Avicennan (d. 1037/428) Peripatetic philosophy (*mashshā'*), Suhrawardī's (d. 1191/587) Illuminationist philosophy (*ḥikmat al-ishrāq*), and the Sufism of Ibn al-'Arabī (d. 1240/638).

There has been a resurgence of scholarship in the writings of Mullā Ṣadrā, particularly after Muḥammad Khwājawī published new editions of all of his commentaries in Arabic in the early 1980s. Muḥsin Bīdārfar's introduction to Khwājawī's edition of commentary on the Qur'ān also drew attention to the other Qur'ān-related texts.[4] Although Henry Corbin (d. 1978/1398), Toshihiko Izutsu (d. 1993/1413), Seyyed Hossein Nasr, and Fazlur Rahman (d. 1988/1408) first opened the door to Ṣadrā's thought in Western Scholarship, most of these studies focused on his philosophical works. It was Nasr who dedicated a chapter to Ṣadrā's Qur'ānic output in his book *Ṣadr al-Dīn Shīr-*

---

1 He was later given the title *Ṣadr al-Muta'allihīn*, the "foremost of the theosophers".

2 Mīr Dāmād was the founder of the philosophical School of Isfahan. Nasr notes, "He was essentially an Avicennan philosopher with an Ishrāqī interpretation of some Peripatetic theses." See Seyyed Hossein Nasr, *Islamic Philosophy from its Origin to the Present* (New York: State University New York Press, 2006), 218.

3 The polymath Bahā' ad-Dīn 'Āmilī was a leading Shi'i theologian and jurist of the Safavid era. He was at once a philosopher, mathematician, architect, astronomer, Sufi and poet. See Sayyid Hossein Nasr, "The School of Isfahan," in *A History of Muslim Philosophy*, ed. M.M. Sharif (Wiesbaden: Otto Harrassowitz, 1966), 2: 910.

4 Mullā Ṣadrā, *Tafsīr al-Qur'ān al-karīm*, ed. Muḥammad Khwājawī (Qum: Intishārāt-i Bīdār, 1987–90).

https://doi.org/10.1515/9783111320083-016

āzī and His Transcendent Theosophy,[5] ushering in a new wave of studies on Ṣadrā's Qur'ānic output. This was followed by Latimah Peerwani's articles and study *On the Hermeneutics of the Light Verse,* where she outlined the principles of Ṣadrā's hermeneutics and his exegetical approach.[6] Christian Jambet has also devoted attention to *Tafsīr āyat an-nūr* reconciling Ṣadrā's philosophical treatment and spiritual approach to the Qur'ān.[7] Mohammed Rustom has written several studies on Ṣadrā's Qur'ānic writings, including "Approaching Mullā Ṣadrā as Scriptural Exegete: A Survey of Scholarship on His Quranic Works,"[8] which surveys the literature on Ṣadra's approach to the Qur'ān. His tour de force, *The Triumph of Mercy: Philosophy and Scripture in Mullā Ṣadrā,* is an in-depth study investigating Ṣadrā's commentary on the Qur'ān's opening chapter, *Tafsīr sūrat al-Fātiḥa.*[9] This work highlights Ṣadrā's esoteric hermeneutical vision, his comprehensive approach to the Qur'ān, and the profound influence of Ibn al-ʿArabī (d. 1240/638) on his thought.[10] Another important study that addresses Ṣadrā's commentaries within the larger backdrop of Sufi commentaries is Annabel Keeler's article, "Mullā Ṣadrā's Commentary on *Sūrat al-Sajda.*"[11]

2 His Works and Influences on his Commentaries

Ṣadrā's writings can be categorized into four main areas: philosophical works, commentaries on key Islamic philosophical texts, polemic pieces, and scriptural commentaries. His philosophical works include *al-Asfār al-ʿarbaʿa* (The Four Journeys), *Shawāhid ar-*

5 Seyyed Hossein Nasr, *Ṣadr al-Dīn Shīrāzī and His Transcendent Theosophy* (Tehran: Institute for Humanities and Cultural Studies, 1997²), 123–35. This article was reprinted in a commemorative volume for Toshihiko Izutsu (d. 1993): "The Qur'ānic Commentaries of Mullā Ṣadrā," in *Consciousness and Reality: Studies in Memory of Toshihiko Izutsu,* ed. Jalāl ad-Dīn Āshtiyānī et al. (Leiden: Brill, 1998), 47–57.

6 Latimah Peerwani, "Qur'ānic Hermeneutics: The Views of Ṣadr al-Dīn Shīrāzī," *Proceedings of the British Society for Middle Eastern Studies* (1991): 468–77; Peerwani, "Mullā Ṣadrā Shīrāzī on the Hermeneutics of the Qur'ān: His Philosophical Meditation," in *Eschatology, Exegesis, Hadith (Islam-West Philosophical Dialogue: The Papers Presented at the World Congress on Mullā Ṣadrā, May 1999, Tehran)* (Tehran: Sadra Islamic Philosophy Research Institute, 2005), 369–90.

7 See Ṣadrā, *Le verset de la lumière commentaire,* trans. Christian Jambet (Paris: Les Belles Lettres, 2009); Christian Jambet, *The Act of Being: The Philosophy of Revelation in Mullā Sadrā,* trans. Jeff Fort (New York: Zone Books, 2006).

8 Mohammed Rustom, "Approaching Mullā Ṣadrā as Scriptural Exegete: A Survey of Scholarship on His Quranic Works," *Comparative Islamic Studies* 4 (2008): 75–96.

9 Mohammad Rustom, *The Triumph of Mercy: Philosophy and Scripture in Mullā Ṣadrā* (Albany: State University New York Press, 2012).

10 See Mohammed Rustom's other works on Ṣadrā's Qur'ānic output: "Mullā Ṣadrā's Prolegomenon to the *Mafātīḥ al-Ghayb,*" *JQS* 9, no. 1 (2007): 128–33.

11 Annabel Keeler, "Mullā Ṣadrā's Commentary on Sūrat al-Sajda," in *Eschatology, Exegesis, Hadith (Islam-West Philosophical Dialogue: The Papers Presented at the World Congress on Mullā Ṣadrā, May 1999, Tehran)* Tehran: Sadra Islamic Philosophy Research Institute, 2005: 343–56.

rubūbiyya (Divine Witnesses), *Kitāb al-Mashāʿir* (Book of Metaphysical Penetrations) and *al-Ḥikma al-ʿarshiyya* (Wisdom of the Throne). His commentaries on Islamic philosophical works cover significant texts such as Ibn Sīnā's *Shifāʾ* (The Healing) and Suhrawardī's *Ḥikmat al-ishrāq* (Philosophy of Illumination). His polemic pieces in which he defends his views against the criticism of the religious scholars, include *Sīh aṣl* (The Three Principles) and *Kasr al-aṣnām al-jāhiliyya* (Shattering the Idols of Ignorance). Lastly, his scriptural commentaries, such as his commentary on Kulaynī's *Uṣūl al-Kāfī*, include treatises on religious topics such as *Risālat al-ḥashr* (Treatise on Resurrection), also known as *al-Ḥashriyya*, and *Risāla fī-l-imāma* (Treatise on the Imamate), along with works on hermeneutics and Qurʾānic exegesis, which will be addressed below.[12]

Although Ṣadrā is primarily known as a philosopher, he is also an exceptional an exceptional exegete whose works on the Qurʾān occupy a central position in his metaphysical worldview. Like his philosophical synthesis, his Qurʾānic commentaries bring together four traditions of Qurʾānic commentary: the Sufi, the Shiʿi the theological and the philosophical.[13] It is worthwhile to assess some of these influences beginning with Sufi commentaries, which have often been attributed to Imams Muḥammad al-Bāqir (d. 732/114) and Jaʿfar aṣ-Ṣādiq (d. 765/148). The earliest of these commentaries preserved the statements of the Imams, such as Sahl at-Tustarī's commentary *Tafsīr at-Tustarī*, which is the oldest, then Abū ʿAbd ar-Raḥmān as-Sulamī's (d. 1021/412) *Ḥaqāʾiq at-tafsīr* (The Realities of Exegesis).[14] This trend continued with Sulamī's student Abū-l Qāsim al-Qushayrī (d. 1072/465) in his *Laṭāʾif al-ishārāt* (The Subtleties of the Signs) and later with Ibn al-ʿArabī's contemporary Ruzbihān Baqlī's (d. 1209/606) *ʿArāʾis al-bayān fī ḥaqāʾiq al-Qurʾān* (The Brides of Explanation on the Realities of the Qurʾān).[15] After the initial wave of Sufi commentaries, the most profound influence on Ṣadrā came from Ibn al-ʿArabī's *al-Futūḥāt al-Makkiyya* (The Meccan Openings), which expounds on Qurʾānic themes throughout. The most important post-Ibn al-ʿArabī commentary to influence Ṣadrā is ʿAbd ar-Razzāq al-Kāshānī's (d. 1329/730) *Tafsīr Ibn al-ʿArabī*.

Shiʿi commentaries are generally traditionist rather than esoteric, explaining the verses according to hadith of the Prophet and the Imams. Even when these statements

12 Ṣadrā, *The Wisdom of the Throne*, trans. James Morris (Princeton: Princeton University Press, 1981), 51.
13 Seyyed Hossein Nasr, *Ṣadr al-Dīn Shīrāzī*, 123–35. See Ibrahim Kalin, "An Annotated Bibliography of the Works of Mullā Ṣadrā with a Brief Account of his Life," *Islamic Studies* 42, no. 1 (2003): 35–41; Sajjad Rizvi, *Mullā Ṣadrā Shīrāzī: His Life and Works and the Sources for Safavid Philosophy* (Oxford: Oxford University Press, 2007), 77–87; Mohammad Rustom, "The Nature and Significance of Mullā Ṣadrā's Qurʾānic Writings," *Journal of Islamic Philosophy* 6 (2012): 109–30; Rustom, "Approaching Mullā Ṣadrā".
14 See Farhana Mayer, *Spiritual Gems: The Mystical Commentary Ascribed to Jaʿfar al-Ṣādiq as contained in Sulamī's Ḥaqāʾiq al-tafsīr from the text of Paul Nwyia* (Louisville: Fons Vitae, 2011).
15 Pieter Coppens, "Sufi Qurʾān Commentaries, Genealogy and Originality," *Journal of Sufi Studies* 7, no. 1–2 (2018): 102–24.

reveal mysteries of the Qur'ān, they are classified as traditionist or standard commentaries, such as Muḥammad b. al-Ḥasan aṭ-Ṭūsī's (d. 1067/460)[16] *at-Tibyān fī tafsīr al-Qur'ān* and Abū ʿAlī l-Faḍl b. al-Ḥasan aṭ-Ṭabrisī's (d. 1144/538) *Majmaʿ al-bayān*, and works of ʿAlī b. Bābawayh al-Qummī (d. 939/328) and Muḥammad b. Masʿūd al-ʿAyyāshī (d. 932/320). Alongside these commentaries, Ṣadrā was deeply engaged with the Shiʿi hadith tradition as evidenced by his *Sharḥ Uṣūl al-Kāfī*, a philosophical commentary on Kulaynī's (d. 940–1/329) hadith compilation, *al-Kāfī*.[17]

Ṣadrā was also familiar with the commentaries written by the Ashʿarī theologians, Fakhr ad-Dīn ar-Rāzī (d. 1209/606) and Abū Ḥāmid al-Ghazālī (d. 1111/505) and Muʿtazilī commentator, Abū l-Qāsim az-Zamakhsharī (d. 1144/538), about whom he unfavourably said,

> That which Zamakhsharī and his likes understand from the Qur'ān is not, in reality, knowledge of the Qur'ān. Rather, it goes back to the sciences of lexicography, grammar, verbal expressions, and dialectical theology (*kalām*). But knowledge of the Qur'ān is other than these sciences, just as the skin and husk of man is not man in reality, but only figuratively.[18]

Ṣadrā was averse to these types of commentaries because they not only distracted from the real aim of seeking the Qur'ān's inner reality, but those who pursued such superficial scholarship were themselves misguided and were misguiding others, keeping them fixated on literalists interpretations.

Although Ṣadrā incorporated all aspects of Islamic knowledge in his style of exegesis, particularly philosophy, he embraced the methodology of the Sufis who relied on unveilings and spiritual experience as the basis for their commentaries. Thus, his hermeneutics or *ta'wīl*, (lit. "returning to the origin") was a return to the prophetic model, which was to learn directly from God: "Be God-conscious and God will teach you" (Q 2:282), and thereafter, "Ask the people of remembrance if you do not know" (Q 16:43). In this regard, after the teachings of the Shiʿi Imams, Ṣadrā was deeply indebted to the philosophical Sufism of the great master, Muḥyī d-Dīn b. al-ʿArabī, whose teachings form the warp and woof of Ṣadran thought in general. But Ṣadrā also drew from earlier masters of Sufism such as Bāyazīd Basṭāmī (d. 874/261), Abū l-Qāsim al-Junayd (d. 910/297), ʿAbd Allāh al-Anṣārī (d. 1089/481) and ʿAyn al-Quḍāt Hamadānī (d. 1131/525).

---

16 Muḥammad b. al-Ḥasan b. ʿAlī b. al-Ḥasan, better known as ash-Shaykh aṭ-Ṭūsī and ash-Shaykh aṭ-Ṭā'ifa, a great Shiʿi jurist and compiler of hadith. His famous compilation *al-Istibṣār fī mā ukhtulifa min al-akhbār* is one of the "four books" or canonical hadith works in Shiʿism.

17 For studies on Ṣadrā's hadith interpretation, see Karim Crow, "Mullā Ṣadrā on the First Intellect in his *Sharḥ Uṣūl al-Kāfī*," in *Eschatology, Exegesis, Hadith*, 571–90; Maria Dakake, "The Origin of Man in Pre-Eternity and his Origination in Time: Mullā Ṣadrā and Imāmī Shīʿite Tradition," in *Eschatology, Exegesis, Hadith*, 147–66.

18 Ṣadrā, *Sīh aṣl*, ed. Seyyed Hossein Nasr (Tehran: University of Tehran Press, 1961), 84, cited in Rustom, *The Triumph of Mercy*, 29.

3 His Works on the Qur'ān

His works on the Qur'ān[19] consist of sixteen pieces, thirteen of which are commentaries on independent suras, and three of which revolve around the theoretical aspects of the Qur'ān. His major exegetical works include: *Mutashābihāt al-Qur'ān* (The Ambiguous Verses of the Qur'ān), *Asrār al-āyāt* (The Mysteries of the Verses), *Mafātīḥ al-ghayb* (Keys to the Unseen) and *Tafsīr al-Qur'ān al-karīm* (Commentary on the Noble Qur'ān). In addition, the following works deal the exegesis of individual suras:[20]

1. *Tafsīr sūrat al-Fātiḥa* (sura 1, The Opening)
This is the last of his complete commentaries, written less than a decade before his death. Ṣadrā's views on ontology, cosmology, psychology, and eschatology closely follow Ibn ʿArabī's doctrines. He also covers topics such as the meaning of the *Basmala* (the phrase: "In the name of God the Merciful, the Compassionate"), and his theory of substantial motion (*al-ḥaraka al-jawhariyya*).

2. *Tafsīr sūrat al-Baqara* (sura 2, The Cow)
Though incomplete, it is the last of Ṣadrā's commentaries. In this work, he addresses the questions of freewill and determinism as held by the other Ashʿarī and Muʿtazalī theological schools. An important discussion on the inimitability of the Qur'ān (*iʿjāz al-Qur'ān*) is provided in relation to the 23rd verse which challenges the unbelievers to produce a verse of equal measure.

3. *Tafsīr sūrat as-Sajda* (sura 32, The Prostration)
This 135-page commentary is of the entire sura. Ṣadrā discusses his eschatology and spiritual psychology, the correspondences between the microcosmic heart and the divine Throne, as well as other concepts such as the Perfect Man, the Muḥammadan Light and the levels of God's folk.

4. *Tafsīr sūrat Yāsīn* (sura 36, Yā-Sīn)
Traditional sources consider this sura to be the "heart" of the Qur'ān. In this 450-page commentary, similar to his *al-Mabda' wa-l-maʿād*, Ṣadrā devotes his attention to key eschatological issues and the posthumous states of the soul, relying heavily on Ibn al-ʿArabī's doctrines.

---

19 This section is derivative of Mohammad Rustom's more complete outline of Ṣadrā's Qur'ānic works in, "The Nature and Significance of Mullā Ṣadrā's Qur'ānic Writings."

20 The following are the most popular editions of these works: *Mafātīḥ al-ghayb*, ed. Muḥammad Khwājawī (Beirut: Mu'assasat at-Tārīkh al-ʿArabī, 1999); "*Mutashābihāt al-Qur'ān*" in Ṣadrā, *Sih risāla-yi falsafī*, ed. Jalāl ad-Dīn Āshtiyānī (Tehran: Markazī-yi Intishārāt-i Daftar-i Tablīghāt-i Islāmī, 2000²), 255–84; *Tafsīr al-Qur'ān al-karīm*, ed. Muḥammad Khwājawī (Qum: Intishārāt-i Bīdār, 1987–90); *Asrār al-āyāt wa-anwār al-bayyināt*, ed. S.M. Mūsawī (Tehran: Intishārāt-i Ḥikmat, 2006).

5. *Tafsīr sūrat al-Ḥadīd* (sura 57, Iron)
The first of his commentaries, this 280-page work connects his philosophical ideas found in the *Asfār* to Qur'ānic symbolism. Ibn al-ʿArabī's influence is evident in his exposition of man's receptivity to the divine names, being the locus of the comprehensive name *Allāh*.

6. *Tafsīr sūrat al-Wāqiʿa* (sura 56, The Event)
In this 120-page commentary, Ṣadrā addresses eschatological issues and also explores the Imaginal world, much in the spirit of Ibn al-ʿArabī.

7. *Tafsīr sūrat al-Jumʿa* (sura 62, Friday)
This 200-page commentary deals with ontology, prophetology, the levels of faith and praxis. Towards the end, there is a discussion on the degrees of remembrance (*dhikr*).

8. *Tafsīr sūrat aṭ-Ṭāriq* (sura 86, The Morning Star)
This short 50-page gnostic commentary discusses some cosmology with respect to the stages of man's development and spiritual journey. According to Ṣadrā, the basis of this commentary is intuition and unveiling.

9. *Tafsīr sūrat al-Aʿlā* (sura 87, The Supreme)
One of the shorter commentaries, this 50-page work explains the nature of God's transcendence, given that the sura begins by a command to glorify the name of God (*sabbiḥi sma rabbika l-aʿlā*).

10. *Tafsīr sūrat az-Zilzāl* (sura 99, The Quaking)
This short, 34-page commentary raises some eschatological points related to the opening of scrolls on the Day of Judgement.

11. *Tafsīr sūrat al-Ikhlās* (sura 112, Sincerity)[21]
This gnostic commentary discusses divine unity, its degrees and the knowledge of God.

The following are commentaries on individual verses:

12. *Tafsīr āyat al-kursī* (The Verse of the Pedestal)
This is a 300-page commentary on verse of the Throne (Q 2:255) and the two verses following it where Ṣadra explains the doctrine of spiritual authority (*walāya*). Beginning with God's *walāya*, Ṣadrā discusses the *walāya* of the Imams and their intercession in the hereafter.

21 While this work is not commonly attributed to Ṣadrā, Kalin has included it in his bibliography. See Ibrahim Kalin, "An Annotated Bibliography".

13. *Tafsīr āyat an-nūr*[22] (The verse of Light)
Completed in 1620/1029, this work draws attention the tradition of commentaries on this verse (Q 24:35), highlighting the sayings of the early Sufis, such as al-Ḥasan al-Baṣrī (d. 728/110), Abū Saʿīd al-Kharrāz (d. 899/286), Dhū n-Nūn al-Miṣrī (d. 860/245), Basṭāmī, Abū Bakr ash-Shiblī (d. 946/334), as well as those of al-Ghazālī, Fakhr ad-Dīn ar-Rāzī, and Naṣīr ad-Dīn aṭ-Ṭūsī (d. 1274/672). The verse of light serves as his point of departure for describing the human microcosm and his ideas on spiritual psychology.

4 Hermeneutical Vision

In this section, we will take a closer look at Ṣadrā's hermeneutical vision in his key theoretical works on the Qurʾān: *Mafātīḥ al-ghayb* (Keys to the Unseen),[23] *Asrār al-ayāt* (Secrets of the Verses),[24] *Mutashābihāt al-Qurʾān* (The Ambiguous Verses).[25] *Mafātīḥ al-ghayb*, which consists of an introduction and twenty chapters, was written in 1632/1042 when he had already composed ten commentaries, making it a mature reflection on his hermeneutical approach. However, only the first two chapters focus specifically on hermeneutics, while the remaining chapters unfold his method in the context of other themes, such as the nature of knowledge, cosmogony, eschatology, wayfaring and others.

The second work of hermeneutics, *Asrār al-āyāt* (Secrets of the Verses), delves into the nature of revelation and the divine books. It also covers a range of theological, philosophical and mystical issues, including proofs for God's existence, the Muḥammadan reality, the Perfect Human the origin of the world, eschatology, the nature of God's speech and the divine book.

In the last and shortest of his hermeneutical works, the *Mutashābihāt al-Qurʾān* (The Ambiguous Verses), Ṣadrā elucidates the nature of the ambiguous or allegorical verses and the necessity of unveiling to interpret them correctly. This work serves as a polemic against the exoteric commentators of the Qurʾān, who attempt to interpret the allegorical verses without the requisite insight nor unveiling.

When we begin to analyse Ṣadrā's approach to the Qurʾān in the *Mafātīḥ al-ghayb*, it becomes evident that Ṣadrā was not merely a traditional exegete, but a gnostic seeking its inward meanings. His study of hermeneutics was driven by spiritual experience

22 Ṣadrā, *Tafsīr*, 4: 345–427; Ṣadrā, *Majmūʿat*, Lithographed by Aḥmad Shīrāzī (Tehran, 1904), 358–75. The work has been translated by Latimah Peerwani, *On the Hermeneutics of the Light Verse of the Qurʾān*, (London: ICAS Press, 2004).
23 The published version with Mullā ʿAlī Nūrī's glosses is 700 pages. See Ṣadrā, *Mafātīḥ al-ghayb*, ed. Muḥammad Khwājawī (Beirut: Muʾassasat at-Tārīkh al-ʿArabī, 1999).
24 Ṣadrā, *Asrār al-āyāt wa-anwār al-bayyināt*, ed. S.M. Mūsawī (Tehran: Intishārāt-i Ḥikmat, 2006).
25 Ṣadrā, *Sih risāla-yi falsafī*, ed. Jalāl ad-Dīn Āshtiyānī (Tehran: Markazī-yi Intishārāt-i Daftar-i Tablīghāt-i Islāmī, 2000), 257–84.

and the very impulse to write on the Qur'ān was not his own but divinely inspired, as he states:

> (a) command has issued from the Lord of my heart (*āmir qalbī*), a spiritual allusion has come forth from my innermost recesses (*waradat ishāra min sirr ghaybī*). His judgement and decision have come to pass and He has decreed that some of the divine symbols (*rumūz ilāhiyya*) be brought forth, and that the matters related to the Qur'anic sciences, the Prophetic allusions, secrets of faith, flashes of wisdom, esoteric glimmerings connected to the wonders of the glorious revelation and the subtleties of Qur'anic interpretation become manifest.[26]

His goal in writing was to expound on the esoteric dimension of the Qur'ān, addressing exoteric and linguistic matters only as they pertain to the esoteric. Ṣadrā aimed to elucidate symbols and subtleties through what he describes as piercing insight: "The Master of the Holy Realm of the Divinity (*ṣāḥib quds al-lāhūt*), the Owner of the Kingdom of the Dominion (*mālik mulk al-malakūt*), granted me a new opening (*fatḥ jadīd*), made the sight of my insight piercing with His light, revealing to my heart an opening which drew me near."[27] With this divine illumination, he began to lay down the foundation of his *taʾwīl* in the *Mafātīḥ*.

The basic structure of his hermeneutics mirrors his cosmology, positing that Being is a graded reality manifesting in three primary planes: the intellectual, the imaginal and the material. The levels of hermeneutics also correspond to these three degrees: the intellectual, symbolic and literal. By "intellectual," Ṣadrā does not mean rational but realities that exist without representation. The first premise of his hermeneutics is that the Qur'ān has both exoteric and esoteric dimensions, quoting the Prophet who states, "God, the Almighty has not revealed a verse except that it possess an outward and inward (dimension),"[28] and Imam ʿAlī b. Abī Ṭālib, "There is no verse in the Qur'ān which does not possess an outward (*ẓāhir*) and inward (*bāṭin*) meaning and a point of departure (*maṭlaʿ*).[29] This principle aligns with the general ethos of Sufi hermeneutic tradition, which appreciates both the literal and symbolic aspects of the Qur'ān. Ṣadra elaborates:

> The Qur'ān has degrees and ranks, just as man has levels and stations. The lowest level of the Qur'ān is like the lowest level of man: the Qur'ān's lowest level is what is contained in the book's binding and covering (*jild wa-aghlāf*), just as the lowest rank of man is what is in the outer covering and skin (*al-ihāb wa-l-bashara*). The husk of man attains nothing but the blackness of the Qur'ān and its sensory form. The man of the outward husk only perceives husk-like meanings (*al-maʿānī al-qishriyya*).[30]

---

26 Ṣadrā, *Mafātīḥ*, 76–77, cited in Rustom, "Mullā Ṣadrā's Prolegomenon," 129.

27 Ṣadrā, *Mafātīḥ*, 78, cited in Rustom, "Mullā Ṣadrā's Prolegomenon," 129.

28 ʿAlī b. ʿAbd al-Malik al-Muttaqī al-Hindī *Kanz al-ʿummāl fī sunan al-aqwāl wa-l-afʿāl* (Beirut: Dār al-Kutub al-ʿIlmiyya, 1998), no. 2461.

29 Ṣadrā, *Mafātīḥ*, 70.

30 Ṣadrā, *Mafātīḥ*, 117, cited in Rustom, *The Triumph of Mercy*, 29.

These various levels of interority are expressed through either the explicit (*muḥkam*) verses or the allegorical (*mutashābih*) ones. The allegorical verses lend themselves to multiple interpretations because of the multivalence of their words. Once one becomes conversant with the literal meaning of the text, they must not be confined by the literal meanings and delve into its inward meanings.

This brings us to another key aspect of his methodology: reading the book of God through its relation to the book of the soul. Ṣadrā posits that there is a correspondence between the outer book of existence and the inner book of the soul. Both Qur'ān and hadith affirm that God's signs manifest on both planes of existence and within the human soul as illustrated by the verse, "We will show him Our signs on the horizons and in the self until it becomes clear that He is the Truth (Q 41:53)" Ṣadrā elaborates:

> Know that the Qur'ān, like man is divided into a manifest (*ʿalan*) and hidden dimension (*sirr*), each of which has an outer (*ẓāhir*) and inner (*bāṭin*) aspect. Its inner aspect has another inner aspect known only to God: and none knows its interpretation but God (Q 3:7). It has also been related in the ḥadīth, "The Qur'ān has an outer and inner aspect." Its inner aspect consists of up to seven inner dimensions (*abṭun*) which are like the levels of man's inner dimensions, such as the soul (*nafs*), heart (*qalb*), intellect (*ʿaql*), spirit (*rūḥ*), innermost mystery (*sirr*), and the hidden and most hidden (*al-khafī wa-l-akhfāʾ*).[31]

There the Book of God can be read on the plane of the intellect or the plane of the heart, each revealing a different reality as it reaches the deepest aspects of the human being.[32] Since the intellect is considered the inner prophet and the inner law, it mirrors the outer prophet and the outer law. Only those who possess true intellect can attain knowledge of the Qur'ān's inner mysteries, and even this understanding is granted through divine opening and assistance. Ṣadrā writes,

> As for the spirit of the Qur'ān, its kernel (*lubb*), and its secret, none but the possessors of deep understanding (Q 12:111) (*ūlū l-albāb*)[33] perceive it. They do not attain this through knowledge

---

31 Ṣadrā, *Mafātīḥ* 115, cited in Rustom, *The Triumph of Mercy*, 27. See also Shigeru Kamada, "Mullā Ṣadrā Between Mystical Philosophy and Qur'ānic Interpretation through His Commentary on the 'Chapter of the Earthquake'," *International Journal of Asian Studies* 2, no. 2 (2005): 280.

32 Imam Jaʿfar aṣ-Ṣādiq (d. 765/148) says, "The book of God Almighty has four aspects: the expression (*ʿibāra*), the allusion (*ishāra*), subtleties (*laṭāʾif*), and realities (*ḥaqāʾiq*). The expression is for the layperson (*ʿawāmm*), the allusion for the elite (*khawāṣṣ*), the subtlety for the saints (*awliyāʾ*), and the reality for the prophets (*anbiyāʾ*)." Muḥammad Bāqir al-Majlisī, *Biḥār al-anwār: al-jāmiʿa li-durar akhbār al-aʾimma al-aṭhar* (Beirut: Muʾassasat al-Wafāʾ, 1983), 92: 95. See Mayer, *Spiritual Gems*, xxi, and Annabel Keeler, *Sufi Hermeneutics: The Qur'an Commentar of Rashīd al-Dīn Maybudī* (Oxford University Press, 2007), 55, citing from as-Sulamī's *Ḥaqāʾiq at-tafsīr*, ed. Paul Nwyia, "'Le tafsîr mystique attribué à Ǧaʿfar Ṣâdiq', Arabic text and introduction," *MUSJ* 43 (1967): 188.

33 *Lubb* means essence or kernel, referring to the intellect.

> acquired by way of learning and thinking. Rather, (they attain this) through God-given (*ladunnī*) knowledge.[34]

However, even if the Qur'ān's true meaning is attained through divinely inspired knowledge, this does not undermine Ṣadrā's synthetic approach of employing reason (*ʿaql*), revelation (*naql*) and gnosis (*ʿirfān*). In other words, the outer and inner meanings are mirrors for each other; one cannot seek the inner by disregarding the outer. The pure heart is like a mirror, clear and free of rust. When it faces the realities of the worlds, it can perceive them according to its receptivity for reflection and perception.

Ṣadrā's commentaries adhere to certain established structures of traditional Qur'ānic exegesis. He often begins with philological concerns, comparing the views of the grammarians and discussing variant readings of a verse. At times, he mentions the occasion of revelation (*asbāb an-nuzūl*) to contextualize the verse before delving deeper into esoteric readings. However, he rejects the methodology of the exoteric commentators insisting that if scripture could be fully understood through the rules of grammar, logic and rational inquiry, then the Arabic scholars and philosophers would have deciphered the entirely of the Qur'ān.[35] Thus, he interprets the verses according to mystical insights, seamlessly integrating them with his metaphysical framework.

5 Conclusion

The reception of Ṣadrā's Qur'ānic corpus differs significantly from his other philosophical writings, especially the *Asfār* which has had the greatest impact in Islamic philosophy since Avicenna. These writings, however, are equally valuable, displaying a thinker vast in scope and prolific in output. In many ways, they represent the culmination of his thought and life's work. The uniqueness of his commentaries lies in the synthetic approach for which he is known. Just as his philosophy is a confluence of various philosophical schools and epistemologies, his hermeneutical vision reflects this syncretic approach to the Qur'ān. However, it is abundantly clear that the thrust of his approach is spiritual and mystical, rather than historical, lexical or legal. Ṣadrā's writings are inexorably rooted in the school of Ibn al-ʿArabī.

Ṣadrā's Qur'ānic works are more praxis-oriented than his philosophical works. He emphasizes the importance of spiritual purification and decorum (*adab*) before embarking on the path of interpretation (*ta'wīl*). He insists that certain conditions must be met, such as presence of heart, reflection on the verses, contemplation on the

34 Ṣadrā, *Mafātīḥ*, 117, cited in Rustom, *The Triumph of Mercy*, 29.

35 Ṣadrā, "Mutashābihāt al-Qur'ān," in *Sih risāla-yi falsafī*, 257–84.

divine mysteries, and the removal of spiritual veils. His method exemplifies an existential reading, reflecting the correspondence between scripture, the book of existence, and the book of the soul.

6 Bibliography

6.1 Sources

Al-Majlisī, Muḥammad Bāqir. *Biḥār al-anwār: al-jāmiʿ li-durar akhbār al-aʾimma al-aṭhar*. 111 Volumes. Beirut: Muʾassasat al-Wafāʾ, 1983.

Mullā Ṣadrā. *Majmūʿat at-tafāsīr*. Lithographed by Aḥmad Shīrāzī. Tehran, 1904.

Mullā Ṣadrā. *Sih aṣl*. Ed. Seyyed Hossein Nasr. Tehran: University of Tehran Press, 1961.

Mullā Ṣadrā. *Kasr aṣnām al-jāhiliyya*. Ed. Muḥammad Taqī Dānishpazhūh. Tehran: University of Tehran Press, 1962.

Mullā Ṣadrā. *Kitāb al-Mashāʿir*. Ed. and trans. Henry Corbin as *Le livre des penetrations métaphysiques*. Tehran: Département d'iranologie de l'Institut franco-iranien, 1964.

Mullā Ṣadrā. *Ash-Shawāhid ar-rubūbiyya*. Ed. Jalāl ad-Dīn Āshtiyānī. Mashhad: Chāpkhāna-yi Dānishgāh-i Mashhad, 1967.

Mullā Ṣadrā. *Al-Mabdaʾ wa-l-maʿād*. Ed. Jalāl ad-Dīn Āshtiyānī. Tehran: Imperial Iranian Academy of Philosophy, 1976.

Mullā Ṣadrā. *Al-Ḥikma al-mutaʿāliya fī l-asfār al-ʿaqliyya al-arbaʿa*. Eds. Riḍā Luṭfī, Ibrāhīm Amīnī, and Fatḥ Allāh Ummīd. Beirut: Dār Iḥyāʾ at-Turāth al-ʿArabī, 1981.

Mullā Ṣadrā. *The Wisdom of the Throne*, trans. James Morris. Princeton: Princeton University Press, 1981.

Mullā Ṣadrā. *Risālat al-ḥashr*. Ed. and trans. into Persian by Muḥammad Khwājawī. Tehran: Intishārāt-i Mawlā, 1983.

Mullā Ṣadrā. *Sharḥ Uṣūl al-Kāfī*. Ed. Muḥammad Khwājawī. Tehran: Muʾassasa-yi Muṭālaʿāt wa-Taḥqīqāt-i Farhangī, 1987.

Mullā Ṣadrā. *Tafsīr al-Qurʾān al-karīm*. Ed. Muḥammad Khwājawī. 8 Volumes. Qum: Intishārāt-i Bīdār, 1987–90.

Mullā Ṣadrā. *Yāddāsht-hā-yi Mullā Ṣadrā hamrāh bā fihrist-i kitābkhāna-yi shakhṣī-yi Mullā Ṣadrā*. Ed. Muḥammad Barakat. Qum: Intishārāt-i Bīdār, 1998.

Mullā Ṣadrā. *Al-ʿArshiyya*. Ed. Fātin Muḥammad Khalīl Labbūn. Beirut: Muʾassasat at-Tārīkh al-ʿArabī, 2000.

Mullā Ṣadrā. "Al-Masāʾil al-qudsiyya." In *Sih risāla-yi falsafī*. Ed. Jalāl ad-Dīn Āshtiyānī. Tehran: Markazī-yi Intishārāt-i Daftar-i Tablīghāt-i Islāmī, 2000^2: 185–254.

Mullā Ṣadrā. "Mutashābihāt al-Qurʾān." In *Sih risāla-yi falsafī*. Ed. Jalāl ad-Dīn Āshtiyānī. Tehran: Markazī-yi Intishārāt-i Daftar-i Tablīghāt-i Islāmī, 2000^2: 257–84.

Mullā Ṣadrā. *Sih risāla-yi falsafī*. Ed. Jalāl ad-Dīn Āshtiyānī. Tehran: Markazī-yi Intishārāt-i Daftar-i Tablīghāt-i Islāmī, 2000^2.

Mullā Ṣadrā. *Mafātīḥ al-ghayb*. Ed. Muḥammad Khwājawī. Beirut: Muʾassasat at-Tārīkh al-ʿArabī, 1999.

Mullā Ṣadrā. *Majmūʿa-yi ashʿār-i Ṣadr ad-Dīn Shīrāzī*. Ed. Muḥammad Khwājawī. Tehran: Intishārāt-i Mawlā, 1998.

Mullā Ṣadrā. *The Elixir of the Gnostics*. Ed. and trans. William Chittick. Provo: Brigham Young University Press, 2003.

Mullā Ṣadrā. *On the Hermeneutics of the Light Verse of the Qurʾān*, trans. Latimah Peerwani. London: ICAS Press, 2004.

Mullā Ṣadrā. *Asrār al-āyāt wa-anwār al-bayyināt*. Ed. Sayyid Muḥammad Mūsawī. Tehran: Intishārāt-i Ḥikmat, 2006.

Mullā Ṣadrā. *Breaking the Idols of Ignorance: Admonition of the Soi-Disant Sufi*, trans. Mahdi Dasht Bozorgi and Fazel Asadi Amjad. London: ICAS Press, 2008.

Mullā Ṣadrā. *Spiritual Psychology: The Fourth Intellectual Journey in Transcendent Philosophy: Volumes VIII & IX of the Asfar*, trans. Latimah Peerwani. London: ICAS Press, 2008.

Mullā Ṣadrā. *Le verset de la lumière commentaire*, trans. Christian Jambet. Paris: Les Belles Lettres, 2009.

Mullā Ṣadrā. *Addenda on the Commentary on the Philosophy of Illumination*. Ed. and trans. Hossein Ziai. Costa Mesa: Mazda, 2010–.

Mullā Ṣadrā. *Tafsīr sūrat al-Fātiḥa*. Ms. Tehran National Library 263.

Nwyia, Paul. "'Le tafsîr mystique attribué à Ǧaʿfar Ṣâdiq', Arabic text and introduction." *MUSJ* 43 (1967): 179–230.

6.2 Secondary Literature

Akbarian, Reza. *The Fundamental Principles of Mullā Sadra's Transcendent Philosophy*. London: Xlibris, 2009.

Coppens, Pieter. "Sufi Qurʾān Commentaries, Genealogy and Originality." *Journal of Sufi Studies* 7, no. 1–2 (2018): 102–24.

Crow, Karim. "Mullā Ṣadrā on the First Intellect in his Sharḥ Uṣūl al-Kāfī." In *Eschatology, Exegesis, Hadith (Islam-West Philosophical Dialogue: The Papers Presented at the World Congress on Mullā Ṣadrā, May 1999, Tehran)*. Tehran: Sadra Islamic Philosophy Research Institute, 2005: 571–90

Dakake, Maria. "The Origin of Man in Pre-Eternity and his Origination in Time: Mullā Ṣadrā and Imāmī Shīʿite Tradition." In *Eschatology, Exegesis, Hadith (Islam-West Philosophical Dialogue: The Papers Presented at the World Congress on Mullā Ṣadrā, May, 1999, Tehran)*. Tehran: Sadra Islamic Philosophy Research Institute, 2005: 147–66.

Ernst, Carl. "Sufism and Philosophy in Mullā Ṣadrā." In *Mullā Ṣadrā and Transcendent Philosophy (Islam-West Philosophical Dialogue: The Papers Presented at the World Congress on Mullā Ṣadrā, May, 1999, Tehran)*. Tehran: Sadra Islamic Philosophy Research Institute, vol. 1, 2001: 173–92.

Hindī, ʿAlī b. ʿAbd al-Malik. *Kanz al-ʿummāl fī sunan al-aqwāl wa-l-afʿāl*. Beirut: Dār al-Kutub al-ʿIlmiyya, 1998.

Jambet, Christian. *The Act of Being: The Philosophy of Revelation in Mullā Sadrā*, trans. Jeff Fort. New York: Zone Books, 2006.

Kalin, Ibrahim. "An Annotated Bibliography of the Works of Mullā Ṣadrā with a Brief Account of his Life." *Islamic Studies* 42, no. 1 (2003): 21–62.

Kalin, Ibrahim. "Between Physics and Metaphysics: Mullā Ṣadrā on Nature and Motion." *Islam and Science* 1, no. 1 (2003): 59–90.

Kalin, Ibrahim. "Mullā Ṣadrā on Theodicy and the Best of All Possible Worlds." *JIS* 18, no. 2 (2007): 183–207.

Kalin, Ibrahim. *Knowledge in Later Islamic Philosophy: Mullā Ṣadrā on Existence, Intellect and Intuition*. Oxford: Oxford University Press, 2010.

Kamada, Shigeru. "Mullā Ṣadrā Between Mystical Philosophy and Qurʾānic Interpretation through His Commentary on the 'Chapter of the Earthquake'." *International Journal of Asian Studies* 2, no. 2 (2005): 275–89.

Keeler, Annabel. "Mullā Ṣadrā's Commentary on Sūrat al-Sajda." In *Eschatology, Exegesis, Hadith (Islam-West Philosophical Dialogue: The Papers Presented at the World Congress on Mullā Ṣadrā, May 1999, Tehran)*. Tehran: Sadra Islamic Philosophy Research Institute, 2005: 343–56.

Keeler, Annabel. *Sufi Hermeneutics: The Qur'an Commentar of Rashīd al-Dīn Maybudī*. Oxford: University Press, 2007.

Kuspinar, Bilal. “Mullā Ṣadrā on the Light-Verse.” In *Eschatology, Exegesis, Hadith (Islam-West Philosophical Dialogue: The Papers Presented at the World Congress on Mullā Ṣadrā, May, 1999, Tehran)*. Tehran: Sadra Islamic Philosophy Research Institute, 2005: 357–68.

Mayer, Farhana. *Spiritual Gems: The Mystical Commentary Ascribed to Ja'far al-Ṣādiq as contained in Sulamī's Ḥaqā'iq al-tafsīr from the text of Paul Nwyia* (Louisville: Fons Vitae, 2011).

Moris, Zailan. *Revelation, Intellectual Intuition and Reason in the Philosophy of Mullā Ṣadrā: An Analysis of the al-Hikmah al-'Arshiyyah*. London: Routledge, 2003.

Morris, James. “Civilization as Dialogue: Spirituality and Philosophy in Mullā Ṣadrā and Today.” In *Mullā Ṣadrā's School and Western Philosophies (Islam-West Philosophical Dialogue: The Papers Presented at the Second World Congress on Mullā Ṣadrā, May, 2004, Tehran)*. Tehran: Sadra Islamic Philosophy Research Institute, vol. 1, 2005: 261–72.

Musadad, Asep N. “Mullā Sadra's Ontological Perspective on the Qur'an.” *Al-Bayān – Journal of Qur'ān and Ḥadīth Studies* 14 (2016): 152–67.

Nasr, Sayyid Hossein. “The School of Isfahan.” In *A History of Muslim Philosophy*, edited by M.M. Sharif. 2 vols. Wiesbaden: Otto Harrassowitz, vol. 2, 1966: 904–32.

Nasr, Seyyed Hossein. *Ṣadr al-Dīn Shīrāzī and His Transcendent Theosophy*. Tehran: Institute for Humanities and Cultural Studies, 1997[2].

Nasr, Seyyed Hossein. “The Qur'ānic Commentaries of Mullā Ṣadrā.” In *Consciousness and Reality: Studies in Memory of Toshihiko Izutsu*, ed. Jalāl ad-Dīn Āshtiyānī et al. Leiden: Brill, 1998: 47–57.

Nasr, Seyyed Hossein. *Islamic Philosophy from its Origin to the Present*. New York: State University New York Press, 2006.

Peerwani, Latimah. “Qur'ānic Hermeneutics: The Views of Ṣadr al-Dīn Shīrāzī.” *Proceedings of the British Society for Middle Eastern Studies* (1991): 468–77.

Peerwani, Latimah. “Mullā Ṣadrā Shīrāzī on the Hermeneutics of the Qur'ān: His Philosophical Meditation.” In *Eschatology, Exegesis, Hadith (Islam-West Philosophical Dialogue: The Papers Presented at the World Congress on Mullā Ṣadrā, May 1999, Tehran)*. Tehran: Sadra Islamic Philosophy Research Institute, 2005: 369–90.

Rizvi, Sajjad. *Mullā Ṣadrā Shīrāzī: His Life and Works and the Sources for Safavid Philosophy*. Oxford: Oxford University Press, 2007.

Rizvi, Sajjad. *Mullā Ṣadrā and Metaphysics: Modulation of Being*. London: Routledge, 2009.

Rustom, Mohammed. “Mullā Ṣadrā's Prolegomenon to the *Mafātīḥ al-Ghayb*.” *JQS* 9, no. 1 (2007): 128–33.

Rustom, Mohammed. “Psychology, Eschatology, and Imagination in Mullā Ṣadrā Shīrāzī's Commentary on the Ḥadīth of Awakening.” *Islam and Science* 5, no. 1 (2007): 9–22.

Rustom, Mohammed. “Approaching Mullā Ṣadrā as Scriptural Exegete: A Survey of Scholarship on His Quranic Works.” *Comparative Islamic Studies* 4 (2008): 75–96.

Rustom, Mohammed. “The Nature and Significance of Mullā Ṣadrā's Qur'ānic Writings.” *Journal of Islamic Philosophy* 6 (2012): 109–30.

Rustom, Mohammed. *The Triumph of Mercy: Philosophy and Scripture in Mullā Ṣadrā*. Albany: State University New York Press, 2012.

Muḥammad Muḥsin Fayḍ Kāshānī

Liyakat Takim

1 Introduction

One of the most prominent Shiite scholars in the Safavid era was Mullā Muḥammad Muḥsin Fayḍ Kāshānī. Born in a family of religious scholars in 1598/1007, Fayḍ started his education at a young age. Although he was born in Kashan, he grew up in Qom and, at the age of twenty he moved to Isfahan, the Shiite center of learning at the time. After residing there for one year he left for Shiraz where he studied hadith and *fiqh* with Sayyid Mājid al-Baḥrānī (d. 1619/1028), one of the outstanding Akhbārī scholars of the time. When al-Baḥrānī died, Fayḍ Kāshānī returned to Isfahan where he studied under Bahā' ad-Dīn al-ʿĀmilī (Shaykh Bahā'ī, d. 1621/1000). He later returned to Qom where he studied under Ṣadr ad-Dīn Muḥammad ash-Shīrazī, also known as Mullā Ṣadrā (d. 1641/1050) for eight years. The latter, who bestowed upon him the title Fayḍ (Generous), became his mentor, teacher, and father-in-law. Mullā Ṣadrā specialized in *ʿirfān* (gnosis) and rationalist philosophy. It was under his supervision that Fayḍ was able to attain visions of the various esoteric sciences.[1] He later returned to Kashan where he taught philosophy and theology and wrote copiously in various fields. He died in Kashan in 1680/1090.

Fayḍ Kāshānī was a prolific author who composed more than one hundred and twenty scholarly works in different fields ranging from hadith, Qur'ānic exegesis, ethics, philosophy, *fiqh*, and gnosis. Of these, about forty have been published. His works include *Abwāb al-jinān* (The Doors of Paradise) and *al-Maḥajja al-bayḍā'* (The Pure Path) which is a Shiite version of al-Ghazāli's *Iḥyā' ʿulūm ad-dīn* (Revival of the Religious Sciences). Fayḍ's other works include *Tafsīr aṣ-Ṣāfī* (The Pure), *al-Wāfī* (The Sufficient), *ash-Shāfī* (The Healer), *Mafātīḥ ash-sharāʾiʿ* (The Keys to the Laws), *Asrār aṣ-ṣalāt* (The Secrets of Prayer), *ʿIlm al-yaqīn* (Knowledge of Certitude), *al-Kalimāt al-maknūna* (The Hidden Words), *Safīnat an-najāt* (The Ark of Salvation), and *Tarjumat aṣ-ṣalāt* (Translation of Prayer). According to his biography *Risāli-yi Sharḥ-i ṣadr*, (Treatise on the Expansion of the Heart), which he wrote himself, Fayḍ Kāshānī also studied other sciences like mathematics, medicine, astronomy and astrolabe.

1 Muḥsin Fayḍ Kāshānī, *al-Maḥajja al-bayḍā' fī tahdhīb al-Iḥyā'* (Tehran: al-Maktaba al-Islāmiyya, 1960), 4: 7. On Fayḍ Kāshānī's gnosticism and ideas on the 'Perfect Man' see Liyakat Takim, "The Concepts of the Absolute and Perfect Man in Mulla Fayd al-Kashani," in *Fayd e Kashani*, ed. Naqībī and Abbās Aḥmadvand (Anṣārīyān, 2013), 33–45.

https://doi.org/10.1515/9783111320083-017

2 Fayḍ Kāshānī and Akhbārīsm

In consonance with the dominant school of thought of the time, Fayḍ espoused the Akhbārī doctrine. The Akhbārīs' main thesis was that the truth can be only attained from the traditions of the Imams. This is because the Imams' hadith engender certitude that they reflect the divine will. This epistemic stance obviated the need for mujtahids or the reliance on their juridical edicts. Many Akhbārīs even claimed that the correct understanding of the Qur'ān is dependent on the hadith since, in itself, the Qur'ān is not a legal or historical document. Thus, despite its divine nature, the Qur'ān could be properly comprehended only through the lens of traditions.[2]

Akhbārī hermeneutical presuppositions also were premised on their assessment that certitude was to be derived solely from the four Shiite books of law composed by Abū Jaʿfar al-Kulaynī (d. 940/329), Ibn Bābawaih aṣ-Ṣadūq (d. 991/381), and Naṣīr ad-Dīn aṭ-Ṭūsī (d. 1067/460). For the Akhbārīs, these works exhibited the most authentic statements of the Imams (*qaṭʿī ṣ-ṣudūr*) and therefore constituted probative force (*hujja*). Due to this, most of the law could be derived from the traditions as preserved in the Shiite canonical sources. There was little, if any, scope for personal reasoning in the law.[3]

Fayḍ Kāshānī 's Akhbarism is evident in his first work on *uṣūl al-fiqh* titled *Naqd al-uṣūl al-fiqhiyya* (The Criticism of the Principles of Jurisprudence). Like other Akhbārīs he refuses to accept any authority outside the sacred sources. Hence, he considers both *ijtihād* (independent reasoning) and *taqlīd* (imitation) to be invalid. Fayḍ's *al-Uṣūl al-aṣliyya* (Original Fundamentals) and *Safīnat an-najāt* are also anti-Uṣūlī tracts. His anti-Uṣūlī stance can be further discerned by the fact that in the *Safīnat an-najāt*, he adduces the "authority verse" in the Qur'ān (Q 4:59) to condemn the office of the mujtahids and their role as the custodians and interpreters of the law. As obedience is due to God, the Prophet, and Imams only, normative law can only be obtained from the traditions (*akhbār*) of the Imams.[4]

Like other Akhbārīs, Fayḍ relied heavily on hadith as a source of legal prescription and legislation. His fourteen-volume work titled *al-Wāfī* comprises of hadiths collected from the four major Shiite hadith works along with his annotations. In the introduction to *al-Wāfī*, Fayḍ further demonstrates his Akhbārī penchant by criticizing basic Uṣūlī concepts and methodology. He rebukes the Uṣūlīs for borrowing and utilizing Sunni tools like *ijmāʿ* (consensus), *istiṣḥāb* (continuity), *ijtihād* (personal rea-

2 On the salient Akhbārī views and beliefs see Mullā Amīn Astarābādī, *al-Fawāʾid al-madaniyya* (Qom: Muʾassasat an-Nashr al-Islāmī, 2003); Robert Gleave, *Scripturalist Islam: The History and Doctrines of the Akhbari Shiʿi School* (Leiden: Brill, 2007).

3 Liyakat Takim, *Shiʿism Revisited: Ijtihad and Reformation in Contemporary Times* (New York: Oxford University Press, 2022).

4 See Said Amir Arjomand, *The Shadow of God and the Hidden Imam: Religion, Political Order and Societal Change in Shiʿite Iran from the Beginning to 1890* (Chicago: Chicago University Press, 1986), 146.

soning) and *ʿaql* (reason) in the Shiite legal system and claims that hadith is the actual and only reliable source of Islamic law.[5] He also abjures the classification and categorization of different genres of hadith, a juristic innovation introduced by ʿAllama al-Hillī (d. 1325/726).

Despite his Akhbārī predilections, Fayḍ Kāshānī disagreed with his Akhbārī compatriots on many issues. For example, in contrast to the founder of the movement Mullā Amīn Astarābādī (d. 1626/1036) Fayḍ stated that one cannot be sure that isolated traditions (*al-khabar al-wāḥid*) are based on certitude and that they necessarily reflect the traditions of the Imams. He also insisted that the Qurʾān should be regarded as an autonomous source from which the law could be derived. Thus, it did not have to be validated by hadith.[6]

For Astarābādī the Qurʾān can be understood only through the hadith. In his *al-Uṣūl al-aṣliyya* Fayḍ disagrees with this stance stating that if it is not possible to discern the meaning of the Qurʾān independently, then the authenticity and application of the Prophet's famous hadith regarding the two weighty things (*thaqalayn*) can be doubted. Indeed, if Astarābādī's argument is correct, then the Prophet has left behind only one rather than two weighty things, namely his family. This view suggests that the Qurʾān does not have any practical value without the hadith. Furthermore, if human beings cannot understand the Qurʾān's meaning directly then a similar claim can be made regarding the Imam's statements. Fayḍ concludes that traditions that prohibit the interpretation of the Qurʾān without recourse to reports from the Imams refer only to the ambiguous (*mutashābih*) verses of the Qurʾān, not its fixed (*muḥkam*) verses.[7]

Fayḍ Kāshānī also disagreed with Astarābādī's rejection of the principle of *barāʾa* (exemption of assessment of duty when the sources are silent on an issue). He agreed with the Uṣūlī position on the subject and stated that traditions from the Imams demonstrated that unless stated otherwise, all things are to be considered permissible.[8] However, despite his more liberal thinking, due to his Akhbārī training, some of Fayḍ's views were in strict compliance with the Akhbārī outlook. For example, he forbids uttering more than five words with a person of the opposite gender to whom one is not related (*ghayr maḥram*)[9] based on a hadith reported from Jaʿfar aṣ-Ṣādiq (d. 765/148), the sixth Imam.[10]

---

5 Muḥsin Fayḍ Kāshānī, *Safīnat an-Najā*, (Tehran: n.p., 1960), 9f.

6 Etan Kohlberg, "Aspects of Akhbari Thought in the Seventeenth and Eighteenth Centuries," in *Eighteenth-Century Renewal and Reform in Islam*, ed. N. Levtzion and J. O. Voll (Syracuse: Syracuse University Press, 1987), 137.

7 Muḥsin Fayḍ Kāshānī, *al-Uṣūl al-aṣliyya* (Tehran: Sāzāman-i Chāp-i Dānishgāh, 1971), 36–38; See also Robert Gleave, *Scripturalist Islam*, 234.

8 Muḥsin Fayḍ Kāshānī, *al-Uṣūl al-aṣliyya*, 143.

9 Muḥsin Fayḍ Kāshānī, *Mafātīḥ ash-sharāʾiʿ* (Qom: Majmaʿ-i Ẓakhāʾir-i Islāmī, 1981), 2: 22. For an online version of this book see: http://lib.eshia.ir/86886/2/22, accessed October 25, 2023.

10 Muḥammad b. al-Ḥasan al-Ḥurr al-ʿĀmilī, *Wasāʾil ash-shīʿa ilā taḥṣīl masāʾil ash-sharīʿa* (Beirut: Dār Ihyāʾ at-Turāth al-ʿArabī, 1967), 20: 212. For an online version of this book see: http://lib.eshia.ir/11025/20/212, accessed October 25, 2023.

3 Hadith Studies

Due to Mājid al-Baḥrānī's fame as a specialist in hadith studies and a great spiritual leader Fayḍ Kāshānī chose to study under him in Shiraz. He studied and memorized hadith through various modes of learning established in the seminaries. Due to his proficiency and expertise in the field, Fayḍ obtained an *ijāza* (permission) for narrating hadith from his teacher.

Significantly, the diffusion of Akhbārī ideology during Fayḍ Kāshānī's time meant that accounts of the lives and traditions of the Imams became ubiquitous and pervasive within the Shiite community. They also became subject to intense personal and devotional attachment. Due to his interest and expertise in hadith Fayḍ compiled a multi-volume collection of the traditions of the Imams called *al-Wāfī*. Along with Muḥammad Bāqir al-Majlisī's (d. 1699/1110) *Biḥār al-anwār*, *al-Wāfī* is considered as one of the greatest Shiite hadith collections.

Fayḍ Kāshānī's methodology in hadith hermeneutics and harmonization frequently differed from those adopted by other scholars. For example, in his discussion on resolving contradictory traditions reported from the Imams, he states that conflicting traditions on a subject cannot be harmonized by resorting to principles of abrogation or the nullification of a ruling. Rather, Fayḍ states that divergent traditions on an issue reflect the disparate social and political circumstances that the Imams lived in.[11] He also maintains that a jurist has to not only examine the discrepancies between traditions but also study the conditions under which those traditions were uttered since changes in circumstances will inevitably lead to the Imams issuing dissimilar statements.[12] On this issue, Fayḍ adopts a distinctly Uṣūlī principle that a legal injunction is connected to a particular *mawḍūʿ* (situation). If it changes then the rulings must be altered to reflect the new circumstances.

4 Fayḍ Kāshānī 's Qurʾānic Exegesis

As is to be expected from an Akhbārī, Fayḍ Kāshānī 's exegetical approach to his work on the Qurʾān titled *Tafsīr aṣ-ṣāfī* goes beyond commenting on grammatical nuances, morphology, syntax, the occasions of revelation, and elaborating the possible meanings of a verse. In fact, for Fayḍ, the understanding of the Qurʾān is framed by the Imams' hadith. It is the latter rather than the former which determines how a verse is viewed and interpreted by the juristic interpretive community. This hermeneutical method is predicated on the postulation that the correct knowledge of the Qurʾān re-

11 Ayatullah Mohaghegh-Damad, "The Role of Time and Social Welfare in the Modification of Legal Rulings," in *The Shiʿite Heritage*, ed. Lynda Clarke (Binghamton: Global, 2001), 217f.
12 Ibid., 218.

sides strictly with the *ahl al-bayt*, the family of the Prophet, and that only they are cognizant of the Qur'ān's esoteric meanings. After all, doesn't the Qur'ān say, "Ask those who remember, if you do not know." (Q 16:43)? For Fayḍ, the Imams are the ones who remember.

Fayḍ's interpretive method is an exteriorization of the subtle and secret teachings of the Imams. His method is based on principles derived from the Imams' traditions. A topos in Shiite hadith literature is that one can only fathom the law and scripture through statements made by an Imam.[13] As I have argued elsewhere many traditions prohibited Shiites from resorting to *ra'y* (personal reasoning) and *qiyās* (analogy). Such traditions are a clear indication of the refractoriness that independent thinking can cause within the Shiite community. Another prominent feature in Shiite hadith literature is the propitious comments made by the Imams in favor of those companions who memorized and preserved their hadith. In contrast, disciples who resorted to independent reasoning and other rational tools were explicitly denounced and excoriated as having deviated from the Imams' teachings and as being followers of Abū Hanīfa (d. 767/150) and his associates.[14]

The dichotomy between personal interpretation and the statements of the Imams can be further discerned from the Shiite world-view which directs the believers to acknowledge and accept the authority of the Imams who, based on the divine guidance they purportedly receive, are able to guide and save their followers from perdition. Personal interpretation and reasoning is a breach of this world-view since it challenges the authority and salvific efficacy of the Imams. In addition, *ra'y* and independent reasoning could create major cleavages within the Shiite community as they encourage Shiites to seek alternative sources of guidance.

There was another reason for condemning *ra'y* and *qiyās*. Shiite scholars were keen to prove that, contrary to their Sunni counterparts, their hermeneutical edifice was constructed on certainty (*qaṭ'*) and not *ẓann* (conjecture). Shiites claimed that Qur'ānic injunctions and prescriptions should not be based on *ra'y* and *qiyās* for these only engender *ẓann*, from which the true intent of the Lawgiver cannot be ascertained. Thus even God's revelation can only be discerned through an Imam. This was a polemical device that Shiites often used against the Sunnis.[15]

Due to these considerations, Fayḍ denied any proper understanding of the Qur'ān without the explicit intervention of the Imam. This is because only the Imam can properly decode the connotation of a verse and provide its correct meaning. Although the Qur'ān can be read at the basic level by anyone, those who wish to gain further

---

13 Muḥammad b. 'Umar al-Kashshī, *Ikhtīyār ma'rifat ar-rijāl* (Mashad: Dānishgāh-i Mashad, 1969), 156.

14 Muḥammad b. Ya'qūb Kulaynī, *al-Kāfī* (Qom: Dār al-Kutub al-Islāmiyya, 1986), 1: 73. See Liyakat Takim, *The Heirs of the Prophet: Charisma and Religious Authority in Shi'ite Islam* (Albany: SUNY, 2006), chapter three for more details.

15 See for example Ja'far b. al-Ḥasan al-Ḥillī, *Ma'ārij al-uṣūl* (Qom: Sayyid ash-Shuhadā', 1983), 179.

insights must seek knowledge from the Imams only.[16] To understand Qurʾānic interpretations, Fayḍ states that a person must also purify his or her soul, which is performed by practising various mystical and spiritual exercises.[17]

The ramifications of Fayḍ's approach transcend the field of hermeneutics. They actually impinge on the soteriological. Those who ignore the Imams' traditions not only resort to their own interpretations but also abandon one of the two weighty things (*thaqalayn*) that is essential for salvation. Stated differently, reliance on the authority of and traditions from the Imams is an indispensable tool of Qurʾānic exegesis as it mitigates the deleterious effects of conjecture or personal opinion. It is only by following the interpretation and authority (*wilāya*) of the Imams that salvation can be ensured.

This view is underlined in the introduction to Fayḍ Kāshānī 's exegetical work titled *aṣ-Ṣāfī*. He reinforces the role of traditions as the second of the two weighty things left by the Prophet. He also prefaces his work with twelve sections that stress the obligation to conjoin the Qurʾān with the hadith of the Imams.[18] The famous Shiite *mufassir* (exegete) Muḥammad b. Masʿūd al-ʿAyyāshī (d. 932/320) quotes Jaʿfar aṣ-Ṣādiq as saying that "(it is) through our *wilāya* that the scriptures are elucidated, and faith becomes clear."[19] Fayḍ also projects the Qurʾān as a nuanced text whose ability to speak to a reader is contingent on his commitment, sincerity, and spiritual preparedness.

Fayḍ Kāshānī cites many hadiths to vindicate the belief that the Imams are the exclusive possessors of Qurʾānic knowledge. He also quotes various traditions emphasizing the close connection between the Qurʾān and the Prophet's family. For example, he cites a tradition reported by Kulaynī, that Sulaym b. Qays al-Hilālī heard ʿAlī, say:

> Every verse that was revealed to the Prophet of Allah was dictated to me by him. I would then write it down with my own hand. He taught me its *taʾwīl* and *tafsīr*, its *nāsikh* (abrogating) and *mansūkh* (abrogated), its *muḥkam* (fixed) and *mutashābih* (ambiguous). The Prophet prayed that Allah should teach me the understanding of the Qurʾān and its preservation. Thereafter, I did not forget a single verse he dictated to me, or any other knowledge that he transmitted to me (. . .).[20]

Traditions such as this not only accentuate the close connection between the Qurʾān and the Imam but also that ʿAlī had memorized the inner meanings of every verse. This knowledge would then be transmitted to succeeding Imams. Hence, in the Shiite

16 Muḥsin Fayḍ Kāshānī, "Towards the Sacred Text: The Importance and Value of the Qur'an in the Life of a Believer," in *al-Sidrah*, trans. Azhar Sheraze, accessed October 25, 2023. http://www.aiseminary.org/al-sidrah/towards-sacred-text-importance-value-qur%ca%bean-life-believer/.

17 Muḥsin Fayḍ Kāshānī, *Spiritual Mysteries Ethical Secrets*, trans. Amina Inloes, Nazmina Virjee, and Muhammed Reza Tajri (London: ICAS Press, 2012), xxii.

18 Muḥsin Fayḍ Kāshānī, *Tafsīr aṣ-ṣāfī*, (Tehran: n.p., 1979), 1: 19f.; 29–34.

19 Muḥammad b. Masʿūd al-ʿAyyāshī, *Kitāb at-Tafsīr* (Tehran: al-Maktaba al-ʿIlmiyya al-Islāmiyya, 1961), 1: 5.

20 See Kulaynī, *al-Kāfī*, 1: 64 (hadith #1). For an online version of this hadith see: http://www.aiseminary.org/al-sidrah/none-know-except-knowing-ahlul-bayt-knowledge-qur%ca%bean/.

view, authority resides not only in God, the scripture and the Prophet but also in the living Imam. It is only the Imam who possesses hermeneutical authority.

Another hermeneutical method that Fayḍ Kāshānī adopts is to interject his commentary on Qur'ānic verses with traditions and to claim that the verses refer to the Imams or their *wilāya* (authority) regardless of the historical or social context of the verses. For example, when discussing Q 39:69 "And the earth will shine by the light of its Lord, and the book will be brought forth, and the Prophets and the witnesses will be brought, and they will be judged based on justice, they shall not be wronged" Fayḍ states that "the light of its Lord" refers to the twelfth Imam, the eschatological messiah. When he emerges people will no longer need sunshine and moonlight. Similarly, verses that speak of the signs of the day of resurrection are often interpreted as referring to the Qā'im, one of the titles of the Mahdi. The Qā'im is also identified as "one of the signs of your Lord" (Q 6:158) who will appear on the day of resurrection.[21] Similarly, Q 3:134 "Those who spend, whether they are prosperous, or live in adversity, those who restrain their anger and forgive people; indeed God loves those who do good deeds" refers to ʿAlī.[22]

Q 13:43 states "Say: Allah is sufficient as a witness between me and you and whoever has knowledge of the Book". Fayḍ claims that this refers to ʿAlī as he had knowledge of the Book'[23] Fayḍ also quotes Muḥammad al-Bāqir (d. 732–736/114–117), the fifth Imam, as stating that 'Surely, this Qur'ān guides to the most straight path' (Q 17:9). The most straight (*aqwam*) path in the verse refers to the *wilāya* of the imams.[24] According to a number of commentators, including Fayḍ, Q 33:71 also refers to *wilāya* of the Imams.[25]

The preceding discussion on Fayḍ Kāshānī's hermeneutics indicates that his methodology transcends the mere explication of the meaning and context of the verses of the Qur'ān. His approach is to valorize and read Shiite beliefs into the scripture by comparing, harmonizing, and inscribing them with the contents of the Qur'ān. For him, hadith works can act as reliable guides for the believers and can even replace the Imams during their absence.[26]

5 *Taḥrīf* in the Qur'ān

Another important dimension to Fayḍ's exegesis on the Qur'ān is the question of the mention of the Imams in it. Not only have the Imams been referred to in the scripture,

21 Kāshānī, *Tafsīr aṣ-ṣāfī*, 2: 173.

22 Ibid., 1: 152. See also Mohammad Ali Amir-Moezzi, *The Spirituality of Shi'i Islam: Beliefs and Practices* (London: Tauris, 2011), 322.

23 Ibid., 1: 880.

24 Ibid., 1: 960.

25 Ibid., 2: 369.

26 Devin J. Stewart, *Islamic Legal Orthodoxy: Twelver Shiite Responses to the Sunni Legal System* (Salt Lake City: The University of Utah Press, 1998), 195.

scholars like Fayḍ believe that they have been explicitly mentioned in it. Stated differently, they believe in the interpolation (*taḥrīf*) or omission of parts of the original Qur᾿ān. Amīr Moezzi cites many instances to demonstrate that, along with other Akhbārīs, Fayḍ Kāshānī believed that the original Qur᾿ān has been interpolated as the names of the Imams and their *wilāya* have been edited. These include verses such as Q 5:67 which was reportedly revealed when the Prophet appointed ʿAlī as his successor at Ghadir Khum.[27] Kāshānī also states:

> The Qur'an which is in our hands is not the entire Qur᾿ān sent down by God to Muḥammad. Rather, there is in it that which contradicts that which God had sent down. There is, moreover, in it that which was altered and changed. There were many things deleted from it, such as the name of ʿAlī in many places and the phrase Āl Muhammad (the family of Muhammad), as well as the names of the 'hypocrites', where they occur (. . .).[28]

Fayḍ vindicates his stance regarding omissions from the Qur᾿ānic text by quoting hadiths regarding the apparent deficiencies in the Qur᾿ān. He concludes that since the Qur᾿ān and the Imams are indissoluble guides, the adulteration and alteration in the Qur᾿ānic text is irrelevant because the *akhbār* can provide the necessary guidance.[29] For Fayḍ, although the Qur᾿ān has been tampered with, since the traditions reflect the Imams' exact statements, the basic essence of the revelation has been protected and transmitted. The focus is therefore more on the hadith than the Qur᾿ān because the former ensures the integrity of the message inscribed in the latter. God's message to humanity is delivered through the Prophet and the Imams. Hence, for Fayḍ, they become part of the revelatory sources.

Fayḍ's stance on *taḥrīf* is not equivocal. In other instances, after mentioning traditions that discuss *taḥrīf*, he states if such traditions are correct then we cannot be sure of anything in the Qur'an since it is quite possible that any verse could have been interpolated or changed. One would therefore lose confidence in the Qur᾿ān as a divine text and that God's commandments in it would lose meaning. It would also nullify God's promise that He would protect the Qur᾿ān (Q 15:9).[30] Here he follows the lines of aṭ-Ṭūsī and Abū ʿAlī al-Faḍl aṭ-Ṭabarsī who rejected the notion of alterations in the Qur᾿ān.

27 Kāshānī, *Tafsīr aṣ-ṣāfī*, 1: 414. See also Amir-Moezzi, *The Spirituality of Shi'i Islam*, 234.

28 Todd Lawson, "Akhbari Approaches to Tafsir," in *Approaches to the Qur'an*, ed. G. R. Hawting and Abdul-Kader A. Shareef (New York: Routledge, 1993), 186f.; Mahmoud Ayoub, "The Speaking and the Silent Qur'an: A Study of the Principles and Development of Imami Shi'i Tafsir," in *Approaches to the History of the Interpretation of the Qur'an*, ed. Andrew Rippin (Oxford: OUP, 1988), 190.

29 Kāshānī, *Tafsīr aṣ-ṣāfī*, 1: 55; Gleave, *Scripturalist Islam*, 219.

30 Rasul Jaʿfariyan, "A Study of Sunni and Shii Traditions Concerning Tahrif," in *al-Islam.org*, accessed October 26, 2023, https://www.al-islam.org/al-tawhid/general-al-tawhid/study-sunni-and-shii-traditions-concerning-tahrif-part-3-rasul-jafariy-0.

Fayḍ further argues that the Imams have explicitly stated that their traditions should be compared with the Qur'ān. Those which agree with the Qur'ān should be accepted, otherwise they should be rejected. If the Qur'ān is altered, then there is no basis for making such a comparison. In his other works, Fayḍ also refutes any adulteration in the Qur'ānic text by dismissing traditions regarding *taḥrīf* as being contrary to the Qur'ān and therefore as invalid.[31]

Although beyond the scope of the present study it should be noted that many traditions cited in Shiite works mention *taḥrīf* in the Qur'ān. This has been admitted by scholars such as ʿAlī b. Ibrāhīm al-Qummī (d. 919/306–7), al-ʿAyyāshī (d. 941/329),[32] Mīrzā Ḥusayn Nūrī (d. 1902/1319),[33] and others. For example, aṣ-Ṣādiq is reported to have stated: "If the Qur'ān is recited as it was revealed, you will find us named (in it)."[34] Nūrī cited more than one thousand traditions to prove his thesis on the interpolation of the ʿUthmanic codex. However, generally speaking, most Shiite scholars have refuted Nūrī's thesis and strenuously denied that there has been any interpolation or alteration in the contents of the Qur'ān. They have also questioned the reliability of such traditions, claiming that the Imams themselves have rejected the view that they were mentioned in the Qur'ān.[35] Shiite scholars have also quoted earlier scholars like aṣ-Ṣadūq, aṭ-Ṭūsī, and al-Murtaḍā (1044/436) who have affirmed the accuracy and authenticity of the present Qur'ān.[36]

6 Juridical Field

Fayḍ Kāshānī was also an accomplished jurist who was capable of inferring laws without having to imitate (*taqlīd*) or follow the edicts of another jurist. To be sure, fourteen commentaries have been written on his book on jurisprudence called *Mafātiḥ ash-sharāʾiʿ*. By exercising his own legal judgements, Fayḍ was able to issue a number of edicts that opposed the *fatāwā* (edicts) of the majority of jurists.

---

31 Ibid.

32 Meir M. Bar-Asher, *Scripture and Exegesis in Early Imami Shiism* (Leiden: Brill, 1999), 16f.; 82f.; 90f.

33 See Mīrzā Ḥusayn Nūrī, *Faṣl al-khitāb fī tahrīf kitāb Rabbi l-Arbāb* (Beirut: Markaz ad-Dirasat al-Fikriyya, 2020. See also Jaʿfar as-Subhani, *Masadir al-fiqh al-islami wa-manabiuhu* (Qum: Mu'assasa al-Imam as-Sadiq, 2007), 40–81.

34 Muḥammad b. Muḥammad al-Mufīd, *al-Masāʾil as-sarawiyya* (Qom: al-Muʾtamar al-ʿAlamī li-Alfiyyat ash-Shaykh al-Mufīd, 1993), 78f.

35 Hossein Modarressi, "Early Debates on the Integrity of the Qur'an: A Brief Survey," *Studia Islamica* 77 (1993): 29f.

36 Ibid., 34f. See also as-Sayyid Abū l-Qāsim al-Mūsawi al-Khuʾī, *Prolegomena to the Qur'an*, trans. Abdulaziz Sachedina (New York: Oxford University Press, 1998), 135–162.

For example, Fayḍ allowed singing and the playing of music in public. He cited traditions to show that they are prohibited only when accompanied by forbidden acts like gambling, or when men can hear women sing. He also considered the people of the book (*ahl al-kitāb*) to be pure (*ṭāhir*) a view that was an anomaly at the time.[37] He also maintained that no Shiite religious tax (*khums*) is payable on savings accumulated from earnings and on agriculture during the absence of the twelfth Imam and that the three shares of the Imam were not payable at all during the occultation.[38] Among his other *fatāwā* which were considered an anomaly at his time was that puberty for a girl is not contingent on her age, rather it is connected to her menstrual cycle. In contrast to the view of the majority of jurists, Fayḍ Kāshānī also maintained that the correct time of *maghrib* prayer starts when the sun sets.[39]

Another controversial juridical edict that Fayḍ Kāshānī issued pertained to the Friday prayers. In contrast to the view of many jurists of his time, he states that the *jum'a* prayer is obligatory on every individual (*wujūb-i 'aynī*) instead of the midday prayers. In his *ash-Shihāb ath-thāqib* (Piercing Light) he argues forcefully that the Friday prayer must be offered by every man unless under special mitigating circumstances. In fact the Safavid king Shah 'Abbas II (d. 1666/1077) issued a royal decree (*farmān*) in 1654/1065 asking Fayḍ to lead the Friday prayers at the 'Atiq Mosque in Isfahan. Later on, due to the heavy criticisms he faced, he resigned from his post and left the city. Fayḍ wrote to the Shah complaining that instead of uniting people and bringing them closer, the Friday prayers had actually created deep rifts within the community.[40]

As mentioned, Fayḍ Kāshānī was not a typical Akhbārī. He also believed in the involvement of the jurist in the administration of both kinds of religious taxes (*zakāt* and *khums*) since the jurist had been appointed by the Imam.[41] He also supported an expansion of juristic practices during the absence of the Imam. This included cooperation with the political entity even if it meant being remunerated for services rendered.[42] Rather controversially, he also maintains that the authority of the *fuqahā'* is

---

37 'Alīriża Fayḍ, *Vīzhegīhāyi ijtihād va fiqh-i Pūyā* (Tehran: Pazhūhishgāh-i 'Ulūm-i Insānī, 1997), 381.

38 Yusuf b. Aḥmad al-Baḥrānī, *Hadā'iq an-nāḍira* (Najaf: Dār al-Kutub, 1957), 12: 462; 'Alīriża Fayḍ, *Vīzhegīhāyi Ijtihād*, 381.

39 'Alīriża Fayḍ, *Vīzhegīhāyi ijtihād*, 426f.

40 Andrew J. Newman, "Fayd al-Kashani and the Rejection of the Clergy/State Alliance: Friday Prayer as Politics in the Safavid Period," in *The Most Learned of the Shi'a: The Institution of the Marja' Taqlid*, edited by Linda Walbridge (New York: Oxford Univ. Press, 2001), 40.

41 Ibid., 40, 42.

42 Ibid., 41. In 1650 at the request of the King he wrote a book on the kingly mirror. In this tract he talks on the significance of kingship and earthly authority and political rule in relation to the spiritual and mundane. See William Chittick "Two Seventeenth-Century Persian Tracts on Kingship and Rulers," in *Authority and Political Culture in Shi'ism*, ed. Said Amir Arjomand (Albany, SUNY, 1988), 269.

the same as that of the Imam except that the jurists cannot enforce an offensive *jihad*. Authority belongs to the jurists, he says, based on 'the right of deputyship' (*ḥaqq an-niyāba).*[43] Fayḍ Kāshānī also maintained that the Perfect Man is the Prophet or the Imam in that they are intermediaries between human and divine.[44] Since he manifests the divine attributes within him, the Perfect Man has destroyed all obstacles that come between the absolute and the temporal. In fact, the world would cease to exist without him.[45] Perhaps surprisingly, in his *al-Wāfī* Fayḍ extends this function to the *'ulamā'* who are also the perfect men and intermediaries between heaven and earth.[46]

7 Conclusion

This chapter has explored different facets of Fayḍ Kāshānī's ideas and views as explicated in his works. It has focused on his hermeneutical methodology in his Qur'ānic exegesis. Here, his approach can be described as Akhbārī since he reads Shiite beliefs and perspectives into the Qur'ān by interjecting his exegesis with traditions from the Imams. He believed that the Imams were explicitly mentioned in the Qur'ān but that their names were later removed.

Fayḍ Kāshānī's academic credentials can be discerned by the fact that he wrote copiously in different fields of Islamic sciences ranging from hadith and Qur'ānic exegesis to the ethical, juridical, and mystical fields. Fayḍ was not a "typical" or strict Akhbārī as can be seen by his differences with his Akhbārī compatriots on many issues and by his deep interest in Shiite theosophy and the influence exerted by Mullā Ṣadrā on him. In fact, it would be more appropriate to categorize him as a "mild" Akhbārī. Despite this, Fayḍ Kāshānī's contribution to Shiite intellectual thought remains profound.

---

43 Seyfeddin Kara & Mohammad Saeed Bahmanpour, "The Legal Authority of the Jurist," *Journal of the Contemporary Study of Islam* 1 (2020): 11; Kāshānī, *Mafātīḥ ash-sharā'i'*, 2: 4f.; See also Liyakat Takim, "From Partial to Complete: Juristic Authority in Twelver Shi'ism, "*Journal of South Asian and Middle Eastern Studies* 43, no. 4 (2020): 6–27.

44 Fayḍ Kāshānī, *Kalimāt Maknūna*, (Tehran: Farāhānī, 1963), 186–92.

45 Shigeru Kamada, "Fayḍ al-Kāshānī's Walāya: The Confluence of Shi'i Imamology and Mysticism," in *Reason and Inspiration in Islam*, ed. Todd Lawson (New York: I.B. Taurus, 2005), 459–63; Takim, "The Concepts," 40.

46 Fayḍ Kāshānī, *al-Wāfī*, 41. See also Ahmad Kazemi Moussavi, *Religious Authority in Shi'ite Islam: From the Office of Mufti to the Institution of Marja'* (Kuala Lumpur: ISTAC, 1996), 123.

8 Bibliography

8.1 Sources

Al-ʿĀmilī, Muḥammad b. al-Ḥasan al-Ḥurr. *Wasāʾil ash-shīʿa ilā taḥṣīl masāʾil ash-sharīʿa*. 20 volumes. Beirut: Dār Ihyāʾ at-Turāth al-ʿArabī, 1967.

Astarābādī, Mullā Amīn. *Al-Fawāʾid al-madaniyya*. Qom: Muʾassasat an-Nashr al-Islāmī, 2003.

Al-ʿAyyāshī, Muḥammad b. Masʿūd. *Kitāb at-Tafsīr*. 2 Volumes. Tehran: al-Maktaba al-ʿIlmiyya al-Islāmiyya, 1961.

Al-Baḥrānī, Yusuf b. Aḥmad. *Hadāʾiq an-nāḍira*. 25 vols. Najaf: Dār al-Kutub, 1957.

Al-Ḥillī, Jaʿfar b. al-Ḥasan. *Maʿārij al-uṣūl*. Qom: Sayyid ash-Shuhadāʾ, 1983.

Kāshānī, Muḥsin Fayḍ. *Al-Maḥajjat al-bayḍāʾ fī tahdhīb al-Iḥyāʾ*. 4 vols. Tehran: al-Maktabat al-Islāmiyya, 1960.

Kāshānī, Muḥsin Fayḍ. *Safīnat an-najā*. Tehran: n.p., 1960.

Kāshānī, Muḥsin Fayḍ. *Kalimāt maknūna*. Tehran: Farāhānī, 1963.

Kāshānī, Muḥsin Fayḍ. *Al-Uṣūl al-aṣliyya*. Tehran: Sāzāman-i Chāp-i Dānishgāh, 1971.

Kāshānī, Muḥsin Fayḍ. *Tafsīr aṣ-ṣāfī*. 5 vols. Tehran: n.p., 1979.

Kāshānī, Muḥsin Fayḍ. *Mafātīḥ ash-sharāʾiʿ*. Qom: Majmaʿ-i Ẕakhāʾir-i Islāmī, 1981.

Kāshānī, Muḥsin Fayḍ. *Spiritual Mysteries Ethical Secrets*, trans. Amina Inloes, Nazmina Virjee, and Muhammed Reza Tajri. London: ICAS Press, 2012.

Kāshānī, Muḥsin Fayḍ. "Towards the Sacred Text: The Importance and Value of the Qurʾan in the Life of a Believer." In *al-Sidrah*, trans. Azhar Sheraze. http://www.aiseminary.org/al-sidrah/towards-sacred-text-importance-value-qur%ca%bean-life-believer/.

Al-Kashshī, Muḥammad b. ʿUmar. *Ikhtīyār maʿrifat ar-rijāl*. Mashad: Dānishgāh-i Mashad, 1969.

At-Tabrisi, Mirza al-Husayn Nūrī. *Faṣl al-khitāb fī tahrīf kitāb Rabbal-l-Arbāb* (Beirut: Markaz ad-Dirāsāt al-Fikriyya, 2020.

8.2 Secondary Literature

Amir-Moezzi, Mohammad Ali. *The Spirituality of Shiʿi Islam: Beliefs and Practices*. London: Tauris, 2011.

Arjomand, Said Amir. *The Shadow of God and the Hidden Imam: Religion, Political Order and Societal Change in Shiʿite Iran from the Beginning to 1890*. Chicago: Chicago University Press, 1986.

Ayoub, Mahmoud. "The Speaking and the Silent Qur'an: A Study of the Principles and Development of Imami Shiʿi Tafsir." In *Approaches to the History of the Interpretation of the Qur'an*, edited by Andrew Rippin. Oxford: OUP, 1988: 177–198.

Bar-Asher, Meir M. *Scripture and Exegesis in Early Imami Shiism*. Leiden: Brill, 1999.

Chittick, William. "Two Seventeenth-Century Persian Tracts on Kingship and Rulers." In *Authority and Political Culture in Shiʿism*, edited by Said Amir Arjomand. Albany: SUNY, 1988: 267–304.

Fayḍ,ʿAlīriża. *Vīzhegīhāyi Ijtihād va Fiqh-i Pūyā*. Tehran: Pazhūhishgāh-i ʿUlūm-i Insānī, 1997.

Gleave, Robert. *Scripturalist Islam: The History and Doctrines of the Akhbārī Shīʿi School*. Leiden: Brill, 2007.

Jaʿfariyan, Rasul. "A Study of Sunni and Shii Traditions Concerning Tahrif." In *al-Islam.org*. Accessed October 26, 2023. https://www.al-islam.org/al-tawhid/general-al-tawhid/study-sunni-and-shii-traditions-concerning-tahrif-part-3-rasul-jafariy-0.

Kamada, Shigeru. "Fayḍ al-Kāshānī's Walāya: The Confluence of Shiʿi Imamology and Mysticism." In *Reason and Inspiration in Islam*, edited by Todd Lawson. New York: I.B. Taurus, 2005: 455–70.

Al-Khuʾī, as-Sayyid Abū l-Qāsim al-Mūsawi. *Prolegomena to the Qur'an*, trans. Abdulaziz Sachedina. New York: Oxford University Press, 1998.

Kohlberg, Etan. "Aspects of Akhbari Thought in the Seventeenth and Eighteenth Centuries." In *Eighteenth-Century Renewal and Reform in Islam*, ed. N. Levtzion and J. O. Voll. New York: 1987: 170–182.

Kulaynī, Muḥammad b. Yaʿqūb. *Al-Kāfī*. 4 vols. Qom: Dār al-Kutub al-Islāmiyya, 1986.

Lawson, Todd. "Akhbari Approaches to Tafsir." In *Approaches to the Qur'an*, edited by G. R. Hawting and Abdul-Kader A. Shareef. New York: Routledge, 1993: 173–210.

Modarressi, Hossein. "Early Debates on the Integrity of the Qur'an: A Brief Survey." *Studia Islamica* 77 (1993): 5–39.

Mohaghegh-Damad, Ayatullah. "The Role of Time and Social Welfare in the Modification of Legal Rulings." In *The Shiʿite Heritage*, edited by Lynda Clarke. Binghamton: Global, 2001: 213–222.

Moussavi, Ahmad Kazemi. *Religious Authority in Shiʿite Islam: From the Office of Mufti to the Institution of Marjaʿ*. Kuala Lumpur: ISTAC, 1996.

Al-Mufīd, Muḥammad b. Muḥammad. *Al-Masāʾil as-sarawiyya*. Qom: al-Muʾtamar al-ʿAlamī li-Alfīyyat ash-Shaykh al-Mufīd, 1993

Newman, Andrew J. "Fayd al-Kashani and the Rejection of the Clergy/State Alliance: Friday Prayer as Politics in the Safavid Period." In *The Most Learned of the Shiʿa*: *The Institution of the Marjaʿ Taqlid*, edited by Linda Walbridge. New York: Oxford Univ. Press, 2001: 34–52.

Nūrī, Mīrzā Ḥusayn. *Faṣl al-Khitāb fī Tahrīf Kitāb Rabbi l-Arbāb*. Beirut: Markaz Dirasat al-Fikriyya, 2020.

Seyfeddin, Kara & Bahmanpour Mohammad Saeed. "The Legal Authority of the Jurist and its Scope in Modern Iran." *Journal of the Contemporary Study of Islam* 1 (2020): 1–27.

Stewart, Devin J. *Islamic Legal Orthodoxy: Twelver Shiite Responses to the Sunni Legal System*. Salt Lake City: The University of Utah Press, 1998.

As-Subhani, Jaʿfar, *Masadir al-fiqh al-islami wa-manabiuhu*. Qum: Muʾassasa al-Imam as-Sadiq, 2007.

Takim, Liyakat. *The Heirs of the Prophet: Charisma and Religious Authority in Shiʿite Islam*. Albany: SUNY, 2006.

Takim, Liyakat. "The Concepts of the Absolute and Perfect Man in Mulla Fayd al-Kashani." In *Fayd e Kashani*, edited by Naqībī and Abbās Aḥmadvand. Anṣārīyān, 2013: 33–45.

Takim, Liyakat. "From Partial to Complete: Juristic Authority in Twelver Shiʿism." *Journal of South Asian and Middle Eastern Studies* 43, no. 4 (2020): 6–27.

Takim, Liyakat. *Shiʿism Revisited: Ijtihad and Reformation in Contemporary Times*. New York: Oxford University Press, 2022.

Bahā' ad-Dīn al-Lāhījī

Mathieu Terrier

1 Introduction

Bahā' ad-Dīn al-Lāhījī, also known as Quṭb ad-Dīn Ashkiwarī or Ishkawarī (d. between 1677 and 1684/1088 and 1095), is a quite marginal philosopher and theologian of seventeenth-century Safavid Iran. His Qur'ānic *tafsīr*, entitled *Tafsīr Sharīf Lāhījī*, composed in both Arabic and Persian and completed in 1675/1086, is at first glance a strictly Shiite traditional work. However, it differs from most traditional Shiite *tafsīrs* by adding philosophical topics to the prevailing sectarian concerns.

2 His Life and His Works

Born in Gīlān in northern Iran around 1600/1010, Muḥammad b. Shaykh 'Alī l-Lāhījī l-Ashkiwarī (or Ishkawarī) belonged to the milieu of Imāmī Shiite clerics in Safavid Iran.[1] After the proclamation of Imāmī Shiism as the state religion in 1501/906, Shāh Ismā'īl and his successor Shāh Ṭahmāsb (r. 1524/930–1576/984) had called upon scholars from the Shiite centers of the Arab countries to immigrate to Iran in order to convert the society to the new faith. However, al-Lāhījī's paternal forefather did not come from Jabal 'Āmil, in present-day southern Lebanon, or from Iraq, as did the great majority of these ulama, but from Yemen, and may have been originally of the Shiite Ismā'īlī faith rather than of the Shiite Imāmī one.[2]

In his early youth, Muḥammad b. Shaykh 'Alī studied in Isfahan under both famous philosophers and Shiite theologians Shayh Bahā'ī (d. 1621/1030) and Mīr Dāmād (d. 1631/1040). However, in 1630/1039 by all accounts, the death of his older brother, which occurred a few years after the death of his father, compelled him to return to his hometown to hold the official position of *shaykh al-islām* of Lāhījān. Yet he seems to have retained familial sympathies towards Sufism, particularly the *Nūrbakhshiyya* brotherhood. He spent most of his life in Lāhījān, made a trip to India where he completed his *Tafsīr* in 1675/1086, returned to Iran and died between 1677/1088 and 1684/1095.

1 On this author, see Mathieu Terrier, *Histoire de la sagesse et philosophie shi'ite: "L'Aimé des cœurs" de Quṭb al-Dīn Ashkevarī* (Paris: Le Cerf, 2016).
2 On the Shiite clerical migration to Iran during the Safavid period, see Rula Jurdi Abisaab, *Converting Persia: Religion and Power in the Safavid Empire* (London/New York: I.B. Tauris, 2004); Andrew J. Newman, *Safavid Iran: Rebirth of a Persian Empire* (London/New York: I.B. Tauris, 2006).

https://doi.org/10.1515/9783111320083-018

Our author composed most his works over a period of twenty years under different pen names. His first work, written in his youth (before 1039/1630) under the name of Quṭb ad-Dīn al-Lāhījī, is a collection of mathematical exercises entitled *Laṭā'if al-ḥisāb*. Around 1659–60/1070, he completed under the name of Quṭb ad-Dīn Ashkiwarī a first version of the *Maḥbūb al-qulūb* ("the Beloved of the Hearts"), an encyclopedia of the wise in three volumes gathering, in the first one, the ancient sages and prophets, in the second one, philosophers, doctors and theologians of Islam, including a section on the Sufi masters, and in the third one, the twelve Imams and some great Shiite authorities; and until 1677/1088, he completed the work with marginal glosses.[3] In 1665/1075, under the same name, he achieved *Thamrat al-fu'ād*, an anthology derived from the previous work, and under the name of Bahā' ad-Dīn al-Lāhījī, *Khayr ar-rijāl*, a work belonging to the "science of men" (*'ilm ar-rijāl*), dealing with the transmitters of hadiths from the canonical work of Muḥammad b. 'Alī b. Bābawayh (d. ca 991/381) *Man lā yaḥḍuruhu al-faqīh*. In 1667/1077, under the name of Quṭb ad-Dīn al-Lāhījī, he wrote *Fānūs al-khayāl fī irā'at 'ālam al-mithāl* ("The Lantern of Imagination. On the Vision of the Imaginal World"), also called *ar-Risāla al-mithāliyya*, an epistle devoted to the concept of the imaginal world (*'ālam al-mithāl*), based on hadiths of the Imams as well as philosophical and mystical sources.[4] In 1675/1086, under the name of Bahā' ad-Dīn al-Lāhījī, he completed his Qur'ānic *Tafsīr*, entitled *Tafsīr ash-Sharīf al-Lāhījī*, in the city of Bīna or Putna in India, as attested by the colophon, although he had probably written most of it in Iran.[5] It was first edited by the Iranian Shiite scholar Muḥaddith Urmawī in 1961/1340 with a comprehensive introduction. Ashkiwarī-Lāhijī is also the author of a Persian commentary on the famous collection of prayers and orations of the fourth Shiite Imam 'Alī Zayn al-'Ābidīn (d. ca. 714/95) *aṣ-Ṣaḥīfa as-sajjādiyya*, and of a commentary on a verse of a *Mathnawī*, probably that of Jalāl ad-Dīn Rūmī (d. 1273/672), whose dates of composition are unknown. Only four of his works have been edited to date, with the third volume of the *Maḥbūb al-qulūb* still unpublished.

The difference between the two nicknames (*laqab*) 'Bahā' ad-Dīn' and 'Quṭb ad-Dīn', coupled with the variety of genres explored by the author, has sown doubt in the minds of prosopography authors. The famous scholar Āqā Buzurg Ṭihrānī (d. 1970/1389) held the opinion that Bahā' ad-Dīn al-Lāhījī, author of the *Khayr ar-rijāl* and the *Tafsīr*, was

3 Quṭb ad-Dīn al-Ashkiwarī, *Maḥbūb al-qulūb*, ed. I. ad-Dībājī and Ḥ. Ṣidqī, 2 vols. (Tehran: Mīrāth Maktūb, 1999–2003), third part unpublished. First part translated in French in Terrier, *Histoire de la sagesse*.

4 Manuscript from the Malik Library in Tehran, n° 1615. See M. Terrier, *Le guide du monde imaginal. Présentation, édition et traduction de la* Risāla mithāliyya *(Épître sur l'imaginal) de Quṭb al-Dīn Ashkevarī* (Turnout: Brepols, 2023).

5 Bahā' ad-Dīn b. Shaykh 'Alī sh-Sharīf al-Lāhījī al-Ashkiwarī, *Tafsīr ash-Sharīf al-Lāhījī*. Ed. Ja'far Pazhūm, 4 vols. (Tehran: 'Ilmī, 2011–12) (abr. *TSL*). Introduction of the first edition by Muḥaddith Urmawī, *TSL*, 1: 57 and 90–91; end of the text, 4: 968.

not the same man as Quṭb ad-Dīn Ashkiwarī, author of the *Maḥbūb al-qulūb*.[6] However, the author of the *Tafsīr* makes several references to his *Fānūs al-khayāl*, in which he refers several times to the *Maḥbūb al-qulūb*. Muḥaddith Urmawī acknowledges the existence of one and the same author but assumes that he wore the two nicknames at two different periods of his life and experienced a radical spiritual evolution: inclined to philosophy at the beginning of his life, he would then have converted to *akhbārī* traditionalism under the influence of Muḥammad Amīn Astarābādī (d. 1624/1033 or 1627/1036)[7]. However, this hypothesis must be rejected or at least tempered. First of all, Ashkiwarī-Lāhījī wrote in the same year *Thamrat al-fu'ād* and *Khayr ar-rijāl*, i.e., a work of philosophy and a work of traditional religious sciences. On the other hand, in 1677/1088, he was still working on the glosses of his *Maḥbūb al-qulūb*. Last but not least, several references to his *Fānūs al-khayāl* are to be found in his *Tafsīr*, on the occasion of philosophical commentaries of various verses. Not only is it certain that the three works are of the same author, but it turns out that the author of the *Tafsīr* did not in any way deny his earlier works and his interest in philosophical ideas, just as the author of the *Maḥbūb al-qulūb* was already well versed in Imāmī traditions.

Being the author of the *Maḥbūb al-qulūb*, the *Fānūs al-khayāl* and the *Tafsīr ash-Sharīf al-Lāhījī*, Ashkiwarī-Lāhījī belongs to a spiritual current which aimed to bring together the esoteric Imāmī tradition, the speculative mysticism of Sufism, especially that of *ash-Shaykh al-akbar* Muḥyī d-Dīn b. ʿArabī (d. 1240/638), the peripatetic philosophy (*mashshā'ī*) and the 'illuminative' philosophy (*ishrāqī*) founded by Shihāb ad-Dīn Suhrawardī (d. 1191/587). This syncretic current, alive in Iran during the Ilkhānid and Safavid periods, has for its great representatives Sayyid Ḥaydar Āmulī (d. after 1385/787), Ibn Abī Jumhūr (d. around 1501/906), and Nūrullāh Shūshtarī (d. 1610/1019).[8] The combination of illuminative philosophy, Akbari Sufism and Imāmī traditionalism is also to be found in an eminent contemporary of our author, Muḥsin al-Fayḍ al-Kāshānī (d. 1679/1090), author of mystical treatises such as the *Kalimāt maknūna*, of a traditionalist *tafsīr*, *aṣ-Ṣāfī*, and of a monumental collection of Imāmī hadiths, *al-Wafī*.[9] Lāhījī actually mentions him in laudatory terms in his *Tafsīr*.

Ashkiwarī-Lāhījī led a marginal and discreet life as a second-rank cleric, far from the great religious centers of Safavid Iran such as Isfahan, Shīrāz or Mashhad, a fact which undoubtedly allowed him to freely develop a thought that went against the tide

6 *TSL*, 1: 17f.

7 Ibid., 18 and 77.

8 See on this Mathieu Terrier, "The Defence of Sufism among Twelver Shi'i Scholars of Early Modern and Modern Times: Topics and Arguments," in *Shi'i Islam and Sufism: Classical Views and Modern Perspectives*, ed. Denis Hermann and Mathieu Terrier (London: I.B. Tauris, 2020), 27–63.

9 For a survey of his works and thought, see Mathieu Terrier, "Anthropogonie et eschatologie dans l'œuvre de Muḥsin Fayḍ Kâshânî: l'ésotérisme shî'ite entre tradition et syncrétisme," in *L'ésotérisme shî'ite, ses racines et ses prolongements. Shi'i Esotericism: its Roots and Developments*, ed. Mohammad Ali Amir-Moezzi (Turnhout: Brepols, 2016), 743–80.

of his time. While his training in Isfahan took place during a period of philosophical and spiritual renaissance under Shāh ʿAbbās the Great (r. 1588–1629/996–1038), his own activity, under the reign of Shāh Sulaymān (r. 1666–1694/1077–1105), coincided with a strong clerical backlash against philosophers and mystics. Although we have no information on his stay in India, it is tempting to explain it by his sympathies for Sufism and fear of persecution, as many Sufis emigrated from Iran to India for this reason at that time. The eclectic and esoteric character of his work takes on its full meaning in this adverse context.

3 His Qurʾānic Commentary (*Tafsīr ash-Sharīf al-Lāhījī*)

The *Tafsīr ash-Sharīf al-Lāhījī* belongs to the neo-traditional (*akhbārī*) Imāmī Shiite movement of the seventeenth/eleventh century, reviving the hermeneutical tradition of early Shiism. According to the original tenet that the impeccable Imams are the only men qualified to interpret the Book of God, it is essentially composed of hadiths attributed to the Imams and includes few personal interventions by the author. The exegetical Imāmī hadiths that constitute the book are borrowed from the earliest Imāmī sources, notably the *Tafsīr*s of ʿAlī b. Ibrāhīm al-Qummī (d. *ca* 919/307) and Muḥammad b. Masʿūd al-ʿAyyāshī (d. *ca* 932/320), and compilations of Imāmī hadiths such as *al-Kāfī* of Muḥammad b. Yaʿqūb al-Kulaynī (d. 940/328), as well as later compilations such as ash-Shaykh Abū ʿAlī aṭ-Ṭabrisī's (d. 1154/548) *al-Iḥtijāj.*

A distinctive feature of this commentary is its bilingual character: each verse and hadith is first quoted in its original Arabic version, then translated and commented on in Persian. This diglossia demonstrates an obvious pedagogical purpose, that of disseminating Shiite early esoteric tradition to a Persian-speaking audience unfamiliar with Arabic and Imāmī sources.

3.1 A Reminder of the Principle of Shiite Exegesis

As a neo-traditional commentary, the *Tafsīr Sharīf Lāhījī* is based on the two fundamental principles of traditional Shiite exegesis. First, just as God is 'the Outward and the Inward' (Q 57:3), the Qur'ān has essentially two levels: its linguistic, outward or exoteric (*ẓāhir*) meaning on the one hand, and its spiritual, inward or esoteric (*bāṭin*) meaning on the other. Second, the Imams are the only legitimate interpreters of the Qurʾān by virtue of their knowledge immediately received from the Prophet (from the first Imam ʿAlī onwards) and their impeccable and infallible nature (*ʿiṣma*), similar to that of the Prophet. This dogma is based on Q 3:7: "It is He who sent down upon thee the Book, wherein are verses clear that are the Essence of the Book, and others ambiguous.

As for those in whose hearts is swerving, they follow the ambiguous part, desiring dissension, and desiring its interpretation, and none knows its interpretation, save only God and those firmly rooted in knowledge," marking the pause after these words and before the rest ("They say: 'We believe in it. . .'") – and not according to the reading generally received by Sunni Muslims: "and none knows its interpretation, save only God (pause). Those firmly rooted in knowledge' say: 'We believe in it. . .". According to hadiths attributed to the sixth Imam, "'Those firmly rooted in knowledge' are 'the Prince of the faithful' (*amīr al-muʾminīn*) (i.e. ʿAlī b. Abī Ṭālib) and the Imams after him," "We (the Imams) are 'those firmly rooted in knowledge,' we know the interpretation (of the Book)," and "The Messenger of God is the greatest of those firmly rooted in knowledge[10]."

As for the "clear" and the "ambiguous" verses, they are also given an esoteric interpretation. According to a tradition reported without reference, the "clear verses" are identified with ʿAlī and the following Imams; and the "ambiguous verses" are "So-and-So" (*fulān wa-fulān*), i.e., Abū Bakr aṣ-Ṣiddīq and ʿUmar b. al-Khaṭṭāb, the two first caliphs considered as the worst enemies of God's allies.[11] This is in line with the oldest kind of Shiite Qurʾānic hermeneutics, the "personalized commentary" to which I will return below. According to the author, the hadith means that the claim of the imamate (i.e. temporal and spiritual authority) of ʿAlī and his descendants is derived from all the clear verses of the Qurʾān, whereas the claim of those to the caliphate is derived from false interpretations of the ambiguous verses. This reflects the dualistic vision typical of early Shiism, in which the history of creation is a permanent fight between forces of Knowledge, Good and Light on the one hand, and forces of Ignorance, Evil and Darkness on the other hand.[12] According to this tenet, the conflict between the allies of God and their enemies is at the core of the Qurʾān.

Even more, according to the doctrine of original Shiism, the sayings of the Imams are not merely an interpretation but a continuation of the Qurʾānic revelation, hence they have the same dual exoteric and esoteric nature as the latter. According to a hadith attributed to the eighth Imam: "Among our traditions (*min akhbārinā*), some are ambiguous like the ambiguous (verses) of the Qurʾān, and some are clear like the clear (verses) of the Qurʾān. Therefore, bring back what is ambiguous to what is clear, and do not follow the ambiguity without the clearness or you will go astray."[13] The

10 *TSL*, 1: 445; from Muḥammad b. Yaʿqūb al-Kulaynī, *Uṣūl al-kāfī*, ed. Ḥasan al-Aʿlamī (Beirut: Muʾassasat al-Aʿlamī, 2005), 122, ḥ. 1, 2 and 3.

11 *TSL*, 1: 444, From Muḥammad b. Masʿūd al-ʿAyyāshī, *Tafsīr al-ʿAyyāshī*, ed. ʿAbdallāh Ṣāliḥī (Qum: Dhawī l-Qurbā, 2016–17), 2: 8.

12 Mohammad Ali Amir-Moezzi, *The Divine Guide in Early Shiʿism: The Sources of Esotericism in Islam*, trans. David Streight (Albany: State University of New York Press, 1994), 7f.; Mohammed Ali Amir-Moezzi and Christian Jambet, *What is Shiʿi Islam? An Introduction*, trans. Kenneth Casler and Eric Ormsby (Abingdon/New York: Routledge, 2018), 13–18.

13 From Muḥammad b. ʿAlī b. Bābawayh, *ʿUyūn akhbār ar-Riḍā*, ed. M. Lājūrdī (Tehran, Nashr-i Jahān, n.d.), 1: 290.

teaching of the Imams therefore has also an exoteric and an esoteric level, which allows a priori to resolve the apparent contradictions or inconsistencies between their traditions.[14]

On the identification of the Imams as the devoted interpreters of the Qur'ān, Lāhījī further reports on Q 6:149: "To God belongs the argument conclusive; for had He willed, He would have guided you all," a hadith of the fifth Imam: "We are the treasurers of God's Science, the interpreters of His revelation, His conclusive argument (*ḥujjatuhu al-bāligha*) for all those between heaven and earth," i.e. men and djinns.[15] And about Q 64:8: "Therefore believe in God and His Messenger, and in the Light which We have sent down," he states that the commentaries of the Sunni Muslims (*al-ʿawāmm*) explain "the light" as the silent Qur'ān (*Qur'ān-i ṣāmit*), whereas according to the Imams, "the light" means the speaking Qur'ān (*Qur'ān-i nāṭiq*), which is the Prince of the faithful and the impeccable Imams[16]. According to the original doctrine indeed, the Qur'ān remains a "silent Book" without the interpretation of the Imam; the Imam is the one who makes the book "speaking," i.e., intelligible, through his interpretation, to such an extent that he is himself called "the speaking Qur'ān," which is in particular a nickname for ʿAlī.[17]

3.2 Allusions to the Falsification of the Qur'ān

Lāhījī reports a large number of traditions, borrowed from the *Tafsīr al-Qummī* and other ancient sources, mentioning the falsification (*taḥrīf*) of the Qur'ān by the enemies of the Imams, and reporting from the Imams certain verses recovered from the authentic Qur'ān. Indeed, the belief that the official Qur'ān had been censored, altered or falsified by the first three caliphs and the Umayyads, and that the original, unadulterated Qur'ān had been preserved only by ʿAlī and the line of his successors, was part of the early doctrine of the Imāmī Shiite Muslims before being progressively suppressed after the Occultation of the twelfth Imam (941/329).[18] However, Lāhījī reports a discussion between early and contemporary scholars, showing that this issue was not settled at his time. Unlike Fayḍ Kāshānī who, in his *Tafsīr aṣ-Ṣāfī*, reports the Imāmī hadiths supporting the claim of the Qur'ān's falsification before discarding them, at least apparently, to 'bet' on the integrity of the Qur'ān, Ashkewarī decides to give credence to these hadiths. In so doing, he does not seem to fear the accusation of

14 *TSL*, 1: 445.

15 Ibid., 988, from Kulaynī, *Uṣūl al-kāfī*, 109, ḥ. 3.

16 *TSL*, 4: 538.

17 See on this Mohammad Ali Amir-Moezzi, *The Silent Qur'an and the Speaking Qur'an: Scriptural Sources of Islam between History and Fervor*, trans. Eric Ormsby (New York: Columbia University Press, 2015).

18 See on this Amir-Moezzi, *The Silent Qur'an and the Speaking Qur'an*.

'exaggeration' (*ghulūw*) commonly made in Shiite circles against the declared supporters of this thesis.

In the commentary of Q 2:143: "Thus We appointed you a midmost nation (*umma wāsiṭa*) that you might be witnesses to the people," Lāhījī states, according to a hadith of the sixth Imam, that this verse is specifically addressed to the impeccable Imams, for the "midmost nation" refers to "the righteous Imams who are the mediators between the Messenger and men." However, according to the *Tafsīr al-Qummī*, this verse was originally revealed as follows: "We appointed you midmost Imams (*a'imma wāsiṭa*)," suggesting that an alteration occurred in the text with the suppression of a single letter. Another hadith of the sixth Imam does not confirm the alteration, but states: "The 'nation' in this verse cannot refer to the people of the *qibla* (i.e. Muslims), for among them there are many whose witness in this world would not even be admissible concerning a sack of dates; how then would it be admissible, at the Resurrection, concerning the actions of men? Rather, this 'nation' refers to us, the chosen Imams. It is still about us that Q 3:110: 'You are the best nation ever brought forth to men,' (was revealed)."[19]

Indeed, in the commentary of Q 3:110, Lāhījī reports a tradition of the sixth Imam according to which this verse was not revealed in this way; "for the best nation would not have killed the Prince of the faithful, al-Ḥasan and al-Ḥusayn," but the original verse was: "You are the best Imams...," with the same omission of a single letter (*a'imma* changed to *umma*). According to another reading, the fifth Imam recited: "You, the best nation. . .," once again with the change of a single letter (*antum* instead of *kuntum*), the personal pronoun "you" referring to the Prophet, ʿAlī and the Imams among their descendants.[20]

In his commentary on Q 4:3: "If you fear that you will not act justly towards the orphans, marry such women as seem good to you, two, three, four," Lāhījī reports a tradition in which a Manichean criticizes the Qur'ān before ʿAlī, saying that there was no link and correspondence between acting justly towards the orphans and marrying women. The first Imam replies: "The cause of this lack of link and correspondence is that the hypocrites have removed a third of the Qur'ān which was between the two verses: 'If you fear that you will not act justly towards the orphans' and 'marry such women as seem good to you.'" This tradition thus affirms that a massive censorship occurred in the Qur'ān and implicitly refers to the early caliphs as "the hypocrites" frequently denounced in the Qur'ān.[21]

19 *TSL*, 1: 265; founded on ʿAlī b. Ibrāhīm al-Qummī, *Tafsīr al-Qummī*, ed. Muḥạmmạd Bāqir al-Abṭaḥī l-Iṣfahānī (Qum: Mu'assasat al-Imām al-Mahdī, 2013–14), 1: 98, who does not present it as the restitution of the authentic verse but as an interpretation.

20 *TSL*, 1: 516f.; al-Qummī, *Tafsīr*, 1: 164.

21 *TSL*, 1: 584; from Abū Manṣūr aṭ-Ṭabrisī, *al-Iḥtijāj*, ed. I. Bahādurī and M. Hādī Beh (Qum: Uswa, 1992–93), 1: 574–609.

Regarding Q 4:166: "But God bears witness to that He has sent down to thee; He has sent it down with His knowledge; and the angels also bear witness; and God suffices for a witness," Lāhījī reports a hadith of the sixth Imam according to which it was originally revealed as follows: "But God bears witness to that He has sent down to thee *about ʿAlī* (underlining added). . .".[22] The meaning of the verse is therefore:

> If the hypocrites do not accept the Divine Alliance and the Caliphate of ʿAlī which has been revealed to you, God will bear witness to that he has revealed to you about it. 'He has sent it down with His knowledge' means that God has revealed the Qur'ān or the Divine Alliance of ʿAlī with full knowledge of the facts.[23]

Paradoxically, despite the allusion to the censorship occurred in the Qur'ān, this verse is considered to be a proof of the miraculous character of the Book (*iʿjāz*), since a work whose author has full knowledge must be a perfect work. Does this mean that the miraculous perfection of the Qur'ān is also to have foiled its own falsification? This idea remains only suggested by the traditions reported by al-Lāhījī, but will be explicitly formulated and developed in a slightly later author, Abū l-Ḥasan al-ʿĀmilī l-Iṣfahānī (d. 1725–26/1138).[24]

In his commentary on Q 15:9: "It is We who have sent down the Remembrance, and We watch over it," taken by the Sunni exegetes as a proof of the integrity of the Qur'ān, Lāhījī reports a discussion on the alteration (*taghyīr*) of the Qur'ān led by his contemporary Mullā Khalīl Qazwīnī (d. 1678/1089), a former student of Mīr Dāmād and a commentator of al-Kulaynī's *al-Kāfī*:

> To deduce from this verse that the Qur'ān has not undergone any change is a weak reasoning, for this verse is in the past tense and belongs to a Meccan sura; after this one, many suras went down to Mecca and even more to Medina. This verse therefore does not prove that the whole Qur'ān has been preserved. Nor does it prove that the Qur'ān is preserved by all men, for perhaps it is preserved only by the Imam of the time and his partisans.[25]

However, aṭ-Ṭabrisī, the same who reported the highly subversive tradition on Q 4:3, stated in his *tafsīr Majmaʿ al-bayān*: "According to the doctrine of our (Imāmī) companions, the truth is that no alteration and diminution has ever taken place in the Qur'ān. Weak traditions (*akhbār ḍaʿīfa*) do not allow us to give up what the truth has firmly established, that is, the absence of alteration and diminution in the Qur'ān."[26] To which Mullā Khalīl replies:

---

22 *TSL*, 1: 745, from al-Qummī, *Tafsīr*, 1: 234.

23 *TSL*, 1: 745.

24 See the chapter devoted to him in this handbook.

25 *TSL*, 2: 663.

26 Quoted in Persian in *Ibid.*, 2: 664; see ash-Shaykh aṭ-Ṭabrisī, *Majmaʿ al-bayān fī tafsīr al-Qur'ān*, ed. n.n., 10 vol. (Beirut: Dār al-ʿUlūm, 1426/2005), 1: 14 (in Arabic).

> Many trustful hadiths have been reported from the elite (*al-khāṣṣ*, i.e. Shiite Muslims) as from the common (*al-ʿāmm*, i.e. Sunni Muslims), which show the suppression of many (matters) from the Qurʾān; to call all these hadiths lies is rather presumptuous (. . .). With all this, the claim that the (original) Qurʾān is identical to what is found in the known folios is difficult to sustain.[27]

Lāhījī also quotes the ambiguous position of Fayḍ Kāshānī in his *Wāfī*: "It may be that the basis of the Qurʾān (*aṣl-i Qurʾān*) is indeed what is read and preserved (in the vulgate), and that the suppressed part is its meanings; the alteration would have affected the explanation of the Qurʾānic verses, which also came to the Prophet by revelation."[28] This confirms by the way that the hadiths of the Imams are indeed the continuation of the revelation received by Muḥammad, which is from the Sunni point of view a claim almost as heterodox as the falsification of the Qurʾān. As for the Sunni exegetes, Fakhr ad-Dīn ar-Rāzī (d. 1209–10/606), in his *Great Commentary*, considers that one cannot argue that this verse is a proof of the absence of change in the Qurʾān, as the Imāmīs could reply that this verse itself was added. On this point, Lāhījī finally takes a stand: "This answer is even weaker than the reasoning it criticizes, but it is still false. For all the Imāmīs who claim that additions and omissions were introduced in the Qurʾān base themselves on the hadiths of the People of the House of Prophethood (*ahl-i bayt-i nubuwwat*) and not on mere speculations."[29]

A final illustration of this heavy controversial claim can be found in the exegesis of Q 37:130, of which Lāhījī mentions the two possible readings: "Peace be upon Elias," and "Peace be upon the family of Yāsīn (*āl yāsīn*)." The latter is the reading of the Imams, who understand by "the family of Yāsīn" the family of Muḥammad, since "Yāsīn" is another name for "Muḥammad" as in Q 36:1–3: "Yā Sīn! By the wise Qurʾān, thou art truly among the Envoys." According to a hadith of the first Imam, God chose to say "Peace be upon the family of Yāsīn (*āl yāsīn*)" because He knew that if He would have said: "Peace be upon the family of Muḥammad," the hypocrites and enemies of Muḥammad's family would have removed it from the Qurʾān."[30] That is to say, more clearly than before, that God had foreknowledge of the falsification of the Qurʾān and knew how to foil it in advance to let its hidden meaning be recognized.

3.3 Personalized Commentaries

As Professor M.A. Amir-Moezzi has shown, the earliest form of Shiite hermeneutics is that of a "personalized commentary," which consists in interpreting the Qurʾānic terms as referring to individual and historical persons, essentially the "People of the

27 Quoted in *TSL*, 2: 664.
28 Quoted in *Ibid.*, 2: 664.
29 Ibid., 664f.
30 Ibid., 3: 902f., from aṭ-Ṭabrisī, *Iḥtijāj*, 1: 597.

Prophet's Holly Family" (*ahl al-bayt*) and their enemies, mainly the three first caliphs and the Umayyads, whose names are supposed to have been censored in the official Qur'ān.[31] This kind of commentary reflects the Shiite dualistic vision opposing the forces of Intelligence and Good to those of Ignorance and Evil with their respective historical representatives, an opposition that Imāmī exegesis finds at the very core of the Qur'ānic revelation.

In the commentary of Q 2:115: "To God belong the East and the West; whithersoever you turn, there is the Face of God (*wajh Allāh*)," Lāhījī quotes a hadith of Imam ʿAlī according to which "The face of God" means, in the exoteric sense, the Qur'ān, and in the esoteric sense, the impeccable Imams.[32] And according to a hadith of the eighth Imam: "In this verse, the meaning of 'the face of God' is ʿAlī b. Abī Ṭālib".[33]

In several cases, the personalized commentary supports or confirms the thesis of the falsification of the Qur'ān. Example giving, on Q 10:15: "And when Our signs are recited to them, clear signs, those who look not to encounter Us say, 'Bring a Qur'ān other than this, or alter it.' Say: 'It is not for me to alter it of my own accord. I follow nothing, except what is revealed to me. Truly I fear, if I should rebel against my Lord, the chastisement of a dreadful day'," Lāhījī reports two hadiths of the sixth Imam: "The meaning of 'alter this' is 'alter (the name of) ʿAlī'"; "'Alter this', that is the Prince of the faithful ʿAlī b. Abī Ṭālib."[34] According to the latter, "Our signs" (*āyātanā*) are the verses that affirm the Divine Alliance of ʿAlī and the Imams. The meaning of the verse is therefore:

> When the verses concerning the Divine Alliance of ʿAlī and the impeccable Imams are read to the polytheists and hypocrites, those of them who do not look to encounter Us, that is to encounter Our allies and the heirs of Our prophet (the Imams), say: 'Bring us a Qur'ān other than this, a Qur'ān in which ʿAlī and his eleven children would not be described, or alter the verses concerning ʿAlī's holiness by other verses, so that we accept your prophecy'.[35]
>
> Thus, "alter this" means "alter what is revealed in the Qur'ān about ʿAlī." However, these words may also refer to the Qur'ān. In this case, they refer to "the speaking Qur'ān" (*al-Qur'ān an-nāṭiq*) which is ʿAlī b. Abī Ṭālib, for he said: "I am the speaking word of God" (*kalām Allāh an-nāṭiq*). And at the end of the verse, again according to the sixth Imam: "I follow nothing, except what is revealed to me" means "what has been revealed about ʿAlī".[36]

All this suggests that the alteration or falsification of the Qur'ān was impossible in substance, on the esoteric level, although it was possible and even ascertained in form, on the exoteric level.

31 Amir-Moezzi, *The Silent Qur'an and the Speaking Qur'an*, chap. 3.
32 Aṭ-Ṭabrisī, *Iḥtijāj*, 1: 594.
33 *TSL*, 1: 239.
34 From al-Qummī, *Tafsīr*, 1: 443.
35 *TSL*, 2: 356f.
36 *Ibid.*, 357.

In other cases, it is difficult to distinguish between the personalized commentary and the restitution of the original Qur'ān after its censorship, an ambiguity probably intended by the Imams – and/or their transmitters – and emphasized by our author. On Q 20:15: "And We made covenant with Adam before, but he forgot, and We found in him no constancy," Lāhījī produces a free commentary, yet inspired by the hadiths of the "people of the Holy Family of Prophethood":

> The covenant made with Adam was the revelation to him of the names of the "people of the cloak" (*aṣḥāb-i kisā'*) (i.e. Muḥammad, 'Alī, Fāṭima, al-Ḥasan and al-Ḥusayn) and the Imams of right guidance. It was from them that he had to ask for the fulfilment of his needs. "But he forgot" to ask for their intercession. This is why Adam is not counted among "the prophets with resolution" (*ulū l-'azm*), for these are the ones who did not fail to ask Muḥammad and the Imams for the fulfilment of their needs. According to a hadith of the sixth Imam, this verse was revealed to Muḥammad as follows: "And We made covenant with Adam before *with the names of Muḥammad, 'Alī, Fāṭima, al-Ḥasan, al-Ḥusayn and the Imams of their lineage*, but he forgot."[37]

However, Lāhījī confesses to doubting whether the mention of Muḥammad and the Imams was part of the revealed verse or only of his explanation, which was also revealed to Muḥammad.[38]

The most relevant personalized commentary is probably that of the famous "verse of light" (Q 24:35) and of another that follows it shortly afterwards.

> God is the Light of the heavens and the earth; the likeness of His Light is as a niche wherein is a lamp (the lamp in a glass, the glass as it were a glittering star) kindled from a Blessed Tree, an olive that is neither of the East nor of the West whose oil wellnigh would shine, even if no fire touched it; Light upon Light; God guides to His Light whom He will.

According to a hadith of the fifth Imam:

> The meaning of 'the niche' is the light of the sacred Knowledge in the heart of the Messenger, and 'the glass' is the heart of 'Alī who received science and knowledge from the Prophet. The light of Knowledge is 'neither of the East nor of the West': it is neither that of the Jews (. . .) nor that of the Christians (. . .), because 'Alī states neither by default (*tafrīṭ*), as the Jews do (by denying Jesus' prophethood), nor by excess (*ifrāṭ*), as the Christians do (by making Jesus the Son of God).[39]

The meaning of "whose oil wellnigh would shine, even if no fire touched it" is that the learned among the Family of Muḥammad is about to speak of knowledge before being questioned. "Light upon Light" means that each Imam receives the light of

37 From Kulaynī, *Uṣūl al-kāfī*, 248, ḥ. 23.

38 *TSL*, 3: 110f.

39 From Muḥammad b. 'Alī b. Bābawayh, *Kitāb at-Tawḥīd*, ed. n.n. (Beirut: Dār al-Aḍwā', 2009), 95, ḥ. 4.

knowledge and wisdom from another Imam of Muḥammad's heirs. This succession of Imams one after the other began with Adam and will last until the great resurrection, for every messenger has an heir and a lieutenant (*khalīfa*). These Imams are the heirs that God has made His proofs for His creatures, and no age can be without one of them.[40]

Lāhījī also reports, from the famous companion of the Prophet and the first Imams Jābir b. ʿAbdallāh al-Anṣārī (d. 697/78), the explanation of this verse by Imam ʿAlī, who would have pronounced it with great rejoicing:

> "The likeness of His Light is as a niche": the niche is Muḥammad the Messenger of God; "wherein is a lamp": I am the lamp; "the lamp in a glass": al-Ḥasan and al-Ḥusayn, because "in a glass" means here "with a glass," as ʿAlī is with al-Ḥasanayn; "the glass as it were a glittering star": ʿAlī b. al-Ḥusayn (the fourth Imam); "kindled from a Blessed Tree": Muḥammad b. ʿAlī; "an olive": Jaʿfar b. Muḥammad (the sixth) ; "that is neither of the East": Mūsā b. Jaʿfar (the seventh); "nor of the West": ʿAlī b. Mūsā (the eighth); "whose oil wellnigh would shine": Muḥammad b. ʿAlī (the ninth); "even if no fire touched it": ʿAlī b. Muḥammad (the tenth); "Light upon Light": al-Ḥasan b. ʿAlī (the eleventh); "God guides to His Light whom He will": al-Qāʾim al-Mahdī (the twelfth).[41]

Here again, the hidden or esoteric meaning of the Qurʾān, as revealed by the Imams, is none other than their own persons.

The interpretation of Q 24:40: "or they are as shadows upon a sea obscure covered by a billow above which is a billow above which are clouds, shadows piled one upon another; when he puts forth his hand, wellnigh he cannot see it. And to whomsoever God assigns no light, no light has he," also belongs to the register of "personalized commentary," but on its negative side. According to the sixth Imam:

> The "shadows" are the first and the second (i.e., caliphs: Abū Bakr and ʿUmar); "a sea obscure covered by a billow," it is Naʿthal, i.e., the third (ʿUthmān b. ʿAffān); "above which is a billow," it is Ṭalḥa and Zubayr; "above which are clouds, shadows piled one upon another," they are Muʿāwiya, Yazīd and all the calamities caused by the Umayyads.[42]

According to another tradition of the same Imam:

> "They are as shadows," they are the first and his companion (Abū Bakr and ʿUmar); "upon a sea obscure covered by a billow," it is the third (ʿUthmān), who is like a wave that confused and blotted out (the Qurʾān); "above which is a billow," it is again the second (ʿUmar), who ordered the third to confuse the Qurʾān; "shadows piled one upon another," they are Muʿāwiya – God damn him – and the calamities of the Umayyads.[43]

---

40 *TSL*, 3: 329–31.
41 *TSL*, 3: 331.
42 Ibid., 334; from al-Qummī, *Tafsīr*, 2: 711.
43 *TSL*, 3: 334; from Kulaynī, *Uṣūl al-kāfī*, 111, ḥ. 5.

3.4 The Return to Life

In several places, Lāhījī supports a major tenet of early Imāmī Shiism, frequently labelled in modern times as "exaggeration" (*ghulūw*): the return to life (*raj'a*), at the time of the reappearance of the Mahdī, of a group of allies and a group of adversaries to take part in the final battle that must redeem all the injustices of history.

Example giving, Q 27:83: "Upon the day when We shall muster out of every nation a troop of those that cried lies to Our signs, duly disposed," is interpretated as follows:

> "Our Signs" are the Prince of the faithful and the Imams. It is well known that according to our companions, this mustering will happen in this life, not in the Great Resurrection, and this verse is about the 'return to life' (*raj'a*). For it makes clear that some men will be mustered and some will not, but at the great resurrection all will be mustered. It is reported in many hadiths that before the great resurrection, when the time of the Redeemer will come, a group of very good men and a group of very bad men will return to this life, the good ones to see the reign of the Imams and enjoy some of the good things of this life, the bad ones to suffer from the pains and endure the reign of the People of the Holy Family whom they did not want to see. So the good ones will take revenge on the bad ones. Meanwhile, all the other men will remain in their tombs until the great resurrection.[44]

The eschatological pattern alluded here is developed by the author in his epistle on the imaginal world: between death and the great resurrection lies the time of the Isthmus and the "retribution of the grave," a minor resurrection of imaginal, not physical, nature. The *raj'a* is a natural and bodily resurrection affecting only a tiny minority of men, the best and the worst, whereas the resurrection in the Isthmus concerns the ordinary mortals averagely advanced in perfection or sunk in corruption.

4 Philosophical Grafts in Imāmī Traditional Exegesis

4.1 The Imaginal World

In several passages of his Qur'ānic commentary, Lāhījī introduces the concept of the imaginal world (*'ālam-i mithāl*), an intermediary world between the material-sensible and the spiritual-intelligible worlds. This concept, inherited from the Ishrāqī ("illuminist") and Akbarī (from the *Shaykh al-akbar* Ibn 'Arabī) schools, is recovered or projected in the esoteric meaning (*bāṭin*) of several Qur'ānic verses and Imāmī hadiths. On each occasion, Lāhījī refers to his own epistle on the subject, *ar-Risāla al-mithāliyya* (*Fānūs al-khayāl*). In doing so, he assumes his project of harmonizing the scriptural tradition (Qur'ān and hadith) with "modern" philosophical and mystical thought, which obviously breaks with the attitude of a typical traditionalist.

44 *TSL*, 3: 510.

Thus, in the commentary of Q 2:154: "And say not of those slain in God's way, 'They are dead'; rather they are living, but you are not aware," Lāhījī states:

> They are living and eternally living, either in their original bodies, or in imaginal bodies (*abdān-i mithālī*), as reported from the sixth Imam: "When God takes (the souls) back from the faithful, they (remain) in frames similar to those of this life." This hadith is commented on, together with the explanation of the imaginal world, in the *Epistle on the Imaginal* (*risāla-yi mithāliyya*) of the author of these lines.[45]

This epistle indeed quotes this hadith, as well as many other Imāmī traditions, describing the states of believers after death and before the great resurrection, in a bodily and visible condition that the author identifies with the imaginal world. By attributing this conception to the Imams, Lāhījī allows himself to apply it to the exegesis of the Qur'ān in an apparently traditional *Tafsīr*.

In the commentary of Q 18:31: "those – theirs shall be Gardens of Eden, underneath which rivers flow; therein they shall be adorned with bracelets of gold, and they shall be robed in green garments of silk and brocade, therein reclining upon couches," Lāhījī quotes a certain Gnostic (*ʿārif*) among his contemporaries, whom we have not been able to identify, according to which:

> 'Green clothes' are a metaphor for imaginal and 'isthmus-nature' bodies (*mithāliyya barzakhiyya*), which are intermediate between the darkness of this world and the whiteness of the higher world, just as green is intermediate between black and white (. . .).The meaning of this verse is therefore that in that world, the well-doing believers will put on imaginal bodies which, like the green color, are intermediate between the blackness of the rough material world and the whiteness of the subtle spiritual world. These bodies have themselves different degrees of subtlety (. . .). Apart from the believers (i.e. the Shiites), the other deceased also possess imaginal bodies and green clothes, but not of silk and brocade.[46]

In other words, non believers will also have imaginal bodies, but of a lesser subtlety and closer to material roughness than those of believers. Ashkiwarī-Lāhījī refers to "ancient sages" who asserted that all things belonging to the world of generation (*ʿālam-i kūnī*) do have an image and a form of apparition (*shabḥ*) in the world of Isthmus (*ʿālam-i barzakh*), and points the reader to his own *Epistle on the Imaginal World* for more information.[47] This identification of the Qur'ānic notion of *barzakh* ("barrier," "limit" or "interval," here translated as "Isthmus") with the intermediary state between this life and the hereafter is already to be found in Imāmī hadiths; however, its interpretation as an imaginal state is mostly due to Sufi thinkers, from Abū Ḥāmid al-Ghazālī (d. 1111/505) to Ibn ʿArabī, and Ishrāqī philosophers such as Shams ad-Dīn Shahrazūrī (d. between 1288/687 and 1305/704).

45 Ibid, 1: 274.
46 Ibid., 2: 900f.
47 Ibid., 2: 901.

This is confirmed in Lāhījī's commentary of Q 23:100: "behind them, is a barrier until the day that they shall be raised up":

> Behind these men, there is an intermediary between this life and the hereafter. This is the tomb in which they will remain until the day they will be resurrected. Many hadiths show that the "barrier" (*barzakh*) means the grave. According to a hadith of (the fourth Imam): "The grave is a garden from among the gardens of Paradise or a pit from among the pits of Hell." According to a hadith of (the sixth Imam): "All our partisans will go to Paradise at the great resurrection, through the intercession of the Prophet or his heirs (i.e., the Imams). However, I fear for you in the Isthmus (*barzakh*)." One asks him: "What is the Isthmus?" He said: "The grave, from the moment of death until the coming of the great resurrection." Some philosophers and religious scholars do believe that the *barzakh* is the imaginal world. They say that this world is intermediate between the worlds of immaterial and material beings; and that it is in this world that men are rewarded or punished before the great resurrection.[48]

4.2 Other Philosophical Topics

On Q 5:64: "The Jews have said, 'God's hand is fettered.' Fettered are their hands, and they are cursed for what they have said. Nay, but His hands are outspread. He expends how He will," Lāhījī introduces the notion of *badā'*. Literally "appearance, emergence" (*ẓuhūr*), it means for the Imāmī Shiites the advent of a divine decree changing a previous one announced to the Prophet or the Imam, i.e., an apparent change of will or resolution in God. This notion, of Kaysanid origin but preserved only by the Imāmī branch of Shiism, was first developed in hadiths of a non-rational (which does not mean irrational) nature attributed to the fifth and sixth Imams, collected in a chapter of al-Kulaynī's *Uṣūl al-kāfī*. It reappears in the works of Imāmī philosophers of the Safavid era, beginning with Mīr Dāmād, the author's teacher, followed by Mullā Ṣadrā (d. 1635–36/1045), who also studied under Mīr Dāmād, in commentaries on these hadiths. This concept allows them to save the tenet of divine Justice in view of the tragic history – from the Shiite point of view – of Islam, i.e. to proceed with a Shiite theodicy, and to defend the possibility of human freedom in a system entirely determined by God.[49]

Lāhījī thus reports this comment found in the *Tafsīr al-Qummī*: "The Jews said that God has brought the order (*al-amr*) to an end and will not bring forth anything new that he had not determined in the first measure. God refuted them by saying: "but His hands are outspread. He expends how He will", that is, He can advance a

48 Ibid., 3: 286.

49 See on this Mathieu Terrier, "The Wisdom of God and the Tragedy of History: the Concept of Appearance (*badā'*) in Mīr Dāmād's *Lantern of Brightness*," in *Philosophy and the Intellectual Life in Shī'ah Islam*, ed. Saiyad Nizamuddin Ahmad and Sajjad Rizvi (Bloomsbury: The Shi'ah Institute Press, 2017), 94–134; Mathieu Terrier, "De l'ésotérisme shi'ite à la philosophie shi'ite. Mullā Ṣadrā, Mīr Dāmād et la question du *badā'*," in *Raison et quête de la sagesse: Hommage à Christian Jambet*, ed. Mohammad Ali Amir-Moezzi (Turnhout: Brepols, 2021), 507–50.

thing or delay it, increase it or decrease it. He possesses *badā'* and Will."[50] He explains it in philosophical terms by stating that the essence of *badā'* is the manifestation of a new will in the world of souls, that is, the natural souls of the heavenly spheres and their powers acting in this world. According to an intriguing tradition of the eighth Imam, quoted in Persian translation and without source, "the creation of the world is one case among many (other) cases of *badā*". This is because God knows all the beneficent measures and conditions of the existence or non-existence of the world and worldly beings, and that these measures and conditions differ according to times and moments. The meaning of *badā'* is thus the renewal, at every moment, of God's determination and will concerning the measures and conditions beneficial to His creation. In this respect, this concept is closely related to that of Providence (*ʿināya*) in Avicenna's (d. 1037/428) philosophy.

Following the Imāmī tradition, Lāhījī finds the most adequate Qur'ānic expression of *badā'* in this meaning in Q 13:39: "God blots out, and He establishes whatsoever He will; and with Him is the Essence of the Book." He quotes a statement by the leader of the *akhbārī* traditionalists, Muḥammad Amīn Astarābādī: "The doctrine of *badā'* is a refutation of the Jews claiming that God has brought the orders to an end because He knows from all eternity the destinies of things and had determined everything in accordance with His knowledge. In short, the answer is that God's decisions and wills are renewed every day according to the benefits He expects." In conclusion, our author refers to the writings of some contemporary scholars on *badā'*, presumably thinking of Mīr Dāmād and Mullā Ṣadrā without mentioning their names.[51] One can see here a discrete attempt to reconcile Imāmī *akhbārī* traditionalism and "modern" Shiite philosophy.

Lāhījī gives a typically philosophical interpretation of Q 41:53: "We shall show them Our signs in the horizons and in themselves, till it is clear to them that it is the truth. Suffices it not as to thy Lord, that He is witness over everything?", an interpretation attributed without precision to "men of realization from among the wise" (*muḥaqqiqīn-i ḥukamā'*). According to it, the first part of the verse ("We shall show them Our signs (. . .) the truth") corresponds to the method of the common men of theoretical examination (*ahl-i naẓar*), those who are accustomed to inferring the cause from the effect, a method labelled as "inductive proof" (*dalīl-i innī*), while the second part of the verse ("Suffices it not (. . .) everything?") corresponds to the method of the elite of men of theoretical examination, those who go from the cause to the effect, a method called "deductive proof" (*dalīl-i limmī*). This is an echo of Plato's distinction between ascending dialectics (*anabasis*) and descending dialectics (*katabasis*). Our author concludes: "Since God is not a cause, but the Cause of all causes, as expressed by Shaykh Abū ʿAlī (Ibn Sīnā): 'There is

50 *TSL*, 1:831, from al-Qummī, *Tafsīr*, 1: 251.
51 *TSL*, 1: 831f.

no demonstration of Him, but He is the demonstration of everything,' the deductive proof of God's existence is harder to the point, since His existence has no cause."[52] We understand that there is only inductive proof of the existence of God, but that such knowledge remains essentially deficient.

A final example of philosophically oriented commentary implicitly may be found on Q 91:5–7: "By the heaven and That which built it, and by the earth and That which extended it! By the soul, and That which shaped it". The author writes:

> It is clear that the world of the Testimony (*ʿālam ash-shahāda*), which is the world of bodies, is divided into simple and compound bodies. The simple ones are again divided into higher and lower. The compounds have many divisions; the noblest among them are the essences of souls. God indicated the higher simple bodies by saying, "By the heaven and That which built it," He indicated the lower simple bodies by saying, "and by the earth and That which extended it," and He indicated the noblest divisions of the compound bodies, which are the essences of souls, by saying, "By the soul, and That which shaped it."[53]

5 Conclusion

In conclusion, Bahā' ad-Dīn al-Lāhījī's commentary on the Qur'ān, stands out in the history of Qur'ānic hermeneutics by the meeting of three main features: the Arabic-Persian diglossia, attesting to his pedagogical purpose; an Imāmī Shiite traditionalist (*akhbārī*) orientation, evidenced by numerous allusions to the falsification of the Qur'ān, the mode of "personalized commentary" and the eschatological theme of the "return to life" (*rajʿa*); and the presence of philosophical topics such as that of the imaginal world. The *Tafsīr ash-Sharīf al-Lāhījī* thus contributed, in a rather confidential way, to the fruitful encounter of the Imāmī Shiite hermeneutical tradition with Islamic philosophy in the modern period.

6 Bibliography

6.1 Sources

Main source

Al-Lāhījī al-Ashkiwarī, Bahā' ad-Dīn b. Shaykh ʿAlī sh-Sharīf. *Tafsīr ash-Sharīf al-Lāhījī*. Ed. Muḥaddith Urmawī, 4 vols. Tehran: ʿIlmī, 1984–85; Reed. Jaʿfar Pazhūm, 4 vols. Tehran: ʿIlmī, 2011–12.

52 Ibid, 4: 40f.
53 Ibid., 865.

Other sources

Al-Ashkiwarī, Quṭb ad-Dīn. *Maḥbūb al-qulūb*. Ed. Ibrāhīm ad-Dībājī and Ḥāmid Ṣidqī, 2 vols. Tehran: Mīrāth Maktūb, 1999–2003, third part unpublished.

Al-ʿAyyāshī, Muḥammad b. Masʿūd. *Tafsīr al-ʿAyyāshī*. Ed. ʿAbdallāh Ṣāliḥī, 4 vols. Qum: Dhawī l-Qurbā, 2016–17.

Ibn Bābawayh, Muḥammad b. ʿAlī. *Kitāb at-Tawḥīd*. Ed. n.n. Beirut: Dār al-Aḍwāʾ, 2009.

Ibn Bābawayh, Muḥammad b. ʿAlī. *ʿUyūn akhbār ar-Riḍā*. Ed. M. Lājūrdī, 2 vols. Tehran: Nashr-i Jahān, n.d.

Al-Kulaynī, Muḥammad b. Yaʿqūb. *Uṣūl al-kāfī*. Ed. Ḥasan al-Aʿlamī. Beirut: Muʾassasat al-Aʿlamī, 2005.

Al-Qummī, ʿAlī b. Ibrāhīm. *Tafsīr al-Qummī*. Ed. Muḥammạd Bāqir al-Abṭaḥī al-Iṣfahānī, 3 vols. Qum: Muʾassasat al-Imām al-Mahdī, 2013–14.

Aṭ-Ṭabrisī, Abū Manṣūr. *Al-Iḥtijāj*. Ed. I. Bahādurī and M. Hādī Beh, 2 vols. Qum: Uswa, 1992–93.

aṭ-Ṭabrisī, Amīn al-islām. *Majmaʿ al-bayān fī tafsīr al-Qurʾān*. Ed. n.n., 10 vols. Beirut: Dār al-ʿUlūm, 1426/2005 – 1427/2006),

The Koran Interpreted, trans. Arthur J. Arberry, 2 vols. London: Allen and Unwin, 1955.

6.2 Secondary Literature

Abisaab, Rula Jurdi. *Converting Persia: Religion and Power in the Safavid Empire*. London/New York: I.B. Tauris, 2004.

Amir-Moezzi, Mohammad Ali. *The Divine Guide in Early Shiʿism: The Sources of Esotericism in Islam*, trans. David Streight. Albany: State University of New York Press, 1994.

Amir-Moezzi, Mohammad Ali. *The Silent Qurʾan and the Speaking Qurʾan: Scriptural Sources of Islam between History and Fervor*, trans. Eric Ormsby. New York: Columbia University Press, 2015.

Amir-Moezzi, Mohammad Ali, and Christian Jambet. *What is Shi'i Islam? An Introduction*, trans. Kenneth Casler and Eric Ormsby. London/New York: Routledge, 2018.

Newman, Andrew J. *Safavid Iran: Rebirth of a Persian Empire*. London/New York: I.B. Tauris, 2006.

Terrier, Mathieu. "Anthropogonie et eschatologie dans l'œuvre de Muḥsin Fayḍ Kâshânî: l'ésotérisme shîʿite entre tradition et syncrétisme." In *L'ésotérisme shîʿite, ses racines et ses prolongements. Shi'i Esotericism: its Roots and Developments*, edited by Mohammad Ali Amir-Moezzi. Turnhout: Brepols, 2016: 743–80.

Terrier, Mathieu. *Histoire de la sagesse et philosophie shi'ite: "L'aimé des cœurs" de Quṭb al-Dīn Ashlevarī*. Paris: Le Cerf, 2016.

Terrier, Mathieu. "The Wisdom of God and the Tragedy of History: the Concept of Appearance (*badā'*) in Mīr Dāmād's *Lantern of Brightness*." In *Philosophy and the Intellectual Life in Shīʿah Islam*, edited by Saiyad Nizamuddin Ahmad and Sajjad Rizvi. Bloomsbury: The Shi'ah Institute Press, 2017: 94–134.

Terrier, Mathieu. "The Defence of Sufism among Twelver Shiʿi Scholars of Early Modern and Modern Times: Topics and Arguments." In *Shiʿi Islam and Sufism. Classical Views and Modern Perspectives*, edited by Denis Hermann and Mathieu Terrier. London: I.B. Tauris, 2020: 27–63.

Terrier, Mathieu. "De l'ésotérisme shi'ite à la philosophie shi'ite. Mullā Ṣadrā, Mīr Dāmād et la question du *badā*.'" In *Raison et quête de la sagesse: Hommage à Christian Jambet*, edited by Mohammad Ali Amir-Moezzi. Turnhout: Brepols, 2021: 507–50.

Terrier, Mathieu. *Le guide du monde imaginal. Présentation, édition et traduction de la* Risāla mithāliyya *(Épître sur l'imaginal) de Quṭb al-Dīn Ashkevarī*. Turnout: Brepols, 2023.

ʿAbd ar-Raʾūf b. ʿAlī l-Fanṣūrī l-Jāwī s-Singkīlī

Peter G. Riddell

1 Introduction

The seventeenth-century Malay commentator ʿAbd ar-Raʾūf b. ʿAlī l-Fanṣūrī l-Jāwī s-Singkīlī[1] (d. 1693/1105) is as foundational to Southeast Asian Qurʾānic exegesis as is Muḥammad b. Jarīr aṭ-Ṭabarī (d. 923/310) to exegesis in the Islamic heartlands. Just as aṭ-Ṭabarī appeared on the scene as the faith of Islam was expanding and flourishing in Arabia and surrounding regions, ʿAbd ar-Raʾūf was born around 1615/1024 into an emerging Islamic context in the Malay world, where Islam had thrown down strong roots and become the faith of significant dynasties but still faced wide-ranging challenges.

2 ʿAbd ar-Raʾūf's Life, Works, and Context

Details of his childhood are scarce but some information can be gleaned from his name. His father's name was ʿAlī, with the *nisba* as-Singkīlī suggesting that ʿAbd ar-Raʾūf was born in the west Sumatran town of Singkil, lying around 450 kilometres to the south of the centre of the Sultanate of Aceh in the seventeenth century. While Singkil itself was a relatively insignificant community in terms of influence, it lay at a distance of 130 kilometres from the active port of Barus, also known as Fanṣūr, to the south – hence ʿAbd ar-Raʾūf's additional *nisba* of al-Fanṣūrī. The extra *nisba*, al-Jāwī, was a common term used, especially in Arabia, to refer to inhabitants of the Malay world.[2]

Much legend surrounds ʿAbd ar-Raʾūf's childhood, with some sources influenced by hagiographical material suggesting that he was related to the famous sixteenth-century Acehnese Sufi Hamza al-Fanṣūrī (d. ca. 1590/998) and met both Hamza and the *Qāḍī Mālik al-ʿĀdil* to the Acehnese court, Shams ad-Dīn as-Samaṭrāʾī (d. 1630/1039).[3]

1 The method of transliterating eleventh/seventeenth century Southeast Asian names poses a dilemma; namely, whether to follow Arabic transliteration conventions or rather whether to assimilate according to common modern Southeast Asian conventions (e.g. ʿAbd ar-Raʾūf or Abdurrauf). Chambert-Loir explains the dilemma well: Cf. Henri Chambert-Loir, "Islamic Law in 17th Century Aceh," *Archipel* 94 (2017): 57, f.n. 6. Given that the present paper is part of a broader study of Qurʾānic exegesis and therefore readership is likely to be broad, we have followed the usual Arabic conventions of transliterating Southeast Asian names based on Arabic nomenclature.

2 Anthony H. Johns, "Islam in Southeast Asia: Reflections and New Directions," *Indonesia* 19 (1975): 50.

3 Ali Hasjmy, "Syekh Abdurrauf Syiah Kuala Ulama Negarawan Yang Bijaksana," in *Universitas Syiah Kuala Menjelang 20 Tahun* (Banda Aceh: Universitas Syiah Kuala, 1980), 367–81.

https://doi.org/10.1515/9783111320083-019

Azyumardi Azra's more sober assessment rings true; ʿAbd ar-Raʾūf's early study of Islam was probably with his father ʿAlī in his native town of Singkil, which he then continued in the larger town of Fanṣūr before moving to the capital of the Sultanate of Aceh to pursue further studies, possibly with the aging Shams ad-Dīn as-Samaṭrāʾī.[4]

While in his late teens and twenties, ʿAbd ar-Raʾūf witnessed the eruption of an Islamic doctrinal polemic in the Acehnese Sultanate. Shams ad-Dīn as-Samaṭrāʾī had dominated the religious stage in Aceh for decades, building on the monistic views of *waḥdat al-wujūd* espoused and taught in Aceh throughout much of the sixteenth century by Hamza al-Fanṣūrī. Shams ad-Dīn had enjoyed the patronage of Sultan Iskandar Muda (r. 1607–36/1016–1046), but with both Shams ad-Dīn's death in 1630/1039 and that of Iskandar Muda six years later, the long-standing centrality of *waḥdat al-wujūd* teaching in the Sultanate was challenged by a newcomer to the religious scene, who won favour with the new Sultan.

The successor to Iskandar Muda as Sultan was his son-in-law, Iskandar ath-Thānī. He engaged a Gujerati scholar, Nūr ad-Dīn ar-Rānīrī, who arrived in 1637/1046 to give spiritual direction in the Sultanate. Ar-Rānīrī took little time in embarking on a campaign to discredit Hamza al-Fanṣūrī, Shams ad-Dīn as-Samaṭrāʾī and their followers, accusing them of embracing *wujūdī* views that were heretical, with accusations that were reminiscent of the infamous trial and execution in 922/309 of Manṣūr al-Ḥallāj, who stood charged with equating Creator and created. Ar-Rānīrī's campaign resulted in book burnings, Inquisition-like interrogations and executions during the period 1637–1644/1046–1053, until the Gujerati scholar was driven out of Aceh by the next Sultan, Tāj al-ʿĀlam Ṣafīyat ad-Dīn Shāh, the daughter of Iskandar Muda and widow of Iskandar ath-Thānī, who had died in 1641/1050.[5]

At the peak of the doctrinal polemic, ʿAbd ar-Raʾūf left Aceh for Arabia in 1642/1052, with the purpose of deepening his study of the Islamic sciences. What may have begun as a voyage of fixed duration became a lengthy sojourn of nineteen years. While in Arabia, ʿAbd ar-Raʾūf wrote a work addressing methods of *dhikr* and various ethical questions, entitled *ʿUmdat al-muḥtājīn ilā sulūk maslak al-mufridīn* (The Issue of Those Needing Direction on the Behaviour of Those who Divide);[6] this work concluded with a colophon which is most helpful for the historian, as it recorded key details of ʿAbd ar-Raʾūf's time in Arabia travelling to various sites of learning, including

4 Azyumardi Azra, *The Origins of Islamic Reformism in Southeast Asia: Networks of Malay-Indonesian and Middle Eastern ʿUlamāʾ in the Seventeenth and Eighteenth Centuries* (Crows Nest, Australia: Allen & Unwin, 2004), 71.

5 Takeshi Ito, "Why did Nuruddin ar-Raniri leave Aceh in 1054 AH?," *Bijdragen tot de Taal-, Land- en Volkenkunde* 134 (1978): 489–91.

6 Damanhuri, "Umdah al-Muhtājīn: Rujukan Tarekat Syattariyah Nusantara," *Ulumuna: Jurnal Studi Keislaman* 17, no. 2 (2013): 303–22.

Mecca, Medina and Yemen.[7] He sat at the feet of some of the greatest scholars of his day, studying virtually the full range of the Islamic sciences: theology, jurisprudence, exegesis, tradition, Sufism and other topics. His two main scholarly mentors during this period were Aḥmad al-Qushāshī (1583–1661/991–1071) and Ibrāhīm al-Kūrānī (1616–1690/1025–1101); between them they instructed ʿAbd ar-Raʾūf in both the exoteric and esoteric sciences. These two scholars were also key in ʿAbd ar-Raʾūf obtaining the *ijāza* qualification to propagate the mystical teachings he had studied. He was thus well-placed to return to Aceh with diverse skills and expertise, which set him up to contribute to a still-fractured society after the polemical campaign of Nūr ad-Dīn ar-Rānīrī.

His return occurred in 1661/1072 following the death of his master Aḥmad al-Qushāshī.[8] There are no records of the reception he received on arrival but it must have been with some fanfare, given both his great expertise and the urgent needs of Acehnese society for reconciliation. After his return, he received the patronage of Sultan Tāj al-ʿĀlam Ṣafīyat ad-Dīn Shāh, continuing to serve under royal favour during the famous period of the four female Sultans in Aceh: Tāj al-ʿĀlam Ṣafīyat ad-Dīn Shāh (1641–1675/1051–1086); Nūr al-ʿĀlam Nakiyyat ad-Dīn Shāh (1675–1678/1086–1089); ʿInāyat Shāh Zakiyyat ad-Dīn Shāh (1678–1688/1089–1099); Kamalat Shāh (1688–1699/1099–1111).[9]

ʿAbd ar-Raʾūf settled at the mouth of the Aceh river and established his religious study centre (Acehnese *dayah*). This location earned him the nickname "Syiah Kuala" (the Shaykh of the mouth of the river), by which he is best known. His *dayah* became the hub for the spread of the Shaṭṭāriyya Sufi order throughout the Malay world.[10]

Given the breadth of his training, it is not surprising that during the last three decades of his life, ʿAbd ar-Raʾūf's prolific literary output covered a wide range of the Islamic sciences. Early in his term as spiritual leader of the Sultanate,[11] he was commissioned by Sultan Ṣafiyyat ad-Dīn Shāh to write three works, together designed to speak into the social challenges faced by Acehnese society and to eradicate the leg-

7 Peter G. Riddell, "The Sufi Path. Travelling in Search of Learning. An Indonesian Case Study," In *Insights into Sufism: Voices from the Heart*, ed. Ruth J. Nicholls and Peter G. Riddell (Newcastle: Cambridge Scholars Publishing, 2020), 151–68.

8 Douwe Adolf Rinkes, *Abdoerraoef Van Singkel: Bidjrage tot de Mystieck op Sumatra en Java* (Heerenveen: Hepkema, 1909), 25.

9 For an excellent detailed study of these female Sultans, cf. Sher Banu A.L. Khan, *Sovereign women in a Muslim kingdom: the sultanahs of Aceh, 1641–1699* (Singapore: NUS Press, 2017).

10 Mohd. Syukri Yeoh Abdullah, "Zawiyah Shaykh Kuala: Pusat Penyebaran Islam di Alam Melayu Abad Ke-17 Masihi," *Sari: International Journal of the Malay World and Civilization* 27, no. 2 (2009): 90f.

11 In the role of *Qāḍī Mālik al-ʿĀdil*, according to Azra, *Origins*, 78. In an important 2017 article, Chambert-Loir argues that evidence is inconclusive about the occupants to the two positions of Acehnese religions leadership, the *Qāḍī Mālik al-ʿĀdil* and the *Shaykh al-Islām*, in eleventh/seventeenth century Aceh. Cf. Chambert-Loir, "Islamic Law," 51–96.

acy of the earlier polemics unleashed by Nūr ad-Dīn ar-Rānīrī.[12] The works in question considered both sharīʿa and *taṣawwuf*, in order to address a charge of the earlier polemic that the *wujūdī* Sufis were neglecting the requirements of the Law.[13] The first work was *Mir'āt aṭ-ṭullāb fi tashīl maʿrifat al-aḥkām ash-sharʿiyya li-l-mālik al-wahhāb* (Mirror of the seekers of knowledge of the law of God),[14] completed in 1663/1074, a large manual of jurisprudence providing guidelines on wide-ranging matters: "social, political and economic life, arranged in sections on commercial, matrimonial and criminal law."[15] The other two works, *Kifāyat al-muḥtājīn* and *Daqā'iq al-ḥurūf*, addressed Sufi topics, with the former affirming the transcendence of God over his creation and rejecting the monistic understanding of the *wujūdī* Sufis,[16] while the latter specifically focused on the teachings of Muḥyī d-Dīn b. al-ʿArabī (d. 1240/638), the pioneer of *wujūdī* thinking, again affirming the separation of Creator and creature, as stated in the following quotation from the work: "This section is to show proofs that the world and all created things are other than Almighty God, and they are very numerous."[17]

The number of ʿAbd ar-Ra'ūf's written works is a matter of debate. Since Voorhoeve's initial list in 1952 of around twenty works,[18] that figure has increased with further scholarship. Fathurahman "associates" thirty-six works with ʿAbd ar-Ra'ūf, choosing his words carefully in pointing out that the authorship of some of the works is disputed.[19] Abdullah goes even further, listing "as many as 56 manuscripts of study in various fields of religious disciplines, including 1 work on exegesis, 1 work on hadith, 4 works on ethics, 8 works on fiqh, 41 works on theology and Sufism and 1 historical work."[20] While most of the early study of ʿAbd ar-Ra'ūf's writings was carried out by Western scholars, it is appropriate and welcome that the pendulum has swung to-

12 Salman Harun, "Hakekat Tafsir Tarjuman al-Mustafid Karya Syekh Abdurrauf Singkel" (PhD diss., Institut Agama Islam Negeri Syarif Hidayatullah, Jakarta, 1988), 27–30.

13 For a study of ʿAbd ar-Ra'ūf's efforts to reconcile sharīʿa and *taṣawwuf*, cf. A.S.M. Shukri and R. Arif, "The role of Shaykh Abd Al-Ra'uf Al-Fansuri in the reconciliation of sufism and shariʿah of 17th century the Malay world," *Al-Shajarah* 23 (2018): 207–45.

14 Cf. Peunoh Daly, *Hukum Perkawinan Islam: Suatu Studi Perbandingan Dalam Kalangan Ahlus-Sunnah Dan Negara-Negara Islam* (Jakarta: Bulan Bintang, 1988); also, Jelani Harun, "Mir'at Al-Tullab By Syeikh Abdul Rauf Singkel: A Preliminary Study of Manuscripts Kept In The Special Collections, Leiden University Library," *Malay Literature* 26, no. 2 (2015): 119–38.

15 Annabel Gallop, "A Malay work on Islamic law from Aceh: Mirat al-tullab," Asian and African studies blog, last modified December 5, 2016, https://blogs.bl.uk/asian-and-african/2016/12/a-malay-work-on-islamic-law-from-aceh-mirat-al-tullab.html?utm_source=feedburner&utm_medium=feed&utm_campaign=Feed%3A+asian-and-african+%28Asia+and+Africa%29.

16 Azra, *Origins*, 83.

17 A. H. Johns, "Daqā'iq al-Hurūf by `Abd al-Ra'ūf of Singkel," *JRAS* 87, no. 3–4 (1955): 151.

18 Petrus Voorhoeve, "Bajan Tadjalli," *Tijdschrift voor Indische Taal-, Land-, en Volkenkunde* 23, no. 1 (1952): 87–115.

19 Oman Fathurahman, *Tanbih al-Masyi: Menyoal Wahdatul Wujud* (Jakarta: Mizan, 1999), 28–30.

20 Abdullah, "Zawiyah Shaykh Kuala," 94.

wards scholarly studies by Southeast Asian researchers in the late twentieth and early twenty-first centuries, with excellent contributions by scholars whose names appear in the bibliography and others whose research output was not accessible for this present study.

3 ʿAbd ar-Raʾūf's *Tarjumān al-mustafīd*

Our attention now turns to one of ʿAbd ar-Raʾūf's greatest works, his commentary on the Qurʾān, entitled *Tarjumān al-mustafīd* (The Interpreter of the Beneficial). Given the breadth of his training in Arabia, it is not surprising that ʿAbd ar-Raʾūf chose to write a commentary in Malay on the entire Qurʾān. However, his reasons for doing so extend beyond that explanation. His extensive writings on Sufism were of considerable value to those readers for whom Sufi thought was integral to their Islamic faith. Furthermore, his works on jurisprudence, with their clear injunctions and guidelines, represented a crucial contribution to the Sultanate of Aceh as it negotiated its pathway out of contestation and polemic. However, for many Muslims in the Sultanate and beyond, neither Sufi thought nor detailed jurisprudence spoke into their daily needs. Furthermore, the religious community leaders, the imams, were in need of guidance on interpreting Islam's sacred text, the Qurʾān, as they gave direction to the masses. *Tarjumān al-mustafīd* spoke into all these needs and arguably represents ʿAbd ar-Raʾūf's greatest legacy to the common Muslim faithful. As Arivaie Rahman comments: "(. . .) the purpose of writing this commentary was to make it easier for ordinary people to understand the Qurʾān because many of them were not proficient in Arabic and there was also no complete written Malay interpretation of the Qurʾān."[21]

A consideration of various contextual factors, relating to ʿAbd ar-Raʾūf's theological maturity, his occupational status, and surviving manuscript evidence, points to *Tarjumān al-mustafīd* having been completed around 1675/1086. It has had a major impact on the development of Islam in Southeast Asia. Over a dozen handwritten manuscripts survive, in various locations in Southeast Asia, Europe and South Africa.[22] The work has been printed many times, with its various editions reflected in the bibliography.

21 Arivaie Rahman, "Tafsir Tarjumân al-Mustafî Karya ʿAbd al-Rauf al-Fanshuri: Diskursus Biografi, Kontestasi Politis-Teologis, dan Metodologi Tafsir," *MIQOT: Journal of Islamic Studies* 41, no. 1 (2018): 9.
22 Ervan Nurtawab, "Discourse on translation in hermeneutics: its application to the analysis of Abdurra'ûf's turjumân al-mustafîd" (master's thesis, Sekolah Pascasarjana UIN Syarif Hidayatullah, Jakarta, 2007), 21.

3.1 The Sources of *Tarjumān al-mustafīd*

Tarjumān al-mustafīd represents the first commentary in Malay upon the whole Qur'ān. Previous efforts at exegesis in Malay appear to have been fragmentary, according to surviving evidence. The most substantial work of Qur'ānic exegesis predating *Tarjumān al-mustafīd* was a commentary upon sura 18, *al-Kahf* (The Cave), of which only one copy survives.[23]

References to *Tarjumān al-mustafīd* in Dutch colonial writings identified the work as a rendering of the authoritative Arabic commentary *Anwār at-tanzīl wa-asrār at-ta'wīl*, by Nāṣir ad-Dīn al-Bayḍāwī (d. 1286/685). Such references were made by the Dutch colonial authority on Islam, Christiaan Snouck Hurgronje, as well as the scholars Douwe Adolf Rinkes and Petrus Voorhoeve. It appears that this association was made on the basis of a reference to al-Bayḍāwī's commentary on the title page of the first printed edition of *Tarjumān al-mustafīd* that appeared in Istanbul in 1884/1302. The association of *Tarjumān al-mustafīd* with al-Bayḍāwī's commentary then reappeared in subsequent editions of the work, and in some later scholarship.[24]

Anthony Johns was an early voice to notice discrepancies between *Tarjumān al-mustafīd* and al-Bayḍāwī, though he did not question the association of the two.[25] In my 1984 PhD thesis I argued, on the basis of a detailed study of *juz'* 16 of *Tarjumān al-mustafīd*, that both the sources and the authorship of the work were composite, with the *Tafsīr al-Jalālayn*, rather than al-Bayḍāwī's *Anwār*, representing the core source, though supplemented by material drawn from al-Bayḍāwī, al-Khāzin's *Lubāb al-ta'wīl fī ma'ānī t-tanzīl*, and several other sources, including inputs by 'Abd ar-Ra'ūf himself. Regarding authorship, the full version of *Tarjumān al-mustafīd* was shown to represent the core commentary by 'Abd ar-Ra'ūf, supplemented by information on the *qirā'āt* and narrative material added by 'Abd ar-Ra'ūf's student, Dā'ūd ar-Rūmī.[26]

The association of *Tarjumān al-mustafīd* with *Tafsīr al-Jalālayn* as core source was borne out by Salman Harun's doctoral research[27] and also by investigation undertaken by Azyumardi Azra[28] and Ervan Nurtawab.[29] In 2018, Mokhtar and Mokhtar provided further evidence of the relative disconnect between the commentaries by 'Abd ar-Ra'ūf and al-Bayḍāwī, examining eighteen suras and reporting on a detailed comparison of the content of Q 1:1–7 and Q 2:1–15. They concluded as follows: "There

23 Peter G. Riddell, *Malay Court Religion, Culture and Language: Interpreting the Qur'ān in 17th Century Aceh* (Leiden: Brill, 2017).

24 For a full discussion of this issue, cf. Peter G. Riddell, *Transferring a Tradition: 'Abd Al-Ra'ūf Al-Singkilī's Rendering into Malay of the Jalālayn Commentary* (Berkeley, CA: University of California, Centers for South and Southeast Asia Studies, 1990), chapter 2.

25 Johns, "Islam in Southeast Asia," 47f.

26 Riddell, *Transferring*, chapter 2.

27 Harun, "Hakekat Tafsir," 27–30.

28 Azra, *Origins*, 83.

29 Nurtawab, "Discourse," 30.

are similarities in some aspects between the two commentaries, but these similarities are only external. A detailed examination reveals very clear differences between these two commentaries."[30]

In further 2018 research, Arivaie Rahman examined the exegesis of suras 1 and 2 in *Tarjumān al-mustafīd* and found the most heavy use in those Qurʾānic chapters to be of al-Khāzin's commentary, not those of al-Bayḍāwī nor the Jalālayn. His conclusion was that "this commentary is not a translation of Tafsîr al-Baidhâwî or the Jalâlain commentary, but an original work compiled by ʿAbd ar-Raʾūf with reference to various books of commentary."[31] This conclusion of composite sourcing is consistent with the earlier-mentioned studies, but the observation regarding core use of al-Khāzin's commentary in suras 1 and 2 is interesting and bears closer examination in other parts of *Tarjumān al-mustafīd*. We shall consider the respective claims of the Jalālayn and al-Khāzin as foundational exegetical source of *Tarjumān al-mustafīd* in the following section.

3.2 The Style of *Tarjumān al-mustafīd*

Tarjumān al-mustafīd presents its exegesis according to the standard order of the suras rather than reflecting a chronological arrangement or taking a thematic approach. A sample of the exegesis of *Tarjumān al-mustafīd* assists us in a reconsideration of the style and sources of the commentary. What follows is the *Tarjumān al-mustafīd* commentary upon Q 38:21–26, a coherent pericope which concerns the prophet David being subjected to a divine test. The text in parentheses represents a rendering into English of the text of the Qurʾān. The text outside the parentheses represents commentary of two main kinds: (a) Phrase-by phrase commentary upon the Qurʾān drawn up by ʿAbd ar-Raʾūf, as discussed above; (b) More discrete exegetical interpolations drawing on a range of commentators. These interpolations have been set apart in the translation which follows under the headings of "anecdote" or "clarification". These interpolations were composed and added to ʿAbd ar-Raʾūf's commentary by his student, Dāʾūd ar-Rūmī.

21. (Have you heard) O Muhammad (the story of the disputants who climbed the walls of the sanctuary) of David's place of prayer?
22. (When they came upon David), i.e., before David (and David was startled by them. They said: "Do not be afraid,) O David. We represent (two groups who are in dis-

30 Ahmad Baha' Mokhtar and Ahmad Effat Mokhtar, "Analisis Pengaruh Tafsir Al-Baydawi Dalam Tafsir Turjuman Al-Mustafid," *Jurnal Al-Sirat* 1, no. 17 (2018): 94
31 Rahman, "Tafsir," 12.

pute and have reached an impasse) between us. (Judge between us in fairness, and do not be unjust, and guide us to the right path.")

23. ("My brother here has ninety-nine wives, while I only have one wife. He said) i.e., the one who had ninety-nine wives said: ('Give her into my care) and I will marry her'. (And he imposed himself upon me in this dispute.")

(*Anecdote*): Khāzīn reports that Ibn ʿAbbās said that "Give her into my care" was actually stated as "Divorce her so that I can marry her."

24. In judgement, David (said: "He wronged you in demanding that your sheep be added to his flock of sheep. Many are those) . . . (who are unjust one to another; in contrast, those who have faith and do good works are few indeed); there lies therein a sign."

Anecdote: This was a test by God of His prophet David, peace be upon him. There were various accounts of the prophets, peace be upon them, and (also) disagreement among the religious scholars regarding the prophets' accounts and the reason for this test.

Ibn Masʿūd related that the sin of God's prophet David was in demanding that another man divorce his wife.

One version relates that this was obligatory in the religion of the Israelite prophets; however Almighty God did not favour David doing this because He loved the world and had added woman (to man) and they had multiplied with gifts bestowed upon them by Him. So Almighty God sent two angels in the form of two men, and they went up before him in his place of prayer, and David did not recognize the two until the two were before him as he prayed.

One version relates that the two were (the archangels) Gabriel and Michael. When the prophet David had passed judgement between them, he saw one look at his companion and laughing as they ascended to Heaven. At that moment the prophet David realised that Almighty God had wished to cleanse him.

And God knows best.

(David realised that We had subjected him to a test) because of his love for the woman. (He begged forgiveness of his Lord and fell upon his knees) and turned back to Almighty God in penitence.

25. (So he was forgiven for this (transgression), and he was nearer to Us) through increasing goodness in the world (and the best placed) in the hereafter.

Clarification: regarding the differences among the three readers in reading *wa-lī*. Abū ʿAmr and Nāfiʿ agree in reading it *wa-lī* with an unvowelled *yāʾ*, while Ḥafṣ reads it with a *fatḥa* on the *yāʾ*.

And God knows best.

26. (O David! We have made you a Viceregent in the world) in order to govern over people, (so judge between people with fairness. Do not follow your lusts, otherwise you will be lost from) the signs which point to *tawḥīd*. (Truly all those who lose) belief in (Almighty God face a terrible punishment for forgetting the day of reckoning) leading to loss of faith.

From the above short pericope of verses, ʿAbd ar-Raʾūf's style of progressive commentary, moving through the verses phrase by phrase, is clear. This is consistent with the style of *tafsīr taḥlīlī*, where the commentator follows the Qur'ānic order of the suras and āyas, with the exegetical approach focusing on the detail, reflecting the interests and expertise of the exegete.[32] Though only representing six verses of a single sura, this excerpt is rich with elements that provide crucial insights into the identity and style of *Tarjumān al-mustafīd*. We shall take each verse in turn, identifying features that open windows into our Malay commentary.

Verse 21 contains little in-verse commentary, mainly representing a rendering by ʿAbd ar-Raʾūf of the text of the Qurʾān. The two small exegetical phrases, "O Muhammad" and "of David's place of prayer", appear to have been taken directly from the *Jalālayn* commentary on this verse, which contains the equivalent and succinct phrases *yā Muḥammad* and *miḥrāb Dāʾūd*.[33] While al-Khāzin's also contains the phrase *yā Muḥammad* as well as reference to the place as David's *miḥrāb*,[34] these references are immersed within lengthy narrative commentary. The succinctness of ʿAbd ar-Raʾūf's commentary reflects that of the Jalālayn, making it the more likely source for this verse.

In verse 22, the very minor exegetical insertions are merely designed to smoothe the flow of the Qur'ānic text in Malay to aid reader comprehension. The Malay text was likely to have been the creation of ʿAbd ar-Raʾūf himself without drawing significant information from the commentators.

The commentary on verse 23 is in two parts. ʿAbd ar-Raʾūf exercises some liberty with the Qur'ānic text, replacing the Arabic reference to "ewes" (*naʿja*) by reference to "wives". This is a rendering of the phrase that occurs in the *Jalālayn* at this point: *yuʿabbir bihā ʿan al-marʾa*.[35] This decision by ʿAbd ar-Raʾūf to exclude the reference to ewes was more bold than the approach taken by the Jalālayn and al-Khāzin, both of whom first translate the Qur'ānic text with ewes, then insert an explanatory phrase specifying the actual reference as a woman. ʿAbd ar-Raʾūf's approach has the effect of significantly lessening the metaphorical sense of the Arabic text, tying it more closely to the Biblical narrative of David and Bathsheba, though he makes no direct reference to Bathsheba or Uriah.[36]

Verse 23 is enhanced by a short exegetical Anecdote, added by Dāʾūd ar-Rūmī. It identifies al-Khāzin as its source, and quotes al-Khāzin's source in turn as being Ibn

32 Fahd ʿAbd ar-Raḥmān ar-Rūmī, *Buḥūth fī uṣūl at-tafsīr wa-manāhijuhu* (Investigation into the foundations and methods of Quranic exegesis) (Riyadh: Maktabat al-Tawba, 1996), 57–59.

33 Jalāl ad-Dīn Muḥammad b. Aḥmad al-Maḥallī and Jalāl ad-Dīn ʿAbd ar-Raḥmān b. Abī Bakr as-Suyūṭī, *Tafsīr al-Qurʾān al-ʿaẓīm wa-huwa tafsīr al-Jalālayn* (Beirut: Dār al-Fikr, 1981), 374.

34 ʿAlī b. Muḥammad Ibrāhīm al-Khāzin al-Baghdādī and ʿAbd Allāh b. Aḥmad Nasafī, *Tafsīr al-Qurʾān al-jalīl al-musammā Lubāb at-taʾwīl fī maʿānī t-tanzīl*, vol. 4 (Beirut: Muḥammad Amīn Damj, n.d).

35 Al-Maḥallī and as-Suyūṭī, *al-Jalālayn*, 374.

36 2 Samuel 11.

ʿAbbās.[37] This short text serves several purposes. It has a didactic function, familiarising the audience with one of the key Companions. It also reinforces ʿAbd ar-Raʾūf's approach to this verse by repeating that this matter concerns wives, not ewes, further cementing the link between the Qurʾānic account and the Biblical account of David and Bathsheba. However, note that al-Khāzin goes further than ʿAbd ar-Raʾūf, actually naming Uriah and identifying him as "the husband of the woman whom David married".[38] In doing so, the Anecdote connects Ibn ʿAbbas with the David and Bathsheba account, potentially making him, al-Khāzin, and indirectly ʿAbd ar-Raʾūf, vulnerable to accusations of importing *Isrāʾiliyyāt* elements which demean the reputation of the prophet David into commentary on this narrative.

Verse 24 serves as an action peak in this short set of verses and comprises three elements. Abd ar-Raʾūf's in-verse exegesis appears in two segments, enfolding an exegetical Anecdote inserted by Dāʾūd ar-Rūmī. In his in-verse commentary, Abd ar-Raʾūf reverts to talking about sheep rather than wives. He has identified the real point of reference in the previous verse; his concern in the new verse is to underscore that the point of this account is that God is testing David. The second part of the in-verse commentary after the Anecdote is drawn from the *Jalālayn*. Abd ar-Raʾūf further consolidates the link with the Biblical David and Bathsheba story by adding the phrase "because of his love for the woman" in explaining the reason for the test.[39]

This section of commentary on verse 24 also affirms God's sovereignty, in a way which speaks into the Acehnese Sultanate's debate about *wujūdī* monism. That contentious teaching is rejected, with a clear statement of dualism evident in David's humble submission to a transcendent God when he "turned back to Almighty God in penitence." That short exegetical phrase could have been taken from the Jalālayn (*ay sājidan*) or al-Khāzin, though the succinctness of the former and the sheer wordiness of the latter, plus the fact that Dāʾūd ar-Rūmī was adding sections from al-Khāzin, makes the Jalālayn the more likely source for such short exegetical phrases.

Dāʾūd ar-Rūmī's lengthy Anecdote in the middle of the verse 24 commentary is taken from al-Khāzin.[40] He is careful to avoid some of the most sordid details of the David and Bathsheba/Uriah narrative that al-Khāzin chooses to include; e.g. "It was said that David wanted to kill Uriah and marry his wife. This was his fault."[41] The Anecdote serves several purposes. Dāʾūd ar-Rūmī takes the opportunity to introduce a famous early Companion scholar by referring to Ibn Masʿūd. Furthermore, he familiarises readers with diverse opinions on the topic at hand, in the process giving an alternative model to the single-mindedness of ar-Rānīrī from earlier times. In this,

37 Al-Khāzin, *Lubāb*, 35.

38 "(. . .) *zawj al-marʾa allatī tazawwajahā Dāwūd*": al-Khāzin, *Lubāb*, 35.

39 "*Ay baliyya bi-maḥabbatihi tilka al-marʾa*: al-Maḥallī and as-Suyūṭī, *al-Jalālayn*, 375.

40 Al-Khāzin, *Lubāb*, 33.

41 "*Wa-qīla inna Dāwūd aḥabba an yaqtula ʿUwriyya fa-yatazawwaju imraʾatahu fa-hādhā kāna dhanbahu*": al-Khāzin, *Lubāb*, 34.

there is a sense of this material being devised for a *dayah* classroom, and it may well be that Dāʾūd ar-Rūmī's additions were in part designed to ensure that *Tarjumān al-mustafīd* was suitable for classroom settings. This Anecdote also reinforces ʿAbd ar-Raʾūf's in-verse commentary affirmation of the sovereignty of God, with the reminder of who is in charge: "Almighty God sent two angels."

In verse 25, the in-verse commentary by ʿAbd ar-Raʾūf is taken straight from the *Jalālayn*.[42] Though short, this verse is important in making clear that this is no *wujūdī* commentary. The Arabic text of the Qurʾān makes reference to David being "nearer to Us," which could have easily been explained in terms of monistic nearness. ʿAbd ar-Raʾūf prefers to allude to the sharīʿa by specifying nearness to God through good deeds: "through increasing goodness in the world".[43] Verse 25 is also enhanced by a "Clarification" paragraph, with information on the *qirāʾāt*. This discussion is brief and does not truly show the significant level of detail devoted to the *qirāʾāt* throughout *Tarjumān al-mustafīd*. Nevertheless, the short Clarification paragraph above does demonstrate that this commentary favoured three of the authoritative reading systems of the Qurʾān, namely those of Abū ʿAmr b. al-Alāʾ (d. 770/154), Nāfiʿ b. ʿAbd ar-Raḥmān b. Abī Nuʿaym al-Madanī (d. 785/169), and Ḥafṣ b. Sulaymān b. al-Mughīra (d. 796/180). This information would seem to be further evidence of an intention to serve the needs of Islamic students in a *dayah*, suggesting that Dāʾūd ar-Rūmī's additions, which post-dated ʿAbd ar-Raʾūf's core commentary, may have been designed to expand the readership of *Tarjumān al-mustafīd*.

Verse 26 serves as the didactic peak of our short pericope of six verses. The content is not only valuable in serving that particular literary function but also because it provides overwhelming evidence of ʿAbd ar-Raʾūf's preference for the *Jalālayn* as the core commentary for his in-verse exegesis, at least in this short pericope. The verse contains four exegetical insertions within the Malay rendering of the Qurʾānic text: English translations are "in order to govern over people," "the signs which point to *tawḥīd*," "belief in" and "leading to loss of faith". These four phrases represent renderings of the following phrases in the *Jalālayn* respectively: "*tudabbiru amr an-nās; ay ʿan ad-dalāʾil ad-dālla ʿalā tawḥīdihi; ay ʿan al-īmān bi-llāh;* and *al-murattab ʿalayhi tarkuhum al-īmān.*"[44]

42 Al-Maḥallī and as-Suyūṭī, *al-Jalālayn*, 375.

43 "*Ay ziyādat khayrin fī d-dunyā*": al-Maḥallī and as-Suyūṭī, *al-Jalālayn*, 375.

44 Al-Maḥallī and as-Suyūṭī, *al-Jalālayn*, 375.

4 Conclusion

The above discussion of our short pericope of six verses from sura 38, *Ṣād*, has provided us with crucial insights into the nature of *Tarjumān al-mustafīd*. They have demonstrated the composite authorship of this work, with the combined efforts of ʿAbd ar-Raʾūf and Dāʾūd ar-Rūmī having the effect of taking the skeletal outline of the Qurʾānic text and converting it into a comprehensive and stand-alone narrative account. The six verses have also demonstrated the composite sources of *Tarjumān al-mustafīd*, with the *Jalālayn* commentary serving as primary (though not sole) source of the in-verse commentary created by ʿAbd ar-Raʾūf and al-Khāzin's *tafsīr* serving as primary source for the interpolated anecdotes. With regard to the commentary of al-Bayḍāwī, its non-appearance in the above six verses points to the nature of its presence within *Tarjumān al-mustafīd*: helpful when it appears, but its inputs are sporadic. This sets in even more stark relief the error underlying the long-standing reputation of *Tarjumān al-mustafīd* as a rendering of al-Bayḍāwī's *Anwār*.

We have also seen ways that the six verses examined speak into ʿAbd ar-Raʾūf's Acehnese context. This is no Sufi commentary, yet ʿAbd ar-Raʾūf takes the opportunity to subtly refute *wujūdī* teaching by affirming dualistic themes consistent with an attempt to reconcile Sufism and sharīʿa. Furthermore, an Acehnese society recovering from a period of polarisation and polemic was in need of unambiguous direction in terms of doctrine and practice. ʿAbd ar-Raʾūf seems to have seen his Qurʾān commentary as contributing to this need, preparing a work that was accessible to Muslims without specialist knowledge but, with Dāʾūd ar-Rūmī's additions, also being suitable for the emerging Islamic schools, the *dayahs*, that were appearing in an increasingly Islamised Malay world.

ʿAbd ar-Raʾūf's reputation is considerable in Islamic Southeast Asia, but he remains largely unknown in other parts of the Muslim world. This is unfortunate and undeserved, as when we consider the crucial pioneering role of *Tarjumān al-mustafīd* in the annals of Southeast Asian Qurʾānic exegesis, supplemented by the huge contribution made by ʿAbd ar-Raʾūf through his prolific writing on the other Islamic sciences, we realise that he can fairly claim to be considered as the aṭ-Ṭabarī of Southeast Asian Islam.

5 Bibliography

5.1 Sources

Al-Fanṣūrī, ʿAbd ar-Raʾūf b. Shaykh ʿAlī l-Jāwī. *Tarjumān al-mustafīd*. 2 vols. Istanbul: Matbaʿa ʿUthmaniyya, 1884.

Al-Fanṣūrī, ʿAbd ar-Raʾūf b. ʿAlī. *Tarjumān al-mustafīd*. Singapore: Sulayman Maraghi, 1951.

Al-Fanṣūrī, ʿAbd ar-Raʾūf b. Muḥammad ʿAlī. *Tafsīr Anwār al-Bayḍāwī*. 3 vols. Penang: Persamar Press, 1961.

Al-Fanṣūrī, ʿAbd ar-Raʾūf b. ʿAlī. *Tarjumān al-mustafīd*. Cairo: Muḥammad Idrīs al-Marbāwī, 1981[4].
Al-Fanṣūrī, ʿAbd ar-Raʾūf b. ʿAlī. *Tarjumān al-mustafīd*. Beirut: Dār al-Fikr, 1990.
Al-Khāzin al-Baghdādī, ʿAlī b. Muḥammad Ibrāhīm and ʿAbd Allāh b. Aḥmad Nasafī, *Tafsīr al-Qurʾān al-jalīl al-musammā Lubāb at-taʾwīl fi maʿānī t-tanzīl*. vol. 4. Beirut: Muḥammad Amīn Damj, n.d.
Maḥallī, Jalāl ad-Dīn Muḥammad b. Aḥmad and Jalāl ad-Dīn ʿAbd ar-Raḥmān b. Abī Bakr as-Suyūṭī. *Tafsīr al-Qurʾān al-ʿaẓīm wa-huwa tafsīr al-Jalālayn*. Beirut: Dār al-Fikr, 1981.

5.2 Secondary Literature

Abbas, Syahrizal. *Corak pemikiran hukum Islam Syeikh Abdurrauf As-Singkili: studi terhadap kitab Mi'rat al-Thullab tentang hakim perempuan*. Banda Aceh: Yayasan PeNA, 2008.
Abdullah, Mohd. Syukri Yeoh. "Zawiyah Shaykh Kuala: Pusat Penyebaran Islam di Alam Melayu Abad Ke-17 Masihi." *Sari: International Journal of the Malay World and Civilization* 27, no. 2 (2009): 87–118.
Adam, Syahrul and Maman Rahman Hakim. "Menelusuri Jalan Sufi (kajian Kitab ʿUmdat al-muḥtājīn ilā Suluk Maslak Al Mufradīn Karya ʿAbd al-Râuf al-Sinkîlî)." *Kordinat: Jurnal Komunikasi Antar Perguruan Tinggi Agama Islam* 16, no. 2 (2017): 367–88.
Arif, Ridwan. *Sheikh ʿAbd Al-Raʾuf Al-Fansuri: Rekonsiliasi Tasawuf dan Syariat Abad Ke-17 di Nusantara*. Jakarta: Buku Kompas, 2021.
Azra, Azyumardi. *The Origins of Islamic Reformism in Southeast Asia: Networks of Malay-Indonesian and Middle Eastern 'Ulamā' in the Seventeenth and Eighteenth Centuries*. Crows Nest, Australia: Allen & Unwin, 2004.
Basri, Fairuzah Haji. "ʿUmdat al-Muḥtājīn ilā Sulūk Maslak al-Mufridīn: Satu Kajian Teks." Master's Thesis, Universiti Kebangsaan Malaysia, 2002.
Chambert-Loir, Henri. "Islamic Law in 17th Century Aceh." *Archipel* 94 (2017): 51–96.
Daly, Peunoh. *Hukum Perkawinan Islam: Suatu Studi Perbandingan Dalam Kalangan Ahlus-Sunnah Dan Negara-Negara Islam*. Jakarta: Bulan Bintang, 1988.
Damanhuri. *Akhlak: perspektif pemikiran tasawuf Syeikh Abdurrauf As-Singkili*. Banda Aceh: ArRijal Publisher, 2011.
Damanhuri. "ʿUmdah al-Muḥtājīn: Rujukan Tarekat Syattariyah Nusantara." *Ulumuna: Jurnal Studi Keislaman* 17, no. 2 (2013): 303–22.
Fathurahman, Oman. *Tanbih al-Masyi: Menyoal Wahdatul Wujud*. Jakarta: Mizan, 1999.
Gallop, Annabel. "A Malay work on Islamic law from Aceh: Mirat al-tullab." Asian and African studies blog. Last modified December 5, 2016. https://blogs.bl.uk/asian-and-african/2016/12/a-malay-work-on-islamic-law-from-aceh-mirat-al-tullab.html?utm_source=feedburner&utm_medium=feed&utm_campaign=Feed%3A+asian-and-african+%28Asia+and+Africa%29.
Harun, Jelani. "Mir'at Al-Tullab By Syeikh Abdul Rauf Singkel: A Preliminary Study Of Manuscripts Kept In The Special Collections, Leiden University Library." *Malay Literature* 26, no. 2 (2015): 119–38.
Harun, Salman. "Hakekat Tafsir Tarjuman al-Mustafid Karya Syekh Abdurrauf Singkel." PhD diss., Institut Agama Islam Negeri Syarif Hidayatullah, Jakarta, 1988.
Hasjmy, Ali. "Syekh Abdurrauf Syiah Kuala Ulama Negarawan Yang Bijaksana." In *Universitas Syiah Kuala Menjelang 20 Tahun*. Banda Aceh: Universitas Syiah Kuala, 1980: 367–81.
Irawati, Marlinda. "Women's Political Communications: Assessing Leadership In Queen Safiatuddin In Aceh Darussalam in the 17th Century." *Inter Komunika: Jurnal Komunikasi* 1, no. 1 (2016): 65–73.
Ito, Takeshi. "Why did Nuruddin ar-Raniri leave Aceh in 1054 AH?." *Bijdragen tot de Taal-, Land- en Volkenkunde* 134 (1978): 489–91.
Johns, Anthony H. "Daqāʾiq al-Ḥurūf by ʿAbd ar-Raʾūf of Singkel." *JRAS* 87, no. 3–4 (1955): 139–58.
Johns, Anthony H. "Islam in Southeast Asia: Reflections and New Directions." *Indonesia* 19 (1975): 33–55.

Johns, Anthony H. "The Qur'an in the Malay World: Reflections on 'Abd al-Ra'uf of Singkel (1615–1693)." *JIS* 9, no. 2 (1998): 120–45.

Khan, Sher Banu A.L. *Sovereign women in a Muslim kingdom: the sultanahs of Aceh, 1641–1699*. Singapore: NUS Press, 2017.

Mokhtar, Ahmad Baha'. "Syarh Latif 'ala Arba'in Hadithan Li Al-Imam Al-Nawawi Karangan Syeikh Abd Rauf Al-Fansuri: Satu Kajian Teks." Master's Thesis, Universiti Malaya, 2008.

Mokhtar, Ahmad Baha' and Ahmad Effat Mokhtar. "Analisis Pengaruh Tafsir Al-Baydawi Dalam Tafsir Turjuman Al-Mustafid." *Jurnal Al-Sirat* 1, no. 17 (2018): 85–95.

Nasution, Ismail Fahmi Arrauf. "Termination Of Wahdatul Wujud In Islamic Civilization In Aceh: Critical Analysis Of Ithaf Al-Dhaki, The Works Of Ibrahim Kurani." *Addin: Media Dialektika Ilmu Islam* 11, no. 2 (2017): 401–32.

Nurtawab, Ervan. "Discourse on translation in hermeneutics: its application to the analysis of Abdurra'ûf's turjumân al-mustafîd." Master's Thesis, Sekolah Pascasarjana UIN Syarif Hidayatullah, Jakarta, 2007.

Nurtawab, Ervan. "The Problems of Translation in Turjuman al-Mustafid: A Study of Theological and Eschatological Aspects." *Studia Islamika: Indonesian Journal for Islamic Studies* 18, no. 1 (2011): 33–65.

Putra, Afriadi. "Khazanah Tafsir Melayu (Studi Kitab Tafsir Tarjuman Al- Mustafid Karya Abd Rauf Al-Sinkili)." *Jurnal Syahadah* 2, no. 2 (2014): 69–88.

Rahman, Arivaie. "Tafsir Tarjumân al-Mustafî Karya 'Abd al-Rauf al-Fanshuri: Diskursus Biografi, Kontestasi Politis-Teologis, dan Metodologi Tafsir." *MIQOT: Journal of Islamic Studies* 42, no. 1 (2018): 1–22.

Riddell, Peter G. *Transferring a Tradition: 'Abd Al-Ra'ūf Al-Singkilī's Rendering into Malay of the Jalālayn Commentary*. Berkeley, CA: University of California, Centers for South and Southeast Asia Studies, 1990.

Riddell, Peter G. *Malay Court Religion, Culture and Language: Interpreting the Qur'ān in 17th Century Aceh*. Leiden: Brill, 2017.

Riddell, Peter G. "The Sufi Path. Travelling in Search of Learning. An Indonesian Case Study." In *Insights into Sufism: Voices from the Heart*, edited by Ruth J. Nicholls and Peter G. Riddell. Newcastle: Cambridge Scholars Publishing, 2020: 151–68.

Rinkes, Douwe Adolf. *Abdoerraoef Van Singkel: Bidjrage tot de Mystieck op Sumatra en Java*. Heerenveen: Hepkema, 1909.

Rivauzi, Ahmad. "Landasan Filosofis Pemikiran Tasawuf Abdurrauf Singkel Tentang Allah, Manusia, Dan Alam." *Jurnal Theologica* 28, no. 2 (2017): 299–328.

Ar-Rūmī, Fahd ʿAbd ar-Raḥmān. *Buḥūth fī uṣūl at-tafsīr wa-manāhijihi* (Investigation into the foundations and methods of Quranic exegesis). Riyadh: Maktabat at-Tawba, 1996.

Shifaturrahmah, Wahyuni. "Mempertimbangkan Tarjuman Al-Mustafid (Telaah Atas Kemurnian Tafsir Karya Abd. Rauf Sinkel)." Wahyunishifaturrahmah's Blog. Last modified February 16, 2010. https://wahyunishifaturrahmah.wordpress.com/2010/02/16/mempertimbangkan-tarjuman-al-mustafid-telaah-atas-kemurnian-tafsir-karya-abd-rauf-sinkel/.

Shukri, A.S.M. and R. Arif. "The role of Shaykh Abd Al-Ra'uf Al-Fansuri in the reconciliation of sufism and shari'ah of 17th century the Malay world." *Al-Shajarah* 23 (2018): 207–45.

Suarni. *Perbedaan Qiraat Dan Pengaruhnya Terhadap Penafsiran Al-Quran: Studi Tafsir Tarjuman Al-Mustafid Karya Syeh Abdurrauf As-Singkili: Laporan Penelitian*. Darussalam, Banda Aceh: Pusat Penelitian dan Penerbitan, Lembaga Penelitian dan Pengabdian kepada Masyarakat Universitas Islam Negeri Ar-Raniry, 2015.

Voorhoeve, Petrus. "Bajan Tadjalli." *Tijdschrift voor Indische Taal-, Land-, en Volkenkunde* 23, no. 1 (1952): 87–115.

Wan Abdullah, Wan Nasyrudin. "Tafsir Tarjuman al-Mustafid: kajian intertekstualiti." PhD diss., Universiti Malaya, 2008.

Abū l-Ḥasan al-ʿĀmilī l-Iṣfahānī

Mathieu Terrier

1 Introduction

Abū l-Ḥasan b. Muḥammad Ṭāhir an-Nabāṭī l-Futūnī l-ʿĀmilī l-Iṣfahānī, commonly known as Abū l-Ḥasan al-ʿĀmilī l-Iṣfahānī, was a Shiite Imāmī theologian and Qurʾān commentator. He was born in Isfahan around 1660/1070, and died and was buried in Najaf in 1725–26/1138 or 1727–28/1140.[1] He was a member of a large family from the village of Futūn in Jabal ʿĀmil (now southern Lebanon), one of the important centers of Twelver or Imāmī Shiism since at least the 10th century. Part of this family had emigrated to Najaf, the main center of Imāmī religious studies around the sanctuary of the first Imam ʿAlī b. Abī Ṭālib (d. 661/40), and another to Isfahan, capital city of the Safavid kingdom since the end of the sixteenth/tenth century. After proclaiming Imāmī Shiism as the state religion in Iran in 1501/906, Shāh Ismāʿīl (ruled 1501–1524/906–930) and his successors called upon Arab Shiite scholars to form a clergy in order to convert Iranian society.[2] Abū l-Ḥasan al-ʿĀmilī l-Iṣfahānī belonged to this class of organic intellectuals of Arab origin. There is scant information about his life. He studied in Baḥrayn, Najaf and Isfahan with renowned Shiite scholars such as Niʿmatullāh al-Jazāʾirī (d. 1701/1112), Muḥammad Ṣāliḥ Khātūnābādī (d. 1704–5 or 1714/1116 or 1126) (his maternal uncle), Aḥmad Muḥammad b. Yūsuf al-Baḥrānī (d. 1690–1/1102), and possibly Muḥsin al-Fayḍ al-Kāshānī (d. 1680/1091). His main teacher and father-in-law was Muḥammad Bāqir al-Majlisī (d. 1699–1700/1111), the compiler of the *Biḥār al-anwār*, the most comprehensive collection of Imāmī hadiths (111 vols., ed. Beirut 1982–83/1403), who exerted a great political influence under the reigns of Shāh Sulaymān (r. /1666–1694/1077–1105) and Shāh Sulṭān Ḥusayn (r. 1694–1722/1105–1134). Al-ʿĀmilī l-Iṣfahānī taught religious sciences at the seminary (*hawzah*) of Najaf where he had numerous pupils. He did not witness the siege of Isfahan by Afghan rebels and the fall of the Safavid kingdom in 1722/1134. In Shiite circles, al-ʿĀmilī l-Iṣfahānī is primarily known for his Qurʾānic commentary entitled *Mirʾāt al-anwār wa-mishkāt al-asrār*, the only work of his to have been edited, which aroused the interest of the philosopher and orientalist Henry Corbin.[3]

1 Muḥsin al-Amīn al-ʿĀmilī, *Aʿyān ash-shīʿa*, ed. Ḥasan al-Amīn (Beirut: Dār at-Taʿāruf li-l-Maṭbūʿāt, 1983), 7: 242; Mohammad Ali Amir-Moezzi, “al-ʿĀmilī al-Iṣfahānī, Abū l-Ḥasan,” in *EI*[3], ed. Kate Fleet et al., Consulted online on 14 November 2023 http://dx.doi.org/10.1163/1573-3912_ei3_COM_22714.

2 On this, see Rula Jurdi Abisaab, *Converting Persia. Religion and Power in the Safavid Empire* (London/New York: I.B. Tauris, 2004); Andrew J. Newman, *Safavid Iran. Rebirth of a Persian Empire* (London/New York: I.B. Tauris, 2006).

3 Henry Corbin, “ʿÂmelî Eṣfahânî, Abû'l-Ḥasan,” in *EIr*, ed. Ehsan Yarshater (London/New York: Routledge, 1985), 1: Fasc. 9, 931–32.

https://doi.org/10.1515/9783111320083-020

2 A Manifesto of Shiite Traditionalist Hermeneutics

Al-ʿĀmilī's intention in *Mirʾāt al-anwār wa-mishkāt al-asrār* was to give a complete Imā mī esoteric interpretation (*taʾwīl*) of the Qurʾān, from the first verse to the last. This included gathering the traditions attributed to the Imams that were relevant to each verse, from the first to the last. However, this monumental project remained largely unfinished; the two preserved manuscript volumes of this work do not contain more than three introductions (*muqaddimāt*) and the commentary to the first and half of the second sura.[4] Only the first volume including the *muqaddimāt* was edited in Iran in 1885–6/1303, and later reedited as an introduction of Hāshim al-Baḥrānī's (d. 1696/1107) *Tafsīr al-burhān*.[5]

As it stands, the work is therefore less a Qurʾānic commentary than a treatise on the interpretation of the Qurʾān, consisting of theoretic prolegomena and a lexicon of Qurʾānic terms. It shall be considered as a manifesto of the *Akhbārī*-traditionalist theological school, which bases its views solely on the traditions attributed to the Imams, as opposed to the *Uṣūlī*-rationalist school, which promotes the authority of the reasoning issued by qualified scholars (*al-mujtahidūn*).

The axiomatic tenet of a Shiite traditionalist Qurʾānic commentary is that the Imams, being the heirs of the Prophet, and impeccable and infallible (*maʿṣūmīn*) just like him, are the only men qualified to interpret the Book of God. This tenet, in turn, is based on a particular reading and interpretation of Q 3:7:

> It is He who sent down upon thee the Book, wherein are verses clear that are the Essence of the Book, and others ambiguous. As for those in whose hearts is swerving, they follow the ambiguous part, desiring dissension, and desiring its interpretation; and none knows its interpretation, save only God and those firmly rooted in knowledge. They say, 'We believe in it; all is from our Lord'; yet none remembers, but men possessed of minds.

This is generally read by Sunni Muslims as "and none knows its interpretation, save only God. And those firmly rooted in knowledge say, 'We believe in it; all is from our Lord' . . . ." According to early Shiite exegesis, "those rooted in knowledge" specifically refers to the "People of the (Prophet's) House" (*ahl al-bayt*), namely Muḥammad, his daughter Fāṭima, his cousin and son-in-law ʿAlī b. Abī Ṭālib, and the eleven following Imams among their descendants, also designated as "the Fourteen Impeccable Ones." The Imams therefore share with God the interpretation of the Book and dispense it to men in their hadith. For this reason, the first Shiite commentators on the Qurʾān, since the "lesser occultation" of the twelfth Imam in 874/260, and before the rationalist turn of the Buyid period (middle of the tenth – middle of the eleventh century),

4 Āqā Buzurg aṭ-Ṭihrānī, *adh-Dharīʿa ilā taṣānīf ash-shīʿa* (Tehran/Najaf: 1970), 20: 264–65, n. 2893.

5 Abū l-Ḥasan b. Muḥammad Ṭāhir al-ʿĀmilī n-Nabāṭī l-Futūnī, *Muqaddimat Tafsīr al-burhān al-musammāt bi-Mirʾāt al-anwār wa-mishkāt al-asrār*, ed. n.n. (Beirut: Muʾassasat al-Aʿlamī li-l-Maṭbūʿāt, 2006).

issued works termed *tafsīr bi-l-maʾthūr*, based exclusively on hadiths attributed to Imams.[6] Such commentaries were produced again in seventeenth-century Safavid Iran by *Akhbārī* scholars, such as Muḥsin al-Fayḍ al-Kāshānī's (d. 1679/1090) *Tafsīr aṣ-Ṣāfī* and Quṭb ad-Dīn Ashkiwarī's (d. between 1677 and 1684/1088 and 1095) *Tafsīr Sharīf al-Lāhījī*. While the *Mirʾāt al-anwār wa-mishkāt al-asrār* is not such a *tafsīr bi-l-maʾthūr*, it summarizes much of the interpretation of the Qurʾān attributed to the Imams in Pre-Buyid Imāmī sources, such as the *Tafsīr*s of Furāt al-Kūfī (d. *circa* 912/300), ʿAlī b. Ibrāhīm al-Qummī (d. *circa* 919/307), and Muḥammad b. Masʿūd al-ʿAyyāshī (d. *circa* 932/320), the earliest compilations of Imāmī hadiths such as *Baṣāʾir ad-darajāt* of Ṣaffār al-Qummī's (d. 902–3/290) and *al-Kāfī* of Muḥammad b. Yaʿqūb al-Kulaynī (d. 940/328), as well as later compilations such as ash-Shaykh Abū Manṣūr aṭ-Ṭabrisī's (d. 1154/548) *al-Iḥtijāj*.

3 A Two-Fold Vision of the World and the Book

The first introduction, divided into three parts (*maqālāt*), presents the first principle of Shiite hermeneutics: the Qurʾān, beneath its literal or exoteric dimension (*ẓuhr*), which professes Divine Oneness and Prophecy, has a hidden or esoteric dimension (*buṭn*), which professes the mission of the Imams (*imāma*) and their Alliance or Friendship with God (*walāya*) (p. 8). According to the dual vision of the world typical of early Shiism, all reality, beginning with God, possesses these two levels: one outer, manifest, or exoteric (*ẓāhir*), and another inner, non-manifest, hidden under the former level, or esoteric (*bāṭin*). In this conception, the Divine Friendship of the Imams, which includes their spiritual and temporal mission on earth, is the very esoteric dimension of Prophecy, namely that of Muḥammad's predication.[7] Having stated that the words of the Qurʾān are not restricted to the men of a single epoch, that of the preaching of Muḥammad, but do have esoteric meanings applied to men of all epochs, the author argues that the esoteric level of the Qurʾān and its interpretation (*taʾwīl*) deals essentially with the Imams, their Divine Alliance and their followers, the Shiites (p. 8–12). Interestingly, in this conception, the Imams are not only the authors of Qurʾānic exegesis, disclosing the hidden truth of the Revelation, but also represent the content of this exegesis, the hidden truth itself, valid for all times. However, ʿĀmilī stresses that the exoteric content of the Book shall not be neglected in favor of its

6 See on this Meir M. Bar-Asher, *Scripture and Exegesis in Early Imāmī Shiism* (Leiden/Jerusalem: Brill and The Magnes Press, 1999).

7 Mohammad Ali Amir-Moezzi, "Shiʿite Doctrine," in *EIr*, ed. Ehsan Yarshater, last updated July 20, 2005, https://iranicaonline.org/articles/shiite-doctrine; Mohammad Ali Amir-Moezzi and Christian Jambet, *What is Shiite Islam? An Introduction*, trans. Kenneth Casler and Eric Ormsby (Abingdon/New York: Routledge, 2018), 13–15.

mere esoteric dimension (p. 21). According to the sixth Imam, the knowledge of the hidden meaning of the verses dealing with religious duties (prayer, alms-giving, pilgrimage) and illicit matters (wine, pork meat, etc.), that is the knowledge of the men alluded to by these symbols (the Imams by the religious duties, their enemies by the illicit matters), does not dispense from obedience to these duties and prohibitions; and that those who assert the contrary, i.e. the partisans of antinomianism, are 'exaggerators' (*ghulāt*) tantamount to infidels – a matter to which the author would return later (p. 22–26).[8] According to the *Akhbārī*-traditionalist view, he asserts that the entire science of Qur'ānic hermeneutics (*at-ta'wīl*) belongs to the Imams and that it is therefore forbidden for the scholar to comment on the Qur'ān according to his opinion (*ra'y*) (p. 26).

The second part of the first introduction develops the view that the Qur'ān, while dealing literally with the oneness of God and the prophecy of Muḥammad, has for esoteric or secret meaning the Imams' Alliance with God (*walāya*) and their leadership over men (*imāma*). The belief in the Imams' Divine Alliance encompasses the prophecy of Muḥammad and the leadership of ʿAlī and his descendants after him, love and obedience to the latter (*tawallā*) and hatred (*tabarra'a*) of their enemies. In early and traditional Shiism, sacred love towards the Imams and sacred hatred towards their enemies, namely the first Caliphs Abū Bakr aṣ-Ṣiddīq (r. 632–634/11–13), ʿUmar b. al-Khaṭṭāb (r. 634–644/13–23) and ʿUthmān b. ʿAffān (r. 644–656/23–35) and most of the Prophet's companions, followed by Umayyad and Abbasid rulers, are two complementary duties of faith.[9] ʿĀmilī thus introduces, in addition to the dual vision previously described, the dualistic vision of the world also typical of early Shiism, in which the history of creation is a permanent battle between forces of Knowledge, Good and light, and forces of Ignorance, Evil and darkness.[10] According to this tenet, the conflict between the Friends of God and their enemies is at the core of the Qur'ān: all the praising words towards believers past (of other religions) and present (Muslims) actually mean, at the esoteric level, the "People of the House" and their followers, the Shiites; likewise, all the reproaches addressed to unbelievers past and present actually concern the Imams' enemies.

ʿĀmilī argues that the belief in the Imams' divine Alliance derives from the belief in the prophecy (*nubuwwa*) of Muḥammad, just as the latter derives from the belief in the Oneness of God (*tawḥīd*). Consequently, professing two of these tenets without the

---

8 Quoted from Ṣaffār al-Qummī, *Baṣā'ir ad-darajāt* (Beirut: Sharikat al-Aʿlamī li-l-Maṭbūʿāt, 2010), 570–80. On this tradition, see Mohammad Ali Amir-Moezzi, "Les Imams et les Ghulāt. Nouvelles réflexions sur les relations entre imamisme 'modéré' et shiʿisme 'extrémiste'," *Shiite Studies Review* 4 (2020): 5–38. Within the limits of this article, I will mention only a few primary sources.

9 See on this Mohammad Ali Amir-Moezzi, *The Divine Guide in Early Shiʿism. The Sources of Esotericism in Islam*, trans. David Streight (Albany: State University of New York Press, 1994), Index s.v. *barā'a* (hatred); Amir-Moezzi, *The Spirituality of Shiʿi Islam. Beliefs and Practices* (London/New York: I.B. Tauris, 2011), 262–67.

10 Amir-Moezzi, *Divine Guide*, 7–8; Amir-Moezzi and Jambet, *What is Shiite Islam?*, 15–18.

third one is of no value for faith, and rejecting the divine Alliance of ʿAlī and his descendants is tantamount to the rejection of God's Oneness and Muḥammad's prophecy. Faith in these three principles is the meaning of 'the original (mankind)' (*fiṭra*) referred to in Q 30:30: "God's original upon which he originated mankind" (p. 40). The divine Alliance of ʿAlī was part of the original Compact (*mīthāq*) of God with the Prophets referred to in Q 3:81: "And when God took compact with the Prophets . . .," that is the very message of all the prophets, the content of all divine Books and the sacred duties of all communities (p. 45).

The third part of the first introduction recalls traditions attributed to the Prophet, holding that the Muslim community would eventually follow step by step the bad example of Jews and Christians. According to this view, Q 84:19: "You shall surely ride stage after stage" means: "You will act like the communities that preceded you in committing betrayal against the heirs of the prophets." This heralds the most controversial part of the book, which immediately follows.

4 The Falsification of the Qurʾān and the Need for Hermeneutics

Indeed, the second introduction supports the thesis that the Qurʾān has suffered alteration in its content (*at-taghyīr fī l-Qurʾān*) during its edition process after the death of Muḥammad, and that this is the reason why the claim of the Imams' divine Alliance is only contained in the hidden level of the Book and its spiritual interpretation (*fī buṭn al-Qurʾān wa-taʾwīlihi*), while being merely alluded in its outer aspect or its literal revelation (*fī ẓāhir al-Qurʾān wa-tanzīlihi*) (p. 62). That is because God, knowing from eternity what would happen to his Book, hid the truth so that it would be unnoticed by the falsifiers but accessible through interpretation. For al-ʿĀmilī, the dual structure of the Qurʾān, exoteric and esoteric, and the necessity of hermeneutics, are thus inextricably linked to an historical and highly polemic claim.

As recent scholarship has shown, early Shiites widely condemned the falsification (*taḥrīf*) of the Qurʾān that occurred during its collection, edition and canonization process initiated under the first three caliphs Abū Bakr, ʿUmar and ʿUthmān, and achieved under the Umayyad ruler ʿAbd al-Malik b. Marwān (r. 685–705/65–86). This claim is supported by most of the ancient Imāmī *tafsīrs* and hadith compilations used by the author. However, since the occultation of the twelfth Imam and the Buyid period, Shiite scholars, likely in search of a *modus vivendi* alongside the Sunni majority, rejected this thesis as "exaggeration" (*ghulūw*) and agreed on the authenticity of the official Qurʾān.[11]

---

11 Hossein Modarressi, "Early Debates on the Integrity of the Qurʾān. A Brief Survey," *SI* 77 (1993), 5–39; Rainer Brunner, *Die Schia und die Koranfälschung* (Würzburg: Ergon Verlag, 2001); Rainer Brun-

Remarkably, al-ʿĀmilī is the first Shiite commentator since the Buyid period to explicitly support the thesis of the falsification of the Qurʾān – although he avoids using the technical term *taḥrīf* in favor of that of *taghyīr* –, while reporting most of the Imāmī traditions confirming this view.[12] For this reason, the *Mirʾāt al-anwār* would eventually serve as a source for Ḥusayn an-Nūrī ṭ-Ṭabrisī's (d. 1902/1320) *Faṣl al-khiṭāb*, the most famous expression of the Shiite criticism of the official Qurʾān in modern times, and a preferred piece of evidence for modern Wahhabi anti-Shiite propaganda.[13]

In the first part of this introduction, al-ʿĀmilī brings together traditions concerning the alteration of the Qurʾān. According to the Imāmī narrative recalled by the author, the original and integral Qurʾān was collected by Imam ʿAlī, who had learned it directly from the Prophet, immediately after the death of the latter, while the matter of his succession was being secretly settled in favor of Abū Bakr. However, after having been prevented by ʿUmar from presenting it to the community, ʿAlī passed his Qurʾān on to his son al-Ḥasan, the second Imam, who passed it on to his brother al-Ḥusayn, the third Imam, and so on up to the twelfth Imam, who took it with him in his occultation (*ghayba*) (in 874/260 for the "lesser occultation," then in 940–1/329 for the "greater" one), and will disclose it, together with all other Books of God in their original content, when returning as the Redeemer (*al-qāʾim*) at the end of history (p. 62). As for the official Qurʾān, it suffered manipulation by the holders of power in order to make the Book of God an instrument for their domination. The author reports that Abū Bakr and ʿUmar, having failed to kill ʿAlī as well as to seize his collection, commissioned Zayd b. Thābit to edit the Qurʾān by deleting what concerned, on one hand, the merits of the "People of the House," and on the other hand, the misdeeds of many companions of the Prophet. As a consequence, if nothing had been removed from or added to the Qurʾān, no one would have ever doubted the divine Alliance of ʿAlī and the following Imams (p. 63). According to al-Kulaynī: "No one claims that the Qurʾān has been collected as it was revealed except the liars; no one has collected and preserved the Qurʾān as it was revealed by God except ʿAlī and the Imams" (p. 64).

A few supposedly censored verses are quoted, whose interpretation intends to restore the original text. Q 3:110: "You are the best nation (*umma*) ever brought forth to men," according to a tradition of the sixth Imam, could not be revealed in this way, "for the best nation would not have killed the Prince of the believers, al-Ḥasan and al-

ner, "La question de la falsification du Coran dans l'exégèse chiite duodécimaine", *AR* 52, no. 1 (2005), 1–41; Etan Kohlberg and Mohammad Ali Amir-Moezzi, *Revelation and Falsification. The* Kitāb al-Qirāʾāt *of Aḥmad b. Muḥammad al-Sayyārī* (Leiden/Boston: Brill, 2009); Mohammad Ali Amir-Moezzi, *The Silent Qur'an and the Speaking Qur'an. Scriptural Sources of Islam between History and Fervor*, trans. Eric Ormsby (New York: Columbia University Press, 2015); M. M. Bar-Asher, *Scripture and Exegesis*, 90–92.

12 R. Brunner, "La question de la falsification du Coran," 17–18.

13 Ibid., 22–38.

Ḥusayn;" but it was: "You are the best *Imams* (*a'imma*) . . ." (emphasis mine), and was altered by the deletion of a single letter. Similarly, in Q 4:166 the mention of ʿAlī has been expunged, since it was originally revealed as follows: "But Allah bears witness that what he has sent down to thee *upon ʿAlī*" (p. 83, emphasis mine).

The second part of this introduction presents traditions issued from Sunni sources like the *Jāmiʿ* of Jalāl ad-Dīn as-Suyūṭī, the *Ṣaḥīḥ* of Muḥammad b. Ismāʿīl al-Bukhārī, and the *Ṣaḥīḥ* of Abū ʿĪsā Muḥammad at-Tirmidhī, mentioning notably ʿUmar's interventions to censor various Qur'ānic collections (*maṣāḥif*). Thus, even the Sunni report confirms that the official Qur'ān is not equal to the original one (p. 70).

The third section explains that the changes made to the Qur'ān are the reason why references to the *walāya* and *imāma* are only allusive, symbolic and indirect (p. 73). In a long tradition,[14] a Manichean, aiming to refute Islam, asked ʿAlī why God, in the Qur'ān, named the prophets who committed minor errors and faults, but did not refer by name to those who committed infidelity (*kufr*) towards the Prophet. The Imam, first, replies that the usurpers of the Caliphate and falsifiers of the Qur'ān are well alluded to by several verses, among which Q 3:78: "And there is a sect of them twist their tongues with the Book;" and that God personally warned the Prophet of what his enemies would do after him in Q 22:52: "We sent not ever any Messenger or Prophet before thee, but that Satan cast into his fancy, when he was fancying; but God annuls what Satan casts, then God confirms his signs" (p. 75–76). As for the Imams, they are referred to in various "ambiguous" verses (*mutashābihāt*) and expressions such as "the face of God" (*wajh Allāh*) (Q 2:115), "the front of God" (*janb Allāh*) (Q 39:56), and "God's remainder" (*baqiyyat Allāh*) (Q 11:86), the latter referring especially to the twelfth Imam (p. 76–77). The Imam states:

> Almighty God has placed in his Book these symbols, which no one knows but Him, His prophets and His proofs on earth (the Imams), because He knew what the usurpers (*al-mubaddilūn*) would do in his Book: to suppress the names of His proofs, to disguise them in the eyes of the people in order to lead them into fallacy. Therefore, God has established in the Qur'ān symbols (*rumūz*) that no one knows except Him, His prophets and His proofs; He has veiled the hearts and eyes of the usurpers over these symbols and other parts of the speech indicating what they have done (p. 77).

According to the same tradition, God divided his word (*kalima*) into three parts: one is known by the learned men as well as the ignorant; another is known only by the one who has purified his spirit and whose breast God has opened to Islam; a third is known only by God and his friends. To the first part belongs the greatness of the Prophet, to the second that of the Imams. In Q 37:130, "Peace be upon the family of

14 Quoted from Abū Manṣūr aṭ-Ṭabrisī, *al-Iḥtijāj*, ed. Ibrāhīm Bahādurī and Muḥammad Hādī Beh (Qum: Uswa, 1992–93), 1: 574–609.

Yāsīn (*āl yāsīn*),"[15] "Yāsīn" refers to the Prophet Muḥammad, as in Q 36:1–3: "Yā Sīn! By the wise Qur'ān, thou art truly among the Envoys". God expressed that in this way only because he knew that if he would have said: "Peace be upon the family of Muḥammad," the falsifiers would have suppressed it. The Imam emphasizes: "This and other similar symbols have been established in the Book so that the falsifiers (*al-muḥarrifūn*) would be unaware of their meaning." However, he claims that he should not expose everything that has been suppressed or altered in the Qur'ān, out of respect for the duty of concealment (*taqiyya*) (p. 77–78). To the Manichean's challenge, he finally replies that the nominal denunciation of the prophets for their minor faults was intended by God precisely to show that the silence on the infidels and hypocrites was due to the misdeeds of the falsifiers. If God had denounced the latter openly, discord would have increased within the community and the men of the truth would have been killed by the men of the false. God thus revealed many verses in metaphorical language (*tajawwuz*), indirect suggestion or indication (*ta'rīḍ*), metonymy (*kināya*) and symbols (*rumūz*), particularly concerning the *walāya*, in order to protect the truth against the falsification of the hypocrites (*taḥrīf al-munāfiqīn*) (p. 80–82). It suggests, in the words of Leo Strauss, that God himself exerted in the Qur'ān an "art of writing under persecution," a "writing between the lines" in order to thwart censorship and falsification.[16] This divine art of writing is the main concern of the Shiite Qur'ānic hermeneutics as summarized by our author.

The fourth part of this introduction confronts the statements of Imāmī scholars supporting or opposing the alteration (*taghyīr*) of the Qur'ān, and points to the invalidity of the opponents' arguments (p. 83). On the one hand, al-Kulaynī, al-Qummī, al-ʿAyyāshī and the other earliest hadith scholars reported many traditions on falsification. According to a statement received privately from his master al-Majlisī, the claim of the falsification of the Qur'ān is one of the necessities of the Shiite doctrine (p. 84). On the other hand, Abū Jaʿfar b. Bābawayh, alias ash-Shaykh aṣ-Ṣadūq (d. 991/381), stated that the belief of the Shiites is that the Qur'ān revealed by God to his Prophet is nothing but the official Qur'ān, and that the claim of the falsification is a hallmark of the 'exaggerators' (*ghulāt*). Similarly, ash-Shaykh Abū Jaʿfar aṭ-Ṭūsī (d. 1068/460) argued that the hadiths dealing with a loss or diminution (*nuqṣān*) in the Qur'ān are nothing but isolated traditions (p. 84). The author replies that it is correct to deny the addition of verses to the Qur'ān, but not any changes and censorship; that the traditions attesting to the alteration of the Qur'ān, even in Sunni sources, are far too numerous to be disqualified; and that this position does not lead to the rejection of the whole Qur'ān (p. 85–86). The argument of the "miraculous integrity" (*iʿjāz*) of the book is also excluded (p. 86). And the verses generally alleged to exclude any falsification,

15 Arberry translates: "Peace be upon Elias," see *The Koran Interpreted*, trans. Arthur J. Arberry, (Oxford: Oxford University Press, 1964), 461.

16 Leo Strauss, *Persecution and the Art of Writing* (Chicago: University of Chicago Press, 1988).

like Q 41:41–42: "Those who disbelieve in the Remembrance when it comes to them – and surely it is a Book Sublime; falsehood comes not to it from before it nor from behind it; a sending down from One All-wise, All-laudable," and Q 15:9: "It is we who have sent down the Remembrance, and we watch over it," are said to refer only to the Qurʾān preserved by the Imams (p. 87).

5 The Balance between Orthodoxy and Exaggeration

The third introduction deals again with the argument that the esoteric level of the Book concerns the Imams' divine Alliance and leadership on men, when its exoteric level concerns God's Oneness and Muḥammad's prophecy. The first part, in seven sections, recalls certain principles of Shiite Qurʾānic hermeneutics and gives a first outline of the following glossary by exposing the esoteric meaning of various Qurʾānic terms and categories. For example, "the unbelievers" (*al-kāfirūn*), "the hypocrites" (*al-munāfiqūn*), "the polytheists" (*al-mushrikūn*), are those who rejected the Imam's divine Alliance, and were hypocrites towards it and associated with the Imam someone who was not Imam. The author then states that many of the verses concerning obviously ancient peoples and communities actually refer to the community of Islam, the Imams, their followers (*shīʿa*) and their enemies (p. 88–89). Thus Q 7:159: "Of the people of Moses there is a nation who guide by the truth, and by it act with justice," esoterically refers to the people of Muḥammad and the Shiites among them. In many verses, God does not address all Muslims, as it may seem, but exclusively the Imams, as in Q 2:136: "Say you: 'We believe in God, and in that which has been sent down on us'," and Q 81:29: "but will you shall not, unless God wills" (p. 90). Personal pronouns (*ḍamāʾir*) can refer to a subject not mentioned before, such as the *walāya* and ʿAlī 'the Prince of the Believers' (*amīr al-muʾminīn*), even without censorship (p. 92–93). Past grammatical tense can be used to refer to future events (p. 94). When God expresses himself by the 'we' of majesty, it often includes God, the Prophet and the Imams altogether. Even "the Almighty" (*al-jalāla*), "the Divinity" (*al-ilāh*) and "the Lord" (*ar-rabb*), in many places, refer esoterically to the Imam (p. 95–96). Q 43:84: "And it is he who in heaven is God and in earth is God;" Q 57:4: "He is with you wherever you are;" and Q 58:7: "Three men conspire not secretly together, but he is the fourth of them, neither five men, but he is the sixth of them," all refer to the Imam as God's representant among his creation, and Q 16:51: "God says: "Take not to you two gods. He is only one God," actually means: "Take not to you two Imams. He is only one Imam" (p. 97).

The author devotes an appendix (*tadhyīl*) to explain and discuss the charges of *tafwīḍ* (attribution of divine powers to Imams) and *ghulūw* (exaggeration, extremism) which might be asserted against the previous developments. After defending the exclusive reliance on the teachings of the Imams as reported in the earliest collections of hadiths, he criticizes the two symmetrical doctrinal faults of excess (*ifrāṭ*) and defect (*tafrīṭ*) in the veneration of the Imams. According to a famous hadith attributed to several of them: "Our teaching is arduous; the only ones who can withstand it are a prophet sent to men, an angel of proximity, or an (initated) believer whose heart has been tested by God for faith," many authentic traditions and beliefs have been wrongly accused of exaggeration by narrow-minded Shiites (p. 101). According to al-Majlisī again, this is an ungodly attitude to reject authentic traditions on the grounds that they conflict with reason (*ʿaql*) (p. 103). Beliefs that are truly exaggeration, according to the Imams' traditions, are nothing but attributing divinity (*ulūhiyya*) or lordship (*rubūbiyya*) to the Imams, confusing them with the prophets or placing them above the latter, supporting the transmigration (*tanāsukh*) of the Imam's soul from one to another, denying the death or martyrdom of an Imam, and claiming that the knowledge of the Imams dispenses from the respect of ritual obligations (prayer, almsgiving, fasting) and food or sexual prohibitions (*maḥārim*), i.e. antinomianism (*ibāḥa*) (p. 104–9). According to the early Imāmī doctrine, the Imams are the representatives (*nuwwāb*) of God, who has delegated (*fawwaḍa*) to them authority, knowledge, intercession in the last life, without the power of creation (*khalq*), sustenance (*rizq*), and abolition of the Law. To outbid these attributions is to sin in excess, to refuse them is to sin by defect (p. 110–16). In any case, attributing to the Imams supra-rational knowledge, supernatural powers and, to a certain extent, divine nature, is by no means exaggeration.[17]

6 An Imāmī Shiite Qur'ānic Lexicon

The second part of this introduction, in fact the largest part of the book, is a lexicon of thirteen thousand Qur'ānic terms in alphabetical order, giving for each one a literal explanation (*tafsīr*) and an esoteric interpretation (*ta'wīl*) according to Imāmī traditions. Many of those interpretations belong to the "personalized commentary" typical of the earliest Imāmī *tafsīr*s, which consists in interpreting abstract terms or general categories as concrete, individual and historical persons, essentially the "People of the House" and their adversaries, whose names are supposed to have been censored by the Caliphs' power.[18] Three major exegetic and dogmatic themes, already identified, run through this technical dictionary: Imamo-centrism, anti-Sunnism and messianism.

---

17 On this, see M.A. Amir-Moezzi, "Les Imams et les Ghulāt".

18 M.A. Amir-Moezzi, *The Silent Qur'an and the Speaking Qur'an*, chapter 3.

6.1 Imamo-centrism

The focus on the Imam – ʿAlī first of all – and his divine Alliance as the hidden center of Revelation is certainly the first axiom of Shiite Qurʾānic exegesis. Many Qurʾānic terms are interpreted as referring specifically to Imam ʿAlī, such as *imām*, "guide" or "pattern," in Q 36:12: "everything we have numbered in a clear pattern (*imām mubīn*)" (p. 138–39), or in association with Prophet Muḥammad, like *īmān*, "faith," in Q 5:5: "Whoso disbelieves in the faith, his work has failed" (p. 143). The Sharia notions of *zakāt*, "alms-giving," and *ṣalāt*, "ritual prayer," in Q 2:43: "And perform the prayer, and pay the alms," are synonymous with ʿAlī (p. 280–81). *An-nafs*, "the person" or "the soul," refers to Muḥammad and ʿAlī together in Q 4:1: "Mankind, fear your Lord, who created you of a single soul," and especially to ʿAlī in Q 89:27: "O soul at piece" (p. 504–7). *An-niʿma*, "(God's) blessing," refers to ʿAlī's divine Alliance in Q 16:83: "They recognize the blessing of God, then they deny it;" while in Q 31:20: "He has lavished on you his blessings, outward and inward," the "outer" blessing means Muḥammad's prophecy and the "inner" one ʿAlī's divine Alliance (p. 513–15). *Al-wālid*, "father," "begetter," in Q 90:3: "by the begetter, and what he begot," means ʿAlī, and "what he begot" the Imams of his descendants. In the dual form, *al-wālidān*, "parents" (father and mother), in Q 17:23: "Thy Lord has decreed you shall not serve any but him, and to be good to parents," means both Muḥammad and ʿAlī (p. 523–24). The eschatological symbol of *al-mīzān*, "the Balance," always refers to ʿAlī, as in Q 55:7: "and heaven – he raised it up, and set the Balance," where "heaven" refers to Muḥammad (p. 529–31). Finally, *al-wajh*, "the face," while referred to God as in Q 28:88: "All things perish, except his Face," esoterically concerns the person of ʿAlī as the divine man, according to the fundamental tenet of early Shiism previously labeled as "exaggeration" (p. 531–32).

Many terms refer to the "People of the House" altogether or separately. For example, *āya*, pl. *āyāt*, "sign," especially "sign of God" (*āyat Allāh*), in Q 20:127: "So we recompense him who is prodigal and believes not in the signs of his Lord" (p. 149–50); and *raḥm*, pl. *arḥām*, "matrix" and by extension "blood relationship," in Q 8:75: "but those related by blood (*ulū l-arḥām*) are nearer to one another in the Book of God," refer to all the Fourteen Impeccable Ones (p. 267–70). *Ash-shuhadāʾ*, "the witnesses" or "the martyrs," in Q 4:69: "Whosoever obeys God, and the Messenger – they are with those whom God has blessed, Prophets, just men, martyrs, the righteous; good companions they!," refers especially to ʿAlī, al-Ḥasan and al-Ḥusayn, while "the righteous" refers to the nine following Imams (p. 315). Derived from *ḍaʿf*, "weakness," "abasement," *al-mustaḍʿafūn* or *alladhīna kānū yustaḍʿafūn (fī l-arḍ)*, "the weak" or "the abased ones" ("on earth" or "in the land"), refers to the Imams as the defeated actors of history promised to final victory, as in the frequently adduced Q 28:5: "Yet we desired to be gracious to those that were abased in the land, and to make them leaders, and to make them the inheritors" (p. 354–55). Likewise, *at-taṭhīr*, "cleaning," "purification," and its derivative *al-muṭahhar*, pl. *-ūn*, "the purified one(s)," always refer to them in their attribute of impeccability (*ʿiṣma*), like in the particularly canonical Q 33:33: "People of the House, God only desires to put away from you abomination and to cleanse you" (p. 360). Derived from *ẓulm*, "evil," "injustice," the category *al-maẓlūm*, pl. *-ūn*, "victim of injustice," also refers to them, as in Q 17:33: "Whosoever is slain unjustly, we have appointed to his next-of-kin authority," where "whosoever is slain unjustly" designates al-Ḥusayn and "his next-of-kin" the twelfth Imam (p. 292). Expressions derived from *al-ʿilm*, God's "knowledge," are identified with the Imams, as "men possessed of knowledge" (*ulū l-ʿilm*) in Q 3:18: "God bears witness that there is no god but he – and the angels, and men possessed of knowledge – upholding justice," and "those rooted in knowledge" in Q 3:7, as seen before (pp. 257, 389–91). Likewise, *kalimāt*, "words (of God)," in Q 2:37: "Thereafter Adam received certain words from his Lord, and God turned towards him," refer to the names of Muḥammad, ʿAlī, Fāṭima, al-Ḥasan and al-Ḥusayn (p. 465–68). Finally, *an-nūr*, "the light," one of God's names, refers, at the esoteric level, to the Imams and ʿAlī in particular, as in Q 57:28: "He will appoint for you a light whereby you shall walk, and forgive you" (p. 502–3).

Other symbolic notions are said to refer to the Imams' divine Alliance and leadership over men, as *amāna*, "trust" or "deposit," in Q 4:58: "God commands you to deliver trusts back to their owners" (p. 143); *al-ḥaqq*, "truth" or "right," in the pericope "It is the truth" (Q 2:144 and 41:53 among others) (p. 210); *ad-dīn*, "religion," as in Q 98:5: "That is the religion of the True" (p. 242); *al-hudā* or *al-hidāya* (p. 544–46), "guidance," as in Q 9:33: "It is he who has sent his Messenger with the guidance and the religion of truth" (p. 535–37). Lastly, Qur'ānic notions, generally considered designating the Muslims as a whole, are said to refer especially to the Shiites, such as *umma*, "community" or "nation", in Q 3:110: "Ye are the best nation ever brought forth to men" (previously mentioned with another interpretation), Q 2:143: "Thus we appointed you a midmost nation," and Q 23:52: "Surely this community of yours is one community" (p. 136–38); and *al-muttaqūn*, "the godfearing," in Q 2:2: "That is the Book, wherein is no doubt, a guidance to the godfearing" (p. 535–37).

6.2 Anti-Sunnism

Opposition to the leaders of the Muslim majority – which became "Sunni" in the third/ninth century – is a hallmark of the traditional Shiite exegesis. Following the method of "personalized commentary," the first three Caliphs Abū Bakr, ʿUmar and, to a lesser extent, ʿUthmān, are never referred to by their names but only by the expression "so-and-so" (*fulān wa-fulān*) or as "the first," "the second" and "the third;" as well as Muʿāwiya b. Abī Sufyān, the adversary of ʿAlī at the battle of Ṣiffīn (657/37) and the founder of the Umayyad dynasty, and his son Yazīd, responsible for the massacre of Imam al-Ḥusayn and his relatives in Karbalāʾ (685/61), are identified with numerous Qurʾānic notions. Generally speaking, all the notions contrary to Islam and faith are said to have as their esoteric meaning the wrongs committed against the Imams and their authors. Thus *ash-shirk*, "associationism" (p. 327–29); *aḍ-ḍalāl*, "doctrinal wandering," "heresy" (p. 356–57);

aṭ-ṭughyān, "insolence," "rebellion," and *aṭ-ṭāghī*, *aṭ-ṭāghūt*, "insolent," as in Q 38:55: "All this; but for the insolent awaits an ill resort," which was revealed for Abū Bakr, ʿUmar and the Umayyad (p. 365); derived from *aẓ-ẓulm*, "injustice," "evil," *aẓ-ẓālim*, *aẓ-ẓālimūn*, "the evildoer," in Q 2:193: "Fight them, till there is no persecution and the religion is God's; then if they give over, there shall be no enmity save for evildoers (*aẓ-ẓālimīn*)," refers to the assassins of Imam al-Ḥusayn (p. 367–68). Likewise, *al-fasād*, "corruption" (p. 410–11), *an-nifāq*, "hypocrisy" (p. 509), *al-kufr*, "unbelief" (p. 459–62), and their related words, are identified with the Imams' enemies.

According to early Shiite exegesis, as seen previously, the holy history of the ancient nations, mainly Jews and Christians, is only mentioned in the Qur'ān in order to deal esoterically with the Islamic nation and the misdeeds of its majority. With regard to the Jews, the author establishes in principle that all that is said about them positively in the Qur'ān under the name of *banū isrāʾīl*, "the sons of Israel," actually refers, at the hidden level, to the Imams, as in Q 2:47: "Children of Israel, remember my blessing wherewith I blessed you, and that I have preferred you above all beings;" and that all that is said about them negatively under the name *al-yahūd*, "the Jews," actually concerns the Imams' enemies (pp. 176 and 540). Likewise, according to the early Shiite dualistic vision, all the negative groups and characters of holy history are interpreted as code names for the Imams' enemies. For instance, *al-jibt*, "the demons," and *aṭ-ṭāghūt*, "the idols," evoked in Q 4:51: "Hast thou not regarded those who were given a share of the Book believing in demons and idols," refer to Abū Bakr and ʿUmar (p. 187); *al-ʿijl*, "the (golden) calf," in Q 2:51: "And when we appointed with Moses forty nights then you took to yourselves the Calf after him and you were evildoers," is identified with Abū Bakr (p. 385). The latter identification goes hand in hand with that of Aaron (*Harūn*) with ʿAlī, according to the Prophet's saying addressed to him: "Your rank compared to mine is like Aaron's rank compared to Moses', except that there is no prophet after me" (p. 543). The negative figure of "the Sāmirī" (*as-sāmirī*), associated with the calf in Q 20:85, 20:87 and 20:95, is identified with ʿUmar (p. 291–92). As for "Pharaoh" (*firʿawn*), he is said to designate Abū Bakr or Muʿāwiya (p. 422), and in the latter option, his accomplice Hāmān (mentioned in Q 28:6, 28:8 and 28:38) is identified with Yazīd (p. 543).

6.3 Messianism

According to the Shiite early doctrine, the twelfth Imam will reappear at the end of times as the Redeemer, in order to "cover the earth with justice as it was covered with injustice." At this occasion, ancient Prophets, previous Imams and some of their followers on one hand, their enemies on the other hand, will return to life in order to take part to the apocalyptic battle. Many Qur'ānic words and verses are interpreted as referring to the manifestation (*ẓuhūr*) of the twelfth Imam and the "return to life" (*rajʿa*) of his allies and enemies. In Q 32:21: "And we shall surely let them taste the nearer chastisement, before the greater; haply so they will return," "the nearer chastisement" (*al-ʿadhāb*) refers to the "return to life" and the defeat of the Imams' enemies (p. 372). *Al-fatḥ*, "victory," in Q 61:13: "help from God and a nigh victory," refers to the Redeemer's victory in this world (p. 409). *Al-fajr*, "dawn," in Q 89:1: "By the dawn!," refers to the Redeemer in person (p. 412). *Al-qatl*, "fighting," "killing," in Q 22:39: "Leave is given to those who fight because they were wronged – surely God is able to help them," refers apparently to the battle of Badr (624/2) and esoterically to that of the Redeemer (p. 438–39). *Al-qiyāma*, "resurrection," notably in Q 75:1 and 75:6: "No! I swear by the Day of Resurrection (. . .) asking, 'When shall be the Day of Resurrection?'," actually refers to the "return to life" (p. 444–48). Likewise, *an-naṣr*, "help," "victory," and *an-nāṣir*, *al-manṣūr* (p. 497–99), "the helper," "the victorious", as in Q. 42:41: "And whosoever helps himself after he has been wronged – against them there is no way," in which "whosoever helps himself" (*al-muntaṣir*) refers to Imam al-Ḥusayn upon his return to life with the Redeemer, or to the Redeemer himself. As for the Great Resurrection, the Imams are identified to *al-aʿrāf*, "the ramparts" or "the battlements" separating Paradise, mentioned in Q 7:46 and 7:48 (p. 382–83).

7 Synthesis and Conclusion

The book ends with a conclusion in two parts. It first deals with the fourteen sequences of isolated letters (*al-ḥurūf al-muqaṭṭaʿāt*) opening twenty-nine suras of the Qurʾān. All are said to hide the formula "ʿAlī is a true path that we hold firmly" (*ʿAlī ṣirāṭ ḥaqq numsikuhu*) or "The path of ʿAlī is a truth that we hold firmly" (*ṣirāṭ ʿAlī ḥaqq numsikuhu*), and each of them is interpreted as a code for a name of God, a nickname of the Prophet, or the evocation of a historical event subsequent to the Qurʾānic preaching. For example, the sequence *k.h.y.ʿ.ṣ*, opening sura 19, *Maryam* (Mary), according to a tradition attributed to the Prophet Zechariah: "The *kāf* means Karbalāʾ, the *hāʾ* the destruction of the (Prophet's) parents (*halāk ʿitra*), the *yāʾ* Yazīd – may God curse him –, the *ʿayn* al-Ḥusayn's thirst (*ʿaṭash*) and the *ṣād* his patience (*ṣabr*)" (p. 559–60).

The second part recalls twelve fundamental lessons of traditional Imāmī exegesis, among which: what has been revealed about the ancient communities, their obedience or disobedience to their prophets, actually concerns the community of Muḥammad, their obedience or disobedience to him and to the Imams of his family; the esoteric meaning of all that God declared illicit in the Qurʾān are the Imams of injustice (*aʾimmat al-jawr*) – beginning with the first three caliphs –, and that of all that God declared licit are the Imams of the True; the meaning of all praising in the Qurʾān is the Imams, their divine Alliance and their partisans, and the meaning of all blaming is the enemies of the Imams and their followers; the knowledge of the esoteric meaning of the Book of God does not dispense with respect to its legislative content; the names of the Divinity, the Lord or the Divine Majesty (*jalāl*), can be interpreted esoterically as referring to the Imam. With the notable exception of the falsification of the Qur'ān, all the main themes of Imāmī hermeneutics, sometimes labelled as "exaggerated," are thus summarized.

In conclusion, the work of al-ʿĀmilī l-Iṣfahānī is of considerable importance in the history of Qur'ānic hermeneutics. It is exemplary of the traditional Imāmī Shiite *tafsīr*, based on the two-fold vision of the world, dual and dualistic, that marks early Imāmī Shiism. It thus demonstrates the resilience of Shiite esotericism in the modern times marked by the rationalization and juridification of doctrine. At the same time, while aiming to be strictly faithful to the non-rational, not to say irrational, teachings of the Imams, it conveys the philosophical idea of an "art of writing between the lines" exerted by God in the Qur'ān, justifying the need for hermeneutics through reference to the violent history of the beginnings of Islam, as well as by purely theological tenets.

8 Bibliography

8.1 Sources

Al-ʿĀmilī n-Nabāṭī l-Futūnī, Abū l-Ḥasan b. Muḥammad Ṭāhir. *Muqaddimat Tafsīr al-burhān al-musammāt bi-Mir'āt al-anwār wa-mishkāt al-asrār*. Ed. n.n. Beirut: Mu'assasat al-Aʿlamī li-l-Maṭbūʿāt, 2006.

The Koran Interpreted. Translated by Arthur J. Arberry. Oxford: Oxford University Press, 1964 (first ed. 2 vols. London: Allen and Unwin, 1955).

8.2 Secondary Literature

Abisaab, Rula Jurdi. *Converting Persia. Religion and Power in the Safavid Empire*. London/New York: I.B. Tauris, 2004.

Al-Amīn al-ʿĀmilī, Muḥsin. *Aʿyān ash-shīʿa*, edited by Ḥasan al-Amīn. 10 vols. Beirut: Dār at-Taʿāruf li-l-Maṭbūʿāt, 1983.

Amir-Moezzi, Mohammad Ali. *The Divine Guide in Early Shiʿism. The Sources of Esotericism in Islam*, trans. David Streight. Albany: State University of New York Press, 1994.

Amir-Moezzi, Mohammad Ali. "Shi'ite Doctrine." In *EIr*, edited by Ehsan Yarshater. Iranica online, 1996–. Last updated July 20, 2005. https://iranicaonline.org/articles/shiite-doctrine.

Amir-Moezzi, Mohammad Ali. "al-ʿĀmilī al-Iṣfahānī, Abū l-Ḥasan." In *EI*[3], edited by Kate Fleet et al. Brill, 2007–. Article published online 2009. Consulted online on 14 November 2023 http://dx.doi.org/10.1163/1573-3912_ei3_COM_22714. http://dx.doi.org.prext.num.bulac.fr/10.1163/1573-3912_ei3_COM_22714>.

Amir-Moezzi, Mohammad Ali. *The Spirituality of Shiʿi Islam. Beliefs and Practices*. London/New York: I.B. Tauris, 2011.

Amir-Moezzi, Mohammad Ali. *The Silent Qur'an and the Speaking Qur'an. Scriptural Sources of Islam between History and Fervor*, trans. Eric Ormsby. New York: Columbia University Press, 2015.

Amir-Moezzi, Mohammad Ali. "Les Imams et les Ghulāt. Nouvelles réflexions sur les relations entre imamisme 'modéré' et shiʿisme 'extrémiste'." *Shiite Studies Review* 4 (2020): 5–38.

Amir-Moezzi, Mohammad Ali, and Christian Jambet. *What is Shi'i Islam? An Introduction*, trans. Kenneth Casler and Eric Ormsby. Abingdon/New York: Routledge, 2018.

Bar-Asher, Meir M. *Scripture and Exegesis in Early Imāmī Shiism*. Leiden/Jerusalem: Brill and The Magnes Press, 1999.

Brunner, Rainer. *Die Schia und die Koranfälschung*. Abhandlungen für die Kunde des Morgenlandes 53. Würzburg: Ergon Verlag, 2001.

Brunner, Rainer. "La question de la falsification du Coran dans l'exégèse chiite duodécimaine." *AR* 52, no. 1 (2005): 1–41.

Corbin, Henry. "ʿÂmelî Eṣfahânî, Abû'l-Ḥasan." In *EIr*, edited by Ehsan Yarshater. 16ff. vols. London/New York: Routledge, vol. 1, 1985: Fasc. 9, 931–32. Online: http://www.iranicaonline.org/articles/ameli-esfahani-abul-hasan-sarif-b.

Kohlberg, Etan, and Mohammad Ali Amir-Moezzi. *Revelation and Falsification. The* Kitāb al-Qirā'āt *of Aḥmad b. Muḥammad al-Sayyārī*. Leiden/Boston: Brill, 2009.

Modarressi, Hossein. "Early Debates on the Integrity of the Qur'ān. A Brief Survey." *SI* 77 (1993): 5–39.

Newman, Andrew J. *Safavid Iran. Rebirth of a Persian Empire*. London/New York: I.B. Tauris, 2006.

Al-Qummī, Ṣaffār. *Baṣā'ir ad-darajāt*. Ed. n.n. Beirut: Sharikat al-Aʿlamī li-l-Maṭbūʿāt, 2010.

Strauss, Leo. *Persecution and the Art of Writing*. Chicago: University of Chicago Press, 1988.

Aṭ-Ṭabrisī, Abū Manṣūr. *Al-Iḥtijāj*. Ed. Ibrāhīm Bahādurī and Muḥammad Hādī Beh, 2 vols. Qum: Uswa, 1992–93.

Aṭ-Ṭihrānī, Āqā Buzurg. *Adh-Dharīʿa ilā taṣānīf ash-shīʿa*. 25 vols. Tehran/Najaf, 1934–1978.

Ismāʿīl Ḥaqqī l-Brūsawī

Nurullah Denizer

1 Introduction

Throughout Islamic history, there have been many scholars who have lived within the tradition of Islamic science and contributed to the development of science with their works and ideas. The contributions made by some of these scholars to science and culture have influenced both the periods they lived in and the periods that have come after them. Ismāʿīl Ḥaqqī l-Brūsawī, who lived in the late seventeenth and early eighteenth centuries, is one of the scholars from the post-classical period whose influence has endured to the present day through his works. Traveling to various regions, including Anatolia, the Balkans, Egypt, and the Hijaz, Al-Brūsawī became a notable figure in the Islamic world both during his lifetime and after his death, especially with his *tafsīr*, titled *Rūḥ al-bayān*.

Al-Brūsawī wrote a number of works in the field of language and literature in addition to Islamic fields such as Qurʾānic exegesis (*tafsīr*), Islamic jurisprudence (*fiqh*), hadith, and Islamic theology (*kalām*). Moreover, he engaged in artistic activities such as music and calligraphy. Although al-Brūsawī was a versatile scholar who worked in an array of scientific disciplines, the dominant aspect of his scientific personality is that he was a Sufi. It is therefore possible to associate almost all the works he wrote about different branches of science with the field of Sufism. Al-Brūsawī, who, after Meḥmed Muḥyiddīn Uftāde (d. 1580/988) and ʿAzīz Maḥmūd Hudāyī (d. 1628/1038), was one of the most famous figures of the Sufi order *Jalwatī ṭarīqa*, always prioritized his identity as a Sufi while writing his works on various scientific fields. This article will discuss the life of al-Brūsawī, his scientific personality as an exegete, and the contributions he made to the science of *tafsīr*.

2 Life, Works, Political, and Intellectual Context of al-Brūsawī

2.1 Al-Brūsawī's Life

Al-Brūsawī gives detailed information about himself in his two works *Kitābu's-Silsileti'l-Jalwatiyye* and *Tamāmü'l-Feyz fī Bābi'r-Ricāl*, as well as in other works. His *wāridāt*[1]-

1 The concept of *wāridāt*, which is the plural of the word *wārid*, expresses the meanings that come from the unseen (from God) to the heart without the intention of the servant.

https://doi.org/10.1515/9783111320083-021

type works also reveal detailed information about his life. The most important and correct source on al-Brūsawī's life is thus his own works. Many studies dealing with his life are likewise essentially based on this information.[2]

Known as the shaykh of the Jalwatī and as an exegete and poet, Ismāʿīl Ḥaqqī l-Brūsawī was born in 1653/1063 in Aytos, which is located within the borders of what is now Bulgaria. He earned the name "al-Brūsawī" due to his prolonged stay in Bursa and "Jalwati" due to his affiliation with the *Jalwati ṭarīqa*. But it was with his al-Brūsawī moniker that he achieved prominence. Al-Brūsawī, received his primary education from Aḥmad Efendi, successor (*khalīfe*) to Atpazarī ʿUthmān Faḍlī Efendi (d. 1691/1102), who would later become his shaykh, before going to Edirne to study Arabic, *tafsīr*, *fiqh*, and *kalam*. After completing this education, he was sent by his teacher to ʿUthmān Faḍlī Efendi in Istanbul, where, he took lessons in *kalām* and *farāʾiḍ*,[3] *ūṣūl al-fiqh*, Persian, and calligraphy, and started preaching on behalf of his shaykh. In 1675, he was appointed *khalīfe* and sent to Skopje. Al-Brūsawī experienced some problems with the Mufti of Skopje and local dignataries due to claims that he was engaging in anti-religious behaviors. He later carried out preaching and religious guidance activities (*irshād*) in Koprulu and Usturumca. Afterward, he stayed in Edirne for some time at the invitation of his spiritual mentor. In 1685, he was sent to Bursa as *khalīfe*. Al-Brūsawī preached in various mosques, in particular in the Bursa Grand Mosque. At the invitation of Sultan Muṣṭafā II (d. 1703/1115), he participated in the first and the second Austrian campaigns between 1695 and 1696 to boost soldier morale. After staying in Tekirdağ and Damascus for a while, he settled in Üsküdar, where he worked as a Friday preacher in Üsküdar Ahmadiyya Mosque. He was prosecuted for allegedly uttering words contrary to the Islamic creed in his preaching. Even though the testimony of many witnesses revealed the accusations were groundless, he returned to Bursa in 1723. After spending his final years dealing with spiritual guidance (*irshād*) and writing, Al-Brūsawī died in Bursa on July 20, 1725/9th Dhū l-Qaʿda 1137 and was buried in the garden of the mosque he had built in Tuzpazarı.[4]

2.2 Al-Brūsawī's Works

Although there is no definitive information about the number of works that al-Brūsawī authored, the studies that have been carried out tell us that he wrote about 120–130 books. He composed most of his works in Turkish and Arabic, wrote some of them in Arabic-Turkish, and occasionally used Persian. Al-Brūsawī's most voluminous

2 For studies on al-Brūsawī, see Ali Namlı, *İsmâil Hakkı Bursevî: Hayatı, Eserleri ve Tarikat Anlayışı* (Istanbul: İnsan Yayınları, 2001), 15–19.

3 The branch of scholarship that studies Islamic inheritance law.

4 Sâkıb Yıldız, "Türk Müfessir İsmaʿīl Ḥaḳḳı Brūsevi'nin Hayatı," *Atatürk Üniversitesi İslamî İlimler Fakültesi Dergisi* 1 (1975): 103–26; Namlı, *İsmâil Hakkı Bursevî*, 33–113.

works are *Rūḥ al-bayān*, *Sharḥ al-Muḥammadiyya (Faraḥ ar-rūḥ)*, *Sharḥ ʿalā Tefsîrī Cüzʾiʾl-Akhīr liʾl-Ḳāḍiʾl-Bayḍāvī*, *Sharḥu Nukhbetiʾl-Fiker*, *Rūḥ al-Mathnawī*, *Kitābüʾn-Netīce*, *Tuḥfe-i khaṣekiyye*, and *Kitābüʾn-Necāt*.[5]

2.3 Political and Intellectual Context of al-Brūsawī

Given the widespread belief that the Treaty of Karlowitz (1699/1110) marked the start of the Ottoman Empire's decline, it is clear that the time in which al-Brūsawī lived (1653–1725/1063–1137) was characterized by the tail end of the stagnation period and the initial years of regression. Consequently, al-Brūsawī had a deep understanding of the political, economic, and social dilemmas that the Ottoman Empire was facing both domestically and internationally.

Because al-Brūsawī resided in Istanbul during the most prosperous phase of the Tulip Era (1718–1730/1130–1142) but refrained from its ostentatious and lavish culture, his works do not contain any explicit condemnation of the lifestyles of the era.[6] However, the following words demonstrate that he strongly opposed the luxury, waste, and worldly life he witnessed during this period:

> O world-addicted people! One day, your eyes will be filled with soil, and you will fill up and find a remedy in it. You've had enough of embracing this body, it is going to be devastated. You have had enough of prioritizing this world, it will remain behind. You are also wasting your other properties, as you waste your life. You leave your households behind and go out and make conversations with the degenerates.[7]

Al-Brūsawī lived in a period in which the quality of the madrasas, the centers of scientific activity, was in decline. The scarcity of scientifically competent scholars coupled with an increase in cases in which privileged individuals who paid bribes were appointed as teachers over more competent candidates were two of the factors disrupting the madrasas.[8] Al-Brūsawī uses the terms "bad scholars" and "shaykhs of time" in *Rūḥ al-bayān* to caution against individuals who fail to abide by principles of the Qurʾān such as by acting upon one's knowledge, preferring the hereafter over this world, practicing asceticism (*zuhd*) and piety (*taqwā*), purifying the soul of negative qualities, and

5 For detailed information about Brūsawī's works, see Mehmed Tâhir Bursalı, *Osmanlı Müellifleri*, ed. Ali Fikri Yavuz and İsmail Özen (Istanbul: Meral Yayınevi, 1971), 1: 121–24; Mehmet Şemseddin Ulusoy, *Yâdigâr-ı Şemsî*, ed. Mustafa Kara and Kadir Atlansoy (Bursa: Uludağ Yayınları, 1997), 182–86; Namlı, *İsmâil Hakkı Bursevî*, 161–219.

6 Yıldız, "Türk Müfessir İsmaʿīl Ḥaḳḳı Brūsevi'nin Hayatı," 121.

7 Mehmet Ali Ayni, *Türk Azizleri: İsmail Hakkı Bursalı ve Ruhü'l-Beyan Müellifi* (Istanbul: Marifet Yayınevi, 1944), 218f.

8 Ismail Hakkı Uzunçarşılı, *Osmanlı Tarihi: XVI. yüzyıl ortalarından XVII. yüzyıl sonuna kadar* (Ankara: Türk Tarih Kurumu, 1988), 3/2: 490.

ultimately turning toward the Creator while turning away from that which has been created.[9] Although the Qāḍīzāde-Sivāsī conflict,[10] which took place over disagreements on a number of religious and social issues just before al-Brūsawī's time, had come to an actual end, with the state taking various measures, these discussions subsequently continued on the intellectual level, and al-Brūsawī, too, scattered his own views on some of these topics of discussion throughout his works.[11]

Influenced by his spiritual mentor ʿUthmān Faḍlī Efendi, as well as Muḥyī d-Dīn b. al-ʿArabī (d. 1240/638), Mawlānā Jalāl ad-Dīn ar-Rūmī (d. 1273/672), Ṣadr ad-Dīn al-Qūnawī (d. 1274/673), and ʿAzīz Maḥmūd Hudāyī in the context of their views on *taṣawwuf*, al-Brūsawī adopted the opinion that the way to reach truth (*ḥaqīqa*) was through the union of *sharīʿa* and *ṭarīqa*. According to him, all the people who pretended to be Sufis, spiritual mentors (*shaykh*), and scholars (*ālim*), and who underestimated compliance with the order of Truth (*Ḥaqq*, God) and behaved in a lazy fashion were ignorant, superstitious people who did not know about the Sufi path (*sulūk*) or Truth (*ḥaqq*).[12] Al-Brūsawī stated that those who were veiled from the Truth (*ḥaqq*) but devoted to *sharīʿa* were closer to the Truth (*ḥaqq*) than the people of *ṭarīqa*, who made groundless claims and were deceived (those who think they are on the right path, even though they are on the wrong path).[13] In this respect, it can be said that al-Brūsawī was against any concept of *ṭarīqa* that excludes *sharīʿa*. Many works by al-Brūsawī, which are widely accepted in almost all Sufi circles, have been published and widely read.[14]

3 The *Tafsīr* of al-Brūsawī: *Rūḥ al-bayān*

Rūḥ al-bayān fī tafsīr al-Qurʾān, which has a special place among the works of al-Brūsawī and is a sign of his scholarly competence, is his most important, well-known, and voluminous work. However, he also wrote other works in the field of *tafsīr*: *Taʿlīqa ʿalā awāʾil Tafsīr al-Bayḍāwī, Sharḥ ʿalā Tafsīrī Cüzʾiʾl-Akhīr liʾl-Ḳāḍiʾl-Bayḍāwī, Kitāb al-Mirʾāt li-ḥaqāʾiq baʿḍ al-aḥādīth wa-l-āyāt, Lawāyiḥ tataʿallaqu bi-baʿḍ al-āyāt wa-l-*

9 Ismāʿīl Ḥaqqī l-Brūsawī, *Rūḥ al-bayān*, ed. Aḥmad ʿUbayd ʿInaya (Beirut: Dār Iḥyāʾ at-Turāth al-ʿArabī, 2001), 2: 67, 87; 3: 89, 453; 4: 151; 8: 12.

10 Uzunçarşılı, *Osmanlı Tarihi*, 3/1: 355–59.

11 Namlı, *İsmail Hakkı Bursevî*, 25.

12 Ismāʿīl Ḥaqqī l-Brūsawī, *Faraḥ ar-rūḥ: Sharḥ al-Muḥammadiyya* (Istanbul: Hacı Muharrem Efendi Matbaası, 1877/1294), 1: 339.

13 Ismāʿīl Ḥaqqī l-Brūsawī, *Sharḥ al-Mathnawī (Rūḥ al-Mathnawī)* (Istanbul: Matbaʿa-i Āmire, 1870/1287), 1: 449.

14 Ali Namlı, "İsmail Hakkı Bursevî," in *TDVIA*, ed. İslâm Araştırmaları Merkezi (Ankara: Türkiye Diyanet Vakfı Yayınları, 2001), 23: 102–4. For more on Ismāʿīl Ḥaqqī l-Brūsawī's life, works, and understanding of *ṭarīqa*, see. Namlı, *İsmail Hakkı Bursevî*.

aḥādīth (printed in *Maqālāt-i Ismāʿīl Ḥaqqī*), and *Mecmūʿatu Āyāti'l-Müntekhabe* (*Mecmūʿatu't-Tefāsīr*). In addition to these, al-Brūsawī wrote some epistles (*rasāʾil*) that contain commentaries on suras and verses. Moreover, in many of his other works, he provides commentary on some short suras and various verses independently.

The following passage from Al-Brūsawī's introduction to *Rūḥ al-bayān* provide significant insight into the purpose of and circumstances surrounding the writing of his commentary:

> During my visits to some parts of Anatolia, I compiled some of the pages from the *tafsīr* and some books on various sciences (*ʿulūm*). These included a little more than a part of the sura *Āl ʿImrān*. But they were scattered in their expression and contained long explanations like the east wind (*Ṣabā*). It was as if some of them were like the west wind and some like the east wind. I wanted to summarize them by purifying them of deficiencies and excesses. I wanted to summarize the pages as the dots, letters, and words permitted. I wanted to add to them my intuitive senses from the *Maʿrifat-i ilāhiyye* (divine knowledge). I wanted to line up all these things properly and put them in a literary form. Although my resources are few and my strength is insufficient, I want to interpret the noble Qur'ān to the end. If the Almighty God gives me the opportunity and possibility, I would like to finish this very important work and write up what I have written for the benefit of the people so that it may be a provision for me on the day of judgment when wealth and children will not benefit me.[15]

According to al-Brūsawī's statements, God inspired the author to write his Qurʾān commentary, guiding him with divine knowledge (*ishārāt*). He thus states that he began his commentary in the month of Shaʿbān in 1685/1096, naming his work *Rūḥ al-bayān fī tafsīr al-Qurʾān*.[16] *Rūḥ al-bayān*, which can be described as a reviewed and textualized version of the sermons he gave in the Great Mosque of Bursa by commenting on the Qurʾān, was completed over a period of approximately twenty-one years. The author's manuscript of the sixteen-book commentary is kept in the Bursa Old Manuscript and Printed Works Library. This volume has been published many times in Istanbul and the Arab world, and has also been the subject of summarization and text criticism studies by various editors.[17] Al-Brūsawī organized his commentary into three volumes, and each volume includes a distinct introductory chapter. This structure results in approximately ten parts (*juzʾ*) of the Qurʾān per volume. The three volumes are organized as follows: volume 1 comprises suras 1 – *al-Fātiḥa* (The Opening) – to 9 – *at-Tawba* (Repentance); volume 2 covers suras 10 – *Yūnus* (Jonah) – to 29 – *al-ʿAnkabūt* (The Spider); and volume 3 consists of suras 30 – *ar-Rūm* (The Byzantines) – to 114, *an-Nās* (Mankind).[18]

15 Al-Brūsawī, *Rūḥ al-bayān*, 1: 21.

16 Al-Brūsawī, *Tamāmü'l-Feyz fī Bābi'r-Ricāl*, ed. Ramazan Muslu and Ali Namlı (Damascus: Dār Nīnawā, 2011), 2: 108.

17 Ali Namlı, "Rūḥu'l-Beyān," in *TDVIA*, ed. İslâm Araştırmaları Merkezi (Ankara: Türkiye Diyanet Vakfı Yayınları, 2008), 35: 211, 213.

18 Namlı, *İsmâil Hakkı Bursevî*, 179.

3.1 Sources of *Rūḥ al-bayān*

3.1.1 Traditional (*naqlī*) Sources of *Rūḥ al-bayān*

Al-Brūsawī drew on multiple sources to write his commentary. Foremost among them were commentaries such as *al-Kashshāf* by az-Zamakhsharī (d. 1144/538), *Mafātīḥ al-ghayb* by Fakhr ad-Dīn ar-Rāzī (d. 1210/606), *al-Jāmiʿ li-aḥkām al-Qurʾān* by al-Qurṭubī (d. 1273/671), *Anwār at-tanzīl* by al-Bayḍāwī (d. 1286/685), *Madārik at-tanzīl* by an-Nasafī (d. 1310/710), and *Irshād al-ʿaql as-salīm* by Abū s-Suʿūd Efendi (d. 1574/982), as well as Sufi commentaries like *Baḥr al-ḥaqāʾiq wa-l-maʿānī* by Najm ad-Dīn Dāye (d. 1256/654), *Tabṣirat al-mutadhakkir* by al-Kawāshī (d. 1281/680), *Taʾwīlāt al-Qurʾān* by al-Kāshānī (d. 1335/736), and *ʿArāʾis al-bayān* by Rūzbihān-i Baqlī (d. 1209/606). In addition to these, *Rūḥ al-bayān* consulted various *fiqh*, hadith, *kalām*, philological, and literary sources. Works by Azīz Maḥmūd Hudāyī and ʿUthmān Faḍlī Efendi were also used as sources for *Rūḥ al-bayān*.[19]

3.1.2 Epistemological Sources of *Rūḥ al-bayān*

The commentator's method of commentary is largely shaped by his approach to acquiring knowledge sources. In Islamic thought, the sources of knowledge are generally categorized as the mind, transmission (*naql/riwāya*), the senses and revelation (*waḥy*), and inspiration (*ilhām*).[20] Al-Brūsawī demonstrates some of his views about these sources of knowledge in *Rūḥ al-bayān*. In this section, we will examine his method of obtaining knowledge and sources, providing an objective analysis of his approach.

3.1.2.1 Mind

Al-Brūsawī believes that the mind, as a means of knowledge acquisition, is not the destination; rather, it is valid only in matters that can be known through the mind.[21] In the explanation he gives of the verse "And such are the parables we set forth for mankind, but only those understand them who have knowledge" (Q 29:43), he remarks that it indicates that science (*ʿilm*) is more virtuous than the mind.[22] Al-Brūsawī believes that those whose faith is based on imitation (*taqlīd*) are sinful because they

19 Nurullah Denizer, *Kur'ân'ın Sûfî Yorumu: İsmail Hakkı Bursevî Örneği* (Ankara: Türkiye Diyanet Vakfı Yayınları, 2019), 132–38.

20 Necip Taylan, "Bilgi," in *TDVIA*, ed. İslâm Araştırmaları Merkezi (Ankara: Türkiye Diyanet Vakfı Yayınları, 1992), 6: 158; Şakir Gözütok, "Tasavvufî Eğitimde Bilginin Elde Edilmesi," *Tasavvuf* 2, no. 6 (2001): 92.

21 Al-Brūsawī, *Rūḥ al-bayān*, 5: 565.

22 Ibid., 6: 601.

have given up using their mind; using the mind is obligatory (*wājib*), and certainty (*yaqīn*) and the unity of God (*tawḥīd*) can only be grasped by abandoning imitation.[23] On the other hand, a person is not liable for issues that cannot be conceived of through the mind. Al-Brūsawī states that the verse "It was never God's (part) that he should send a folk astray after He had guided them until He had made clear unto them what they should avoid" (Q 9:115) is evidence of this.[24] According to this remark by al-Brūsawī, the mind must obey revelation (*waḥy*) in religious matters. Without revelation (*waḥy*) elucidating certain religious matters, it is not possible for the mind to acquire knowledge of those subjects on its own. Al-Brūsawī is thus of the opinion that the mind alone cannot comprehend everything, because in order to grasp the secrets of meaning and truths, one must have wisdom (*ḥikma*). "Wisdom, as a divine gift (*mawhība*), is granted to the hearts of prophets and *awliyāʾ* when God's attributes – his beauty (*jamāl*) and magnificence (*jalāl*) – manifest. In that moment, the attributes of the Creator become manifest, while the attributes specific to the creation of the servant disappear."[25] For this reason, says al-Brūsawī, the knowledge that is obtained through the mind is suspicious and predictable compared to the knowledge of those who possess certainty (*yaqīn*) and know the Truth (*ḥaqq*). What comes to the surface of the heart (*zāhir*) as a result of knowledge gained through reason is faith, and what comes to the heart's interior (*bātin*) is certainty (*yaqīn*).[26] Al-Brūsawī's approach to the mind and to the knowledge obtained through the mind shows that he regards the mind as acceptable only under the guidance of the revelation (*waḥy*) that God has granted to people through the *sharīʿa* and the prophets.[27]

3.1.2.2 Transmission (*naql*/*riwāya*)

Dividing science into two parts as the *ʿilm al-bayān* obtained by means of religion and the *ʿilm al-ʿiyān* obtained through removing the invisible veil (*kashf*),[28] According to al-Brūsawī, the acquisition of knowledge through narration is used only for the sciences related to the *sharīʿa*.[29] According to him, the path to attaining Truth (*ḥaqīqa*) involves acquiring knowledge through divine inspiration (*al-ʿilm al-ladunī*), which is obtained by achieving a mystical state (*ḥāl*), rather than through the mind and transmission. The Prophets' connection with God is an example of this because God taught

---

23 Ibid., 4: 250.
24 Ibid., 3: 664.
25 Ibid., 1: 527.
26 Ibid., 4: 63.
27 Ibid., 3: 666.
28 Al-Brūsawī, *Rūḥ al-bayān*, 6: 419. *Kashf*: According to the Sufis, the influence of the material world (soul desires) creates a veil that prevents the heart from seeing the invisible world. When the soul is purified from its bad features through various rituals, this veil is removed and the invisible world is seen clearly. The removal of this veil, that is, the opening of the heart's eye, is called *kashf*.
29 Ibid., 2: 501.

the Prophets the truth of things by showing it to them, not by narrating and transmitting it.[30] Moreover, al-Brūsawī, who attaches certain importance to the acquisition of knowledge by means of transmission on the way to Truth (*ḥaqīqa*), explains the characteristics of the person preferred for the task of guidance (*irshād*) as follows: "What is necessary for an intelligent person is to protect his soul from his desires and to think about leading it to the right path. The one who is preferred for the task of guidance (*irshād*) is the one who knows best the ways of the mind, transmission, and *kashf*."[31]

3.1.2.3 The Senses

According to al-Brūsawī, only tangible knowledge belonging to the world of objects can be obtained through the five senses. Beyond this, it is not possible for the five senses to make any contribution to obtaining metaphysical knowledge. According to him, the information obtained by the five senses plays a role that leads the person to doubt and distracts him from certainty in obtaining and evaluating the knowledge of truth. For exact knowledge is the knowledge that comes directly from the divine source to the heart. When the soul inhabits a physical body in this world, it has the potential to establish a direct connection with a divine source, providing access to truth and spiritual knowledge. The senses only provide knowledge about the physical dimension, which raises uncertainty about the knowledge that comes from a divine source and sows doubts about it.[32] According to al-Brūsawī, the five senses not only fail to provide knowledge about Truth but also drive the person away from the spiritual and otherworldly dimension, that is, his essence. The senses cannot grasp many aspects of *ma'rifa* within themselves, nor can they fully comprehend the truth of things.[33] Therefore, sensory knowledge alone is neither sufficient nor reliable for understanding the physical or metaphysical world.

3.1.2.4 Revelation-Inspiration (*Kashf-Ilhām*)

Al-Brūsawī considers *kashf* and *ilhām* to be the exclusive guidance granted by God to his superior servants. In his citation of the hadith "Whoever acts according to what he knows, God will make him heir to the sciences (*ʿulūm*) which he does not know,"[34] he states that the first of the sciences mentioned here is *ʿilm al-muʿāmala*, which means knowledge of how to perform daily work and worship. The second is *ʿilm al-*

30 Ibid., 539f.

31 Ibid., 5: 565.

32 Ibid., 1: 59, 90f.

33 Al-Brūsawī, *Rūḥ al-bayān*, 5: 233.

34 Abū ʿAbd Allāh Muḥammed b. ʿAlī b. Muḥammed ash-Shawkānī, *al-Fawāʾid al-majmūʿa fī l-aḥādīth al-mawḍūʿa*, ed. Abd ar-Raḥmān Yaḥyā l-Muʿallimī (Beirut: al-Maktaba al-Islāmiyya, 1986), 286. The sources mention that this *hadith* is weak.

mukāshafa, which is a gift from God that can only be obtained by purifying one's heart and by receiving inspiration from knowledge instilled by God Himself into one's heart. The *ladun* knowledge acquired in this way is higher, brighter, and more valuable than other knowledges. The knowledge obtained by people themselves (*al-ʿilm al-kasbī*) is of comparatively little value, like the shell of a walnut compared to the walnut itself.[35] According to al-Brūsawī, *ladun* knowledge is not knowledge that can be obtained by reading or studying. In order to obtain it, it is necessary to turn away from everything except God (*mā siwā*),[36] and the knowledge gained through reading and learning from those who have knowledge, such as *dhikr* und *muraqaba*, must be reflected in practice.[37] Although al-Brūsawī places the knowledge obtained through *kashf* and *ilhām* on a higher level than the knowledge obtained through the mind,[38] he believes that such knowledge is likewise limited. According to him, only a certain portion of knowledge of the hereafter can be achieved through *kashf* and *ilhām*.[39]

The *awliyā'*, the Friends of God,[40] are obliged to behave in accordance with the knowledge they have obtained through divine inspiration (*ilhām*). The Prophets and the *awliyā'* internalize divine revelation and inspiration (*waḥy* and *ilhām*) in the place designated by God, and they wait and observe God's orders in all circumstances. Unless they are given orders and informed by God, they do not approve of or follow *waḥy* or *ilhām*.[41] These statements show that al-Brūsawī also regards *ilhām* as being at the level of *waḥy*. Just as the prophets place revelation (*waḥy*) in its proper place, preach it, and act in accordance with it, so too must the *awliyā'* position the inspiration (*ilhām*) they have received where God wills, and act upon it – with the exception that, if *ilhām* is not supported by another divine order, there is no obligation to approve of or follow it. On the other hand, these statements by al-Brūsawī suggest that acting in accordance with the information that comes from inspiration (*ilhām*) is only necessary for the person who is the subject of the inspiration. Accordingly, it is possible to say that al-Brūsawī considers the knowledge obtained through *kashf* and *ilhām* to be subjective.

35 Al-Brūsawī, *Rūḥ al-bayān*, 4: 29.

36 This is a *taṣawwuf* term meaning everything but Allāh.

37 This is a *taṣawwuf* term meaning that the servant is conscious and aware that he is under the constant surveillance of Allāh. Al-Brūsawī, *Rūḥ al-bayān*, 1: 91.

38 Ibid., 3: 424f.

39 Ibid., 1: 70.

40 Plural; singular: *walī*. This is a term that means "friend of God."

41 Al-Brūsawī, *Rūḥ al-bayān*, 4: 388.

3.2 Exegetical-Hermeneutical Method of *Rūḥ al-bayān*

Non-textual factors shaped the commentator's purpose and preferred method for writing commentary, stemming from his personal and scientific experiences prior to writing. These factors are not directly stated in the text, but they significantly influence it. In this context, given that al-Brūsawī devoted his life to teaching, preaching, mentoring, and writing, these qualities also arguably shaped his approach to commentary. From this standpoint, it is unsurprising that the majority of al-Brūsawī's works, most notably his commentary, exhibit clear signs of guidance (*irshād*) aligned with the core principles of Sufism (*taṣawwuf*) and the Sufi path (*ṭarīqa*), where he held the highest position as a leader and guide.[42]

It is evident that al-Brūsawī's Sufi identity is reflected in *Rūḥ al-bayān*, which is regarded as an excellent book of religious teaching that provides explanations of verses following the commentary rules.[43] Because the work was created within the framework of sermons in the Bursa Grand Mosque, it is a work for the public. For this reason, it includes some stories unrelated to direct exegesis, exemplary events, and poems with Sufi content, and even stories that are clearly fictitious.[44] Thus, the stories are intended to guide the people. As a matter of fact, as almost every researcher who has studied al-Brūsawī emphasizes, al-Brūsawī, who was extremely knowledgeable of classical commentary literature, sometimes included religious teachings, parables, wise words, poems, and mystical allegories[45] that go beyond the literal meaning of the interpreted verse. This is not an acceptable method of exegesis within the generally accepted discipline of *tafsir*.[46] Al-Brūsawī explains the place accorded to religious teachings in his work in the following quote:

> When I was assigned to preach and sermonize, I paid great attention to subjects of religious teaching (*maw'iẓa*), and I collected them from many commentaries. I lined them up like a line of silk. I opened and solved the parts of the Qur'ān that are closed to people and whose meanings are locked with them. I explained them, trying to speak at a level that people can understand, without entering into different meanings that not everyone can understand and bear in the beginning. I also added some encouraging (*targhīb*) and intimidating (*tarhīb*) information appropriate to each verse, as well as some Sufi interpretations that would not be hidden or closed to anyone who is of mind (i.e., those who will read this commentary).[47]

---

42 Ömer Başkan, "Halkı İrşâd Çabasının Tefsire Yansımış Biçimi: Bursevî'nin Rûhu'l-Beyân'ı ve Tefsir Yöntemine Dair Bir Değerlendirme," *Din bilimleri Akademik Araştırma Dergisi* 8, no. 3 (2008): 128, 130f.

43 Ömer Nasuhi Bilmen, *Büyük Tefsir Tarihi* (Istanbul: Bilmen Yayınevi, 1974), 2: 714.

44 Süleyman Ateş, *İşârî Tefsir Okulu* (Istanbul: Yeni Ufuklar Neşriyat, 1998), 244; see, e.g., al-Brūsawī, *Rūḥ al-bayān*, 2: 113; 3: 585; 4: 197, 424; 9: 65; for poems, see al-Brūsawī, *Rūḥ al-bayān*, 1: 116; 2: 66; 6: 45, 7: 258; 10: 62.

45 Commentary based on knowledge obtained from *kashf* and *ilhām*/revelation and inspiration.

46 Başkan, "Halkı İrşâd Çabasının Tefsire Yansımış Biçimi: Bursevî'nin Rûhu'l-Beyân'ı ve Tefsir Yöntemine Dair Bir Değerlendirme," 138.

47 Al-Brūsawī, *Rūḥ al-bayān*, 1: 524f.

Based on al-Brūsawī's explanations, it is possible to define *Rūḥ al-bayān* as a Sufi commentary written in accordance with the people's level of understanding in order to guide them. It also has classical commentary features.

Aside from being recognized as a Sufi commentary, *Rūḥ al-bayān* is a multifaceted commentary in terms of the sources it references and its approach to commentary, encompassing features of transmission (*tafsīr bi-r-riwāya*), commentary based on opinion (*tafsīr bi-d-dirāya*), interpretation of the Qur'ān through the Qur'ān itself (*tafsīr al-Qur'ān bi-l-Qur'ān*), interpretation of the Qur'ān through *sunna*, and interpretation of the Qur'ān through the sayings of Muḥammad's companions. The commentary includes Qur'ānic sciences such as the reasons for revelation (*asbāb an-nuzūl*), abrogation (*nāsikh wa-mansūkh*), quotations from past commentators, philological explanations, poems, and mystical interpretations.

Many of the hadiths mentioned in *Rūḥ al-bayān* belong to the category of sound traditions (*sahīḥ*); however, the work has been criticized for containing narratives that are weak and fabricated.[48] *Rūḥ al-bayān* also includes hadiths without mentioning their chain of transmission (*isnād*), comparing hadith with the Qur'ān and pointing out that they also have esoteric meanings (*bāṭinī*).[49]

Al-Brūsawī uses the phrase "know that" (*iʿlam*)[50] at the beginning of passages when he wants to draw attention to a point in his commentary, and also deploys the phrase "if you say this, I will say this" (*fa-in qulta . . . qultu . . .*"),[51] which can be found in a great number of commentaries. While approaching the judicial verses (*aḥkām*) from the standpoint of the Ḥanafī school, Al-Brūsawī, mentions the *qirā'āt* differences,[52] even if they minor, and occasionally criticizes groups such as the Muʿtazila, Qadariyya, and Jabriyya.[53]

Traditional exegesis (*naql*) is predominant in this work, which was written for the public to be used by the public, as Al-Brūsawī himself states. The author has carefully selected symbolic Sufi comments (*ishārī*) and clearly explains them. He does not discuss any controversial issues at length. Whereas al-Brūsawī speaks about his own Sufi experiences in his specifically Sufi works, in *Rūḥ al-bayān* he is a transmitter of others' reports.[54] The symbolic (*ishārī*) nature of *Rūḥ al-bayān* is evident almost throughout the work. Thus, after providing explanations from classical commentaries, al-Brūsawī mentions symbolic comments made by previous Sufis and attempts to explain the esoteric (*bāṭin*) meaning of the verses in a "softened" way.

---

48 Ateş, *İşârî Tefsir Okulu*, 244.
49 Al-Brūsawī, *Rūḥ al-bayān*, 9: 401.
50 Ibid., 2: 59; 4: 181; 5: 76; 6: 206; 9: 397.
51 Ibid., 1: 236; 4: 87; 7: 403; 8: 184; 10: 630.
52 Ibid., 1: 113, 143, 503; 3: 185; 6: 172.
53 Ibid., 1: 113, 143, 503; 3: 185; 6: 172.
54 Namlı, *İsmail Hakkı Bursevî*, 213.

While interpreting the verses symbolically, al-Brūsawī validates his explanations with Qur'ānic verses[55] and hadiths.[56] In addition, he assigns symbolic meanings to some hadiths[57] and derives additional symbolic meanings by referencing *isrā'īliyyāt*.[58] He occasionally assigns more than one symbolic meaning to the same verse[59] and uses the order of words in the verse[60] as well as the etymological and morphological similarities between words as a source for such meanings.[61]

In general, it is remarkable that the symbolic (*ishārī*) comments in *Rūḥ al-bayān* do not contradict the literal (*ẓāhir*) meaning of the verses. As previously mentioned, the author combines the outer meaning of the Qur'ān (*sharī'a*) and the inner meaning (*haqī,qa*), in his commentary, stating that there cannot be any contradiction between the two. However, it should be noted that the symbolic (*ishārī*) comments in some parts of the commentary are made by removing the verses from their historical and textual context and by changing their grammatical structures.[62] In *Rūḥ al-bayān*, symbolic (*ishārī*) comments are related to almost every subject matter, such as worship, the rules of mutual social relationships (*mu'āmalāt*), *kawnī* verses,[63] doctrines (*'aqā'id*), and *kalām*, narratives and proverbs. However, symbolic (*ishārī*) comments are used more widely in ambiguous (*mutashābih*) verses, narratives, and proverbs.[64]

3.3 Theological Influence of *Rūḥ al-bayān* in Past and Present

Rūḥ al-bayān, which is the result of an intense and diligent endeavor, has been an important source for works written after it and continues to exert influence today.

Ya'qūb Afwī (d. 1736/1149), the master (*postnishīn*)[65] of Āsitāne-i Hudāyī[66] and one of al-Brūsawī's contemporaries, quotes from *Rūḥ al-bayān* in his work *Tuḥfat al-wu''āẓ* (Hacı Selim Ağa Library, 'Azīz Maḥmūd Hudāyī, no. 623). The influence of *Rūḥ al-*

55 Al-Brūsawī, *Rūḥ al-bayān*, 1: 38, 476; 2: 260f.; 3: 38.

56 Ibid., 1: 46f., 3: 199.

57 Ibid., 1: 402; 2: 155; 4: 232.

58 Al-Brūsawī, *Rūḥ al-bayān*, 1: 83, 510f.

59 Ibid., 1: 93; 4: 46; 5: 305, 314, 338; 6: 30;, 170; 7: 58; 8: 411, 591; 9: 59.

60 Ibid., 1: 418; 2: 328; 3: 115.

61 Ibid., 1: 75; 2: 86; 3: 217.

62 Ibid., 1: 130, 268; 2: 124, 226; 3: 270, 678f.; 4: 223.

63 Verses that show the existence, unity, and lofty attributes of the Creator who connects the innumerable varieties and differences in the universe into a continuous law and order, and that mention the evidence derived from the characteristics of creatures.

64 Nurullah Denizer, "İşârî Tefsirlerdeki Kıssa ve Mesel Yorumlarında İ'tibâr/Analoji Yöntemi: İsmail Hakkı Bursevî Örneği," *Gümüşhane Üniversitesi İlahiyat Fakültesi Dergisi* 6, no. 11 (2017): 95–110.

65 This is the term used in Sufi *takkas* to designate a person who has the *shaykh* authority, especially within the *Mawlawī* tradition.

66 Āsitāne is one of the names given to the great *takkas* that are the center of a *tarīqa* or a branch of a *tarīqa*.

bayān is visible in both the method and content of *Anīs al-jinān fī tafsīr al-Qurʾān* (Bursa Old Manuscript and Printed Works Library, General, no. 976–977) by Ashrafzādeh ʿIzzeddīn Efendi (d. 1740/1153), another one of al-Brūsawī's contemporaries. One of the main sources of the commentary written by Ashrafzādeh's son Abdulqadīr Necīb (d. 1788/1202), titled *Zubdat al-bayān* (Bursa Old Manuscript and Printed Works Library, General, no. 217–218), is *Rūḥ al-bayān*. Ghazzīzādeh Abdullaṭīf Efendi (d. 1832/1247) benefited from *Rūḥ al-bayān* while preparing his Turkish commentary *Futūḥāt kanz al-Qurʾān*. Almost half of the quotations in *Tafsīr-i Sūre-i Wāqıʿa* (Istanbul 1310), written by Meḥmed b. ʿUthmān al-Ḥilmī l-ʿAdavī in 1851, were taken from *Rūḥ al-bayān*. Awliyāzādeh Ismāʿīl Ḥaqqī's (d. 1865/1282) work, titled *Miftāḥ at-tafāsīr* (Istanbul 1286), mainly consists of subjects chosen from *Rūḥ al-bayān*. Hasan Basri Çantay (d. 1964/1384) benefited from *Rūḥ al-bayān* in the footnotes he added to the meanings of the verses. Alongside the commentaries by Qāḍī Nāṣir ad-Dīn Bayḍāwī, Fakhr ad-Dīn ar-Rāzī, and Ebussuʿūd Efendi, Ömer Nasuhi Bilmen (d. 1971/1391) also mentions *Rūḥ al-bayān* as one of the works he benefited from while preparing his commentary. *Rūḥ al-bayān* has also been summarized by various scholars.[67]

One of the most important academic studies on Ismāʿīl Ḥaqqī l-Brūsawī and *Rūḥ al-bayān* is a doctoral thesis presented by Sakıb Yıldız (d. 1995/1416) at Sorbonne University in 1972 titled *L'exégète turc Ismāʿil Hakkī Burūsawī, sa vie, ses oeuvres et la méthode dans son tafsīr* Rūh al-Bayān. Approximately one-third of this study is devoted to analyzing various aspects of *Rūḥ al-bayān*. Since then, scholars have continued to study *Rūḥ al-bayān* and a number of other works by al-Brūsawī since then, with several master's and doctoral theses being written. *Rūḥ al-bayān* was translated into Turkish by a delegation led by Hasan Kamil Yilmaz under the title *Rûhu'l-beyân: Kur'an Meâli ve Tefsiri*.

4 Conclusion

In addition to his Sufi Qurʾān commentary *Rūḥ al-bayān*, al-Brūsawī wrote many other works in the fields of Sufism (*taṣawwuf*), *tafsīr*, hadith, *fiqh*, and *kalām*, which are characterized as Sufi writings. In addition to these original works by al-Brūsawī, the glosses (*ḥāshiye*) at the beginning and end of his commentary, and his explanations (*sharḥ*) of some important previous works, have secured his fame and influence from his time until the present day. *Rūḥ al-bayān*, the most famous and voluminous of al-Brūsawī's works, remains popular today and has been frequently used in preaching and teaching activities. It has also been the subject of academic studies. While the works al-Brūsawī wrote in Arabic enabled him to be recognized throughout the Islamic world, his Turkish works written in a simple language also enabled him to

67 Namlı, "Rūḥu'l-Beyān," 212f.

reach all sections of society, from members of Sufi orders and madrasas to statesmen and ordinary people.

5 Bibliography

5.1 Sources

Al-Brūsawī, Ismāʿīl Ḥaqqī. *Sharḥ al-Mathnawī (Rūḥ al-Mathnawī)*. 2. vols. Istanbul: Matbaʿa-i Āmire, 1870.

Al-Brūsawī, Ismāʿīl Ḥaqqī. *Faraḥ ar-rūḥ: Sharḥ al-Muḥammadiyya*. 2 vols. Istanbul: Hacı Muharrem Efendi Matbaası, 1877.

Al-Brūsawī, Ismāʿīl Ḥaqqī. *Rūḥ al-bayān*. Ed. Aḥmad ʿUbayd ʿInaya, 10 vols. Beirut: Dār Iḥyāʾ at-Turāth al-ʿArabī, 2001.

Al-Brūsawī, Ismāʿīl Ḥaqqī. *Tamāmü'l-Feyz fī Bābi'r-Ricāl*. Ed. Ramazan Muslu and Ali Namlı, 2 vols. Damascus: Dār Nīnawā, 2011.

5.2 Secondary Literature

Ateş, Süleyman. *İşârî Tefsir Okulu*. Istanbul: Yeni Ufuklar Neşriyat, 1998.

Ayni, Mehmet Ali. *Türk Azizleri: İsmail Hakkı Bursalı ve Ruhü'l-Beyan Müellifi*. Istanbul: Marifet Yayınevi, 1944.

Başkan, Ömer. "Halkı İrşâd Çabasının Tefsire Yansımış Biçimi: Bursevî'nin Rûhu'l-Beyân'ı ve Tefsir Yöntemine Dair Bir Değerlendirme." *Din bilimleri Akademik Araştırma Dergisi* 8, no. 3 (2008): 127–44.

Bilmen, Ömer Nasuhi. *Büyük Tefsir Tarihi*. 2. vols. Istanbul: Bilmen Yayınevi, 1974.

Bursalı, Mehmed Tâhir. *Osmanlı Müellifleri*. Ed. Ali Fikri Yavuz and İsmail Özen. 2. vols. Istanbul: Meral Yayınevi, 1971.

Denizer, Nurullah. "İşârî Tefsirlerdeki Kıssa ve Mesel Yorumlarında İ'tibâr/Analoji Yöntemi: İsmail Hakkı Bursevî Örneği." *Gümüşhane Üniversitesi İlahiyat Fakültesi Dergisi* 6, no. 11 (2017): 92–114.

Denizer, Nurullah. *Kur'ân'ın Sûfî Yorumu: İsmail Hakkı Bursevî Örneği*. Ankara: Türkiye Diyanet Vakfı Yayınları, 2019.

Gözütok, Şakir. "Tasavvufî Eğitimde Bilginin Elde Edilmesi." *Tasavvuf* 2, no. 6 (2001): 91–104.

Namlı, Ali. "İsmail Hakkı Bursevî." In *TDVIA*, edited by İslâm Araştırmaları Merkezi, 44 vols. Ankara: Türkiye Diyanet Vakfı Yayınları, vol. 23, 2001: 102–4.

Namlı, Ali. *İsmâil Hakkı Bursevî: Hayatı, Eserleri ve Tarikat Anlayışı*. Istanbul: İnsan Yayınları, 2001.

Namlı, Ali. "Rūḥu'l-Beyān." In *TDVIA*, edited by İslâm Araştırmaları Merkezi, 44 vols. Ankara: Türkiye Diyanet Vakfı Yayınları, vol. 35, 2008: 211–13.

Ash-Shawkānī, Abū ʿAbd Allāh Muḥammad b. ʿAlī b. Muḥammad. *Al-Fawāʾid al-majmūʿa fī l-aḥādīth al-mawḍūʿa*. Ed. Abd ar-Raḥmān Yahyā l-Muʿallimī. Beirut: al-Maktaba al-Islāmiyya, 1986.

Taylan, Necip. "Bilgi." In *TDVIA*, ed. İslâm Araştırmaları Merkezi, 44 vols. Ankara: Türkiye Diyanet Vakfı Yayınları, vol. 6, 1992: 157–61.

Ulusoy, Mehmet Şemseddin. *Yâdigâr-ı Şemsî*. Ed. Mustafa Kara and Kadir Atlansoy. Bursa: Uludağ Yayınları, 1997.

Uzunçarşılı, İsmail Hakkı. *Osmanlı Tarihi: XVI. yüzyıl ortalarından XVII. yüzyıl sonuna kadar*. 6 vols. Ankara: Türk Tarih Kurumu, vol. 3/1–2, 1988.

Yıldız, Sâkıb. "L'exégète turc Ismāʿil Hakkī Burūsawī, sa vie, ses oeuvres et la méthode dans son tafsīr *Rūh al-Bayān*." PhD. diss., Sorbonne University, 1972.

Yıldız, Sâkıb. "Türk Müfessir İsmaʿīl Ḥaḳḳı Brūsevi'nin Hayatı." *Atatürk Üniversitesi İslamî İlimler Fakültesi Dergisi* 1 (1975): 103–26.

Shāh Walī Allāh

Kamran Bashir

1 Introduction

In any historiography of *tafsīr* literature and Qur'ānic hermeneutics, Shāh Walī Allāh (1703–1762/1115–1176) cannot be overlooked. For the historians of Muslim religious thought in South Asia, he and his family at times provide a basis for periodizing intellectual trends.[1] No revisionist history of eighteenth-century Muslim thought in the wider Muslim world can ignore the creativity of this Delhi polymath.[2] As this chapter will highlight, the significance of Walī Allāh's contribution to Muslim religious thought stems not only from his influence in his own times but is also due to his widespread, favorable reception in South Asia, particularly from the latter part of the nineteenth/thirteenth century onward. Muslim scholars of all stripes, including those who otherwise behaved belligerently to each other when it came to the question of "authentic" Islam, were often seen to appropriate Walī Allāh in order to impress their particular viewpoints upon their audiences. This universal admiration also owes to the nature of his work. His writings and thought often lend themselves to multiple interpretations. For the architects of modernism in modern Islamic thought, he was thus a proto-modernist in Muslim intellectual history who experimented with new approaches to older questions. For traditionalists, on the other hand, he was not dismissive of centuries-old Muslim tradition and therefore crafted his thought within the dictates of Islamic tradition.[3] These features of his personality and work invite new researchers to revisit his times and oeuvre in a fresh light.[4] This chapter aims to determine his contribution to Muslim Qur'ānic sciences and to the larger sub-field of Qur'ānic studies.

1 On Walī Allāh's place and influence in the evolution of modern Qur'ān commentaries and hermeneutics, see Johanna Pink, "Striving for a New Exegesis of the Qur'ān," in *The Oxford Handbook of Islamic Theology*, ed. Sabine Schmidtke (Oxford: Oxford University Press, 2016).

2 See Ahmad S. Dallal, *Islam without Europe: Traditions of Reform in Eighteenth Century Islamic Thought* (Chapel Hill: The University of North Carolina Press, 2018); Francis Robinson, "Islamic Reform and Modernities in South Asia," *Modern Asian Studies* 42, no. 2–3 (March 2008): 259–81.

3 In modern Islam, traditionalists are seen as Muslim scholars who uphold the centuries-long Muslim intellectual tradition; modernists, by contrast, are perceived as the ones who stress understanding Islam directly through the Qur'ān and hadith. On traditionalism and modernism, see Muhammad Qasim Zaman, *'Ulama in Contemporary Islam: Custodians of Change* (Princeton: Princeton University Press, 2002), 1–16.

4 For instance, on the significance of Walī Allāh's work in recognizing the complexity and creativity in Islamic legal thought, see Ebrahim Moosa, "Introduction," in Shāh Walī Allāh, *Shāh Walī Allāh's Treatises on Juristic Disagreement and Taqlīd*, trans. Marcia Hermansen (Louisville, KY: Fons Vitae, 2010), vii–xxi.

https://doi.org/10.1515/9783111320083-022

2 Life, Times, and Works of Walī Allāh

Walī Allāh was born in Mughal India near the end of the long rule of the great Emperor Aurangzeb (r. 1658–1707/1068–1119). Never did the Mughal Empire regain its past glory after the passing away of this sovereign. Much of the eighteenth century after Aurangzeb was characterized by weak Mughal governance, internecine wars, foreign invasions (including the sacking of Delhi by the Iranian king Nādir Shāh (r. 1736–1747/1149–1160)), and the flexing of imperial muscles by would-be colonial rulers from the British Empire. In true historical terms, the question of whether the eighteenth century was a period of "decline" or not is open to fresh inquiry.[5] However, in the context of his times, Walī Allāh's intellectual work and activism aimed at some kind of revival encompassing both religious and political spheres. This is why he is said to have contributed toward thwarting foreign invasions and non-Muslim incursions into the strongholds of Muslim rule.[6]

In terms of his broader development as a religious savant, he belonged to a family of Arab migrants who had settled in India somewhere in the thirteenth century.[7] His father, ʿAbd ar-Raḥīm (d. 1719/1132), was a holy man, a Sufi who was prominent among the religious leaders of his time. One of his enduring contributions was the establishment of a religious seminary, the Madrasa Raḥīmiyya in Delhi, which Walī Allāh came to steer before the demise of his father. The seminary remained active until the mid-nineteenth century. Another key feature of Walī Allāh's life was his 1730/1143 journey to the holy places in the Hijaz region of Arabia, where he stayed for more than a year, attending lectures by well-known Arab teachers. His biographers and historians of the eighteenth-century Muslim world disagree on the significance of what Walī Allāh learned in the Hijaz for his later work. Whatever he learned there, he applied it to his intellectual situation in new ways and transformed it for his own projects. It is therefore difficult to say that he borrowed some kind of Wahhabism from the Hijaz. In any case, his journey to the Hijaz has provided his biographers with a basis for periodizing many of his writings.

---

5 For a revisionist history of this "decline" process, see Satish Chandra, *Parties and Politics at the Mughal Court, 1707–1740* (New Delhi/New York: Oxford University Press, 2002). See also Freeland Abbott, "The Decline of the Mughal Empire and Shāh Walī Allāh," *MW* 52 (January 1962): 115–23.

6 For a recent analysis of Walī Allāh's political thought and its context, see Muhammad Qasim Zaman, "Political Power, Religious Authority, and the Caliphate in Eighteenth-Century Indian Islamic Thought," *JRAS* 30, no. 2 (2020): 313–40.

7 For his biography and the intellectual context of eighteenth-century Delhi, see Saiyid Athar Abbas Rizvi, *Shāh Walī-Allāh and his Times: A Study of Eighteenth Century Islām, Politics and Society in India* (Canberra: Maʿrifat Publications, 1980). Other important works on his life and times are: J.M.S. Baljon, *Religion and Thought of Shāh Walī Allāh Dihlawī 1703–1762* (Leiden: Brill, 1986); M. Ikram Chaghatai, ed., *Shah Waliullah (1703–1762): His Religious and Political Thought* (Lahore: Muḥammad Ashraf, 2005).

When it comes to Walī Allāh's intellectual output, even just a quick glimpse at his sizeable oeuvre is impressive.[8] He worked on almost all the Islamic sciences, spanning philosophical theology, hadith studies, jurisprudence, mysticism, and Qur'ānic sciences. Scholarly analysts of his work have characterized his intellectual labor as an attempt to reconcile Muslim philosophical theology with Muslim jurisprudence. In addition, Walī Allāh's larger contributions are seen in terms of his attempts to limit the differences among Muslim schools of thought on matters of Qur'ānic interpretation, legal rulings, and mystical learning. Some of the well-known writings that draw attention when taking an initial glance at his work include:[9] 1) *Ḥujjat Allāh al-bāligha* and *al-Budūr al-bāzigha*, which explain the rationales behind *sharī'a* rulings;[10] 2) *at-Tafhīmāt al-ilāhiyya*, which includes a series of essays on philosophical and mystical topics;[11] 3) *Fuyūḍ al-Ḥaramayn*, which presents the visions that Walī Allāh had during his stay in the Hijaz;[12] 4) *al-Muṣaffā* and *al-Musawwā*, Persian and Arabic glosses on Mālik b. Anas's *Muwaṭṭa'*;[13] 5) *al-Inṣāf fī bayān sabab al-ikhtilāf*, which discusses the backgrounds to intellectual conflicts among Muslim schools of thought;[14] and 6) his Qur'ānic works *Fatḥ ar-Raḥman*, a Persian translation of the Qur'ān with exegetical glosses, and *al-Fawz al-kabir fī uṣūl at-tafsīr*, a treatise dealing with Qur'ānic hermeneutics.[15]

3 Understanding Walī Allāh's Qur'ānic Hermeneutics

Interest in engaging with Walī Allāhī hermeneutics has never subsided in academic circles. Academic studies on modern Muslim Qur'ān commentaries and hermeneutics have shown particular interest in his famous booklet entitled *al-Fawz al-kabīr fī uṣūl*

8 For a periodization of his works, see Baljon, *Shāh Walī Allāh Dihlawī.*

9 Marcia Hermansen's recent bibliography on Walī Allāh is a useful online resource: Marcia Hermansen, "Shah Wali Allah Bibliography," last modified 2022, https://www.academia.edu/2163744/Shah_Wali_Allah_Bibliography.

10 Shāh Walī Allāh, *Ḥujjat Allāh al-bāligha* (Arabic), ed. 'Uthmān Jum'a ad-Dumayriyya (Riyadh: Maktabat al-Kawthar, 1999). For a partial English translation, see Shāh Walī Allāh, *The Conclusive Argument from God*, trans. Marcia Hermansen (Leiden: Brill, 1996). For the other text, see Shāh Walī Allāh, *al-Budūr al-bāzigha* (Arabic) (Hyderabad: Shāh Walī Allāh Academy, 1970).

11 Shāh Walī Allāh, *at-Tafhīmāt al-ilāhiyya* (Arabic/Persian), 2 vols. (Hyderabad: Shāh Walī Allāh Academy, 1967).

12 Shāh Walī Allāh, *Fuyūḍ al-Ḥaramayn* (Arabic with Urdu translation) (Karachi: Muḥammad Sa'īd, n.d.).

13 Shāh Walī Allāh, *Musawwā Muṣaffā* (Arabic/Persian), 2 vols. (Karachi: Muḥammad 'Alī Karkhāna-i-Islāmī Kutub, 1980).

14 Shāh Walī Allāh, *al-Inṣāf fī bayān sabab al-ikhtilāf* (Cairo: Muḥibb ad-Dīn al-Khaṭīb, 1965).

15 The works on the Qur'ān are discussed below. Secondary works also mention another small piece of writing by Walī Allāh, *Uṣūl dar tarjamat al-Qur'ān*, a pamphlet outlining principles of Qur'ānic translation.

at-tafsīr, whose translations from the original Persian abound, especially in Arabic and Urdu.[16] However, despite this enduring interest, there has been no systematic academic study that investigates Walī Allāh's works on the Qur'ān in any great detail. In particular, no work has made his Persian translation of the Qur'ān, *Fatḥ ar-Raḥmān*, and the accompanying exegetical glosses a special object of research, particularly within the intellectual context of eighteenth-century Delhi.

In order to make sense of the substantial regional literature on the Qur'ān, Walī Allāh's seminal works on the Qur'ān provide a framework for periodizing the Qur'ānic exegetical and hermeneutical literature produced in South Asia. The underlying rationale for this periodization is not based on any claim that Walī Allāh brought about any significant transformations in Qur'ānic hermeneutics in South Asia. Given his considerable Qur'ānic output, it appears that he did initiate some new thinking about the Qur'ān, but the extent to which this new approach to the scripture had an impact in the ensuing decades would require a separate historical study. Nonetheless, in the wake of his widespread, favorable reception in South Asia from the latter half of the nineteenth century onward, his ideas on interpreting the Qur'ān did gain prominence in British India (see the section below on his impact).[17] He is considered the prototype of a "modern" reader of the Qur'ān in several respects, for example, in terms of his motivation for translating the Qur'ān into the Persian vernacular; his movement to return to the Qur'ān (*rujūʿ ilā l-Qur'ān*) to interpret its meaning and understand Islam; and his encapsulation of tendencies to both revisit and preserve traditional positions on Qur'ānic exegesis.[18]

Walī Allāh's Persian translation of the Qur'ān, entitled *Fatḥ ar-Raḥmān*, was an important milestone in his larger Qur'ān project. In the introduction, Walī Allāh announces that his new translation is meant for both the common man and specialists.[19] He does not elucidate what exactly he had in mind when he defined this large audience. At a simpler level, we can speculate that anybody who could read Persian, par-

16 For the study of *al-Fawz al-kabīr*, I have used the Arabic translation Shāh Walī Allāh, *al-Fawz al-kabīr fī uṣūl at-tafsīr*, trans. Saʿīd Aḥmad al-Bālanfūrī (Damascus: Dār al-Ghawthānī li-d-Dirāsāt al-Qur'āniyya, 2008).

17 In their studies of modern *tafsīr*, historians recognize Walī Allāh as a milestone in Qur'ānic hermeneutics, particularly in South Asia. Cf. Pink "Striving for a New Exegesis," 3–6.

18 On the significance of Walī Allāh for South Asian Islam, see Baljon, *Shāh Walī Allāh Dihlawī*. Some viewed Walī Allāh as a revivalist on the lines of Taqī d-Dīn Aḥmad b. Taymiyya (d. 1328/728). See Rizvi, *Shāh Walī-Allāh*.

19 For his introduction to the translation of the Qur'ān, I used Shāh Walī Allāh, *Tarjamat al-Qur'ān: Fārsī, al-mawsūm ba-Fatḥ ar-Raḥmān* (Bombay: ʿAlī Bhā'ī Sharaf ʿAlī, n.d.). For the text of his translation, I consulted Shāh Walī Allāh, *Qur'ān-i majīd mutarjam: Tarjama-yi fārsī* (Lahore: Pāk Company, 2004). On the reception of his translation during his own times, the studies on Walī Allāh mention a general opposition from the fundamentalist theologians. It seems that the older idea that the Qur'ān cannot be rendered into any other language has prevailed to some extent. See Rizvi, *Shāh Walī-Allāh*, 229–31.

ticularly in eighteenth-century Delhi, was considered a member of his audience. Moreover, he mentions that there are other Persian translations available and that he is dissatisfied with them. His biographers have also noted some kind of animosity toward his project of Qur'ānic translation, which was seen as an attempt to put the Qur'ān, which required specialist interpretation skills, in the hands of lay people. One can see this work as one of the early steps toward diminishing of the status enjoyed by Arabic in the disciplines that made up the Islamic sciences in South Asia. It is clear that, within a few decades of his demise, vernacular translations of the Qur'ān, in particular Urdu translations, were making inroads into the sphere of Qur'ānic learning. However, despite the importance of the Qur'ānic scholarship produced by Walī Allāh and later by his three sons, one must practice caution in identifying changes and continuity in Qur'ānic hermeneutics in South Asia. The question of Walī Allāh's influence is a formidable one. Even though his impactful work, *al-Fawz al-kabīr* and *Fatḥ ar-Raḥmān*, can be perceived as marking a break with the past scholarship produced in the Islamic tradition, the actual growth of his intellectual influence is a separate historical question. Exercising caution in relation to the above observations stemming from the current state of scholarship on Walī Allāh's Qur'ānic works, we can briefly ascertain the concrete contributions made by Walī Allāh in terms of his Qur'ānic hermeneutics. Considering his overall thought, it seems that he was revisiting the history of Islamic thought, both Qur'ānic and hadith-based.[20] The following will examine *al-Fawz al-kabīr* in more detail.

3.1 A New Investigation of Walī Allāh's *al-Fawz al-kabīr*

Our aim here is to evaluate the hermeneutical potential latent in Walī Allāh's principles of Qur'ānic interpretation, which he discusses in detail in *al-Fawz al-kabīr*. We can begin by noting three aspects of this work that are key to understanding the core of Walī Allāh's Qur'ānic thought: First, Walī Allāh helps his readers to make sense of the whole Qur'an. To do so, he categorizes various types of Qur'ānic narrative and discourse, through which the scripture provides guidance to its believers. These are the different types of knowledge (*ʿilm*) that God has given to mankind through the Qur'ān. Walī Allāh claims that this realization of different kinds of knowledge is a divine gift given to him. There are five types of *ʿilm*: 1) the *ʿilm al-aḥkām* in the verses that deal with legal issues; 2) the *ʿilm al-jadal* (or *ʿilm al-mukhāṣama*) in the verses that contain debates with groups who opposed believers, including *mushrikūn* (polytheists), *yahūd* (Jews), *naṣārā* (Christians), and *munāfiqūn* (hypocrites); 3) the *ʿilm tadhkīr*

20 On Walī Allāh's review of the Islamic juristic tradition and its disagreements, see Jonathan A.C. Brown, *Misquoting Muḥammad: The Challenge and Choices of Interpreting the Prophet's Legacy* (London: Oneworld Publications, 2014), 15–68. Brown's analysis is based on his study of Walī Allāh's *al-Inṣāf fī bayān sabab al-ikhtilāf*.

bi-ʿalāʾ Allāh in the verses that discuss Allāh's omnipotence, omnipresence, and providence; 4) the *ʿilm tadhkīr bi-ayyām Allāh* in the Qurʾānic stories of peoples and nations meant to guide the reader; and 5) the *ʿilm tadhkīr al-mawt wa-mā baʿd al-mawt* in the verses that address matters related to eschatology, death, and the hereafter.

Second, Walī Allāh reviews the evolution of the Muslim intellectual tradition from its formative period to the later period. In doing so, he engages with all the key issues of debate and controversy in the field of *ʿulūm al-Qurʾān* (traditional Qurʾānic sciences), issues that occupied Muslim scholars throughout Muslim intellectual history. What seems conspicuous in Walī Allāh's thought in this regard is his forthright acknowledgement of the challenges underlying any effort to interpret the Qurʾān given the scripture's particular linguistic style and diction. In addition to the linguistic peculiarities of the Qurʾān, he raises the issues of *naskh* (abrogation) and *asbāb an-nuzūl* (occasions of revelation)[21] as debated in the genres of the traditional Qurʾānic sciences. However, what is striking here is the particular linguistic manner in which he discusses these issues in his work. Walī Allāh's articulation is somewhat equivocal in the sense that his approach prompted his academic observers to label him both a "traditionalist" and "modernist" when they saw him being appropriated by Muslim scholars of various stripes.

Third, despite taking the form of a systematic approach to the question of how to understand the scripture, Walī Allāh's Qurʾānic hermeneutics is clearly based on the idea of a divinely bestowed understanding (*wahbī*) of the Qurʾān. What he emphasizes from the beginning of his work was that understanding the text of the Qurʾān is actually a special gift from the Almighty and that he has been graced with this ability. Walī Allāh stresses that it is only through divine grace that a reader can understand the Qurʾān as he should.[22] The implication of this feature of his Qurʾānic hermeneutics is that his emphasis on the *wahbī* nature of Qurʾānic understanding leaves open the possibility that a particular verse may provide a class of selected believers with some kind of access to what Allāh actually means. This implies that they may be eligible for divine guidance in matters of interpretation. This thinking has had a direct and an indirect impact on new ways of establishing authority in interpretive matters.

The above three aspects of Walī Allāh's approach to the Qurʾān can be seen as the larger theoretical framework within which we can attempt to make sense of his complex Qurʾānic hermeneutics. The main question about Walī Allāh remains, namely: What exactly was Walī Allāh's opinion on the correct method of interpreting the Qurʾān? We can draw certain conclusions from *al-Fawz al-kabīr* in order to answer

---

21 Andrew Rippin explains that by *asbāb an-nuzūl*, "we mean reports, transmitted generally from the Companions of Muḥammad, detailing the cause, time and place of the revelation of a portion (usually a verse) of the Qurʾān." Andrew Rippin, "Occasions of Revelation," in *EQ*, ed. Jane Dammen McAuliffe, accessed May 31, 2021 from https://referenceworks.brillonline.com/entries/encyclopaedia-of-the-quran/occasions-of-revelation-EQSIM_00305.

22 Walī Allāh, *al-Fawz al-kabīr*, 145f.

this question. It appears that his aim in this short tract was, in fact, a step toward his larger project of putting the Qur'ān at the center of Muslim reform, although he never articulated that project in any clear detail. Despite his focus on the Qur'ān, he continued to reconcile the Qur'ān with other sources from the Muslim tradition, such as *sīra*, hadith, and *asbāb an-nuzūl*. His own tract *Fatḥ al-Khabīr* is a proof of this.[23] *Fatḥ al-Khabīr* is a short piece of writing in which Walī Allāh compiles important exegetical hadiths and *asbāb an-nuzūl* for some selected Qur'ānic verses. But despite the equivocality of his articulation, Walī Allāh, as a figure who had considerable influence on the late nineteenth century, definitively and vocally sets a somewhat new precedent of revisiting the use of unreliable materials by earlier generations of exegetes and *sīra* writers, such as Muḥammad b. 'Umar al-Wāqidī. This has even led some modern academic observers to place him in the category of the earliest modernists of Muslim scholarship.[24] This trend of dismissing the record of early exegetical materials in the Muslim tradition reached its logical climax at the turn of the twentieth century. A good example of this is Shiblī Nu'mānī, a renowned British Indian Muslim theologian and historian of the late nineteenth and early twentieth centuries. Nu'mānī criticized the very basis of the Muslim exegetical tradition in his rebuttal of the earliest premodern *tafsīr* works, such as those by Mujāhid b. Jabr, Ismā'īl b. 'Abd ar-Raḥmān as-Suddī, and Muqātil b. Sulaymān, on whose reports later generations of exegetes relied.[25]

Moreover, Walī Allāh openly admits the complexity of the task facing any interpreter of the Qur'ān and acknowledges the apparent problems posed by the text, particularly its repetitions, breaks, and philological challenges. He refers to the *naẓm* ("coherence") of the Qur'ān in *al-Fawz al-kabīr*, but he does not use the term *naẓm* to establish the text's coherence along the lines of premodernists in the South Asian exegetical tradition like 'Alī b. Aḥmad al-Mahā'imī (d. 1431/835)[26] or in the manner of later generations of Deobandī *mufassirūn* ("commentators"). The concept of *naẓm* in Walī Allāh is related to the clear meaning of the Qur'ān or the relationship between a Qu-

23 For the text of *Fatḥ al-Khabīr*, I have used Shāh Walī Allāh, *al-Fawz al-kabīr fī uṣūl at-tafsīr: Ma' Fatḥ al-Khabīr* (Delhi: Maṭba'-yi Muḥammadī, n.d.). The work *Fatḥ al-Khabīr* was completed around 1756. Baljon, *Shāh Walī Allāh Dihlawī*, 12.

24 See Pink, "Striving for New Exegesis," 3–8.

25 Shiblī Nu'mānī, "'Ulūm al-Qur'ān," in *Maqālāt-i Shiblī*, ed. Sayyid Sulaymān Nadwī (Azamgarh: Dār al-Muṣannifīn, 1999), 1: 33.

26 Al-Mahā'imī, who flourished in fifteenth-century Gujarat, is a concrete specimen of a medieval thinker who attempted to chart new exegetical ground, but within the limits of traditional resources of Qur'ānic exegesis. In his *Tabṣīr ar-Raḥmān wa-taysīr al-mannān*, al-Mahā'imī offered a detailed technique of making connections between verses of the Qur'ān and presented a somewhat developed concept of the unity of the sura. In other words, his was a sophisticated attempt to read the Qur'ān as a coherent text.

r'ānic word or phrase and its exact and understandable meaning.[27] Thus, Walī Allāh claims that his hermeneutical principles could lead to a clear understanding of Qur'ānic vocabulary and verses. In other words, committing to his linguistic principles, he says, will reveal the philological *naẓm* or clarity of the Qur'ān, including its eloquence[28]:

> It should be known that the Qur'ān was revealed in the true idiom language of the Arabs. (Those Arabs) possessed the capacity to appreciate (the Qur'ānic diction) and understood the implied meanings without any difficulty, as God has said about the Qur'ān, "And the clear Book," "The Qur'ān in Arabic, so that you may understand," and "Its verses are unambiguous and clear." The aim of the Law-Giver was to avoid delving into the interpretation of allegorical verses (*al-mutashābihāt*) of the Qur'ān, in drawing the picture of the truth of Allāh's attributes, in ascertaining the ambiguous, and in the telling of the stories in (their) minuteness. That was the reason why the Companions asked the Prophet very few questions; very little is traceable (to them) in this connection. But when that era (of the Companions) ended and the non-Arabs (*al-ʿajam*) mixed up with the Arabs, the original language became obsolete (. . .). The method used to overcome this difficulty is to accept the meaning (of the Qur'ān) that was understood by the Companions, their immediate followers, and other men of learning.[29]

It is important to note that Walī Allāh had a particular approach to solving the complexity involved in interpreting the Qur'ān. Two features of his method stand out. First, like many other Muslim interpreters, he emphasized the indispensability of understanding the Arabic rhetoric and idiom as employed in the genres of pre-Islamic Arabic. For example, the Qur'ānic use of ellipsis (*ḥadhf*) could be understood by an interpreter only if he was familiar with this particular aspect of pre-Islamic writing genres. Moreover, he disagreed with the high frequency of *naskh* in the Qur'ān, a notion that is visible in the history of Qur'ānic thought in Muslim intellectual tradition. Walī Allāh opined that a careful philological and contextual analysis of the Qur'ān could reduce *naskh* instances to as low a number as only five. In all other instances, the Qur'ānic text could be explained meaningfully without considering the relevant verses as abrogated:

> A great reason for this difficulty (of settling the questions of *naskh* in the Qur'ān) is the difference between the opinion of the ancients (*al-mutaqaddimīn*) and the moderns (*al-muta'akhkhirīn*) about the technical meaning of the word *naskh*. What we know from the research on the (use of) the language of the Companions and the Followers is that they used the word "abrogation" in its very grammatical sense, that is, "the replacement of one thing by another," and not in the sense taken by the people who developed a certain terminology (around the Qur'ān).[30]

---

27 On the question of how premodern Muslim scholars thought about the concept of *naẓm*, see Mustansir Mir, *Coherence in the Qur'ān: A Study of Iṣlāḥī's Concept of Naẓm in* Tadabbur-i Qur'ān (Indianapolis: American Trust Publications, 1986).
28 Walī Allāh, *al-Fawz al-kabīr*, 53, 89, 103.
29 Ibid., 54f.
30 Ibid., 57.

Second, for Walī Allāh, the key to an authentic understanding of the Qur'ān, in addition to God's grace, lies in the sayings of the early generations of Muslims, which included the Prophet's Companions, Successors, and Successors to the Successors. However, in his approach to the Qur'ān, Walī Allāh proves to be a thinker of considerable complexity. He stresses that in order to understand Qur'ānic vocabulary and the other terms used in the traditional disciplines of the Qur'ānic sciences, it is important to understand them the way they were understood by the earliest generations of Muslims. This feature of his Qur'ānic hermeneutics has led him to be interpreted in a number of ways. The case of *asbāb an-nuzūl* provides us with a representative example, also highlighting the nature of assumptions that observers of modern *tafsīr* carry with them about premodern Qur'ānic hermeneutics. Premodern Qur'ān commentaries are generally understood to have employed *asbāb* materials uncritically. In contrast, the prevailing perception is that there was a discernible shift toward a more cautious approach in the modern period. In this regard, Walī Allāh's Qur'ānic thought is routinely analyzed as an example of an obvious disparaging of *asbāb* material.[31] This assumption is problematic on several counts. The very understanding of what is meant by *asbāb* seems to have been a controversial issue in the Muslim scholarly tradition right up to the present day. Therefore, if Walī Allāh was seen as rethinking the meaning and importance of *asbāb* in the interpretation of the Qur'ān, it is difficult to consider his thinking as marking a necessary shift toward a modern interpretation of the scripture. There are two reasons for this: First, Walī Allāh engaged with a long tradition of understanding the nature of *asbāb an-nuzūl*, voices of which can be heard in premodern works of *'ulūm al-Qur'ān* (the traditional disciplines of the Qur'ānic sciences)[32] such as Abū l-Qāsim az-Zarkashī's *al-Burhān* and Jalāl ad-Dīn as-Suyūṭī's *al-Itqān*. As Az-Zarkashī writes,

> It is known as the common practice of the Companions and their Successors that when one of them says that the verse was revealed about a given matter (*nazalat hādhihi al-āya fī kadhā*), he means that the verse encompasses (or deals with) that matter or injunction; not that the matter was the reason for its revelation.[33]

Az-Zarkashī's statement makes it clear that the very definition of what is meant by *sabab* was a controversial issue in the Muslim tradition of his day. Walī Allāh appropriates az-Zarkashī's understanding without directly mentioning him. In addition, the

---

31 Pink, "Striving for a New Exegesis," 3–6.

32 For an introduction to the development of the Muslim sciences based on the Qur'ān, see Claude Gilliot, "Traditional Disciplines of Qur'ānic Studies," in *EQ*, ed. Jane Dammen McAuliffe, accessed May 31, 2021 from https://referenceworks.brillonline.com/entries/encyclopaedia-of-the-quran/traditional-disciplines-of-quranic-studies-EQCOM_00207. The discussions on *asbāb an-nuzūl* form an important part of *'ulūm al-Qur'ān* works.

33 Badr ad-Dīn Muḥammad b. 'Abd Allāh az-Zarkashī, *al-Burhān fī 'ulūm al-Qur'ān*, ed. Muḥammad Abū al-Faḍl Ibrāhīm (Cairo: Dār al-Ḥadīth, 2006), 33. Walī Allāh revived the discussion anew around az-Zarkashī's Arabic phrase of *nazalat hadhihi al-āya fī kadhā*.

modern, "established opinion" in the academy that Walī Allāh considered *asbāb* as having no or less value for *tafsīr* is in itself a matter of contention.[34] This dispute throws light on how an early scholar like Walī Allāh was received and interpreted in later times. The issue of the multiple interpretations of Walī Allāh was demonstrated when one of the most notable scholars in the South Asian Deoband tradition, Muḥammad Taqī ʿUthmānī (b. 1943/1362), was asked about Walī Allāh's opinions on *asbāb*'s role in the interpretation of Qurʾānic verses. Reminding us of the additional aspects of Walī Allāh's position on *asbāb*, ʿUthmānī wrote the following:

> The people have not understood correctly what Shāh Walī Allāh has written about *asbāb an-nuzūl* in his book, *al-Fawz al-kabīr fī uṣūl at-tafsīr* (. . .). The fact of the matter is that he considers *asbāb an-nuzūl* to be *a pre-requisite to interpreting the Qurʾān* (emphasis added). Shāh Ṣāḥib himself notes: "The knowledge of two things is strictly necessary for a commentator to know. One is those events that are alluded to in the verses. We cannot understand those verses unless we know those events. The other is the fact that in some (Qurʾānic) stories, the words used are too generic in nature. With the help of *shān-i nuzūl* (occasions of revelation), these words acquire specific meaning; or verses sometimes have an apparent (*ẓāhir*) meaning but *asbāb an-nuzūl* determine a different meaning. Without the knowledge of these *asbāb* traditions, it is difficult to understand Qurʾānic verses."[35]

All of the above discussions about Walī Allāh's Qurʾānic hermeneutics suggests that his approach to understanding the Qurʾān had the potential to provide guidelines for how to pursue different avenues in the field of Qurʾānic interpretation. Consequently, his Qurʾānic thought could be a source of inspiration for even two opposing Muslim schools of thought. This meant that Walī Allāh was held in high esteem by all. The rich history of Muslim exegetical thought in British India corroborates this feature of his Qurʾānic thought. Walī Allāh functioned as a patron scholar for many competing schools of Muslim exegesis.

34 Walī Allāh discussed his position on *asbāb* material in one of his seminal Persian tracts that dealt with principles of Qurʾānic interpretation. Translations of the text in Arabic and Urdu seemed to become more popular with the passage of time in South Asia. The Arabic translation used in the current study is Walī Allāh, *al-Fawz al-kabīr*, trans. al-Bālanfūrī. Baljon places this work by Walī Allāh in his writings in the period after his return from the Hijaz in 1732/1144. An earlier Arabic translation is attributed to Muḥammad Munīr ad-Dimashqī. See Baljon, *Shāh Walī Allāh Dihlawī*.

35 Muḥammad Taqī ʿUthmānī, *ʿUlūm al-Qurʾān*. Originally published in 1976. Karachi: Maktabat Dār al-ʿUlūm, 2010), 79–82. In the above quote, ʿUthmānī was referring to Walī Allāh, *al-Fawz al-kabīr*, 73f.

4 The Reception and Influence of Walī Allāh's Qur'ānic Hermeneutics

We have already mentioned that Walī Allāh's influence during his lifetime and the period since his death is a complex question. Sweeping claims should be avoided. While Walī Allāh's sons followed in his footsteps in their works on the Qur'ān,[36] we only see a clear borrowing from and engagement with his Qur'ānic thought from the latter part of the nineteenth century. From the second half of the nineteenth century onward, there was a widespread acceptance of his Qur'ānic thought among most schools of thought in British India. Key Qur'ānic exegetes in British India such as Sayyid Aḥmad Khān (d. 1898/1316), Ashraf ʿAlī Thānawī (d. 1943/1362), and Ḥamīd ad-Dīn Farāhī (d. 1930/1349) demonstrated their admiration of Walī Allāh's work.[37] Most schools, whether labelled traditionalist, modernist, or fundamentalist, seemed to embrace Walī Allāh as the authoritative representative of their hermeneutical principles. These groups also included those who went down the path of political activism, like ʿUbayd Allāh Sindhī (d. 1944/1363). Aḥmad Khān actively engaged with Walī Allāh in his small booklet on Qur'ānic hermeneutics, *at-Taḥrīr fī uṣūl at-tafsīr*. He agreed with him on some points and disagreed on others. Similarly, Thānawī, a patron of Deoband, also consumed the ideas of Walī Allāh and his sons in the commentary, *Tafsīr bayān al-Qur'ān*. For Thānawī, there was no need, in principle, for his own Urdu exegesis when there were already available translations and commentaries written by Walī Allāh and his family. Walī Allāh had a noticeable impact on other scholars as well. For example, Ṣiddīq Ḥasan Khān (d. 1880/1297) of the Ahl-i Ḥadīth school wrote his *Iksīr fī uṣul at-tafsīr* as an extension of Walī Allāh's *al-Fawz al-kabīr* and, in many respects, structured his work according to the division of topics in the Qur'ān suggested by Walī Allāh and based on different types of *ʿilm*. In the same vein, we can read another important Qur'ān commentator, Farāhī, as evidently arguing for his

36 Immediately after Walī Allāh, the next major milestone in Qur'ānic exegesis appeared in the shape of the works written by his sons: the Persian commentary *Fatḥ al-ʿAzīz* written by his eldest son, ʿAbd al-ʿAzīz (d. 1824/1239); and *Muḍiḥ al-Qur'ān*, the Urdu translation and exegetical glosses by his son ʿAbd al-Qādir. As seen in their introductions, these commentaries and translations did not chart any special program of introducing new Qur'ānic hermeneutics. For ʿAbd al-ʿAzīz, his major motivation seemed to be to expand on the exegetical project initiated by his father. For the brothers of ʿAbd al-ʿAzīz, in particular ʿAbd al-Qādir, the main reason to pen their work was similar to that of their father, but they chose a different language as their medium of expression. They exemplified the new trend of rendering the Qur'ān into Urdu for a larger audience in Delhi and other Urdu speaking towns, but their translations could also be presented by their followers as "authentic" renderings of the Qur'ān in situations where the Qur'ān was being translated into regional varieties of Urdu such as Awrangābādī, Gujarātī, and Dakkanī.

37 For a detailed study of these Qur'ān commentators, see Kamran Bashir, *The Qur'an in South Asia: Hermeneutics, Qur'an Projects, and Imaginings of Islamic Tradition in British India* (London and New York: Routledge, 2021).

own hermeneutical positions by showcasing Walī Allāh's views on *asbāb an-nuzūl*. Farāhī's position on the occasions of revelation was somewhat similar to Walī Allāh's, though Farāhī developed it further.

There are three main features of the reception history of the literary output produced in the tradition of Walī Allāh and his family. First, their translations motivated others to render the divine Word in various vernacular languages, not just Urdu. This historical moment can be interpreted as marking the beginning of the dilution of an Arabic tradition of exegesis in South Asia and the launching of some kind of "Protestantization" of the Qur'ānic text. Moreover, the opportunity to translate the Qur'ān that now presented itself paved the way for controversies surrounding what it meant to translate the scripture authentically. Many religious scholars of various post-1857 schools of thought, notably including the Dār al-ʿUlūm at Deoband, considered the translations by Walī Allāh's family to be a benchmark for the "correct" translation of the Qur'ān. Finally, in the imagined Walī Allāhī tradition, authors such as Thanā Allāh Pānīpatī (d. 1810/1125) were seen by traditionalists as "custodians" of an "authentic" premodern exegetical tradition even though they occasionally introduced innovations. The impact of the Walī Allāhī tradition seems to have been more pronounced in the way it has been used to seek support in South Asian inter-school polemics than in serving as a concrete basis to develop a sound methodology to interpret the Qur'ān. Ultimately, Walī Allāh's impact becomes a question of both sociology and epistemology.

5 Conclusion

The Qur'ānic hermeneutics envisaged by Walī Allāh enabled modern Muslim scholars of various schools of thought to come to grips with the intellectual modernity emerging around them. His thoughts on the Qur'ān were universally appropriated because his exegetical principles can be interpreted in a number of ways. A representative case in this regard is the issue of *asbāb an-nuzūl*. Secular and so-called modernist Muslim scholars read his work as dismissing such occasions of revelation. By contrast, Deobandi scholars interpret him differently on this issue. Any two sides in a Muslim polemic on the Qur'ān can draw upon him to derive support for their particular positions. Walī Allāh's own approach to the Qur'ān was not based on dismissing traditional resources on the interpretation of the Qur'ān. His aim was to resolve conflicts among Muslims regarding the understanding of the Qur'ān. This trend toward uniting Muslims around the Qur'ān continued to grow in the nineteenth and twentieth centuries.

6 Bibliography

6.1 Sources

Walī Allāh, Shāh. *Al-Inṣāf fī bayān sabab al-ikhtilāf*. Cairo: Muḥibb ad-Dīn al-Khaṭīb, 1965.

Walī Allāh, Shāh. *At-Tafhīmāt al-ilāhiyya*. 2 vols. Hyderabad: Shāh Walī Allāh Academy, 1967.

Walī Allāh, Shāh. *Al-Budūr al-bāzigha*. Hyderabad: Shāh Walī Allāh Academy, 1970.

Walī Allāh, Shāh. *Musawwā Muṣaffā*. 2 vols. Karachi: Muḥammad ʿAlī Karkhāna-i-Islāmī Kutub, 1980.

Walī Allāh, Shāh. *The Conclusive Argument from God*, trans. Marcia Hermansen. Leiden: Brill, 1996.

Walī Allāh, Shāh. *Ḥujjat Allāh al-bāligha*. Ed. ʿUthmān Jumʿa ad-Dumayriyya. Riyadh: Maktabat al-Kawthar, 1999.

Walī Allāh, Shāh. *Qur'ān-i majīd mutarjam: Tarjama-yi fārsī*. Lahore: Pāk Company, 2004.

Walī Allāh, Shāh. *Al-Fawz al-kabīr fī uṣūl at-tafsīr*, trans. Saʿīd Aḥmad al-Bālanfūrī. Damascus: Dār al-Ghawthānī li-d-Dirāsāt al-Qur'āniyya, 2008.

Walī Allāh, Shāh. *Al-Fawz al-kabīr fī uṣūl at-tafsīr: Maʿ Fatḥ al-Khabīr*. Delhi: Maṭbaʿ-yi Muḥammadī, n.d.

Walī Allāh, Shāh. *Fuyūḍ al-Ḥaramayn*. Karachi: Muḥammad Saʿīd, n.d.

Walī Allāh, Shāh. *Tarjamat al-Qur'ān: Fārsī, al-mawsūm ba-Fatḥ ar-Raḥmān*. Bombay: ʿAlī Bhā'ī Sharaf ʿAlī, n.d.

Az-Zarkashī, Badr ad-Dīn Muḥammad b. ʿAbd Allāh. *Al-Burhān fī ʿulūm al-Qur'ān*.Muḥammad Abū al-Faḍl Ibrāhīm. Cairo: Dār al-Ḥadīth, 2006.

6.2 Secondary Literature

Abbott, Freeland. "The Decline of the Mughal Empire and Shāh Walī Allāh." *MW* 52 (January 1962): 115–23.

Baljon, J.M.S. *Religion and Thought of Shāh Walī Allāh Dihlawī 1703–1762*. Leiden: Brill, 1986.

Bashir, Kamran. *The Qur'an in South Asia: Hermeneutics, Qur'an Projects, and Imaginings of Islamic Tradition in British India*. London and New York: Routledge, 2021.

Brown, Jonathan. *Misquoting Muhammad: The Challenge and Choices of Interpreting the Prophet's Legacy*. London: OneWorld, 2014.

Chaghatai, M. Ikram, ed. *Shah Waliullah, (1703–1762): His Religious and Political Thought*. Lahore: Muḥammad Ashraf, 2005.

Chandra, Satish. *Parties and Politics at the Mughal Court, 1707–1740*. New Delhi/New York: Oxford University Press, 2002.

Dallal, Ahmad S. *Islam without Europe: Traditions of Reform in Eighteenth Century Islamic Thought*. Chapel Hill: The University of North Carolina Press, 2018.

Gilliot, Claude. "Traditional Disciplines of Qur'ānic Studies." In *EQ*, edited by Jane Dammen McAuliffe. Brill, 2001–06. Accessed May 31, 2021, from ttps://referenceworks.brillonline.com/entries/encyclopaedia-of-the-quran/traditional-disciplines-of-quranic-studies-EQCOM_00207.

Hermansen, Marcia. "Shah Wali Allah Bibliography." Last modified 2022. https://www.academia.edu/2163744/Shah_Wali_Allah_Bibliography.

Mir, Mustansir. *Coherence in the Qur'ān: A Study of Iṣlāḥī's Concept of Naẓm in* Tadabbur-i Qur'ān. Indianapolis: American Trust Publications, 1986.

Moosa, Ebrahim. "Introduction." In Shāh Walī Allāh. *Shāh Walī Allāh's Treatises on Juristic Disagreement and Taqlīd*, trans. Marcia Hermansen. Louisville, KY: Fons Vitae, 2010: vii–xxi.

Nuʿmānī, Shiblī. "ʿUlūm al-Qur'ān." In *Maqālāt-i Shiblī*, edited by Sayyid Sulaymān Nadwī. 4 Vols. Azamgarh: Dār al-Muṣannifīn, vol. 1, 1999: 25–33.

Pink, Johanna. "Striving for a New Exegesis of the Qur'ān." In *The Oxford Handbook of Islamic Theology*, edited by Sabine Schmidtke. Oxford: Oxford University Press, 2016.

Rippin, Andrew. "Occasions of Revelation." In *EQ*, edited by Jane Dammen McAuliffe. Brill, 2001–06. Accessed May 31, 2021, from https://referenceworks.brillonline.com/entries/encyclopaedia-of-the-quran/occasions-of-revelation-EQSIM_00305.

Rizvi, Saiyid Athar Abbas. *Shāh Walī-Allāh and his Times: A Study of Eighteenth Century Islām, Politics and Society in India*. Canberra: Maʿrifat Publishing, 1980.

Robinson, Francis. "Islamic Reform and Modernities in South Asia." *Modern Asian Studies* 42, nos. 2–3 (March 2008): 259–81.

ʿUthmānī, Muḥammad Taqī. *ʿUlūm al-Qurʾān*. Originally published in 1976. Karachi: Maktabat Dār al-ʿUlūm, 2010.

Zaman, Muhammad Qasim. *ʿUlama in Contemporary Islam: Custodians of Change*. Princeton: Princeton University Press, 2002.

Zaman, Muhammad Qasim. "Political Power, Religious Authority, and the Caliphate in Eighteenth-Century Indian Islamic Thought." *JRAS* 30, no. 2 (2020): 313–40.

Aḥmad b. ʿAjība

Ruggero Vimercati Sanseverino

1 Introduction

Since the work of Jean-Louis Michon,[1] the Moroccan scholar and Sufi Aḥmad b. ʿAjība al-Ḥasanī (1746–1809/1159–1223) is known to historians of Muslim spirituality as an eminent representative of the Shādhiliyya-Darqāwiyya order and as the author of still nowadays widely read treatises on Sufism. However, his most substantial work is a complete exegesis of the Qur'ān, *al-Baḥr al-madīḍ fī tafsīr al-Qurʾān al-majīd* (The extended Ocean in the Commentary of the Glorious Qur'ān),[2] which combines the expressive (*ʿibāra*) or religious with the allusive (*ishāra*) or spiritual interpretation and represents a synthesis of his vast knowledge of Islamic sciences and his experience as a spiritual master. A second exegetical work is his *Tafsīr al-Fātiḥa al-kabīr* (The great commentary of The Opening), which presents itself as a handbook of Islamic doctrine and practice from a spiritual perspective structured according to the themes of the sura.

2 Life, Works, Political and Intellectual Context

Born in 1747–48/1160–61 in the region of Tetouan in northern Morocco, Aḥmad b. ʿAjība[3] became one of the most respected religious scholars in his homeland. According to his

1 Aḥmad b. ʿAjība al-Ḥasanī, *Le Soufi Marocain Aḥmad Ibn ʿAjība et son Miʿrāj*, trans. Jean-Louis Michon (Paris: J. Vrin, 1990); ders, *Deux Traités sur l'Unité de l'Existence*,, trans. Jean-Louis Michon (Marrakech: Al Qoubba Zarqua, 1998); ders., *L'autobiographie (Fahrasa) du Soufi Marocain Ahmad Ibn ʿAgība*, trans. Jean-Louis Michon (Milan: Arché, 1982).

2 I use the edition of Dār al-Kutub al-ʿIlmiyya edited by ʿUmar Aḥmad al-Rāwī in eight volumes in 2002, which includes references to hadith and seems reliable all in all despite occasional omissions. The Cairo edition of 1998 in 6 volumes by Ḥasan ʿAbbās Zakkā is difficult to find. Some extracts of the work are translated in Aḥmad b. ʿAjība al-Ḥasanī, *The Immense Ocean: Al-Bahr Al-Madid. A Thirteenth Century Quranic Commentary on the Chapters of the All-Merciful, the Event, and Iron*, trans. Mohamed F. Aresmouk and Michael A. Fitzgerald (Louisville, Ky: Fons Vitae, 2009). For a thematic study of the *tafsīr*, see Ruggero Vimercati Sanseverino, "Commentaire coranique, enseignement initiatique et renouveau soufi dans la Darqāwiyya. Le *Baḥr al-madīd fī tafsīr al-Qurʾān al-majīd* d'Aḥmad Ibn ʿAjība (m. 1223/1809)," *Studia Islamica* 107, no. 2 (2012): 209–34, and the PhD research, Mohamed Fadel Ebada Elsayed, "Le commentaire du Coran par le soufi marocain Aḥmad Ibn ʿAjība (1160–1224/1747–1809) – Méthodes et themes" (PhD thesis. Strasbourg, 2020).

3 See Tayeb Chouiref, "Ibn ʿAjība," in *EI*[3], ed. Kate Fleet et al., accessed on October 27, 2023, http://dx.doi.org/10.1163/1573-3912_ei3_COM_30684.

https://doi.org/10.1515/9783111320083-023

autobiography,[4] he experienced in his forties a spiritual crisis that led him to abandon his duties as a professor of religious sciences and enter the Sufi path under the direction of Muḥammad al-Būzīdī[5] (d. 1814/1229) and his master, the founder of the Shādhiliyya-Darqāwiyya, Mawlāy al-ʿArabī ad-Darqāwī[6] (d. 1823/1239). After a time of trial, Ibn ʿAjība became himself a major spiritual master of the order and a prolific author.

The end of the eighteenth century marks the foundation of the great Moroccan Sufi orders,[7] notably the Ṣaqalliyya, the Darqāwiyya, the Tijāniyya, later the Kattāniyya, and corresponds to a period in which Sufism underwent a harsh treatment from the political authorities. The Sultan Muḥammad b. ʿAbd Allāh (1757–1790/1170–1204/) seeks to stabilize the Alawite power by claiming his legitimacy as a religious leader and initiating a reform of Islamic education. As a learned sovereign, he claims an authority superior to that of the *ʿulamāʾ*, an authority that allows him to eliminate the study of Sufi texts from the religious curriculum and denounce forms of Sufism that go beyond what he considers "moderate Sufism". The Sufi circles' reaction is formally heterogeneous, but a certain resentment of a sclerosis of the spiritual life seems to be generally widespread, not unique to Morocco. The foundation of new orders testifies to the will of Sufi circles to renew the organizational and doctrinal framework of their teaching and practice in the face of the socio-religious changes that began with the weakening of the political structures of the Muslim world. However, for the Sufis, the challenge is not in social or political terms, but rather to responds to what is felt to be the weakening of spiritual aspirations and moral decay. Therefore, the renewal's main focal point was to attenuate the antagonistic multitude of legal traditions and Sufi orders by reunifying and concentrating various aspects of Islamic tradition through the formulation of a spiritual conception of its prophetic reference.

The writings of Ibn ʿAjība can be distinguished between those concerning conventional Islamic sciences only which he wrote before his engagement in Sufism, and those which are written from a Sufi perspective. These latter writings constitute the

4 *L'autobiographie (Fahrasa).* For his biography, see Florian A. G. Lützen, *Sufitum und Theologie bei Aḥmad Ibn ʿAjība* (Tübingen: Mohr-Siebeck, 2020), 19–46.

5 Cf. Muḥammad b. Muḥammad al-Mahdī at-Tamsamānī, *al-Imām Sayydī Muḥammad b. Aḥmad al-Būzīdī – tarjumatuhu wa-baʿḍ āthāruhu* (Beirut: Dār al-Kutub al-ʿIlmiyya, 2007).

6 Cf. J. Spencer Trimingham, *The Sufi Orders in Islam* (Oxford: Clarendon Press, 1971), 110–14; Mostafa Zekri, "La tarîqa Shādhiliyya-Darqāwiyya – Les empreintes du shaykh al-ʿArabī al-Darqāwī," in *Une voie soufi dans le monde: la Shādhiliyya*, ed. Éric Geoffroy (Paris: Maisonneuve & Larose, 2005), 229–36; Abdelbaqî Meftah, "L'initiation dans la Shādhiliyya-Darqāwiyya," in *Une voie soufi dans le monde: la Shādhiliyya*, ed. Éric Geoffroy (Paris: Maisonneuve & Larose, 2005), 237–48; Itzchak Weismann, "The Shâdhiliyya-Darqâwiyya in the Arab East," in *Une voie soufi dans le monde: la Shādhiliyya*, ed. Éric Geoffroy (Paris: Maisonneuve & Larose, 2005), 255–70; Frederick De Jong, "Materials relative to the History of the Darqâwiyya Order and its Branches," *AR* 26, no. 2 (1979): 126–43.

7 Cf. Aḥmad Būkārī, *al-Iḥyāʾ wa-t-tajdīd fī l-Maghrib*, 3 vols. (Rabat: Wazārat al-Awqāf wa-sh-Shuʾūn al-Islāmiyya, 2006).

fame of its author and are regularly republished and have found wide distribution throughout the Maghreb and the Middle East. These writings have to be considered within the context of the revivification of Sufism, particularly in its Maghrebian Shādhilite form. Consequently, most of them are commentaries (*sharḥ*) on the classical texts of the Shādhilite tradition.[8] Some of his writings have gained considerable popularity, such as the *Miʿrāj at-tashawwuf ilā ḥaqāʾiq at-taṣawwuf*,[9] a treatise on Sufi terminology, or the *Īqāẓ al-himam*, his commentary on the *Ḥikam* of Ibn ʿAṭāʾ Allāh al-Iskandarī. However, Ibn ʿAjība is also known for another category of writings that reflect his background and the particularity of his approach, in particular in Qurʾānic exegesis, namely those related to "the science of spiritual allusion" (*ʿilm al-ishāra*), which involves interpreting texts of Islamic science according to a spiritually experienced understanding. Concerning Quranic exegesis, Ibn ʿAjība authored a complete commentary, *al-Baḥr al-madīd fī tafsīr al-Qurʾān al-majīd* (The extended Ocean in the Commentary of the Glorious Quran), and three commentaries of sura 1, *al-Fātiḥa* (The Opening), of variable length of which the largest is published as *Tafsīr al-Fātiḥa al-kabīr* (The great commentary of the *Fātiḥa*).[10] Since his master ad-Darqāwī left only one collection of letters,[11] Ibn ʿAjība's *tafsīr* represents not only one of the most significant sources for the study of this Sufi order but also the doctrinal culmination of a whole spiritual tradition whose history begins in Morocco in the fourteenth century with Ibn ʿAbbād ar-Rundī and then in the fifteenth century with Aḥmad Zarrūq and Muḥammad b. Sulaymān al-Jazūlī. Like his predecessors, the author of the *Baḥr al-madīd* is also committed to the apologetics of Sufism while more explicitly asserting the superiority of the Sufi magisterium and thus the pre-eminence of "interior knowledge" (*ʿilm al-bāṭin*) over any other form of religious knowledge[12] while at the same time affirming the necessity of both.

---

8 Among those published: Aḥmad b. ʿAjība al-Ḥasanī, *Iqaz al-Himam fī Sharḥ al-Ḥikam*, ed. Khalīl Manṣūr (Beirut: Dār al-Kutub al-ʿIlmiyya, n.d.); *Kitāb Sharḥ Ṣalāt al-Quṭb bin Mashīsh*, ed. ʿAbd as-Salām al-ʿUmrānī al-Khālidī (Casablanca: Dār ar-Rashād al-Ḥadītha, 2003); ders., *Lawāqiʿ al-Qudusiyya fī Sharḥ al-Waẓīfa az-Zarrūqiyya. al-Jawāhīr al-ʿAjībiyya*, eds. by ʿAbd as-Salām al-ʿUmrānī al-Khālidī. Beirut: Dār al-Kutub al-ʿIlmiyya, 2004.

9 The work is translated from Michon in Ibn ʿAjība, *Le Soufi Marocain*.

10 Aḥmad b. ʿAjība al-Ḥasanī, *Tafsīr al-Fātiḥa al-kabīr*, ed. ʿĀṣim Ibrāhīm al-Kayyālī (Beirut: Dār al-Kutub al-ʿIlmiyya, 2004).

11 *Majmūʿa ar-rasāʾil Mawlāy al-ʿArabī ad-Darqāwī al-Ḥasanī*, ed. Bassām Muḥammad Bārūd (Abu Dhabi: al-Mujammaʿ ath-Thaqāfī, 1999).

12 See Kenneth Honerkamp, "Ibn ʿAbbad, Model of the Shādhiliyya," in *Une voie soufi dans le monde: la Shâdhiliyya*, ed. Éric Geoffroy (Paris: Maisonneuve & Larose, 2005): 159–71.

3 Ibn ʿAjība's Hermeneutical Method and the Analysis of his Major Exegetical Works

Ibn ʿAjība explains his hermeneutical method in various works, which shows how central the relationship between the exterior and the interior, the exoteric dimension of Islam represented by divine law while the esoteric dimension related to divine reality, is for his thought. The two levels of commentary, *ʿibāra* (clear expression) and *ishāra* (allusion), are part of what Ibn ʿAjība in the introduction to his Qurʾānic commentary calls *tafsīr ahl aẓ-ẓāhir*, "the exegesis of the people of the exterior", in contrast to "the exegesis of the people of the interior", the *tafsīr ahl al-bāṭin*. According to Ibn ʿAjība, the latter mode of exegesis "transcends the categories of reason" and "is not apprehended through the continuous transmission of conventional teachings."[13] Ibn ʿAjība severely criticizes those who seek to limit the meaning of the Qurʾānic text to the literal and historical, but he also firmly affirms the need for an exoteric and explicatory exegesis without which the internal meaning must not be approached. If exoteric exegesis is relevant for both conventional scholars and Sufis, only the latter can undertake the interior and spiritual exegesis; as Ibn ʿAjība explains, it is a commentary drawing its explanations from the realization (*taḥqīq*) "of spiritual experience (*dhawq*), states (*ḥāl*) and stations (*maqām*), as a result of the company of accomplished men who themselves possess this experience"[14] of which only the saints are gratified.[15]

As for the meaning that the author gives to *ʿibāra* and *ishāra*, it is in his commentary on the *Ḥikam* that we find these precisions reported and translated by J.-L. Michon: "The expression explains, the allusion indicates (. . .). The Sufis have said: 'Our science is all allusion; when it becomes expression, it becomes occulted'. (. . .) In fact, allusions are understood only by those initiated in this art (*ahl al-fann*)."[16] Ibn ʿAjība widely remains in the tradition of its predecessors as far as the definition of *ʿibāra* and *ishāra* is concerned. What is interesting is the systematic way in which Ibn ʿAjība applies it. He explains three modes of articulating spiritual meanings, leading him to classify four levels of understanding of the allusion. Here *ishāra* takes an intermediate place between clear expression (*ʿibāra*) and symbol (*ramz*), the first being understood by everyone and the second being understood only by those directly concerned. As for the *ishāra*, for Ibn ʿAjība, the lowest level of understanding corresponds to those who only grasp the literal meaning of the allusion, i.e., they do not perceive the allusion as an allusion, that is, as a verbal indication of a meaning that is inexpressible in words. But for the beginners in the spiritual path who are still subject to the experience of distance and separation from the divine presence, the allusion acts as a

13 Ibn ʿAjība, *Le Soufi Marocain*, 108.

14 Ibid., 108.

15 Ibn ʿAjība, *Baḥr*, 1: 108.

16 Ibn ʿAjība, quoted in Ibn ʿAjība, *Le Soufi Marocain*, 105–7.

vehicle carrying them into the experience of this presence. Those, on the other hand, whose intimate being has been extinguished in the divine essence are closer to the object of the allusion, that is God, than the allusion is itself so that they no longer need the allusion, for "he who makes the allusion has identified with Him to whom the allusion is made."[17]

In this perspective, the *tafsīr ishārī* has the function of a hermeneutical key, opening the doorway to the experience of the spiritual truths of the Qurʾānic verses. As the allusion claims to be based on direct experience, not on discursive thought, it allows especially those who are not yet firmly established in the contemplation of the divine presence to apprehend this presence through the divine discourse. Ibn ʿAjība's exegesis is truly intended as a work of spiritual pedagogy that seeks to show to the aspirant how to experience the Qurʾān as God's educating and guiding Self-disclosure. As Michon explains, "his point of view is truly initiatory (*initiatique*): It is that of a master educator accustomed to taking souls at the beginning of the path and leading them towards what they are capable of attaining, without forcing their capacities."[18] The *ishārī* commentary, by transposing the historical dimension of the sacred text and relating it to the reader's spiritual experience, aims to actualize the revelation; the reader is invited to relive the Qurʾānic discourse as being addressed to himself and as alluding to his own spiritual becoming. It is important to understand that the *ishāra* does not claim to articulate an ultimate inner meaning of the Qurʾān, since this meaning is considered to be inexpressible and, at the same time, gives rise to unlimited possibilities of understanding. Hence the necessity of the allusion, *al-ishāra*, which articulates the exegetes' experience of this inner meaning and indicates to the reader a way to accede himself to the spiritual dimension of the Qurʾān.

3.1 His Tafsīr: *Al-Baḥr al-madīd fī tafsīr al-Qurʾān al-majīd*

Ibn ʿAjība's complete Qurʾānic commentary, which proceeds by thematic passages of one or more verses, is presented on three levels, introduced respectively by "I have said" (*qultu*), "God the True says, exalted be His majesty" (*yaqūlu al-Ḥaqq jalla jalāluh*) and "the allusion" (*al-ishāra*). The first level concerns the linguistic aspect and is not always mentioned; the second relates to the literal meaning of words, sacred history, and the commentaries and explanations of the exegetical tradition. The third level concerns spiritual allusions. The first and second parts form the exoteric commentary and the third the esoteric commentary. In his introduction, Ibn ʿAjība states, in interpreting the prophetic word in relation to the exterior and interior of the Qurʾān, that:

17 Ibid., 105.
18 Ibid., 147.

> The exterior (*aẓ-ẓāhir)* concerns the one who studies the exterior of the word, like grammarians and people of language and morphology; the interior *(al-bāṭin)* regards those interested in the meaning of the word, the discourse on order and prohibition, prophetic stories, the doctrine of divine oneness and other subjects that are part of the Qur'ānic sciences; this is the point of view of exegetes (*al-mufassirūn*). The limit (*al-ḥadd*) concerns those who extract the normative rules, i.e., the scholars of sacred law (*al-fuqahā')* (. . .). The point from which one rises (*al-muṭṭala'*) concerns the people of spiritual truths *(ahl al-ḥaqā'iq)* among the great Sufis, for they rise *(yaṭla'ūna)* from the outside of the verse to its inside.[19]

Thus, we see that Ibn ʿAjība distinguishes between four levels, but that linguistic analysis and semantic commentary (*maʿnā l-lafẓ*) are the two correlative aspects of the same thing.

According to Ibn ʿAjība, the three levels correspond to the three dimensions of the word, that is to say, the one that concerns its form, then its literal meaning, and then its inner meaning. It remains to be pointed out that the author gives great importance to this triple dimensionality in all his work because it is a distinction that corresponds to the three levels of the human being (*jism*/body, *nafs*/soul, *rūḥ*/spirit), of worlds (*mulk*/visible dominion, *malakūt*/dominion of transcendence, *jabarūt*/dominion of omnipotence), of religion (*islām*/surrender, *imān*/faith, *iḥsān*/excellence or *sharīʿa*/law, *ṭarīqa*/path, *ḥaqīqa*/reality*)*, of certitude (*ʿilm al-yaqīn, ʿayn al-yaqīn, ḥaqq al-yaqīn*/science, vision and truth of certainty), etc.

3.1.1 The Exoteric and Explanatory Exegesis

In his *tafsīr* Ibn ʿAjība specifies at the beginning of each sura in which of the two periods, traditionally described as Mecca and Medina by Muslim authors, the sura was revealed, before or after the Hegira, but without citing the sources. Precise historical information, such as the names of the people involved, seems not of great relevance to Ibn ʿAjība, as it is more of explanatory commentary. What interests him is to make people understand what the literal text conveys as an explanation of the events that took place between the Prophet and the community to which he was sent. It presents less the historical aspect of these events and its actors than its meaning directly related to the human experience. In fact, it seeks to present the sacred history of revelation as a vehicle for ethical teaching. In this is exemplified the pedagogic personality of Ibn ʿAjība, who seeks with his *tafsīr* not to teach the religious sciences in order to train scholars in Islamic law, but rather to transmit an enriching understanding of the exoteric aspect of Islam, capable of giving access to more interior meanings and experiences.

Ibn ʿAjība identifies a leitmotiv or thematic threads in each sura which are for him in continuity with the sura that precedes and follows. Thus, he specifies at the

---

19 Ibn ʿAjība, *Baḥr*, 1: 19.

beginning of each sura the "relationship" (*al-munāsaba*) of the verse announcing the leitmotif with the previous sura. For Ibn ʿAjība, the coherence of the Qurʾān is expressed by a certain fundamental continuity that is discernible at the literal level. His use of the method of *munāsaba* pursued an exegetical tradition going back to Fakhr ad-Dīn ar-Rāzī (d. 1210/606), Abū Ḥayyān al-Gharnāṭī (d. 1344/744), and Badr ad-Dīn az-Zarkāshī (d. 1391/793). Another element that indicates the author's concern for the presentation of the Qurʾānic text as a coherent whole is the link he establishes between the different passages into which he has divided the sura and according to which he structures his commentary. By introducing them with a "*thumma fassara/dhakara/ . . .*". (lit. "then He (God) explains/reminds/ . . .") he emphasizes the continuity with the previous passage. For Ibn ʿAjība, these arguments follow one another according to explicit rhetoric and wisdom that is not arbitrary. In fact, it is the task of the commentator to unveil this argumentative structure and succession of Qurʾānic discourse.

The linguistic remarks of Ibn ʿAjība often aim at clarifying difficult words and passages that surpass the level of classical Arabic known by non-specialists. In his commentary, the grammatical analysis of the Qurʾānic text has an explanatory value that highlights the multiplicity of meanings and makes it possible to broaden the spectrum of meanings that the text can convey. Concerning the Qurʾānic commentaries used by Ibn ʿAjība,[20] the most frequently cited *tafsīr* in the exoteric commentary on a wide variety of subjects such as linguistics, jurisprudence, prophetic words, the circumstances of revelation seems to be the *Anwār at-tanzīl wa-asrār at-taʾwīl* from the theologian Abū l-Khayr al-Bayḍāwī (d. 1316/715) who relies on rational exegesis in the tradition of Abū l-Qāsim az-Zamakhsharī (d. 1144/538) and Fakhr ad-Dīn ar-Rāzī. The very style of *tafsīr iʿtibārī* in *al-Baḥr al-madīd* is curiously very similar to that of al-Bayḍāwī in the sense that it is an explanatory commentary, which seeks to highlight the message of the verse, i.e., what is called the *murād* or *maqṣūd* (lit. what is intended or aimed at) of a text in Islamic science. Nevertheless, more generally, Ibn ʿAjība uses al-Bayḍāwī to develop interpretations in the exoteric commentary, on which the allusive interpretation of the esoteric commentary can be based. The second most frequently cited *tafsīr* is *al-Wajīz fī tafsīr al-kitāb al-ʿazīz* of the Andalusian ʿAbd al-Ḥaqq b. ʿAṭiyya (d. 1147/541). He often quotes Ibn ʿAṭiyya to give an alternative interpretation of the verse, where he has commented on it differently. He likely takes advantage of the fact that this author, as Ibn ʿAjība himself points out, relates the interpretations of all those before him in a concise form, which allows him to draw from the classical exegetical tradition easily and to integrate into his *tafsīr* certain alternative opinions.

It appears that Ibn ʿAjība uses classical commentaries for various purposes, depending on their specific utilities. It seems that the author intended to show that the

---

20 In this presentation, we will limit ourselves to the most important *tafsīr-s in* relation to the frequency with which Ibn ʿAjība quotes them because the aim is to understand in a global way the use that Ibn ʿAjība makes of secondary sources. This is valid for the study of exoteric and esoteric comments.

exoteric interpretation of the Qur'ān constitutes a basis for the esoteric interpretation, and therefore that the two dimensions are not in opposition, but that, on the contrary, esotericism always relies on the external and literal aspect of the religion, or more precisely, that Sufism deals with nothing else than the internal aspect of the Islam that is practiced by every Muslim.

The author has written his commentary for a wider audience than the narrow circle of scholars. While he insists on the necessity of conforming to divine law, expertise in the religious sciences is not a condition for the practice of Sufism. His commentary is directed primarily to Muslims seeking spiritual access to the divine book without detouring the subtleties of exoteric science. This general approach is also visible in Ibn ʿAjība's method and thematic focus. He does not deviate from the tradition referring to intertextuality, *ḥadīth*, and the sayings of the Companions. Normative and dogmatic topics are addressed in some places. But even if Ibn ʿAjība was himself an accomplished scholar in these fields, he pays relatively little attention to them and often prefers to quote other *tafsīr-s*. For example, what concerns speculative theology or *Kalām*, Ibn ʿAjība quotes in particular al-Bayḍāwī on this subject. The passages that may give rise to theological considerations are generally treated more in detail in the *ishāra* section. Divine attributes for instance, are presented either as the object of contemplation of the Sufi rather than the theologian's reflection and argumentation or as the metaphysical cause of a spiritual virtue related to this divine attribute. Ibn ʿAjība does not wish to place too much emphasis on the dogmatic themes of the Qur'ān and seeks keeping with the spiritual vocation of his *tafsīr*, a vocation that largely sets aside the rational approach to divine reality.

3.1.2 The Characteristics of Ibn ʿAjība's Esoteric Commentary

As the preceding considerations have shown, the exoteric exegesis of Ibn ʿAjība revolves around the possibility of esoteric exegesis. Thus, the different thematic elements of each sura are explained in a relatively open way, in the sense that they are related to the sura's driving motive, but without engaging too much in historical details that would limit the understanding of the verse to this dimension alone. He uses traditional sciences such as grammar and Qur'ānic readings, for example, to prepare the passage from the external meaning to the spiritual and internal meaning. However, the esoteric and allusive commentary always refers to a literal explanation of the Qur'ānic text, showing how a spiritual reading operates a temporal transposition that engages the reader in a shift from the history of salvation to his own personal journey to God and the circumstances in which this happens. Thus, the precise historical moment of the mission of the Prophet Muḥammad becomes the universal model of the situation that every saint and his disciples, as the heir of the Prophet and his Companions, must face. The analogical transposition thus realized is, moreover, following the principle of *ishāra*, because of the spiritual allusion "stages" the external meaning of the text in the immediate present and thus "alludes" to the direct and actualized experience that emerges through a spiritual reading of the text.

It is obvious that Ibn ʿAjība's main concern in the *al-Baḥr al-madīd* is the esoteric commentary since it is there that he develops his interpretation. That is why he cites far fewer secondary sources in the esoteric commentary than in the exoteric commentary. If he cites secondary sources, it illustrates his commentary or presents an interpretation from an alternative perspective. The main sources come from classical Shādhilī literature, in particular, the works of Ibn ʿAṭāʾ Allāh (d. 1309/709) such as the *Ḥikam*[21] and the *Laṭaʾif al-minan.*[22] It is only natural, given the context of the revival of Shādhilī spirituality engaged by ad-Darqāwī, that Ibn ʿAjība refers to these sources. Ibn ʿAjība seems to want to present his commentary as a Shadhilī *tafsīr,* uniting the mastery of the exoteric commentary with an esoteric commentary permeated by the teachings of the Shādhiliyya.

The author includes, of course, sources that are not Shadhilī, too. The *Iḥya ʿulūm ad-dīn* of the famous scholar and Sufi Abū Ḥāmid al-Ghazālī (d. 1111/505) is one of the most cited works dealing with spirituality. Ibn ʿAjība often develops the remarks he draws from al-Ghazālī to give them a more initiatory meaning.[23] He thus continues the tradition of the Shadhilī masters who use the works of al-Ghazālī as a basis for more explicitly esoteric teaching. To illustrate some doctrinal points of his esoteric commentaries, the author of *al-Baḥr al-madīd* frequently quotes extracts from the Sufi poetry of Ibn al-Fāriḍ (d. 1235/632) and Abū l-Ḥasan ash-Shushtarī (d. 1269/667), both known for expressing contemplative experiences of divine unity.

The two *tafsīr-s* of Abū l-Qāsim al-Qushayrī (d. 1072/465) and Sahl at-Tustarī (d. 869/283) are frequently cited in the *al-Baḥr al-madīd.* Both are among the earliest Sufi commentaries of the Qurʾān. Ibn ʿAjība quotes them to give a classical interpretation of certain Qurʾānic passages that have an important role in Sufi teachings. The same holds for the commentary of Abū ʿAbd ar-Rahmān as-Sulamī (d. 1021/412), or the one of Ruzbehān al-Baqlī (d. 1021/412) whom he cites as al-Wartajibī for interpretations of a more contemplative nature. The important question is about the influence of the Andalusian Sufi Muhyī d-Dīn b. ʿArabī (d. 1240/638).[24] If it appears that the *al-Baḥr al-madīd* provides some indications affirming an Akbarian influence,[25] it does, however, not appear explicitly. Ibn ʿAjība's *tafsīr* is more a product of the Shadhilī tradition of

21 See Aḥmad b. ʿAṭāʾ Allāh al-Iskandarī, *The Book of Wisdom,* trans. Victor Danner (New York: Paulist Press, 1978).

22 See Aḥmad b. ʿAṭāʾ Allāh al-Iskandarī, *La sagesse des maîtres soufis,* trans. Éric Geoffroy (Paris: Grasset, 1998).

23 See, for example, Ibn ʿAjība, *Baḥr,* 3: 141, where Ibn ʿAjība affirms the possibility of divine contemplation before physical death.

24 Claude Addas was able to identify doctrinal notions that belong to Akbarian terminology or at least show a strong resemblance, see her preface to Ibn ʿAjība, *Deux traités,* 5–9.

25 See Ibn ʿAjība, *Baḥr,* 1: 133; 2: 50; 4: 96; 6: 8; 7: 35f.

Fez,[26] which has been imbued with certain Akbarian elements such as the notions of "God's self-disclosure" (*tajallī*), of "Muḥammadan heritage" or of the spiritual and metaphysical reality of Muḥammad. Ibn ʿAjība integrates this tradition in the framework of the Darqāwian renewal and develops it in the formulation of his Sufi teaching. Thus, the spiritual master's figure takes on capital importance in the allusive exegesis; he represents the instrumental cause that makes the transformative process of spiritual realization operative. The abandonment of one's self-management (*isqāṭ at-tadbīr*) and the satisfaction of divine decrees (*ar-riḍā*) appear as cardinal attitudes that characterize the state of the saints. As for the practical aspect of the path, it is the *dhikr* (invocation of God's name) and the *ṣuḥba* (company with a spiritual master) that occupy the first rank, then the *tajrīd* (stripping off of everything that hinders consecration to God) for the purgative aspect. One of the most striking features of Ibn ʿAjība's exegesis is how the intrinsic link between the realization of servitude (*al-ʿubūdiya)* and the contemplative experience of God's lordship (*ar-rubūbiya*) synthesizes the different facets of the spiritual path. Conformity to the Muḥammadan model of servitude and realization of the Prophet's experience of extinction in *tawḥīd* thus represents the fulfillment of spiritual perfection.

By taking the sacred text out of its historical dimension and relating it to the reader's spiritual experience, the *ishāra* exegesis is a way to actualize the Qurʾānic revelation.[27] Ibn ʿAjība alludes to it at the very beginning of his *tafsīr* commenting on the first passage of sura 2, *al-Baqara* (The Cow):

> They (people of pious fear) do not cease to ascend through it (the Qurʾān) through the spiritual degrees and states, until they hear (the Qurʾān) from the Greatest without intermediary (. . .). Jaʿfar aṣ-Ṣādiq said: 'By God, God discloses Himself to His creatures through His speech, but they are not aware of it!' (. . .). Indeed, at this level of reading[28] one perceives in the discourse the One who speaks and one does not consider oneself, nor one's own reading, but one is annihilated from oneself, absorbed in the vision of the Lord.[29]

Through the *ishāra*, Ibn ʿAjība wants to offer the reader access to the truly spiritual dimension of the divine book, which he considers "the key to hearts."[30] In this perspective, the Qurʾān, as says Ibn ʿAjība, paraphrasing the beginning of the sura *al-Baqara*, is

---

26 See Ruggero Vimercati Sanseverino, *Fès et sainteté, de la fondation à l'avènement du Protectorat (808–1912). Hagiographie, tradition spirituelle et héritage prophétique dans la ville de Mawlāy Idrīs* (Rabat: Éditions Centre Jacques Berque, 2014), 298–390.

27 Cf. on rendering the Quran accessible as an experience of revelation by the allusive interpretation, see William C. Chittick, *The Sufi Path of Knowledge* (New York: SUNY, 1989): 246–50; Michel Chodkiewicz, *Un océan sans rivage* (Paris: Seuil, 1992), 46.

28 Ibn ʿAjība evokes three levels of reading the Qurʾān, the highest of which is the one cited here.

29 Ibn ʿAjība, *Baḥr*, 1: 53.

30 Ibid., 2: 283.

"a guide to the vision of Our Essence; it indicates the way to reach Our Presence."[31] From the background of this hermeneutical principle, it is possible to distinguish three thematic axes in Ibn ʿAjība's esoteric commentary.[32] These three leitmotivs correspond to some extent to the themes of the Qur'ānic text. Thus, the passages dealing with divine reality, creation, revelation, and eschatology – in other words, everything that implies a kind of intervention and the manifestation of the divine – are generally interpreted as allusions to contemplation and spiritual enlightenment. The Qur'ānic passages concerning prophetic stories, the affirmation of the Prophet's mission, and the events of his life are interpreted as referring to sainthood, the function of the spiritual master, and the spiritual reality of the Prophet. As for the *ayāt al-aḥkām*, verses concerning ritual and law, and all the verses setting out ethical precepts, dealing with virtues, sins, and vices, i.e., everything that determines religious practice and experience, they are most often commented on as regards the spiritual path, its states, and its rules, as well as the obstacles and traps that the disciple must overcome.

At the same time, there is an apologetic aspect to the exegesis of Ibn ʿAjība aiming to show how Sufi practices such as *dhikr* or *ṣuḥba* are inspired by the inner aspect of the divine scripture. One perceives a desire to reformulate the classical teaching of Sufism to adapt it to the demands of a new era. The theme of contestation of sainthood (*inkār al-khuṣūṣiyya*), set in analogy to the contestation of Muḥammad's prophecy and of the revelation he claimed to receive, is particularly significant in this regard.

3.2 The *Tafsīr al-Fātiḥa al-kabīr*

Ibn ʿAjība wrote his great commentary of the sura *al-Fatiḥa* originally as an introduction to the *al-Baḥr al-madīd*, but it finally became an independent work. He bases his commentary on the hermeneutical principle that the sura *al-Fātiḥa* "encompasses all the meanings of the Qur'ān; it is as if it were a summary of the Qur'ān."[33] In his "Great Commentary" Ibn ʿAjība offers extensive commentary on it and uses the thematic structure of the sura to develop a concise encyclopedia of Islamic belief and practice with a spiritual interpretation. Beginning with an extensive introductive section about sciences in general and Qur'ānic sciences in particular, the commentary proceeds by explaining the meanings of the major notions of a verse and then develops the themes related to these. In the commentary of the first and second verses of the *al-Fātīha*, Ibn ʿAjība presents an interpretation of God's ninety-nine "beautiful names" according to the scheme of *taʿalluq* (relation), *takhalluq* (assimilation) and *taḥaqquq*

31 Ibid., 1: 55.
32 See Vimercati Sanseverino, *Commentaire coranique*, 209–34.
33 Ibn ʿAjība, *Baḥr*, 1: 47.

(realization) of each name, developing further the methods of al-Ghazālī, al-Qushayrī, and Zarrūq. The commentary of the third verse, “the King of the Day of Judgement” exposes basic principles of belief and eschatology, emphasizing the eschatological role and hierarchy of saints. The commentary of the first part of the fourth verse, “Thee we worship” is dedicated to Islamic ritual practice and its spiritual meaning, while the second part “and from Thee we seek help” concerns the stages of the path of spiritual realization as treated classically in Sufi literature. This part constitutes a complete treatise of Sufi doctrine and practice developing the notions, themes, and schemes characteristic of Ibn ʿAjība’s writings. The triple *islām-īmān-iḥsān* (surrender-faith-excellence) of the so-called “ḥadīth of Jibrīl” serves as a fundamental structure for the spiritual progression of the disciple, showing thereby how spiritual realization is related to three dimensions of the Islamic religion (*ad-dīn*).[34] The last part of the sura concerning the “straight path” gives rise to developing a theology of religions and a heresiology that follows classical lines.

4 Conclusion

Ibn ʿAjība’s Qurʾānic hermeneutics is founded on a Sufi understanding of divine revelation, and his exegesis aims at indicating how the Qurʾān can be read for the spiritual path, as envisaged in the teaching of the Shādhiliyya-Darqāwiyya. Synthesizing North African Sufism with the tradition of classical Sufi exegesis, Ibn ʿAjība offers interpretive keys presenting the Qurʾānic revelation as the divine basis of the path of spiritual realization.

The concepts of “theophany” (*at-tajallī*) and that of “the prophetic heritage” (*al-wirātha an-nabawiyya*) constitute the actual foundations of the author’s hermeneutics. Both notions are applied and developed to demonstrate, on the one hand, the reality of the contemplative enlightenment to which the Sufis aspire and, on the other hand, to demonstrate the legitimacy of the initiatory transmission by the spiritual masters. The leitmotif of theophany, leading everything back to God, is always related to spiritual education. *Adab*, in its double meaning of formation and behavior, unites in some way the two dimensions – horizontal of the initiatory transmission and vertical of the theophany – because *adab* corresponds to the just attitude towards the various theophanies and is therefore identified with the prophetic heritage transmitted by the saints.

Within this framework, the spiritual allusion (*al-ishāra*) acts as a mediator between the historical and the spiritual meaning of revelation. The aim is to make the reader aware of the function that the Qurʾān must have in a Sufi context and, even more so, to instill a state of mind receptive to its spiritual meaning. By introducing the aspirant

34 For this notion in Ibn ʿAjība’s thought, see Lützen, *Sufitum und Theologie*.

into the experience of an interiorized reading of the Qurʾān, Ibn ʿAjība's Qurʾānic hermeneutics aims to sensitize the reader to the spiritual dimension of revelation by living it as a theophany, an instrument of enlightenment and divine education. In this way, the Sufi reading of the Qurʾān is presented as an initiatory experience that goes hand in hand with the aspirant's inner sanctification. According to this essentially contemplative perspective, the Qurʾānic revelation unveils the spiritual meaning of existence as the place of divine Self-disclosure. The existential phenomena to which the Qurʾān draws attention, such as creation, historical events, eschatology, cosmic phenomena, but also the divine commandments and the rites structuring religious life, represent for the aspirant as many allusions to his journey to God.

Seeking to demonstrate the orthodoxy of Sufism in general and to reinterpret certain constitutive elements within the framework of the Darqāwian revival, such as the importance of the spiritual master and metaphysical conception of *tawḥīd,* are undoubtedly objectives of Ibn ʿAjība's Qurʾānic exegesis. However, Ibn ʿAjība primarily addresses his exegesis to the followers of the Sufi path, and it is to this end that he applies his vast knowledge of Islamic sciences, making his *tafsīr* a valuable source for the study of Muslim spirituality and Qurʾānic exegesis in the early modern era and one of the major works of Sufi exegesis which continues to be widely read today.

5 Bibliography

5.1 Sources

Ad-Darqāwī, al-ʿArabī. *Majmūat ar-rasāʾil Mawlāy al-ʿArabī ad-Darqāwī al-Ḥasanī.* Ed. Bassām Muḥammad Bārūd. Abu Dhabi: al-Mujammaʿ ath-Thaqāfī, 1999.

Ibn ʿAjība al-Ḥasanī, Aḥmad. *Al-Fahrasa, L'autobiographie (Fahrasa) du Soufi Marocain Ahmad ibnʿAjība (1747–1809),* trans. Jean-Louis Michon. Milan: Arché, 1982².

Ibn ʿAjība al-Ḥasanī, Aḥmad. *Le Soufi Marocain Aḥmad ibnAjība (1746–1809) et son Miʿrāj,* trans. Jean-Louis Michon. Paris: J. Vrin, 1990.

Ibn ʿAjība al-Ḥasanī, Aḥmad. *Deux Traités sur L'unité de L'existence,* edited and translated by Claude Addas, and Jean-Louis Michon. Marrakech: Al-Quobba Zarqua, 1998.

Ibn ʿAjība al-Ḥasanī, Aḥmad. *Futūḥāt al-Ilāhīya fī Sharḥ al-mabāḥith al-asṣliyya.* Ed. ʿAbd ar-Raḥmān Ḥasan Maḥmūd. Cairo: ʿĀlam al-Fikr, 2000.

Ibn ʿAjība al-Ḥasanī, Aḥmad. *Al-Baḥr al-madīd fī tafsīr al-Qurʾān al-majīd.* Ed. ʿUmar Aḥmad al-Rāwī. Beirut: Dār al-Kutub al-ʿIlmiyya, 2002.

Ibn ʿAjība al-Ḥasanī, Aḥmad. *Kitāb Sharḥ ṣalāt al-Quṭb b. Mashīsh.* Ed. ʿAbd as-Salām al-ʿUmrānī al-Khālidī. Casablanca: Dār ar-Rashād al-Ḥadītha, 2003.

Ibn ʿAjība al-Ḥasanī, Aḥmad. *Lawāqiʿ al-qudusiyya fī Sharḥ al-waẓīfa az-zarrūqiyya.Al-Jawāhīr al-ʿajībiyya.* Ed. ʿAbd as-Salām al-ʿUmrānī al-Khālidī. Beirut: Dār al-Kutub al-ʿIlmiyya, 2004.

Ibn ʿAjība al-Ḥasanī, Aḥmad. *Tafsīr al-Fātiḥa al-kabīr.* Ed. ʿĀṣim Ibrāhīm al-Kayyālī. Beirut: Dār al-Kutub al-ʿIlmiyya, 2004.

Ibn ʿAjība al-Ḥasanī, Aḥmad. *Iqaz al-himam fi Sharḥ al-Ḥikam.* Ed. Khalīl Manṣūr. Beirut: Dār al-Kutub al-ʿIlmiyya, n.d.

Ibn ʿAjība al-Ḥasanī, Aḥmad. *A Thirteenth/Eighteenth Century Quranic Commentary on the Chapters of the All-Merciful, the Event, and Iron from the Immense Ocean*, trans. Mohamed F. Aresmouk, Michael A. Fitzgerald. Louisville, Ky: Fons Vitae, 2009.
Ibn ʿAṭāʾ Allāh al-Iskandarī, Aḥmad. *The Book of Wisdom*, trans. Victor Danner. New York: Paulist Press, 1978.
Ibn ʿAṭāʾ Allāh al-Iskandarī, Aḥmad. *La sagesse des maîtres soufis*, trans. Éric Geoffroy. Paris: Grasset, 1998.

5.2 Secondary Literature

ʿAzūzī, Ḥasan. *Shaykh Aḥmad Ibn ʿAjība wa minhājuhu fī at-tafsīr*. 2 vols. Rabat: Wizārat al-Awqāf wa-sh-Shuʾūn al-Islāmiyya, 2001.
Būkārī, Aḥmad. *Al-Iḥyāʾ wa-t-Tajdīd as-Ṣūfī fī l-Maghrib, 1204–1330/1790–1912*. Rabat: Al-Mamlaka al-Maghribiyya, Wizārat al-Awqāf wa-sh-Shuʾūn al-Islāmiyyah, 2006.
Chittick, William C. *The Sufi Path of Knowledge*. Albany (N.Y.): State University of New York Press, 1989.
Chodkiewicz, Michel. *Un Océan sans Rivage*. Paris: Seuil, 1992.
Chouiref, Tayeb. "Ibn ʿAjība." In *EI*[3], edited by Kate Fleet et al. Brill, 2007–. Article published online 20016. Accessed on October 27, 2023. http://dx.doi.org/10.1163/1573-3912_ei3_COM_30684.
De Jong, Frederick. "Materials Relative to the History of the Darqâwiyya Order and its Branches." *AR* 26, no. 2 (1979): 126–43.
Elsayed, Mohamed Fadel Ebada. "Le commentaire du Coran par le soufi marocain Aḥmad Ibn ʿAjība (1160–1224/1747–1809) – Méthodes et themes." PhD thesis. Strasbourg, 2020.
Honerkamp, Kenneth. "Ibn ʿAbbad, Model of the Shādhiliyya." In *Une voie soufi dans le monde: La Shâdhiliyya*, edited by Éric Geoffroy. Paris: Maisonneuve & Larose, 2005: 159–71.
Lützen, Flroian A. *Sufitum und Theologie bei Aḥmad Ibn ʿAǧība (gest. 1223/1809) – Eine Studie zur Methode des Religionsbegriffs*. Tübingen: Mohr-Siebeck, 2020.
Meftah, Abdelbaqî. "L'initiation dans la Shādhiliyya-Darqāwiyya." In *Une voie soufi dans le monde: La Shādhiliyya*, edited by Éric Geoffroy. Paris: Maisonneuve & Larose, 2005: 237–48.
Nwyia, Paul. *Ibn ʿAbbād of Ronda, a Mystical Preacher at Qarawīyīn of Fez*. Beirut: The Institute of Oriental Letters of Beirut, 1956.
At-Tamsamānī, Muḥammad b. Muḥammad al-Mahdī. *Al-Imām Sayydī Muḥammad b. Aḥmad al-Būzīdī – tarjumatuhu wa baʿḍ āthāruhu*. Beirut: Dār al-Kutub al-ʿIlmiyya, 2007.
Trimingham, John. *The Sufi Orders in Islam*. Oxford: Clarendon Press, 1971.
Vimercati Sanseverino, Ruggero. "Commentaire coranique, enseignement initiatique et renouveau soufi dans la Darqāwiyya. Le *Baḥr al-madīd fī tafsīr al-Qurʾān al-majīd* d'Aḥmad Ibn ʿAjība (m. 1223/1809)." *Studia Islamica* 107, no. 2 (2012): 209–34.
Vimercati Sanseverino, Ruggero. *Fès et Sainteté, De La Fondation à l'avènement du Protectorat (808–1912)*. Hagiogaphie, tradition spirituelle et héritage prophétique dans la ville de Mawlāy Idrīs. Rabat: Centre Jacques-Berque, 2014.
Weismann, Itzchak. "The Shâdhiliyya-Darqâwiyya in the Arab East." In *Une voie soufi dans le monde: La Shâdhiliyya*, edited by Éric Geoffroy. Paris: Maisonneuve & Larose, 2005: 255–70.
Zekri, Mostafa. "La Tarîqa Shâdhiliyya-Darqâwiyya – Les Empreintes du Shaykh al-'Arabî al-Darqâwî." In *Une voie soufi dans le monde: La Shâdhiliyya*, edited by Éric Geoffroy. Paris: Maisonneuve & Larose, 2005: 229–36.

Thanā' Allāh Pānīpatī

Kamran Bashir

1 Introduction

Thanā' Allāh Pānīpatī (d. 1810/1225), the famed author of the voluminous Arabic Qur'ān commentary *at-Tafsīr al-maẓharī*, is an understudied luminary of eighteenth-century South Asia. Despite his substantial contributions to the various disciplines of the Islamic sciences and his social role as a Muslim judge (*qāḍī*), there has been no systematic study that investigates his life and works in the historical context of his times, except for a few studies focusing on his Qur'ān commentary.[1] Pānīpatī not only left behind a sizeable number of writings, many of which still exist in the form of manuscripts, but also actively learned from and engaged with many established figures of his time, including the renowned Naqshbandī Sufi leader and his mentor Mirzā Maẓhār Jan-i Jānāṇ (d. 1781/1195), as well as Shāh Walī Allāh (d. 1762/1175),[2] under whose tutelage Pānīpatī received training in the traditional Islamic disciplines. Moreover, the reception and impact of his work, especially his *tafsīr*, was widespread in South Asia. This influence can be seen even today in the continuous references that are made to his work, particularly in traditionalist[3] circles like that of the Deobandīs in South Asia. The main focus of this chapter is identifying the key hermeneutical principles that Pānīpatī followed in his interpretation of the Qur'ān.

2 Life, Time, and Works

Pānīpatī[4] (sometimes called Muḥammad al-Hindī in non-Indian circles) was born in 1730/1142 during the reign of the Mughal Emperor, Muḥammad Shāh Rangīlā

1 See Ḥāfiẓ 'Abd ar-Rashīd Aẓhar, "Ad-Dirāsa al-muqārana bayna at-Tafsīr al-maẓharī wa-Fatḥ al-bayān" (PhD diss., University of the Punjab, 1997). Aẓhar tends to characterize Pānīpatī's work as being written along the lines of Taqī d-Dīn Aḥmad b. Taymiyya (d. 1328/728). Also see Muḥammad 'Ādil Khān Āfrīdī, "Lamḥāt at-tarbiya fī t-*Tafsīr al-maẓharī*," *Journal of Islam in Asia* 8, no. 2 (2012): 33–52.

2 See the chapter on Shāh Walī Allāh in this Handbook.

3 Elsewhere, I have challenged the typologies of "modernist" and "traditionalist" as they reflect polemical battles among modern scholars about the issue of the "authenticity" of their positions more than what individual scholars think about themselves in terms of these categories. I contend that modernists revere their Islamic tradition no less than their traditionalist counterparts. These typologies owe their origin to the polemics among modern scholars of various stripes. See Kamran Bashir, *The Qur'an in South Asia: Hermeneutics, Qur'an Projects, and Imaginings of Islamic Tradition in British India* (London and New York: Routledge, 2021).

4 Pānīpat is a historic city in North India, lying some 95 kilometers north of Delhi. Pānīpatī was born there and ultimately performed his duties as *qāḍī* in the same town.

https://doi.org/10.1515/9783111320083-024

(r. 1719–1748/1131–1161), and died in 1810/1125 when British rule had acquired a firm footing in South Asia.[5] Historians often mark this period as one marked by a decline in Mughal administration, accompanied by foreign invasions and non-Muslim incursions into earlier strongholds of Muslim rule. In this regard, Pānīpatī's own will (*waṣiyya*) and many letters to him by his Sufi mentor, Jan-i Jānāṇ, lament the weakening of Muslim rule in India and hope for a revival in Muslim camps.[6] It is in this larger context of the perceived decline in Muslim affairs that Pānīpatī's career unfolded. Like Walī Allāh, he too seemed to be interested in bringing intellectual harmony to Muslim camps by resolving doctrinal and interpretive conflicts in Muslim circles, including disputes among Sufis and 'Ulamā'. Moreover, in a vein similar to Jān-i Janāṇ and Walī Allāh, Pānīpatī contributed to the South Asian discursive tradition of Sunni-Shiite polemics and attempted to dismiss the tenets of the latter sect.

Well-versed in Arabic, Persian, Hindi, and Sanskrit, Pānīpatī produced numerous writings that covered several traditional Islamic scientific disciplines. In addition to his famous Qur'ān commentary, he penned a number of writings that dealt with diverse subjects such as the Sufi path and debates on Sufi beliefs, matters of Islamic jurisprudence, criticism of Shiism, and moral guidance for Muslims.[7] His other works addressed contentious issues among Sufis and 'Ulamā', as well as topics pertaining to the life of the Prophet and his family. Pānīpatī also left a small number of letters that throw light on his life and times. Observers of his religious writings, including his Qur'ān commentary, often characterize his work as upholding the teachings of the Ḥanafī legal school.[8] However, upon closer analysis, this characterization can be seen as a retrospective projection by Pānīpatī's late nineteenth-century admirers, in particular the members of the larger Deobandī school who were embroiled in intellectual conflicts with the so-called modernists and other rival groups in British India.[9] As previously mentioned, Pānīpatī's work appears to have been driven by the desire to eliminate the sites of religious conflict among Sunni groups.

Given the above, brief overview of his life, times, and works, the most striking feature to note is Pānīpatī's multifaceted profile and credentials as a thinker. It is this particular profile that can prompt us to revisit current scholarship on the eighteenth-century Muslim world. A recent work invites us to reimagine this era in a

---

5 For a biography and guide to his works, see 'Ārif, Maḥmūd al-Ḥasan, *Tadhkirat Qāḍī Muḥammad Thanā' Allāh Pānīpatī* (Lahore: Institute of Muslim Culture, 1995). The work is based on al-Ḥasan's doctoral dissertation, submitted at the University of Punjab, Lahore, Pakistan.

6 Al-Ḥasan, *Tadhkira*, 8, 12.

7 For an guide to Pānīpatī's works, see al-Ḥasan, *Tadhkira*, 507–617.

8 M.G. Zubaid Ahmad, *The Contribution of Indo-Pakistan to Arabic Literature: From Ancient Times to 1857* (Lahore: Sh. Muhammad Ashraf, 1968).

9 For a discussion of intra-faith contestations in British India, see Sher Ali Tareen, *Defending Muhammad in Modernity* (Notre Dame, IN: Notre Dame Press, 2020). Also see note 3 above.

new light.[10] In this regard, there is a great need to formally study the works of scholars like Pānīpatī in their specific historical contexts, including the intellectual atmosphere in which their careers progressed.

3 Pānīpatī's Qur'ānic Hermeneutics

Pānīpatī lived in an intellectual environment with an initial inclination toward reconciling (*taṭbīq*)[11] the competing religious insights of the various Islamic scientific disciplines that emerged in Muslim intellectual tradition, including Sufism. Walī Allāh's larger oeuvre can also be understood as heading in the same direction, concerned with resolving intellectual conflicts in Muslim tradition. Pānīpatī seems to have made *at-Tafsīr al-maẓharī* the cornerstone of this program of reconciliation.[12] He engaged with the full tradition of Islamic sciences dealing with the Qur'ān, hadith, and Muslim jurisprudence. In addition to this underlying impulse in his work, he attempted to apply the principles of *at-tafsīr al-ishārī* (esoteric commentary) in order to help travelers on their path to Sufism without ignoring other principles of exegesis that found support in Muslim tradition.

According to Walid Saleh, all of the characteristics mentioned in Pānipatī's *tafsīr* are the basis of its categorization as an encyclopedic commentary.[13] However, this encyclopedic commentary was not without its own an exegetical agenda; it reflects Pānīpatī's underlying hermeneutical approach to the Qur'ān in a number of respects. Since he did not leave behind any tract on his principles of exegesis, his Qur'ānic hermeneutics must be determined through the study of his actual practice of understanding the Qur'ānic text. His commentary is sometimes understood as an expansion of the Qur'ān project outlined in the writings of Walī Allāh,[14] specifically in the latter's work *al-Fawz al-kabīr fī uṣūl at-tafsīr*.[15] However, this assessment, though arguable,

10 See Ahmad S. Dallal, *Islam without Europe: Traditions of Reform in Eighteenth Century Islamic Thought* (Chapel Hill: The University of North Carolina Press, 2018).

11 *Taṭbīq* can be seen as a process where an interpreter applies rules and insights to particular circumstances in order to resolve differences of opinions. It appears that *taṭbīq* literature was in the air in eighteenth-century South Asia, particularly North India. Based on a hypothesis that the eighteenth century saw a tendency toward reconciliation between competing schools of thought in Muslim intellectual tradition, there is a need for a full-length separate study on this area of South Asian intellectual history.

12 Thanā' Allāh Pānīpatī, *at-Tafsīr al-maẓharī*, ed. Aḥmad `Izzu `Ināyat. 10 vols. (Beirut: Dār Iḥyā' at-Turāth al-'Arabī, 2004). For the publication history of this seminal work, see al-Ḥasan, *Tadhkira*, 286–89.

13 Walid Saleh, *The Formation of the Classical* Tafsīr *Tradition* (Leiden: Brill, 2003), 16f.

14 Al-Ḥasan, *Tadhkira*, 269.

15 Shāh Walī Allāh, *al-Fawz al-kabīr fī uṣūl at-tafsīr*, trans. Sa'īd Aḥmad al-Bālanfūrī (Damascus: Dār al-Ghawthānī li-d-Dirāsāt al-Qur'āniyya, 2008).

seems more conjectural than rooted in Pānīpatī's actual practice of textual interpretation. Despite this caveat, there are certain notable features of Pānīpatī's interpretive approach that converge with that of Walī Allāh and are discussed below.

Keeping these points in mind, there are three ways in which we can make sense of the hermeneutical thinking underlying Pānīpatī's Qur'ān commentary. First, his work provides evidence that he was building on Walī Allāh's work on the Qur'ān to a certain extent. Second, Pānīpatī was studying the Qur'ān with a critical outlook on the traditional disciplines of the Islamic sciences. In this regard, he was not dismissive of the centuries-long Muslim tradition. Rather, he produced his commentary by critically engaging with the resources of Muslim intellectual heritage, including *fiqh* (jurisprudence) and *ʿulūm al-Qur'ān* (the traditional sciences of the Qur'ān). As mentioned above, his commentary can be seen as contributing to what we can provisionally refer to as the emerging *taṭbīq* literature of South Asia, particularly Delhi. This literature included Qur'ān commentaries as well. Third, Pānīpatī's work also depicted his sophisticated approach to writing mystical or esoteric commentary (*at-tafsīr al-ishārī*). However, this mystical interpretation was executed within the bounds of the classical *tafsīr* tradition that emerged over time in the larger Muslim world. The following sections provide details of these three perspectives.

3.1 In the Footsteps of Shāh Walī Allāh

Pānīpatī studied under Walī Allāh when the latter was producing his late work on the Qur'ān.[16] This is evidenced in Pānīpatī's *tafsīr* when he builds on his teacher's ideas. A few examples highlight what he borrowed from Walī Allāh's *al-Fawz al-kabīr*. Like Walī Allāh, Pānīpatī tends to see the historical reports on *asbāb an-nuzūl* (occasions of revelation) as not necessarily capturing the events that occasioned the revelation of particular verses. Employing the same vocabulary that Walī Allāh used, Pānīpatī emphasizes that the reports that we commonly view as *asbāb* may not necessarily report the historical reasons for the sending down of certain verses. Instead, such reports can hint at how certain verses were applied to certain circumstances during the time of the Companions of the Prophet or events that might have occurred during or after the lifetime of the Messenger. In order to explain this, Pānīpatī often used the phrase "*al-ʿibra li-ʿumūm al-lafẓ lā makhṣūṣ al-mawrid*" (important is the generic meaning (applied to relevant circumstances), and not the specific use of a word).[17]

In addition to the possible convergence between Walī Allāh and Pānīpatī on the issue of *asbāb an-nuzūl*, Pānīpatī's approach toward the use of *isrāʾīliyyāt* (Jewish legends) and the question of *naskh* (abrogation) tends to be in line with that of his

---

16 Al-Ḥasan, *Tadhkira*, 83–85.
17 Pānīpatī, *at-Tafsīr*, 1: 221.

teacher. Like Walī Allāh, Pānīpatī was reluctant to use the stories of Jewish origins when interpreting the Qur'ān. This is also somewhat linked to the dismissing of some of the extra-Qur'ānic sources of interpretation. For instance, in the case of Q 2:101–2,[18] he does not merely rely on traditional explanations of the names Hārūt and Mārūt as referring to two angels; instead, Pānīpatī holds that these words mean *al-qalb* (heart) and *ar-rūḥ* (soul) respectively.[19] This method of interpreting the text was also in line with the older South Asian tendency to interpret Qur'ānic verses for their esoteric and hidden meanings (which will be discussed in detail below).

On the question of *naskh*, Pānīpatī attempts to reduce the instances of perceived abrogation in the Qur'ān and tries to provide a philological explanation on such occasions. For example, he does not explain verses 183 or 187 of sura 2, *al-Baqara* (The Cow), through the lens of *naskh*.[20]

In addition to the above, Pānīpatī attempts to strike a balance between the use of exegetical traditions (*riwāyāt*) and human reason in explaining Qur'ānic verses. In this regard, Walī Allāh uses the phrase *riwāya wa-tawjīh* (tradition and interpretation) in order to discuss the possible use of human reason to explain the Qur'ān. He explains that a commentator could explore additional aspects of a verse by looking beyond the literal meaning provided by traditional reports. Pānīpatī employs a similar terminology. However, he uses the phrase *tafsīr wa-ta'wīl* (commentary and interpretation) in order to settle this debate in Muslim tradition. He explains that an exegete of the Qur'ān can determine the literal meaning of verses through such sources as *asbāb an-nuzūl* and other exegetical sayings, but after that, a commentator can provide additional interpretation of those verses. However, this additional interpretation should not be made against the injunction of the Qur'ān and sunna. For example, while explaining verse Q 2:193, Pānīpatī writes: "I say (*qultu*) that the verse can also mean that if unbelievers restrain themselves from bloodshed, the sin for transgression will fall upon those who commit injustice, meaning that (since) you enforce your action on those who, (in fact), want to avoid warfare, it is you who will be the oppressor."[21]

Viewing Walī Allāh and Pānīpatī together, we can conclude that they were not offering anything new in this regard. They were simply trying to explain, in their own manner, the same debate between the proponents of *tafsīr bi-l-mā'thūr* (traditionist

18 The verses Q 2:101–2 read: "When God sent them a messenger confirming the Scriptures they already had, some of those who had received the Scripture before threw the Book of God over their shoulders as if they had no knowledge, and followed what the evil ones had fabricated about the Kingdom of Solomon instead. Not that Solomon himself was a disbeliever; it was the evil ones who were disbelievers. They taught people witchcraft and what was revealed in Babylon to the two angels Harut and Marut. Yet these two never taught anyone without first warning him, 'We are sent only to tempt – do not disbelieve.'"

19 Pānīpatī, *at-Tafsīr*, 1: 123.

20 Ibid., 211.

21 Pānīpatī, *at-Tafsīr*, 1: 241.

exegesis) and *tafsīr bi-r-rā'y* (deductive exegesis) that Muslim exegetical tradition encapsulated.

3.2 A Critical Perspective on Premodern Exegetical Tradition

As mentioned above, Pānīpatī was not simply expanding on Walī Allāh's Qur'ān project. His own creative reading of the Qur'ān originates from his innovative application of traditional Islamic sciences and his borrowing from key Sunni sources. This broader approach to reading Muslim scripture is evidenced in Pānīpatī's extensive use of traditional books dealing with *tafsīr*, hadith, *ʿilm al-kalām*, and other traditional disciplines. However, Pānīpatī approaches these traditional sources with a critical eye and with the aim of finding out how to reconcile the apparent conflicts among competing Muslim schools found in Muslim tradition. For example, he uses hadiths and variant readings (*qira'āt*) in order to stress the complexity underlying the interpretation of the Qur'ānic verses. All this points to the fact that Pānīpatī, through his sophisticated use of traditional sources in his commentary, was attempting to anchor the interpretation of the Qur'ān in a centuries-old Muslim intellectual tradition rather than try out something innovative. This approach was also meant to offer an otherwise perplexed reader of the scripture a smooth understanding of the Qur'ān. As the later impact of his work shows, it was this feature of his *tafsīr* that appealed to Muslim reformers, in particular Deobandī traditionalists like Ashraf ʿAlī Thānawī (d. 1943/1362),[22] who were active in the late nineteenth century in British India.

One example of Pānīpatī's premodern traditional approach is his understanding of the story connected with the Satanic verses. The history of the Muslim response to this issue throws light on changing Muslim responses to the story and what it means in terms of Qur'ānic interpretation.[23] Premodern Muslim scholarly tradition had actually been divided on the nature and historicity of this incident. Pānīpatī accepted this tradition from Abū Jaʿfar aṭ-Ṭabarī in line with some early Muslim opinions on the issue. He was trying to reconcile the traditions regarding Q 17:73–74 and Q 22:52, and was not altogether denying every report. He thus tended to interpret the related Qur'ānic verses based on the medieval exegetical tradition.[24]

On the one hand, Pānīpatī was upholding the premodern tradition of Qur'ānic exegesis while, on the other, offering the Qur'ān's readers a smooth reading. This is manifested in Pānīpatī's hermeneutical approach, which tended to read Qur'ānic verses as inter-connected pieces of text. This Muslim preoccupation with reading the Qur'ān as a coherent text would gain momentum from the latter part of the nine-

22 See Ashraf ʿAlī Thānawī, *Tafsīr bayān al-Qur'ān* (Lahore: Maktaba Raḥmāniyya, 2011), 1: 23.
23 See Kecia Ali, *Lives of Muhammad* (Cambridge, MA: Harvard University Press, 2014), 62–64.
24 Pānīpatī, *at-Tafsīr*, 5: 308f.; 6: 254–57.

teenth century, when Muslim scholars reacted to Christian criticism on the Qur'ān.[25] This practice in Pānīpatī is demonstrated when the subject of a particular verse shifts to another subject. An example can be seen in his commentary on sura 2, *al-Baqara* (The Cow), where he connects the first five verses with the later ones, writing, "(W)hen God has mentioned His favored individuals and saints in the Book (at this point), He now mentions the very opposite of these former persons as well. However, since the context of the two groups is different, the text does not explain all this in detail."[26]

Similarly at Q 2:27, he connects the topic of giving glad tidings to believers and warnings to unbelievers with the subsequent topic. Pānīpatī explains the shift to the new topic in the following words: "When the traits and wicked ideas of unbelievers have been explained, now God directs His speech to these people again through negative interrogation."[27] The same interpretive technique is used at Q 2:40, where the Qur'ān begins its speech directed toward the Jews.[28]

Moreover, Pānīpatī did not belive that the practice of *tafsīr* writing should always be focused on identifying only one meaning of the Qur'ān. This is how Walid Saleh has characterized classical *tafsīr*, which, in his opinion, is semantically polyvalent.[29] The following commentary on Q 50:16 reveals Pānīpatī's hermeneutical thinking:

> The *'ulamā'* differ on what this closeness (*al-iqrābiyya*) means. Some say that this *qurb* (closeness) means "intellectual closeness." Bayḍāwī says that it means "We know the condition of our follower because We are closer to him than his jugular vein." But this invites us to think about the person of God (. . .). Baghawī says that it means God knows us better than us (. . .). The revered Sufis say that these verses mean that God is closer to us than our very persons.[30]

Despite his reverence for traditional Muslim resources for the interpretation of foundational texts, Pānīpatī continued to apply creativity throughout his explanations of the Qur'ān. The same can be said of his legal understanding of the Qur'ān. Although, he was inclined toward Ḥanafī jurisprudence, it is difficult to characterize his commentary as decisively shaped by the Ḥanafī school on legal matters. On many occasions, he expresses his intention to enable readers to recognize the underlying complexity in any given legal dilemma, which involves appreciating arguments from other schools such as Shāfi'ī and Ḥanbalī as well.

In summary, it is evident that Pānīpatī utilized premodern exegesis for a specific purpose, which was to familiarize his audience with an imagined underlying unity

---

25 For the emergence of the concept of *naẓm* in its modern garb, see Kamran Bashir, *The Qur'an in South Asia: Hermeneutics, Qur'an Projects, and Imaginings of Islamic Tradition in British India* (London and New York: Routledge, 2021).

26 Pānīpatī, *at-Tafsīr*, 1: 29.

27 Ibid., 51.

28 Ibid., 69.

29 Saleh, *Classical Tafsīr Tradition*, 151f.

30 Pānīpatī, *at-Tafsīr*, 9: 36.

within the Muslim tradition, despite its diversity. For this reason, his work appealed to traditionalists in late nineteenth-century British India.[31]

3.3 Mystical Commentary

On occasions, Pānīpatī applies the principles of *at-tafsīr al-ishārī* (esoteric commentary) in his own creative ways. This is understandable given the Sufi training he underwent under his master Jan-i Jānāṇ. Pānīpatī seems cognizant of the controversy over the validity of mystical interpretation in Muslim tradition and, due to his program of reconciliation, attempts to follow a careful path in his approach towards *at-tafsīr al-ishārī*. Early readings of Pānīpatī's commentary maintain that he seems to have drawn upon ideas of earlier scholars such as Andalusian jurist Abū Isḥāq ash-Shāṭibī (d. 1388/790), for whom mystical interpretation should be based on Qur'ānic verses and not on the whims of Sufis. Pānīpatī also follows the principles outlined by Jalāl ad-Dīn as-Suyūṭī (d. 1505/910) for writing mystical exegesis.[32] For example, Pānīpatī subscribes to the traditional opinion that a mystical interpretation should be carried out according to the evident word-meaning relationship found in Qur'ānic verses and not something beyond the dictates of language. He explains this as follows:

> The revered Sufis say that in verse Q 18:24, *wa-dhkur rabbaka idhā nasayta* means "you should remember God when you forget everyone except Him." Sufis say that the continuous, uninterrupted remembering of God is not possible unless one forgets everything except Him (. . .). Thus, that kind of remembrance is not possible without forgetting everything, and this (human) condition is called the annihilation of the heart (*fanā' al-qalb*).[33]

He further clarifies that his aforementioned interpretation reflects the Qur'ānic vocabulary, the dictates of human reason, and what can be learned from classical Arabic idiom.[34] Keeping the above points in mind, Pānīpatī summarizes his Sufi approach to the understanding of the Qur'ān:

> For every verse of the Qur'ān, there is an evident feature and a hidden feature, and for every limit of a specific knowledge, there is a (new) "horizon" (*maṭla'*); it is also reported that for every letter, there is a limit, and for every limit, there is a (new) horizon. This has been reported by al-Baghawī from Ibn Mas'ūd. Hence, (on the one hand,) the Arabic alphabet apparently acts as the elements (or ingredients) of the Qur'ān and (provides) the larger landscape of the Qur'ān; much Qur'ānic discourse is arranged through this alphabet. (Likewise) the Qur'ān offers us a variety of delicacies (*laṭā'if*) and inimitable elements. (On the other hand,) the very same letters depict the

31 For a discussion of the intellectual atmosphere in British India, in particular surrounding the Qur'ān, see Bashir, *The Qur'an in South Asia*, 43–75.
32 Al-Ḥasan, *Tadhkira*, 471f.
33 Pānīpatī, *at-Tafsīr*, 5: 373.
34 Ibid.

"all-encompassing nature" of the Qur'ān and connote the springs brimming with Allah's blessings. These letters are also, in fact, secrets between God and His Messenger. Nobody can know this except by the one whom God (wills) to bestow this (blessing).[35]

4 Impact of Pānīpatī's Qur'ānic Hermeneutics

Pānīpatī's work on the Qur'ān has been consistently received favorably, in particular among what are usually labelled traditionalist schools, for example, the Deobandi tradition in South Asia. The precise impact of his work during his lifetime and immediately after his death is a question that is open to further inquiry. Pānipatī's biographers have determined that he finished his Qur'ān commentary in 1793/1207. His letters reveal that there was some interest in his work during his lifetime. As in the case of other Qur'ān commentaries and translations, this interest and favorable reception grew with the spread of print technology in British India.[36] The first print edition of his first volume was published in 1855/1271 and was also mentioned by Carl Brockelmann in his *Geschichte der Arabischen Litteratur*. It gained the attention of scholars outside British India, like in the Ottoman regions of Syria and modern-day Turkey, where Pānīpatī was called Muḥammad al-Hindī.[37] Later volumes were printed as well, resulting in the gradual recognition and influence of Pānīpatī's commentary. His biographers identify the later widespread recognition of *at-Tafsīr al-maẓharī* in the admiration it received from many scholarly groups and the translation of the commentary into Urdu. It prompted others to write supercommentary (*ḥāshiya*) on the text as well.[38] Deoband luminary, Ashraf 'Alī Thānawī praised the work for its comprehensiveness and said that he had continuously read the commentary from an early age.[39] Thānawī cited Pānīpatī's work in a number of passages in his Qur'ān commentary.

5 Conclusion

What can be inferred from the various aspects of Pānīpatī's consistently positive reception in traditionalist circles is that his work served as a useful tool for his readers to utilize for their own purposes. It can be viewed as both a traditionalist and modernist piece. Its traditionalist elements derived from Islamic intellectual traditions,

35 Pānīpatī, *at-Tafsīr*, 1: 22.

36 On the impact of print technology on the publishing of religious literature, see Ulrike Stark, *An Empire of Books: The Naval Kishore Press and the Diffusion of the Printed Word in Colonial India* (New Delhi: Orient Blackswan, 2009).

37 Al-Ḥasan, *Tadhkira*, 487.

38 Ibid., 493f.

39 Ibid., 488.

while it may be considered modernist as it tackled the challenges of interpreting the Qur'ān during Pānīpatī's time with creativity and innovation. However, from an epistemological perspective, the tafsīr and associated hermeneutics were intricate and creative. The key lesson is to differentiate epistemological insights from sociological ones when analyzing a work of Qur'ānic interpretation. Constructed categories such as "Ḥanafī" and "traditionalist" are inadequate as they designate the disagreements among Muslim scholars and not any concrete categories into which Muslim scholars can be placed. In fact, as Stanley Fish argues,[40] Pānīpatī was reacting to the text of the Qur'ān in his own way.

6 Bibliography

6.1 Sources

'Ārif, Maḥmūd al-Ḥasan, *Tadhkirat Qāḍī Muḥammad Thanā' Allāh Pānīpatī*. Lahore: Institute of Islamic Culture, 1995.

Aẓhar, Ḥāfiẓ 'Abd ar-Rashīd. "Ad-Dirāsa al-muqārana bayna at-Tafsīr al-maẓharī wa-Fatḥ al-bayān." PhD diss., University of the Punjab, 1997.

Pānīpatī, Qāḍī Thanā' Allāh. *At-Tafsīr al-maẓharī*. Edited by Aḥmad `Izzu `Ināyat. 10 vols. Beirut: Dār Iḥyā' at-Turāth al-'Arabī, 2004.

Thānawī, Ashraf 'Alī. *Tafsīr bayān al-Qur'ān*. 3 vols. Lahore: Maktaba Raḥmāniyya, 2011.

Walī Allāh, Shāh. *Al-Fawz al-kabīr fī uṣūl at-tafsīr*, trans. by Sa'īd Aḥmad al-Bālanfūrī. Damascus: Dār al-Ghawthānī li-d-Dirāsāt al-Qur'āniyya, 2008.

6.2 Secondary Literature

Āfrīdī, Muḥammad 'Ādil Khān. "Lamḥāt at-tarbiya fī *t-Tafsīr al-maẓharī*." *Journal of Islam in Asia* 8, no. 2 (2012): 33–52.

Ahmad, M.G. Zubaid. *The Contribution of Indo-Pakistan to Arabic Literature: From Ancient Times to 1857.* Lahore: Sh. Muhammad Ashraf, 1968.

Ali, Kecia. *Lives of Muhammad*. Cambridge, MA: Harvard University Press, 2014.

Bashir, Kamran. *The Qur'an in South Asia: Hermeneutics, Qur'an Projects, and Imaginings of Islamic Tradition in British India*. London and New York: Routledge, 2021.

Dallal, Ahmad S. *Islam without Europe: Traditions of Reform in Eighteenth Century Islamic Thought*. Chapel Hill: The University of North Carolina Press, 2018.

Fish, Stanley. *Is There a Text in This Class? The Authority of Interpretive Communities*. Cambridge, MA: Harvard University Press, 1982.

Ingram, Brannon D. *Revival from Below*. Oakland, California: The University of California Press, 2018: 128.

40 Stanley Fish, *Is There a Text in This Class? The Authority of Interpretive Communities* (Cambridge, MA: Harvard University Press, 1982).

Jackson, William Kesler. “A Subcontinent’s Sunni Schism: The Deobandi-Barelvi Rivalry and the Creation of Modern South Asia.” PhD diss., Syracuse University, 2013.

Saleh, Walid. *The Formation of the Classical* Tafsīr *Tradition*. Leiden: Brill, 2003.

Stark, Ulrike. *An Empire of Books: The Naval Kishore Press and the Diffusion of the Printed Word in Colonial India*. New Delhi: Orient Blackswan, 2009.

Tareen, Sher Ali. *Defending Muhammad in Modernity*. Notre Dame, IN: Notre Dame Press, 2020.

Qur'ānic Hermeneutics in the Ottoman Period

Halis Albayrak and Esra Gözeler

1 Introduction

The Ottoman State (1300–1922/700–1340), the longest-lasting state of the Turkish-Islamic world, boasted a social and scholastic structure formed by different languages and cultures over a period of 622 years. In terms of scholarship, the Ottoman period was characterized by a continuation of the period of Anatolian Seljuks and Principalities.[1] Research into scholarly activities during the Ottoman era have divided the Ottoman scholarly tradition into three periods.[2] In the first period, under the influence of the Sufi Dāvūd-i Qayṣarī (d. 1350/751) – head *mudarris* of the Iznik Madrasa and the first *mudarris* of Ottomans – Sufism, theology, and philosophy were the main fields of the early Ottoman scholastic platform from the outset.[3] In the second period, there was the Mullā Shams ad-Dīn Fanārī (d. 1431/834) school, which excelled in the fields of logic, *tafsīr*, *uṣūl al-fiqh*, and metaphysics.[4] Within Mullā Fanārī's scholarly paradigm, "logic, the methodology of the Islamic sciences, and Sufi metaphysics" coexisted without contradicting one another.[5] The third period began with Ali Kushcu (d. 1474/879), the *mudarris* of Hagia Sophia Madrasa and an astronomer and mathematician.[6] Later, a mathematical dimension was incorporated into the Ottoman scholarly tradition, which had essentially been based on Sufi theology until that point. So, how did Qur'ānic hermeneutics develop during these periods?

1 Mehmet Paçacı, "Osmanlı Medresesinde Tefsir Öğretiminin Yeri ve Tefsirin Çağdaşlıkla Karşılaşması," in *Osmanlı Toplumunda Kur'an Kültürü ve Tefsir Çalışmaları*, ed. Bilal Gökkır et al. (Istanbul: İlim Yayma Vakfı Kur'an ve Tefsir Akademisi, 2011), 1: 73; Mustafa Öztürk, *Osmanlı Tefsir Mirası* (Ankara: Ankara Okulu Yayınları, 2012), 16; Susan Gunasti, "Political Patronage and the Writing of Qur'ān Commentaries among the Ottoman Turks," *JIS* 24, no. 3 (2013): 337.

2 İhsan Fazlıoğlu, "Osmanlılar: Düşünce Hayatı ve Bilim," in *TDVIA*, ed. İslâm Araştırmaları Merkezi (Ankara: TDV, 2007), 33: 549; Öztürk, *Osmanlı Tefsir Mirası*, 16.

3 Öztürk, *Osmanlı Tefsir Mirası*, 16; Mehmet Bayrakdar, "Dâvûd-i Kayserî," in *TDVIA*, ed. İslâm Araştırmaları Merkezi (Ankara: TDV, 1994), 9: 33.

4 Öztürk, *Osmanlı Tefsir Mirası*, 17.

5 Tahsin Görgün, "Molla Fenârî," in *TDVIA*, ed. İslâm Araştırmaları Merkezi (Ankara: TDV, 2005), 30: 247.

6 Öztürk, *Osmanlı Tefsir Mirası*, 17.

Note: All English translations of the Qur'ān in this chapter are from Arthur John Arberry, *The Koran Interpreted* (London: George Allen and Unwin Ltd., 1955). All transcriptions of the Qur'ānic verses, with small modifications, are taken from the *Corpus Coranicum* Project: "Corpus Coranicum," accessed February 12, 2021, https://corpuscoranicum.de.

https://doi.org/10.1515/9783111320083-025

2 One Perspective on the Structure of Qurʾānic Hermeneutics in the Ottoman Period

Most research on Ottoman *tafsīr* begins with an introduction similar to the following: "Among the studies on the history of *tafsīr*, the Ottoman era has not been given sufficient coverage." It is even suggested that this period was left "dark."[7] The term "missing link" has also been used to describe Ottoman *tafsīr*.[8] Qurʾānic hermeneutics is defined as the "missing link" in the Ottoman period, either because the relevant period is not included or because it was inadequately evaluated, especially in classical works and modern research into the history of *tafsīr*. In short, Ottoman *tafsīr* activities are characteristic of a period that has been neglected in *tafsīr* historiography. For this reason, studies in this field have a common purpose, that is, to fill a gap in the history of *tafsīr*.

Reciting, memorizing, and attempting to understand the Qurʾān were some of the religious and scholarly activities that characterized Ottoman society. Some Ottoman scholars memorized the text of a *tafsīr* together with the Qurʾān. On the other hand, the science of *tafsīr* was taught in Ottoman madrasas at the highest level, while *tafsīr* lessons were also given in mosques and palaces. In the month of Ramadan, a certain *tafsīr* was read in the presence of the sultan, and discussions were held concerning the text. These lessons, which continued from 1759/1172 until 1924/1342, when the caliphate was abolished, were known as *Huzur Dersleri* (Imperial Sessions). *Huzur Dersleri* which began on the first day of Ramadan, were usually completed in eight sessions. In these sessions, participants generally read the *tafsīr* of Nāṣir ad-Dīn Abū Saʿīd ʿAbdullāh b. ʿUmar b. Muḥammad al-Bayḍāwī (d. 1286/685), *Anwār at-tanzīl*.[9]

Abū l-Qāsim Maḥmūd b. ʿUmar b. Muḥammad al-Khwārizmī az-Zamakhsharī (d. 1144/538) should be cited among the leading names that supervised *tafsīr* activities in the Ottoman era. His *tafsīr*, *al-Kashshāf*, was one of the main textbooks used in Ottoman madrasas. The reason for this was that the principal source of his Qurʾānic hermeneutics was rational evidence.[10] However, in view of the number of *sharḥ* and *ḥāshiyas* (commentary and glosses) written on *al-Kashshāf* in the Ottoman period, az-Zamakhsharī

---

7 Muhammed Abay, "Osmanlı Döneminde Yazılan Tefsir ile İlgili Eserler Bibliyografyası: Tefsirler, Haşiyeler, Sûre Tefsirleri, Tercümeler," *Dîvân İlmî Araştırmalar* 6, no. 1 (1999): 249.

8 Dücane Cündioğlu, "Çağdaş Tefsir Tarihi Tasavvurunun Kayıp Halkası: 'Osmanlı Tefsîr Mirası' – Bir Histografik Eleştiri Denemesi," *İslâmiyât* 2, no. 4 (1999): 64–72; Halil Şimşek, "The Missing Link in the History of Quranic Commentary" (PhD diss., University of Toronto, 2018), 1.

9 Ziya Demir, *Osmanlı Müfessirleri ve Tefsir Çalışmaları: Kuruluştan X/XVI: Asrın Sonuna Kadar* (Istanbul: Ensar Neşriyat, 2006), 99; Abay, "Osmanlı Döneminde Yazılan Tefsir ile ilgili Eserler Bibliyografyası," 250; Öztürk, *Osmanlı Tefsir Mirası*, 48–53; Mehmet İpşirli, "Huzur Dersleri," in *TDVIA*, ed. İslâm Araştırmaları Merkezi (Ankara: TDV, 1998), 18: 441.

10 Öztürk, *Osmanlı Tefsir Mirası*, 21.

clearly played a significant role.[11] Another person that influenced *tafsīr* works in the Ottoman period was Fakhr ad-Dīn Muḥammad b. ʿUmar b. Ḥusayn ar-Rāzī (d. 1210/606). His *tafsīr, Mafātīḥ al-ghayb,* was one of the main sources for many *tafsīr* works written in the Ottoman period. However, when it comes to determining the most influential exegete on Ottoman *tafsīr* literature, there is no doubt that it was al-Bayḍāwī.[12] Philological explanations, theological and judicial discussions, and a clarity of preferences comprise the common *tafsīr* methodology of these three exegetes.

Qur'ānic hermeneutics during the Ottoman era can be divided into four groups: first, the translation and interpretation of the entire Qur'ān; second the translation and interpretation of some verses or suras of the Qur'ān; third, the commentaries and *ḥāshiyas* of the Qur'ānic *tafsīrs*; and fourth, *tafāsīr* translations. A bibliographic study on Ottoman *tafsīr* conducted in 1999 examined 583 *tafsīr* works by 403 different authors. According to the study, sixty-seven of them were complete *tafāsīr*, 273 were *tafsīr* of *juz'* or suras, 228 were *ḥāshiyas*, and the remaining fifteen were *tafsīr* translations.[13] Although Arabic was preferred in the composition of these works, Turkish and Persian were also used.[14] The most frequently translated *tafsīr* work in the Ottoman era was *Tafsīr al-Qur'ān al-karīm* by Abū l-Layth as-Samarqandī (d. 983/373), a jurist and Sufi exegete.[15] This was a *riwāya* (report)-based commentary and included Sufistic and judicial explanations, which indicates why this *tafsīr* was widely preferred in the Ottoman period.

Tibyān Tafsīri was another classical work that shaped the *tafsīr* culture of the Ottoman period and became popular among the people. It is widely accepted that this *tafsīr* was written by ʿAyinṭābī Maḥmad Efendī (d. 1699/1111). It was the first Turkish *tafsīr* published in the Ottoman period and is one of the first examples of "translation-commentary (*maāl-tafsīr*)"-style works, as it also includes a Turkish translation of the Qur'ān. The background to the composition of this work is as follows: ʿAyinṭābī Maḥmad Efendī was invited to Istanbul to attend the *Huzur Dersleri* held during the period of Mehmed IV (d. 1693/1104). At these sessions, Mehmed IV asked ʿAyinṭābī Maḥmad Efendī to translate the Qur'ān into Turkish, upon which ʿAyinṭābī Maḥmad

11 See Walid Saleh, "The Gloss as Intellectual History: *Hāshiyahs* on *al-Kashshāf*," *Oriens* 41 (2013): 217–59; Taha Boyalık, *el-Keşşâf Literatürü: Zemahşerî'nin Tefsir Klasiğinin Etki Tarihi* (Istanbul: İSAM, 2019).

12 Abay, "Osmanlı Döneminde Yazılan Tefsir ile İlgili Eserler Bibliyografyası," 250f.

13 Ibid., 256f. For *tafāsīr* in Turkish, see Hidayet Aydar, "Türkler'de Kur'an Çalışmaları," *İstanbul Üniversitesi İlahiyat Fakültesi Dergisi* 1 (1999): 159–235; Halil Şimşek, "Turks and Their Translations/Commentaries on the Qur'ân: An Historical and Bibliographical Survey," *Kilis 7 Aralık Üniversitesi İlahiyat Fakültesi Dergisi* 6, no. 11 (2019): 395–420.

14 Abdulhamit Birışık and Recep Arpa, "Osmanlı Dönemi Tefsir Çevirileri," *Türkiye Araştırmaları Literatür Dergisi* 9, no. 18 (2011): 192.

15 Abdulhamit Birışık, "Osmanlı Döneminde Türkçe Tefsirler," in *Başlangıçtan Günümüze Türklerin Kur'an Tefsirine Hizmetleri: Tebliğler ve Müzakereler*, ed. Muhammed Abay, prepared by İsmail Kurt and Seyit Ali Tüz (Istanbul: Ensar Neşriyat, 2012), 202–3; Birışık and Arpa, "Osmanlı Dönemi Tefsir Çevirileri," 195.

Efendī, according to the general opinion, wrote his own *maāl-tafsīr* based on Khiḍr b. ʿAbd ar-Raḥmān al-Azdī's work, *at-Tibyān fī tafsīr al-Qur'ān*. This work has the characteristics of both *at-tafsīr ad-dirāya* and *at-tafsīr ar-riwāya*.[16]

In terms of its sources and methods, the Ottoman *tafsīr* writing tradition can be divided into two main schools: (1) *at-Tafsīr bi-d-dirāya* and (2) *Sharḥ*s and *hāshiya*s. Below, we will examine these schools according to the commentaries of their leading figures.

3 *At-Tafsīr bi-d-Dirāya*/Rational Exegesis

During the Ottoman era, the main framework of *dirāya tafāsīr*, which include various topics or discussions of the explanation of a certain verse, were the fields of *kalām* and *taṣawwuf*. The theological approach interprets the verses through the filter of reason, considering information provided by the *riwāya*s insufficient. In this method, elucidation is replaced by interpretation. The Sufi approach attempts to derive the inner interpretations of the verses as well as their apparent meanings. Attaining these interpretations requires not only knowledge but also moral maturity. Accordingly, those who have knowledge of Arabic can come to know the literal, apparent meaning (*ẓāhir*) of the verses, and the Gnostics (*ʿārif*) can comprehend the hidden meaning (*bāṭin*). We will now evaluate this school through the *tafsīr* theories of Mullā Fanārī, Ismāʿīl Ḥaqqī Bursawī, and Abū s-Suʿūd al-ʿImādī.

3.1 Mullā Fanārī (d. 1431/835): *Taṣawwuf* and *Tafsīr*

Mullā Fanārī, whose real name was Shams ad-Dīn Muḥammad b. Ḥamza, was appointed by Yıldırım Bayezid (d. 1403/805) to the position of *mudarris* at the Manastır Madrasa and *qāḍī* (judge) of Bursa in 1393/795. He was one of the Ottoman State's leading scholars studying *taṣawwuf*. His interest in Sufi culture is clearly apparent in his works. Muḥyī d-Dīn b. al-ʿArabī (d. 1240/638) influenced his interest in Sufi thought even further. But he was a theologian who followed the tradition of ar-Rāzī. He is known as a scholar who transmitted the philosophy of Abū ʿAlī b. Sīnā (d. 1037/428) into the Ottoman culture of Islamic studies.[17]

16 Recep Arpa, "İlk Matbu Türkçe (Osmanlıca) Tefsir: Tibyan Tefsiri," in *Osmanlı Toplumunda Kur'an Kültürü ve Tefsir Çalışmaları*, ed. Gökkır et al., 1: 241, 243f.; Recep Arpa, "Tibyân Tefsiri," in *TDVIA*, ed. İslâm Araştırmaları Merkezi (Ankara: TDV, 2012), 41: 127f.

17 İbrahim Hakkı Aydın, "Molla Fenârî," in *TDVIA*, ed. İslâm Araştırmaları Merkezi (Istanbul: TDV, 2005), 30: 245; Görgün, "Molla Fenârî," 247; Mustafa Öztürk, "Molla Fenari ve Tefsirde Senkretizm," in *Uluslararası Molla Fenari Sempozyumu: Bildiriler*, ed. Tevfik Yücedoğru et al. (Bursa: Bursa Büyükşehir Belediyesi Yayınları, 2010), 401f.

Mullā Fanārī was inspired by traditions of language, methodology, *fiqh*, philosophy, logic, theology, and Sufism. His *tafsīr*, *'Ayn al-a'yān*, is the first example of the *Sufi* perspective of *tafsīr* written in Arabic during the Ottoman period. The work consists of an introduction (*muqaddima*), an explanation of *isti'ādha* (seeking refuge in Allah) and *Basmala*, and an interpretation of sura 1, *al-Fātiḥa* (The Opening). Mullā Fanārī explains sura 1 under the subheadings "*qirā'āt* (variant readings), language (*i'rāb balāgha*), hadith, *kalām*, *aḥkām*, *al-ḥaqā'iq* (realities), *al-ma'ārif* (gnostic knowledge), *at-tadhkīr* (reminding), *at-targhīb* (encouraging good), and *at-tarhīb* (discouraging evil)."[18] Among these titles, *al-ḥaqā'iq*, *al-ma'ārif*, *at-tadhkīr*, *at-targhīb*, and *at-tarhīb* ascribe *'Ayn al-a'yān* the characteristics of a Sufi *tafsīr*. He interprets the verses under the title of *al-ḥaqā'iq* by conveying opinions from, in particular, Ibn al-'Arabī and Ṣadr ad-Dīn al-Qūnawī (d. 1274/672), without discussing them in Sufi terminology. Under the titles of *al-ma'ārif*, *at-targhīb*, and *at-tarhīb*, he determines the moral values contained in the verses and the responsibilities of human beings.[19]

The *muqaddima* of *'Ayn al-a'yān* can be classified as a guide for the science of *tafsīr*. This *muqaddima* consists of four chapters. In the first chapter (1st *bāb*) entitled "The Definition of *Tafsīr* Science and Related Subjects," Mullā Fanārī examines the following topics: defining *tafsīr* (1st section), dividing the science of *tafsīr* into *tafsīr* and *ta'wīl* (2nd section), the permissibility of *tafsīr* and *ta'wīl* (3rd section), knowing the various aspects of *tafsīr* and *ta'wīl* (4th section), the collective obligation (*farḍ al-kifāya*) of learning the science of *tafsīr* (5th section), the companions of the Prophet Muḥammad (*ṣaḥāba*), and the successors (*tābi'ūn*) from whom the *tafsīr* was learned (6th section). This chapter concludes with an epilogue (*ḥātima*) in which he discusses the truths, divisions, and means of the reason of knowledge and perception (*idrāk*).[20] In the second chapter (2nd *bāb*), titled "The Need for the Science of *Tafsīr* and Related Subjects," he examines the following subjects: the need for *tafsīr* (1st section), the superiority of this science (2nd section), and the suras of the Qur'ān, their recitation and virtues of people of the Qur'ān (3rd section).[21] This chapter is followed by the third chapter (3rd *bāb*) "The Subject of the Science of *Tafsīr*." In this chapter, Mullā Fanārī deals with the following points: the description of the Qur'ān (1st section); the descending (*nuzūl*) of the Qur'ān and the revelation, the first and last verse revealed, recitation (2nd section); the codification of the Qur'ānic text (3rd section); the mentioning of names in the Qur'ān (4th section); and the chapters of the Qur'ān and Makkī-Madanī, *nāsikh-mansūkh*, sura, verses, words, and letters (5th section).[22] The fourth chapter (4th *bāb*),

18 Taha Boyalık, "Molla Fenârî'nin Tefsir İlminin Mahiyetine Dair Tartışmasının Tahlili," *İslâm Araştırmaları Dergisi* 18 (2007): 93.

19 Boyalık, "Molla Fenârî'nin Tefsir İlminin Mahiyetine Dair Tartışmasının Tahlili," 94.

20 Shams ad-Dīn Muḥammad b. Ḥamza Mullā Fanārī, *'Ayn al-a'yān* (Istanbul: Dar Sa'āda, 1907/1325), 4–16.

21 Ibid., 17–41.

22 Ibid., 41–85.

entitled "Knowledge Required for *Tafsīr* Science,"[23] completes the *muqaddima*. We can examine Mullā Fanārī's Qur'ānic hermeneutics by making the following comments, based on the titles of *qirā'āt, kalām, at-tadhkīr, al-ḥaqā'iq*, and *al-ma'ārif*:

I. *māliki yawmi d-dīni*
the Master of the Day of Doom (Q 1:4)

Mullā Fanārī points out the difference in the way of recitation (*qirā'a*) by stating that the word *mālik* is also pronounced as *malik*, noting that his preference is *mālik*, which he explains as follows: "As the number of letters of a word increases, so does its meaning and reward. Accordingly, the number of letters in the word *mālik* is more. Allah praised him as *mālik al-mulk* (Q 3:26), but did not praise him as *malik al-mulk*. Therefore, the word *mālik* is more comprehensive."[24]

II. *iyyāka na'budu wa-iyyāka nasta'īnu*
Thee only we serve; to Thee alone we pray for succor (Q 1:5)

In the context of this verse, some of the reminders (*tadhkīr*) made by Mullā Fanārī for readers regarding the meaning of worshiping Allah and asking His help are as follows:

> Worship is to affirm everything Allah informs, to obey his predestination, to obey his commands and prohibitions and to trust him.
>
> Everyone wants, knowingly or unknowingly; some of them win. All this happens only with the help of Allah. Therefore, say *we only ask you for help*.
>
> We only ask you for help, we do not ask anyone else. Because without your permission and assistance, no one else can help me. For this, I ask you for help, directly from the main source of help, not from the intermediaries, *O Lord*![25]

III. *ihdinā ṣ-ṣirāṭa l-mustaqīma*
Guide us on the straight path (Q 1:6)

Mullā Fanārī demonstrates a philological interest in the interpretation of this verse and explains the meaning of the word *hidāya* (guidance) in three stages. This is also an example of his Qur'ānic semantics. In the first stage, he presents the literal meaning by stating that the word means "to signify, to achieve what is desired." In the second stage, he lists the meanings of the word *hidāya* in the Qur'ān. In the third stage, he reports that Junayd al-Baghdādī (d. 909/297), one of the most prominent representatives of early Sufism, explained that the word *ihdinā* meant "to make our hearts lean toward you; let our goal be to appear before you; be your evidence for everything that comes to us from you, guide us."[26] Mullā Fanārī creates a semantic system by identifying the

23 Ibid., 85–125.
24 Ibid., 205f.
25 Ibid., 300.
26 Ibid., 302f.

meaning of the word *hidāya* in the Arabic language, its meanings in the Qur'ān, and its meaning in the post-Qur'ānic Sufi terminology.

One of the realities (*ḥaqā'iq*) that Mullā Fanārī draws attention to in the context of the same verse is as follows:

> *Ihdinā* means "guide us" and is a prayer in the imperative form. The reason why the prayer phrase is plural is that it comes after the phrase "we only worship you." As if it means "we pray with the tongue of each person, follow us!" Every individual who prays in this way wants to benefit from the abundance of the community.[27]

According to one of the branches of knowledge (*ma'ārif*) mentioned by Mullā Fanārī in his interpretation of this verse, guidance is divided into three parts. The first is the guidance of all beings in general. According to this, guidance is every living creature turning toward the things that will benefit him and away from the things that will harm him. The second is reserved for believers and leads them to paradise. The third is the guidance of distinguished/elite (*khawāṣṣ*) believers. This leads them directly to Allah.[28] This last definition of guidance by Mullā Fanārī points to the transmission of Sufi culture to Qur'ānic hermeneutics.

3.2 Ismā'īl Ḥaqqī Bursawī (d. 1725/1137): *Maw'iẓa* and *Tafsīr*

Bursawī achieved renown with the sermons and *tafsīr* lessons he gave in the Grand Mosque of Bursa[29] and other mosques. The title of his *tafsīr* is *Rūḥ al-bayān fī tafsīr al-Qur'ān*. It is a complete interpretation (from sura 1, *al-Fātiḥa* (The Opening) to sura 114, *an-Nās* (Mankind)) of the Qur'ān. He completed this work in twenty-three years, equivalent to the period in which the Qur'ān was revealed. He wrote his *tafsīr* in Arabic by adding Sufi commentaries and poems to his sermons, in which he interpreted the Qur'ān in the Grand Mosque of Bursa and other mosques, beginning with sura 1.[30]

Because it was first introduced as a sermon, *Rūḥ al-bayān* contains advice and recommendations. Moreover, it was a significant source of methods for *maw'iẓa* (guidance) in later times. Bursawī's theory of *tafsīr* comprises Sufi interpretations and poems. Persian poems and explanations are frequently mentioned in the *tafsīr*, in which Bursawī uses both the method of *riwāya* and *dirāya*. He enriches this method with his *Sufi* interpretations. He applies the method of interpretating the Qur'ān with the Qur'ān (*tafsīr al-Qur'ān bi-l-Qur'ān*) by first explaining the verses by means of other verses. However, he also refers to hadiths in the explanation of Qur'ānic verses.

27 Ibid., 323.

28 Ibid., 354.

29 Bursa is a city located in northwestern Turkey.

30 Ali Namlı, "İsmail Hakkı Bursevi'nin *Rûhu'l-Beyân fî Tefsîri'l-Kur'ân* Adlı Tefsiri," *Türkiye Araştırmaları Literatür Dergisi* 9, no. 18 (2011): 368.

He performs grammatical analysis to determine the meaning of words and also uses rhetorical information when necessary.[31] We can evaluate his method of providing Sufi interpretations and recommendations, which is characteristic of his Qurʾānic hermeneutics, by looking at the following examples:

I. . . . *matā naṣru llāhi a-lā inna naṣra llāhi qarībun*
When comes God's help? Ah, but surely God's help is nigh. (Q 2:214)

In this verse, believers are warned to not be ungrateful to the Prophet Muḥammad for their hardships. This warning is a reminder that those who believed in the previous prophets asked the following question because of the difficulties they faced: "O God! When will you help?" Bursawī explains this question from a Sufi perspective: "It points out that reaching Allah and achieving a superior degree in His presence will be achieved by abandoning desires, passions, pleasures and enduring hardships with devotion."[32] The interpretation of gaining devotion/asceticism in the sense of reaching God (*waṣl*), giving up worldly desires and passions, eating and drinking less, sleeping less, worshipping more frequently, and thus improving morals points to an interpretation that is far removed from the textual and historical context of the verse.

II. *wa-mathalu lladhīna yunfiqūna amwālahumu btighāʾa marḍāti llāhi wa-tathbītan min anfusihim . . .*
But the likeness of those who expend their wealth, seeking God's good pleasure, and to confirm themselves . . . (Q 2:265)

This verse was revealed in the context of the hypocrites who helped others out of pride in order to show off. Believers are warned not to be like the hypocrites who simply spend their wealth ostentatiously. Bursawī defines those who heed the Prophet Muḥammad's call and help others only for the sake of God as follows:

> They asked God to correct their souls in faith and obedience, in order to save them from the disgrace of the stinginess in their souls, from the love of worldly goods and the inferiority of keeping these to themselves. The more you release the soul, the more you spoil it. After all, the person gets used to laziness and stinginess. He is fond of property; he does not spend his wealth on anything good. However, if you fulfill physical and financial worship and assign various duties to the soul, it will obey you and give up its previous habits.[33]

Bursawī explains the spending (*infāq*), which is one of the basic tenets of the Qurʾān, by associating it with the disciplining of the soul, one of the basic principles of Sufism.

III. *wa-thiyābaka fa-ṭahhir*
thy robes purify (Q 74:4)

31 Ibid., 377–81.

32 Ismāʿīl Ḥaqqī Bursawī, *Rūḥ al-bayān fī tafsīr al-Qurʾān* (Istanbul: Dar Saʿāda, 1912/1330), 1: 330.

33 Ibid., 424.

This verse is one of the early verses revealed to the Prophet Muḥammad. Within the Sufi paradigm, Bursawī explains this statement as follows: "There is a transition from inner purification to outer purification. Because, in general, the person who purifies the soul avoids impurity and wants to spread purification in everything. Indeed, religion is based on cleanliness."[34] The verse specifically points to the Prophet Muḥammad's early experience of the revelations. However, Bursawī expands further on the interpretation by applying the meaning of the verse to the dimension of cleansing the soul.

3.3 Abū s-Suʿūd al-ʿImādī (d. 1574/982): Reason and *Tafsīr*

Abū s-Suʿūd was a *mudarris* in various madrasas. He served as the *qāḍī* (judge) of Istanbul (1533/940) and Kazasker of Rumelia (1533/940). He was the fourteenth Ottoman *Shaykh al-Islām* (1545–1574/952–982).[35] The most famous *tafsīr* among the *dirāya tafāsīr* of the Ottoman era is Abū s-Suʿūd's work, titled *Irshād al-ʿaql as-salīm ilā mazāyat al-kitāb al-karīm*. This *tafsīr* is a complete commentary on the Qur'ān and was presented by Abū s-Suʿūd to the Ottoman Sultan Suleiman the Magnificent (d. 1566/974).

Abū s-Suʿūd wrote a short *muqaddima* at the beginning of his *tafsīr*. He assigned *al-Kashshāf* and *Anwār at-tanzīl* a special position among the *tafāsīr* he read and described them as the main works of their time written in this field.[36] We can summarize Abū s-Suʿūd's Qur'ānic hermeneutics as follows: He gives grammatical explanations and refers to pre-Islamic poetry. He explains the judicial verses with reference to the Ḥanafī school by using the expression "according to us (*ʿindanā*)," makes comparisons with other judicial schools, and in some cases even criticizes them. Hadiths and narratives of *asbāb an-nuzūl* (occasions of revelation) are among Abū s-Suʿūd's main sources. One of the features of this *tafsīr* is the way that it identifies differences in the recitals.[37] We can examine his theory of exegesis through two examples:

I. *yā-ayyuha lladhīna āmanū idhā qumtum ilā ṣ-ṣalāti fa-ghsilū wujūhakum wa-aydiyakum ilā l-marāfiqi wa-msaḥū bi-ruʾūsikum wa-arjulakum ilā l-kaʿbayni . . .*
O believers, when you stand up to pray wash your faces, and your hands up to the elbows, and wipe your heads, and your feet up to the ankles . . . (Q 5:6)

This verse states that ablution is obligatory (*farḍ*). Abū s-Suʿūd begins the interpretation of this verse by establishing a semantic link between the phrase "*idhā qumtum*

34 Ismāʿīl Ḥaqqī Bursawī, *Rūḥ al-bayān* (Istanbul: ʿUthmān Bey Maṭbaʿasī, 1928), 10: 225.
35 Ahmet Akgündüz, "Ebüssuûd Efendi," in *TDVIA*, ed. İslâm Araştırmaları Merkezi (Ankara: TDV, 1994), 10: 365f.
36 Abū s-Suʿūd al-ʿImādī, *Irshād al-ʿaql as-salīm ilā mazāyā l-kitāb al-karīm*, ed. ʿAbd al-Qādir Aḥmad ʿAṭāʾ (Riyadh: Maktabat ar-Riyāḍ al-Ḥadītha, n.d.), 1: 3.
37 Süleyman Ateş, "İrşâdü'l-Akli's-Selîm," in *TDVIA*, ed. İslâm Araştırmaları Merkezi (Ankara: TDV, 2000), 22: 456f.

ilā ṣ-ṣalāti fa-ghsilū" ("when you stand up to pray wash") and the 98th verse of sura 16, *an-Naḥl* (The Bee), *"fa-idhā qara'ta l-qur'āna fa-sta'iẓ bi-llāhi mina sh-shayṭāni r-rajīmi"* ("When thou recitest the Koran, seek refuge in God from the accursed Satan"). The connection he establishes between the verses sets an example for the interpretation of the Qur'ān through the Qur'ān. Accordingly, one of the conditions of prayer (*ṣalāt*) is intention (*niyya*) and the other is ablution (*wuḍu'*). After this introduction, we can follow Abū s-Su'ūd's judicial approach in interpreting the verses. Abū s-Su'ūd notes that according to the literal, apparent meaning of the verse, it is necessary (*wājib*) for anyone who intends to perform prayers to perform ablution, but that this issue is also controversial among scholars. Accordingly, it is not necessary for a person who already has ablution. Abū s-Su'ūd supports this judicial approach with a narrative, mentioning the report that the Prophet Muḥammad performed the five daily prayers with one ablution on the day of the conquest of Mecca. Abū s-Su'ūd continues his judicial statements on performing ablution as follows: "It is not valid to interpret the command in the verse as recommended (*mandub*) only for those who have ablution. Rather, the command specifically addresses those who do not have ablution."[38] Another point that reflects the Qur'ānic hermeneutics of Abū s-Su'ūd regarding this verse is the differences between the recital of the word *arjulakum* (your feet). In this context, he states that the word *arjulakum* refers to the word *wujūhakum* (your faces), while the common *sunna*, practice of companions, and the opinion of the majority of scholars (*mujtahid imāms*) confirm the *naṣb* recital of the word as *arjulakum*, the ruling of washing (feet). He supports this view with the statement "The limitation of *ilā l-ka'bayni* (up to the ankles) confirms this, since no limitation (*taḥdīd*) is used for wiping." In this context, Abū s-Su'ūd mentions that the word can also be read as *arjulikum*. According to him, the wisdom of this is as follows: "While washing the feet during ablution, the use of water should be minimized and washed in a manner close to wiping."[39] Abū s-Su'ūd's interpretation of this verse also includes an interdenominational comparison: he discusses how much of the head should be wiped by reporting the preferences of Imām ash-Shāfi'ī (the minimum amount of the head), Abū Ḥanīfa (the front of the head), and Imām Mālik (the whole head).[40] The judicial aspect determines the framework and limits of this interpretation.

II. *innī ana rabbuka fa-khla' na'layka . . .*
I am thy Lord; take off thy shoes . . . (Q 20:12)

This is one of the verses in which the terminology of the Sufi *tafsīr* is clearly apparent. The verse describes the following scene: the Prophet Moses and his family lose their way on the journey from Median to Egypt. When the Prophet Moses notices a fire on the way and goes over to it, thinking there will be someone there to give them direc-

38 Abū s-Su'ūd, *Irshād al-'aql as-salīm*, 2: 14f.
39 Ibid., 16.
40 Ibid.

tions, God calls out to him, "I am thy Lord; take off thy shoes." Abū s-Su'ūd cites four different explanations in his interpretation of this verse: First, the Prophet Moses was ordered to take off his shoes because being barefoot is more befitting of humility and good manners. This is why the Ka'ba is circumambulated barefoot. Second, his feet must touch the ground in order to be blessed by the Sacred Valley (the Valley of Tuwa). Third, Moses was ordered to remove his shoes because they were made of tanned donkey skin and therefore impure. The fourth and final reason is the command to cleanse the heart of family and property.[41] Although Abū s-Su'ūd does not prioritize any one of these views, he prefers an interpretation that emphasizes spiritual cleansing as a common ground. Thus, although *Irshād al-'aql as-salīm* is not categorized as a Sufi *tafsīr*, its Qur'ānic hermeneutics reflect the Sufistic approach, which is the main character of the Ottoman scholarly tradition.

4 *Sharḥ* and *Ḥāshiyas*/Commentary and Marginal Notes

A *ḥāshiya* (plural *ḥawāshī*) is an explanation or annotation attached to the margins of a work. It contains explanations about both the annotated main text and the *sharḥ* text. The language of *ḥawāshī* is Arabic, and expressions such as *qāla* (said) / *aqūlu* (I say) and *qawluhu* (the author's following word) are widely used in these texts. This school, which was developed to foster understanding of all classical works, was founded by Dāvūd-i Qayṣarī, the first Ottoman *mudarris*. It emerged as the scholarly tradition of the Ottoman State was taking shape and became widespread in the second half of the fifteenth century.[42]

In order to maintain the classical scholarly tradition, Ottoman scholars preferred to write commentaries and *ḥāshiya* on influential *tafāsīr*. Therefore, the *tafāsīr* of az-Zamakhsharī and al-Bayḍāwī, two of the texts used as textbooks in Ottoman madrasas, are the works on which *sharḥ-ḥāshiya* were most commonly written.[43] So why was writing *sharḥ* and *ḥāshiya* preferred among Ottoman *tafsīr* works? The answer to this question is more or less as follows: to provide a better understanding of a work, to add supplementary information, and to generate discussions on the Islamic texts.[44] Writing *sharḥ* and *ḥāshiya* did not mean repeating classical texts. On the contrary, these texts helped to develop a culture of criticism, introducing new opinions not in-

41 Ibid., 3: 618.

42 Abay, "Osmanlı Döneminde Yazılan Tefsir ile ilgili Eserler Bibliyografyası," 254.

43 Muhammed Abay, "Osmanlı Döneminde Tefsir Haşiyeleri," in *Başlangıçtan Günümüze Türklerin Kur'an Tefsirine Hizmetleri: Tebliğler ve Müzakereler*, ed. Muhammed Abay, prepared by İsmail Kurt and Seyit Ali Tüz, 177, 179.

44 Ibid., 171f., 175.

cluded in the main text. They were original texts that made the traditional texts more understandable by applying methods of investigation, analysis, and criticism.[45] On the other hand, the *sharḥ-ḥāshiya* texts also served to adapt and localize the main/original texts to the locations in which they were read.[46]

We can learn more about this school by looking at two examples found in *Futūḥ al-ghayb*, the most extensive *ḥāshiya* text written by Abū Muḥammad Sharaf ad-Dīn Ḥusayn b. ʿAbd Allāh b. Muḥammad aṭ-Ṭībī (d. 1343/743) on *al-Kashshāf.* Az-Zamakhsharī records that the word *ghurfatan* ("a handful," functioning as the *mafʿūl*, the object) in Q 2:249 is read with a *fatḥa* as *gharfatan* (interpreted as "grasp in the hand," derived from the *maṣdar*, the verbal noun).[47] In his *ḥāshiya*, aṭ-Ṭībī adds that Abū l-Ḥasan al-Kisāʾī, Ḥamza al-Kūfī, ʿĀṣim b. Abī n-Najūd, and Abū ʿUmarb b. ʿĀmir preferred the *qirāʾa* of *ghurfatan*, while Nāfiʿ al-Madanī, Ibn Kathīr al-Makkī, and Abū ʿAmr b. ʿAlāʾ al-Baṣrī preferred the *qirāʾa* of *gharfatan.*[48] This *ḥāshiya* by aṭ-Ṭībī is an additional explanation that is not included in *al-Kashshāf.*

Another feature of writing *ḥāshiya* is related to the interpretation of verse 150 of sura 7, *al-Aʿrāf* (The Heights). When explaining this verse, az-Zamakhsharī records that, according to one narrative, the Torah consisted of seven parts. When the Prophet Moses threw the tablets on the ground, they broke, and six parts were lost; only one part remained. The parts that were lost had contained detailed instructions, while the remaining part emphasized guidance and mercy.[49] In the *ḥāshiya*, aṭ-Ṭībī reveals the contradiction between the two narratives by determining that az-Zamakhsharī elsewhere reports that the Torah weighed seventy camel loads, that a part of it was read in a year, and that no one but Moses, Joshua, Ezra, and Jesus could read the Torah.[50] This *ḥāshiya* by aṭ-Ṭībī takes the form of a critique.

---

45 Ibid., 172; Mesut Kaya, "Şerh ve Hâşiyelerin Tefsir İlmine Katkıları: Teftâzânî'nin *el-Keşşâf* Şerhi Örneği," in *İslam İlim ve Düşünce Tarihinde Şerh Geleneği*, eds. Mesut Kaya and Sezai Engin (Istanbul: Endülüs Yayınları, 2020), 206.

46 Mesut Kaya, "Osmanlı İlim Geleneğinde Şerh ve Hâşiye Yazıcılığı – Ahmed el-Karamânî'nin Tefsiri/Zemahşerî Şerhi Örneği," *Marife Dini Araştırmalar Dergisi* 15, no. 1 (2015): 15.

47 Az-Zamakhsharī, *al-Kashshāf ʿan ḥaqāʾiq ghawāmiḍ at-tanzīl wa-ʿuyūn al-aqāwīl fī wujūh at-taʾwīl*, ed. ʿĀdil Aḥmad ʿAbd al-Mawjūd and ʿAlī Muḥammad Muʿawwaḍ (Riyadh: Maktabat al-ʿUbaykān, 1418/1998), 1: 475.

48 Aṭ-Ṭībī, *Futūḥ al-ghayb fī l-kashf ʿan qināʿ ar-rayb: wa-huwa Ḥāshiyat aṭ-Ṭībī ʿalā l-Kashshāf*, ed. ʿUmar Ḥasan al-Qiyyām (Dubai: Jāʾizat Dubai ad-Dawliyya li-l-Qurʾān al-Karīm, 1434/2013), 3: 469; Muhammet Sacit Kurt, *Muʿtezile'ye Karşı Bir Ehl-i Sünnet Müdafiʿi: eṭ-Ṭībī'nin (743/1343) Futūḥu'l-Ġayb Adlı el-Keşşāf Haşiyesi ve Eserinde ez-Zemaḫşerī'ye (538/1144) Yönelttiği Eleştiriler* (Ankara: İlâhiyât, 2018), 96.

49 Az-Zamakhsharī, *al-Kashshāf*, 2: 512.

50 Aṭ-Ṭībī, *Futūḥ al-ghayb*, ed. Jamīl Muḥammad Banī ʿAṭā, 6: 589; Kurt, *Muʿtezile'ye Karşı Bir Ehl-i Sünnet Müdafiʿi*, 102f.

5 Conclusion

Ottoman-era *tafsīr* works were developed in accordance with the periods of the Ottoman scholarly tradition. In the beginning or early period, the fields of theology, Sufism, and philosophy played a founding role. By adding the paradigms of logic, jurisprudence, and metaphysics to these fields, Mullā Fanārī also influenced the content of the works written after him. In the last period, the influence of mathematics and astronomy in particular on Ottoman scholarly culture was likewise reflected in Qur'ānic hermeneutics. For example, *Sarā'ir al-Qur'ān*, a work by Ghāzī Aḥmad Mukhtār Pāshā (d. 1918/1335), an Ottoman grand vizier, is considered to have been a pioneering volume of the scientific *tafsīr* (*at-tafsīr al-'ilmī*) movement in the Ottoman era. This work is not a complete *tafsīr*; Aḥmad Mukhtār Pāshā prepared it during his lifetime by combining religious knowledge with the latest developments in astronomy. Moreover, Ottoman *tafsīr* culture was reflected in the first Turkish *tafsīr*s of modern Turkey. In the *tafsīr* of Elmalılı Hamdi Yazır (d. 1942), for example, explanations and interpretations based on language, *kalām*, jurisprudence, *uṣūl al-fiqh*, and Sufism are evident.

There were two main schools of Qur'ānic hermeneutics in the Ottoman period. The first school was a style of *at-tafsīr ad-dirāya* that developed on the basis of classical *tafsīr*, adopting the scholarly systems of language, *kalām*, jurisprudence, and Sufism as references. *Al-Kashshāf* and *Anwār at-tanzīl* are based on the multidimensional features of *at-tafsīr bi-d-dirāya*, reflecting this style of *tafsīr* tradition. In the second school of Qur'ānic hermeneutics in the Ottoman period, *shurūḥ* and *ḥawāshī* were more widely accepted than *tafsīr* compilations. The reason for and function of writing *sharḥ* and *ḥāshiya* on *tafsīr* was "to enrich and convey the existing heritage and culture to future generations."[51] In the writing of *tafsīr*, *sharḥ*, and *ḥāshiya*, Ottoman scholars explained the Qur'ānic verses based on the methods, interpretations, and preferences of classical *tafsīr*. Generally speaking, Qur'ānic hermeneutics in the Ottoman period is no longer a "missing link." However, there are still certain aspects of it waiting to be discovered.

6 Bibliography

6.1 Sources

Abū s-Su'ūd al-'Imādī. *Irshād al-'aql as-salīm ilā mazāyā l-kitāb al-karīm*. Ed. 'Abd al-Qādir Aḥmad 'Aṭā', 5 vols. Riyadh: Maktabat ar-Riyāḍ al-Ḥadītha, n.d.

Bursawī, Ismā'īl Ḥaqqī. *Rūḥ al-bayān fī tafsīr al-Qur'ān*. 10 vols. Istanbul: Dar Sa'ādat, vol. 1, 1912.

Bursawī, Ismā'īl Ḥaqqī. *Rūḥ al-bayān*. 10 vols. Istanbul: 'Uthmān Bey Maṭba'asī, vol. 10, 1928.

Mullā Fanārī, Shams ad-Dīn Muḥammad b. Ḥamza. *'Ayn al-a'yān*. Istanbul: Dar Sa'āda, 1907.

51 Öztürk, *Osmanlı Tefsir Mirası*, 210.

Aṭ-Ṭībī, al-Ḥusayn ibn Muḥammad. *Futūḥ al-ghayb fī l-kashf ʿan qināʿ ar-rayb: wa-huwa Ḥāshiyat aṭ-Ṭībī ʿalā l-Kashshāf*. 17 vols. Ed. ʿUmar Ḥasan al-Qiyyām, (vol. 3), Jamīl Muḥammad Banī ʿAṭā (vol. 6). Dubai: Jāʾizat Dubai al-Dawliyya li-l-Qurʾān al-Karīm, 2013.

Az-Zamakhsharī, Maḥmūd ibn ʿUmar. *Al-Kashshāf ʿan ḥaqāʾiq ghawāmiḍ at-tanzīl wa-ʿuyūn al-aqāwīl fī wujūh at-taʾwīl*. Ed. ʿĀdil Aḥmad ʿAbd al-Mawjūd and ʿAlī Muḥammad Muʿawwaḍ. 6 vols. Riyad: Maktabat al-ʿUbaykān, 1998.

6.2 Secondary Literature

Abay, Muhammed. "Osmanlı Döneminde Yazılan Tefsir ile İlgili Eserler Bibliyografyası: Tefsirler, Haşiyeler, Sûre Tefsirleri, Tercümeler." *Dîvân İlmî Araştırmalar* 6, no. 1 (1999): 249–303.

Abay, Muhammed. "Osmanlı Döneminde Tefsir Haşiyeleri." In *Başlangıçtan Günümüze Türklerin Kur'an Tefsirine Hizmetleri: Tebliğler ve Müzakereler*, edited by Muhammed Abay, prepared by İsmail Kurt and Seyit Ali Tüz. Istanbul: Ensar Neşriyat, 2012: 167–94.

Akgündüz, Ahmet. "Ebüssuûd Efendi." In *TDVIA*, edited by İslâm Araştırmaları Merkezi, 44 vols. Ankara: TDV, vol. 10, 1994: 365–71.

Arberry, Arthur John. *The Koran Interpreted*. London: George Allen and Unwin Ltd., 1955.

Arpa, Recep. "İlk Matbu Türkçe (Osmanlıca) Tefsir: Tibyan Tefsiri." In *Osmanlı Toplumunda Kur'an Kültürü ve Tefsir Çalışmaları-I*, edited by Bilal Gökkır, Necdet Yılmaz, Necmettin Gökkır, Ömer Kara, Muhammed Abay, Mustafa Karagöz. Istanbul: İlim Yayma Vakfı Kur'an ve Tefsir Akademisi, vol. 1, 2011: 241–60.

Arpa, Recep. "Tibyân Tefsiri." In *TDVIA*, edited by İslâm Araştırmaları Merkezi, 44 vols. Ankara: TDV, vol. 41, 2012: 127–28.

Ateş, Süleyman. "İrşâdü'l-Akli's-Selîm." In *TDVIA*, edited by İslâm Araştırmaları Merkezi, 44 vols. Ankara: TDV, vol. 22, 2000: 456–58.

Aydar, Hidayet. "Türkler'de Kur'an Çalışmaları." *İstanbul Üniversitesi İlahiyat Fakültesi Dergisi* 1 (1999): 159–235.

Aydın, İbrahim Hakkı. "Molla Fenârî." In *TDVIA*, edited by İslâm Araştırmaları Merkezi, 44 vols. Ankara: TDV, vol. 30, 2005: 245–47.

Bayrakdar, Mehmet. "Dâvûd-i Kayserî." In *TDVIA*, edited by İslâm Araştırmaları Merkezi, 44 vols. Ankara: TDV, vol. 9, 1994: 32–35.

Birışık, Abdulhamit. "Osmanlı Döneminde Türkçe Tefsirler." In *Başlangıçtan Günümüze Türklerin Kur'an Tefsirine Hizmetleri: Tebliğler ve Müzakereler*, edited by Muhammed Abay, prepared by İsmail Kurt and Seyit Ali Tüz. Istanbul: Ensar Neşriyat, 2012: 199–240.

Birışık, Abdulhamit, and Recep Arpa. "Osmanlı Dönemi Tefsir Çevirileri." *Türkiye Araştırmaları Literatür Dergisi* 9, no. 18 (2011): 191–232.

Boyalık, Taha. "Molla Fenârî'nin Tefsir İlminin Mahiyetine Dair Tartışmasının Tahlili." *İslâm Araştırmaları Dergisi* 18 (2007): 73–100.

Boyalık, Taha. *El-Keşşâf Literatürü: Zemahşerî'nin Tefsir Klasiğinin Etki Tarihi*. Istanbul: İSAM, 2019.

Corpus Coranicum: "Corpus Coranicum." Accessed February 12, 2021. https://corpuscoranicum.de.

Cündioğlu, Dücane. "Çağdaş Tefsir Tarihi Tasavvurunun Kayıp Halkası: 'Osmanlı Tefsîr Mirası' – Bir Histografik Eleştiri Denemesi." *İslâmiyât* 2, no. 4 (1999): 64–72.

Demir, Ziya. *Osmanlı Müfessirleri ve Tefsir Çalışmaları: Kuruluştan X/XVI: Asrın Sonuna Kadar*. Istanbul: Ensar Neşriyat, 2006.

Fazlıoğlu, İhsan. "Osmanlılar: Düşünce Hayatı ve Bilim." In *TDVIA*, edited by İslâm Araştırmaları Merkezi, 44 vols. Ankara: TDV, vol. 33, 2007: 548–56.

Görgün, Tahsin. "Molla Fenârî." In *TDVIA*, edited by İslâm Araştırmaları Merkezi, 44 vols. Ankara: TDV, vol. 30, 2005: 247–48.

Gunasti, Susan. "Political Patronage and the Writing of Qur'ān Commentaries among the Ottoman Turks." *JIS* 24, no. 3 (2013): 335–57.

İpşirli, Mehmet. "Huzur Dersleri." In *TDVIA*, edited by İslâm Araştırmaları Merkezi, 44 vols. Istanbul: TDV, vol. 18, 1998: 441–44.

Kaya, Mesut. "Osmanlı İlim Geleneğinde Şerh ve Hâşiye Yazıcılığı – Ahmed el-Karamânî'nin Tefsiri/ Zemahşerî Şerhi Örneği." *Marife Dini Araştırmalar Dergisi* 15, no. 1 (2015): 9–31.

Kaya, Mesut. "Şerh ve Hâşiyelerin Tefsir İlmine Katkıları: Teftâzânî'nin *el-Keşşâf* Şerhi Örneği." In *İslam İlim ve Düşünce Tarihinde Şerh Geleneği*, edited by Mesut Kaya and Sezai Engin. Istanbul: Endülüs Yayınları, 2020: 201–42.

Kurt, Muhammet Sacit. *Mu'tezile'ye Karşı Bir Ehl-i Sünnet Müdafi'i: eṭ-Ṭîbî'nin (743/1343) Futūḥu'l-Ġayb Adlı el-Keşşāf Haşiyesi ve Eserinde ez-Zemaḫşerî'ye (538/1144) Yöneltti̇ği Eleştiriler*. Ankara: İlâhiyât, 2018.

Namlı, Ali. "İsmail Hakkı Bursevi'nin *Rûhu'l-Beyân fi Tefsîri'l-Kur'ân* Adlı Tefsiri." *Türkiye Araştırmaları Literatür Dergisi* 9, no. 18 (2011): 365–85.

Öztürk, Mustafa. "Molla Fenari ve Tefsirde Senkretizm." In *Uluslararası Molla Fenari Sempozyumu: Bildiriler*, edited by Tevfik Yücedoğru, Orhan Ş. Koloğlu, U. Murat Kılavuz and Kadir Gömbeyaz. Bursa: Bursa Büyükşehir Belediyesi Yayınları, 2010: 399–421.

Öztürk, Mustafa. *Osmanlı Tefsir Mirası*. Ankara: Ankara Okulu Yayınları, 2012.

Paçacı, Mehmet. "Osmanlı Medresesinde Tefsir Öğretiminin Yeri ve Tefsirin Çağdaşlıkla Karşılaşması." In *Osmanlı Toplumunda Kur'an Kültürü ve Tefsir Çalışmaları-I*, edited by Bilal Gökkır Necdet Yılmaz, Necmettin Gökkır, Ömer Kara, Muhammed Abay, Mustafa Karagöz. Istanbul: İlim Yayma Vakfı Kur'an ve Tefsir Akademisi, vol. 1, 2011: 63–89.

Saleh, Walid. "The Gloss as Intellectual History: *Ḥāshiyahs* on *al-Kashshāf*." *Oriens* 41 (2013): 217–59.

Şimşek, Halil. "The Missing Link in the History of Quranic Commentary." PhD diss., University of Toronto, 2018.

Şimşek, Halil. "Turks and Their Translations/Commentaries on the Qur'ân: An Historical and Bibliographical Survey." *Kilis 7 Aralık Üniversitesi İlahiyat Fakültesi Dergisi* 6, no. 11 (2019): 395–420.

List of Contributors

Halis Albayrak
Professor emeritus of Qur'ānic Studies
Department of Qur'ānic Exegesis
Faculty of Divinity, Ankara University
Türkiye

Mukhtar Ali
Assistant Professor
Islamic Studies
University of Illinois Urbana-Champaign
USA

Kamran Bashir
Scholar-in-Residence
Centre for Comparative Muslim Studies
Simon Fraser University
Burnaby, British Columbia
Canada

Mehmet Taha Boyalik
Associate Professor
Faculty of Divinity
Istanbul 29 Mayis University
Türkiye

Angelika Brodersen
Research Associates
Institute of Arabic and Islamic Studies
Ruhr-Universität Bochum
Germany

Nurullah Denizer
Associate Professor
Department of Social Sciences
Faculty of Islamic Sciences
Usak University
Türkiye

Ralf Elger
Professor of Arabic Studies
Seminar for Arabic and Islamic Studies
Martin-Luther-Universität Halle
Germany

Esra Gözeler
Associate Professor
Department of Qur'ānic Exegesis
Faculty of Divinity, Ankara University
Türkiye

Jon Hoover
Professor of Islamic Studies
Department of Philosophy
University of Nottingham
United Kingdom

Morteza Karimi-Nia
Islamic Studies Department
Academy of Sciences of the Islamic Republic of
Iran

Serdar Kurnaz
Professor of Islamic Jurisprudence
Berlin Institute for Islamic Theology
Humboldt Universität zu Berlin
Germany

Giovanni M. Martini
Postdoctoral Researcher
Centro Studi sul Mondo Islamico
Istituto Universitario Orientale
Università di Napoli L'Orientale
Italy

Toby Mayer
Senior Research Associate
Qur'ānic Studies Unit
The Institute of Ismaili Studies
London, United Kingdom

Roy Michael McCoy III
Independent scholar
USA

https://doi.org/10.1515/9783111320083-026

Robert Morrison
George Lincoln Skolfield
Jr. Professor of Religion and Middle Eastern and North African Studies
Director of Middle Eastern and North African Studies Program
Bowdoin College
Brunswick, USA

Mohammed Nekroumi
Professor of Islamic-Religious Studies
Department of Islamic-Religious Studies
FAU Friedrich-Alexander-Universtät
Erlangen-Nürnberg
Germany

Miriam Ovadia
Independent scholar and Postdoctoral Researcher in Islamic Studies
Department of Arabic
Bar Ilan University
Ramat Gan, Israel

Peter Riddell
ACT Senior Research Fellow
Department of Languages
Cultures and Linguistics
SOAS, University of London
United Kingdom

Nimet Seker
Professor of Islamic Textual Studies
Berliner Institut für Islamische Theologie
Humboldt Universität zu Berlin
Germany

Farid Suleiman
Lecturer (Academic Council) for Religious Studies with a Focus on Islamic Theology in the Past and Present
Faculty of Theology
Universität Greifswald
Germany

Liyakat Takim
Professor of Religious Studies
Sharjah Chair in Global Islam
Department of Religious Studies
McMaster University Hamilton
Ontario, Canada

Georges Tamer
Professor of Oriental Philology and Islamic Studies
Friedrich-Alexander-Universität
Erlangen-Nürnberg
Germany

Mathieu Terrier
Postdoctoral Researcher at CNRS and Institut d'études de l'Islam et des sociétés du monde musulman
Paris, France

Gregory Vandamme
Postdoctoral research fellow in Religious Studies
Chargé de recherche FNRS
UCLouvain, Belgium

Ruggero Vimercati Sanseverino
Professor of Hadith Studies and Prophetic Tradition
Center for Islamic Theology
Universität Tübingen
Germany

www.ingramcontent.com/pod-product-compliance
Lightning Source LLC
Chambersburg PA
CBHW060847020425
24478CB00006B/41